Introduction to Modern Virology

N. J. Dimmock
A. J. Easton
K. N. Leppard
School of Life Sciences
University of Warwick
Coventry

SEVENTH EDITION

WILEY Blackwell

Library of Congress Cataloging-in-Publication Data

Dimmock, N. J.
   Introduction to modern virology / N. J. Dimmock, A. J. Easton, K. N. Leppard, School of Life Sciences, University of Warwick, Coventry. – Seventh edition.
        pages cm
   Includes index.
   ISBN 978-1-119-97810-7 (pbk.)
   1. Virology. 2. Virus diseases. I. Easton, A. J. (Andrew J.) II. Leppard, K. N. (Keith N.) III. Title.
   QR360.D56 2016
   579.2–dc23

                                                                                        2015031818

A catalogue record for this book is available from the British Library.

Cover design by Jeremy Tilston

Set in 10/12.5pt MeridienLTStd-Roman by Thomson Digital, Noida, India
Printed and bound in Singapore by Markono Print Media Pte Ltd

1 2016

Contents in brief

Contents

Preface

As before, our aim in this 7th edition of *Introduction to Modern Virology* is to provide a broad introduction to virology, which includes the nature of viruses, the interaction of viruses with their hosts, and the consequences of those interactions that lead to the diseases we see. In doing so, we have focused predominantly on viruses that infect humans, with some examples of viruses of other animals where they illustrate a specific point. However, in the sections covering general principles and processes of virology, we have also included bacterial and plant viruses. The revised text is aimed at undergraduate students at all levels and postgraduates who are learning virology as a new subject.

We have retained the four thematic sections that were introduced in the previous edition. These cover the fundamental nature of viruses, their growth in cells, their interactions with the host organism, and their role as agents of human disease. To complement these, we have added a fifth section that incorporates material relating to virology in a wider context. Each section contains a series of chapters that are typically focused on a topic rather than concentrating on a single virus. Inevitably, some of these topics relate to information in different parts of the book and we have included extensive cross-referencing to allow the reader to explore a broader picture than is possible within a single chapter.

The pace of discovery in the field of virology has continued unabated since the last edition. Our knowledge of the molecular detail of viruses, including their interaction with the host, has increased considerably and continues to grow. We have tried to explore the breadth of this new information while retaining a concise style. Inevitably, this has meant that we have had to choose specific examples while leaving out many others of interest, but we have tried to use examples which demonstrate broad principles as well as specific detail. There is suggested reading for those who want to follow up a subject in more depth.

The study of viruses is as topical and important as ever. The global impact of HIV and chronic hepatitis virus infections continues to be severe and, as we completed this edition, we are seeing hopeful indications of the ending of the most devastating Ebola virus outbreak ever recorded. Beyond these direct impacts on our health, viruses also continue to threaten us through effects on food supplies and our economies. Thus, a good basic understanding of viruses is important for generalists and specialists alike. Our aim in writing this book has been to try to make such understanding as accessible as possible, allowing students across the biosciences spectrum to improve their knowledge of these fascinating entities.

New to this edition

This edition contains a number of important changes and innovations. A major change has been the expansion of the consideration of immunology which now covers two chapters, one on innate immunity and the other on adaptive immunity. This reflects the growing understanding of the importance of the immune system in determining the outcome of virus infection and the contribution of the immune system to viral diseases. These chapters also consider some of the ways that viruses evade the immune response. The consideration of vaccines and antivirals has been expanded and separated into two new chapters to reflect the importance of these approaches to prevention and treatment. Virus evolution is considered in more detail than previously, and we have added new chapters on viral hepatitis, influenza, vector-borne diseases, and exotic and emerging viral infections. Finally, in the last section we have introduced three new chapters on the broader aspects of the influence of viruses on our lives, focusing on the economic impact of virus infections, the ways we can use viruses in clinical and other spheres, and the impact that viruses have on the planet and almost every aspect of our lives.

The text is supplemented throughout by information boxes of two types. These are now distinguished by different colours. One type of box provides supporting information or additional detail about the subject matter of the chapter while the other provides the experimental evidence by which selected key points were established. The aim is to assist the reader in understanding the facts but to also allow them to appreciate the nature of the evidence that underpins them.

We very much hope that the 7th edition of *Introduction to Modern Virology* will enrich the virology experience of students and teachers alike.

Finally, we would like express our thanks the staff at Wiley for their generous support throughout the production of this book.

Nigel Dimmock, Andrew Easton
and Keith Leppard
University of Warwick, October 2015

About the companion website

This book is accompanied by a companion website:

www.wiley.com/go/dimmock/virology

The website includes powerpoints of all figures from the book for downloading

Part I

The nature of viruses

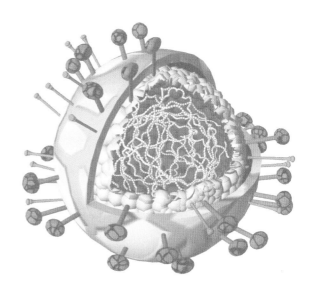

Chapter 1

Towards a definition of a virus

Viruses occur universally, but they can only be detected indirectly. Viruses are obligate intracellular parasites that require a host within which they replicate. Although they are well known for causing disease, most viruses coexist peacefully with their hosts.

Chapter 1 Outline

1.1 Discovery of viruses
1.2 Multiplication of viruses
1.3 The virus multiplication cycle
1.4 Viruses can be defined in chemical terms
1.5 Multiplication of bacterial and animal viruses is fundamentally similar
1.6 Viruses can be manipulated genetically
1.7 Properties of viruses
1.8 Origin of viruses

Viruses are arguably the most ubiquitous and widespread group of organisms on the planet, with every animal, plant and protist species susceptible to infection. The efficiency of replication demonstrated by viruses is such that the infection of a single host can generate more new viruses than there are individuals in the host population. For example, a single human infected with influenza virus can shed sufficient virus particles to be theoretically capable of infecting the entire human population. While not every species has been examined for the presence of viruses, those that have been tested have all yielded up new virus isolates. Further, not only do viruses occur universally but each species has its own specific range of viruses that, by and large, infects only that species. In recent years, the application of new nucleic acid sequencing techniques has demonstrated that a vast array of previously unknown viruses remains to be studied.

Current estimates of the number of individual viruses on earth suggest that they considerably exceed the total number of stars in the known universe, i.e. more than 10^{23} (100 sextillion). This vast number raises questions as to what the viruses are doing there, and what selective advantage, if any, they afford to the species that host them. The answer to the first of these is the same as if the question was posed about any organism – it is simply occupying a particular environmental niche which, in the case of a virus, is another species. The answer to whether or not any benefit accrues for hosting a virus is usually not known, though the adverse effects of virus infections are all too well known. However, it is clear that, despite their adverse effects and the dramatic depictions of viruses in popular media and cinema, viruses have not made their hosts extinct.

Introduction to Modern Virology, Seventh Edition. N. J. Dimmock, A. J. Easton and K. N. Leppard.
© 2016 John Wiley & Sons, Ltd. Published 2016 by John Wiley & Sons, Ltd.

1.1 Discovery of viruses

Although much is known about viruses (Box 1.1), it is instructive and interesting to consider how this knowledge came about. It was only just over 100 years ago, at the end of the 19th century, that the germ theory of disease was formulated, and pathologists were then confident that a causative micro-organism would be found for each infectious disease. Further, they believed that these agents of disease could be seen with the aid of a microscope, could be cultivated on a nutrient medium, and could be retained by filters. There were, admittedly, a few organisms which were so fastidious that they could not be cultivated in the laboratory but the other two criteria were satisfied. However, in 1892, Dmitri Iwanowski was able to show that the causal agent of a mosaic disease of tobacco plants, manifesting as a discoloration of the leaf, passed through a bacteria-proof filter, and could not be seen or cultivated. Iwanowski was unimpressed by his discovery, but Beijerinck repeated the experiments in 1898, and became convinced that this represented a new form of infectious agent which he termed *contagium vivum fluidum*,

what we now know as a virus. In the same year, Loeffler and Frosch came to the same conclusion regarding the cause of foot-and-mouth disease. Furthermore, because foot-and-mouth disease could be passed from animal to animal, with great dilution at each passage, the causative agent had to be reproducing and thus could not be a bacterial toxin. Viruses of other animals were soon discovered. Ellerman and Bang reported the cell-free transmission of chicken leukaemia in 1908, and in 1911 Rous discovered that solid tumours of chickens could be transmitted by cell-free filtrates. These were the first indications that some viruses can cause cancer (see Chapter 25).

Finally, bacterial viruses were discovered. In 1915, Twort published an account of a glassy transformation of micrococci. He had been trying to culture the smallpox agent on agar plates but the only growth obtained was that of some contaminating micrococci. Following prolonged incubation, some of the colonies took on a glassy appearance, and once this occurred no bacteria could be subcultured from the affected colonies. If some of the glassy material was added to normal colonies, they too took on a similar appearance, even if the glassy material was first passed through very fine filters to

---

Box 1.1

Properties common to all viruses

- Viruses have a nucleic acid genome of either DNA or RNA.
- Compared with a cell genome, viral genomes are small, but genomes of different viruses range in size by over 100-fold (ca 3000 nt to 1,200,000 bp)
- Small genomes make small particles – again with a 100-fold size range.
- Viral genomes are associated with protein that at its simplest forms the virus particle, but in some viruses this nucleoprotein is surrounded by further protein or a lipid bilayer.
- The outermost proteins of the virus particle allow the virus to recognise the correct host cell and gain entry.
- Viruses can only reproduce in living cells: they are obligate parasites.

exclude all but the smallest material. Among the suggestions that Twort put forward to explain the phenomenon were either the existence of a bacterial virus or the secretion by the bacteria of an enzyme which could lyse the producing cells. This idea of self-destruction by secreted enzymes was to prove a controversial topic over the next decade. In 1917, d'Hérelle observed a similar phenomenon in dysentery bacilli. He observed clear spots on lawns of such cells, and resolved to find an explanation for them. Upon noting the lysis of broth cultures of pure dysentery bacilli by filtered emulsions of faeces, he immediately realized he was dealing with a bacterial virus. Since this virus was incapable of multiplying except at the expense of living bacteria, he called his virus a *bacteriophage* (literally a bacterium eater), or *phage* for short.

Thus the first definition of these new agents, the viruses, was presented entirely in negative terms: they could not be seen, could not be cultivated in the absence of cells and, most important of all, were not retained by bacteria-proof filters. However, these features define key characteristics of viruses: they are small parasites that require a host in which they replicate.

1.2 Multiplication of viruses

Early studies focused on establishing the nature of viruses. d'Hérelle believed that the infecting phage particle multiplied within the bacterium and that its progeny were liberated upon lysis of the host cell, whereas others believed that phage-induced dissolution of bacterial cultures was merely the consequence of a stimulation of lytic enzymes endogenous to the bacteria. Yet another school of thought was that phages could pass freely in and out of bacterial cells and that lysis of bacteria was a secondary phenomenon not necessarily concerned with

the growth of a phage. It was Delbruck who ended the controversy by pointing out that two phenomena were involved: lysis from within and lysis from without. The type of lysis observed was dependent on the ratio of infecting phages to bacteria (referred to as the *multiplicity of infection*). At a low multiplicity of infection (with the ratio of phages to bacteria no greater than 2:1), then the phages infect the cells, multiply and lyse the cells from within. When the multiplicity of infection is high, i.e. many hundreds of phages per bacterium, the cells are lysed directly, and rather than an increase in phage titre there is a decrease. Lysis is due to weakening of the cell wall when large numbers of phages are attached.

Convincing support for d'Hérelle's hypothesis was provided by the one-step growth experiment of Ellis and Delbruck (1939). A phage preparation such as bacteriophage λ (lambda) is mixed with a suspension of the bacterium *Escherichia coli* at a multiplicity of infection of 10 infectious phage particles per cell, ensuring that virtually all cells are infected. Then, after allowing 5 minutes for the phage to attach, the culture is centrifuged to pellet the cells and attached phage. Medium containing unattached phage is discarded. The cells are then resuspended in fresh medium. Samples of the culture are withdrawn at regular intervals, cells and medium are separated and assayed for infectious phage. The results obtained are shown in Fig. 1.1. After a latent period of 17 minutes during which no phage increase is detected in cell-free medium, there is a sudden rise in the detection of infectious phage in the medium. This 'burst' size represents the average of many different bursts from individual cells, and can be calculated from the total virus yield/number of cell infected. The entire growth cycle here takes around 30 minutes, although this will vary with different viruses and cells. The amount of cell-associated virus is determined by taking the cells pelleted from the medium, disrupting

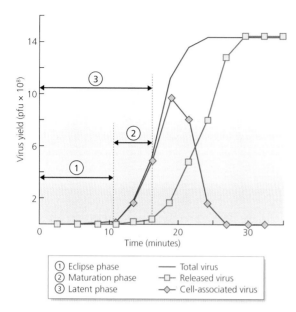

Fig. 1.1   A one-step growth curve of bacteriophage λ following infection of susceptible bacteria (*Escherichia coli*). During the *eclipse phase*, the infectivity of the cell-associated, infecting virus is lost as it uncoats; during the *maturation phase* infectious virus is assembled inside cells (cell-associated virus), but not yet released; and the *latent phase* measures the period before infectious virus is released from cells into the medium. Total virus is the sum of cell-associated virus + released virus. Cell-associated virus decreases as cells are lysed. This classic experiment shows that phages develop intracellularly. A consideration of the methods used to determine the yield of viruses is given in Chapter 5.

them and assaying for virus infectivity as before. The fact that virus appears inside the cells before it appears in the medium demonstrates the intracellular nature of phage replication. It can be seen also that the kinetics of appearance of intracellular phage particles are *linear*, not exponential. This is consistent with particles being produced by assembly from component parts rather than by binary fission.

1.3 The virus multiplication cycle

We now know a great deal about the processes which occur during the multiplication of viruses within single cells. The precise details vary for individual viruses but have in common a series of events marking specific phases in the multiplication cycle. These phases are summarized in Fig. 1.2 and are considered in detail in Part II of this book. The first stage is that of **attachment** when the virus attaches to the potential host cell. The interaction is specific, with the virus attachment protein(s) binding to target receptor molecules on the surface of the cell. The initial contact between a virus and host cell is a dynamic, reversible one often involving weak electrostatic interactions. However, the contacts quickly become much stronger with more stable interactions which in some cases are essentially irreversible. The attachment phase determines the specificity of the virus for a particular type of cell or host species. Having attached to the surface of the cell, the virus must effect entry to be able to replicate in a process called **penetration** or **entry**. Once inside the cell, the genome of the virus must become available. This is achieved by the loss of many, or all, of the proteins that make up the particle in a process referred to as **uncoating**. For some viruses, the entry and uncoating phases are combined in a single process. Typically, these first three phases do not require the expenditure of energy in the form of ATP hydrolysis. Having made the virus genome available it is now used in the **biosynthesis** phase when genome replication, transcription of mRNA and translation of the mRNA into protein occur. The process of translation uses ribosomes provided by the host cell and it is this requirement for the translation machinery, as well as the need for molecules for biosynthesis, that makes viruses obligate intracellular parasites. The newly-synthesized

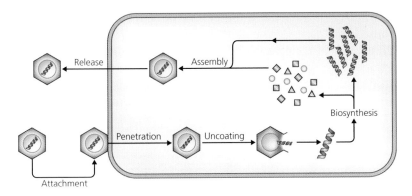

Fig. 1.2   A diagrammatic representation of the six phases common to all virus multiplication cycles. See text for details.

genomes may then be used as templates for further rounds of replication and as templates for transcription of more virus mRNA in an amplification process which increases the yield of virus from the infected cells. When the new genomes are produced, they come together with the newly-synthesized virus proteins to form progeny virus particles in a process called **assembly**. Finally, the particles must leave the cell in a **release** phase after which they seek out new potential host cells to begin the process again. The particles produced within the cell may require further processing to become infectious and this **maturation** phase may occur before or after release.

Combining the consideration of the steps which make up a virus multiplication cycle with the information in the graph of the results of a single step growth curve, it can be seen that during the eclipse phase the virus is undergoing the processes of attachment, entry, uncoating and biosynthesis. At this time, the cells contain all of the elements necessary to produce viruses but the original infecting virus has been dismantled and no new infectious particles have yet been produced. It is only after the assembly step that we see virus particles inside the cell before they are released and appear in the medium.

1.4 Viruses can be defined in chemical terms

The first virus was purified in 1933 by Schlessinger using differential centrifugation. Chemical analysis of the purified bacteriophage showed that it consisted of approximately equal proportions of protein and deoxyribonucleic acid (DNA). A few years later, in 1935, Stanley isolated tobacco mosaic virus in paracrystalline form, and this crystallization of a biological material thought to be alive raised many philosophical questions about the nature of life. In 1937, Bawden and Pirie extensively purified tobacco mosaic virus and showed it to be nucleoprotein containing ribonucleic acid (RNA). Thus virus particles may contain either DNA or RNA. However, at this time it was not known that nucleic acid constituted genetic material.

The importance of viral nucleic acid

In 1949, Markham and Smith found that preparations of turnip yellow mosaic virus comprised two types of identically sized spherical particles, only one of which contained nucleic acid. Significantly, only the particles containing nucleic acid were infectious. A few years later, in 1952, Hershey and Chase demonstrated the independent functions of viral protein and nucleic acid using the head-tail virus, bacteriophage T2 (Box 1.2).

In another classic experiment, Fraenkel-Conrat and Singer (1957) were able to confirm by a different means the hereditary role of viral RNA. Their experiment was based on the earlier discovery that particles of tobacco mosaic virus can be dissociated into their protein and RNA components, and then reassembled to give

Box 1.2

Evidence that DNA is the genetic material of bacteriophage T2. The Hershey-Chase experiment

Bacteriophage T2 was grown in *E. coli* in the presence of ^{35}S (as sulphate) to label the protein moiety, or ^{32}P (as phosphate) to mainly label the nucleic acid. Purified, labelled phages were allowed to attach to sensitive host cells and then given time for the infection to commence. The phages, still on the outside of the cell, were then subjected to the shearing forces of a Waring blender. Such treatment removes any phage components attached to the outside of the cell but does not affect cell viability. Moreover, the cells are still able to produce infectious progeny virus. When the cells were separated from the medium, it was observed that 75% of the ^{35}S (i.e. phage protein) had been removed from the cells by blending but only 15% of the ^{32}P (i.e. phage nucleic acid) had been removed. Thus, after infection, the bulk of the phage protein appeared to have no further function and this suggested (but does not prove – that had to await more rigorous experiments with purified nucleic acid genomes) that the nucleic acid is the carrier of viral heredity. The transfer of the phage nucleic acid from its protein coat to the bacterial cell upon infection also accounts for the existence of the eclipse period during the early stages of intracellular virus development, since the nucleic acid on its own cannot normally infect a cell (Fig. 1.3).

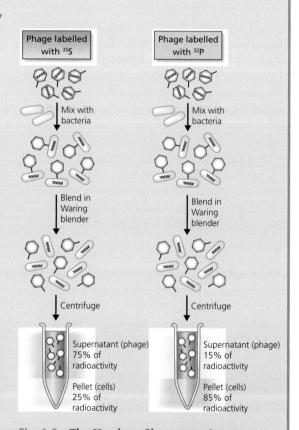

Fig. 1.3   The Hershey-Chase experiment proving that DNA (labelled with ^{32}P) is the genetic material of bacteriophage T2.

particles which are morphologically mature and fully infectious (see Chapter 12). When particles of two different strains (differing in the symptoms produced in the host plant) were each disassociated and the RNA of one reassociated with the protein of the other, and vice versa, the properties of the virus which was propagated when the resulting 'hybrid' particles were used to infect host plants were always those of the parent virus from which the RNA was derived (Fig. 1.4).

The ultimate proof that viral nucleic acid is the genetic material came from numerous observations that, under special circumstances, purified viral nucleic acid is capable of initiating infection, albeit with a reduced efficiency. For example, in 1956 Gierer and Schramm, and Fraenkel-Conrat independently showed that the purified RNA of tobacco mosaic virus can be infectious, provided precautions are taken to protect it from inactivation by ribonuclease. An extreme example is the causative agent of potato spindle tuber disease which lacks any protein component and consists solely of RNA. Because such agents have no protein coat, they cannot be called viruses and are referred to as *viroids*.

Synthesis of macromolecules in infected cells

Following introduction of the virus genetic material into the cell, the next phase of the replication cycle is the synthesis of new macromolecules that play a role in the replication process and/or find their way into the next generation of virus particles. The discovery in 1953, by Wyatt and Cohen, that the DNA of the T-even bacteriophages T2, T4 and T6 contains hydroxymethylcytosine (HMC) instead of cytosine made it possible for Hershey, Dixon and Chase to examine infected bacteria for the presence of phage-specific DNA at various stages of intracellular growth. DNA was

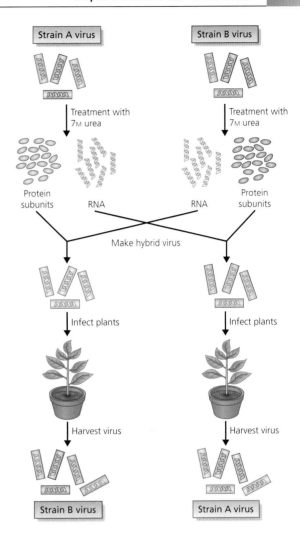

Fig. 1.4 The experiment of Fraenkel-Conrat and Singer which proved that RNA is the genetic material of tobacco mosaic virus.

extracted from T2-infected *E. coli* at different times after the onset of phage growth, and analyzed for its content of HMC. This provided an estimate of the number of phage equivalents of HMC-containing DNA present at any time, based on the total nucleic acid and relative HMC content of the intact T2 phage particle. The results showed that, with T2, synthesis of phage DNA commences about 6 minutes after infection and the amount present then rises sharply, so that by the time the first infectious particles begin to appear 6 minutes later there are 50–80 phage equivalents of HMC. Thereafter, the numbers of phage equivalents of DNA and of infectious particles increase linearly and at the same rate up until lysis, and continue to rise even if lysis is delayed beyond the normal burst time.

Hershey and his co-workers also studied the synthesis of phage protein, which can be distinguished from bacterial protein by its interaction with specific antibodies. During infection of *E. coli* by T2 phage, protein can be detected about 9 minutes after the onset of the latent period, i.e. after DNA synthesis begins and before infectious particles appear. A few minutes later there are approximately 30–40 phages inside the cell. Whereas the synthesis of viral protein starts about 9 minutes after the onset of the latent period, it was shown by means of pulse–chase experiments that the uptake of ^{35}S into intracellular protein is constant from the start of infection. A small quantity (a pulse) of ^{35}S (as sulphate) was added to the medium at different times after infection and was followed shortly after by a vast excess of unlabelled sulphate (chase) to stop any further incorporation of label. When the pulse was made from the 9th minute onward, the label could be chased into material identifiable by its reaction with antibody (i.e. serologically) as phage coat protein. However, if the pulse was made before 9 minutes of infection, although it could still be chased into protein and was non-bacterial, it did not react with antibodies to phage structural proteins. This early protein comprises mainly virus-specified enzymes that are concerned with phage replication but are not incorporated into phage particles. The concept of early and late, non-structural and structural viral proteins is discussed in Chapter 10.

These classical experiments are typical only of head-tail phages (see Section 2.5) infecting

E. coli under optimum growth conditions. *E. coli* is normally found in the anaerobic environment of the intestinal tract, and it is doubtful that it grows with its optimal doubling time of 20 minutes under natural conditions. Other bacterial cells grow more slowly than *E. coli* and their viruses have longer multiplication cycles.

1.5 Multiplication of bacterial and animal viruses is fundamentally similar

The growth curves and other experiments described above have been repeated with many animal viruses with essentially similar results. Both bacterial and animal viruses attach to their target cell through specific interactions with cell surface molecules. Like the T4 bacteriophage, the genomes of some animal viruses (e.g. HIV-1) enter the cell and leave their coat proteins on the outside. However, with most animal viruses, some viral protein, usually from inside the particle, enters the cell in association with the viral genome. In fact, it is now known that some phage protein enters the bacterial cells with the phage genome. Such proteins are essential for genome replication. Many other animal viruses behave slightly differently, and after attachment are engulfed by the cell membrane and taken into the cell inside a vesicle. However, strictly speaking this virus has not yet entered the cell cytoplasm, and is still outside the cell. The virus genome gains entry to the cytoplasm through the wall of the vesicle, when the particle is stimulated to uncoat. Again, the outer virion proteins stay in the vesicle – i.e. outside the cell. Animal viruses go through the same stages of eclipse, and virus assembly from constituent viral components with linear kinetics, as bacterial viruses. Release of progeny virions may happen by cell lysis (although this is not an enzymatic process as it is with some bacterial viruses), but frequently virus is released without major cell damage. The cell may die later, but death of the cell does not necessarily accompany the multiplication of all animal viruses. One major difference in the multiplication of bacterial and animal virus is that of time scale – animal virus growth cycles take in the region of 5–15 hours for completion.

1.6 Viruses can be manipulated genetically

One of the easiest ways to understand the steps involved in a particular reaction within an organism is to isolate mutants which are unable to carry out that reaction. Like all other organisms, viruses sport mutants in the course of their growth, and these mutations can affect all properties including the type of plaque formed, the range of hosts which the virus can infect, and the physico-chemical properties of the virus. One obvious caveat is that many mutations will be lethal to the virus and remain undetected. This problem was overcome in 1963 by Epstein and Edgar and their collaborators with the discovery of *conditional lethal mutants*. One class of these mutants, the *temperature-sensitive mutants*, was able to grow at a lower temperature than normal, the *permissive temperature*, but not at a higher, *restrictive* temperature at which normal virus could grow. Another class of conditional lethal mutants was the *amber* mutant. In these mutants a genetic lesion converts a codon within transcribed RNA into a triplet which terminates protein synthesis. They can only grow on a *permissive* host cell, which has an amber-suppressor transfer RNA (tRNA) that can insert an amino acid at the mutation site during translation.

The drawback to conditional lethal mutants is that mutation is random, but the advent of *recombinant DNA technology* has facilitated controlled mutagenesis, known as reverse genetics, at least for those viruses for which infectious particles can be reconstituted from cloned genomic DNA or cDNA (DNA that has been transcribed from RNA) inserted into a plasmid vector. This process is described in Section 5.7.

1.7 Properties of viruses

With the assumption that the features of virus growth described above for particular viruses are true of all viruses, it is possible to compare and contrast the properties of viruses with those of their host cells. Whereas host cells contain both types of nucleic acid, DNA and RNA, each virus contains only one type. However, just like their host cells, viruses have their genetic information encoded in nucleic acid. Another difference is that the virus is reproduced solely from its genetic material, whereas the host cell is reproduced from the integrated sum of its components. Thus, the virus never arises directly from a pre-existing virus, whereas the cell always arises by division from a pre-existing cell. The experiments of Hershey and his collaborators showed quite clearly that the components of a virus are synthesized independently and then assembled into many virus particles. In contrast, the host cell increases its constituent parts, during which the individuality of the cell is continuously maintained, and then divides and forms two cells. Finally, viruses are incapable of synthesizing ribosomes, and depend on pre-existing host cell ribosomes for synthesis of viral proteins. These features clearly separate viruses from all other organisms, even *Chlamydia* species, which for many years were considered to be intermediate between bacteria and viruses.

1.8 Origin of viruses

The question of the origin of viruses is a fascinating topic; as so often happens when hard evidence is scarce, discussion can be heated but often not illuminating. There are two popular theories: viruses are either degenerate cells or vagrant genes. Just as fleas are descended from flies by loss of wings, viruses may be derived from pro- or eukaryotic cells that have dispensed with many of their cellular functions (*degeneracy*). Alternatively, some nucleic acid might have been transferred accidentally into a cell of a different species (e.g. through a wound or by sexual contact) and, instead of being degraded as would normally be the case, might have survived and replicated (*escape*). Despite decades of discussion and argument there are no firm indications if either, or both, of these theories are correct. Rapid sequencing of viral and cellular genomes is now providing data for computer analysis that are giving an ever-better understanding of the relatedness of different viruses. However, while such analyses may identify, or more commonly infer, the progenitors of a virus, they cannot decide between degeneracy and escape. It is unlikely that all currently-known viruses have evolved from a single progenitor. Rather, viruses have probably arisen numerous times in the past by one or both of the mechanisms outlined above.

Key points

- Viruses are obligate intracellular parasites.
- It is likely that every living organism on this planet is infected by a species-specific range of viruses.
- Viruses multiply by assembling many progeny particles from a pool of virus-specified components, whereas cells multiply by binary fission.
- Viruses have probably originated independently many times.

Further reading

Flint, S. J., Enquist, L. W., Racaniello, V. R., Skalka, A. M. 2008. *Principles of Virology: Molecular Biology, Pathogenesis, and Control*, 3rd edn. ASM Press, Washington DC.

Hull, R. 2013. *Plant Virology*, 5th edn. Academic Press, San Diego.

Knipe, D. M., Howley, P. M. (2007). *Field's Virology*, 5th edn. Lippincott Williams & Wilkins, Philadelphia.

Maclachlan, N. J., Dubovi, E. J. 2010. *Fenner's Veterinary Virology*, 4th edn. Academic Press, San Diego.

Mahy, B. W. J., van Regenmortel, M. H. V. 2008. *Encyclopedia of Virology*, 3rd edn. Academic Press, San Diego.

Richman, D. D., Whitley, R. J., Hayden, F. G. 2009. *Clinical Virology*, 3rd edn. ASM Press, Washington DC.

Zuckerman, A. J., Banatvala, J., Griffiths, P. D., Schoub, B., Mortimer, P. 2009. *Principles and Practice of Clinical Virology*, 6th edn. John Wiley & Sons, Chichester.

Chapter 2

The structure of virus particles

All virus genomes are surrounded by proteins which:

- *protect nucleic acids from nuclease degradation and shearing*
- *contain identification elements that ensure a virus recognizes an appropriate target cell (but plant viruses do not, and enter the cell directly by injection or injury)*
- *contain a genome-release system that ensures that the virus genome is released from a particle only at the appropriate time and location*
- *include enzymes that are essential for the infectivity of many, but not all, viruses*
- *are called* structural proteins, *as they are part of the virus particle.*

All viruses contain protein and nucleic acid with at least 50%, and in some cases up to 90%, of their mass being protein. At first sight it would appear that there are many ways in which proteins could be arranged round the nucleic acid. However, viruses use only a limited number of designs. The limitation on the range of structures is due to restrictions imposed by the considerations of efficiency and stability.

2.1 Virus particles are constructed from subunits

While proteins may have regular secondary structure elements in the form of α helix and β sheet structure, the tertiary structure of the protein is not symmetrical. This is a consequence of the fundamental asymmetry of the amino acid chain making up the protein, hydrogen bonding, disulphide bridges and the incorporation of proline in the secondary structure. Although it may be naïvely thought that the nucleic acid could be covered by a single, large protein molecule, this cannot be the case since proteins are irregular in shape, whereas most virus particles have a regular morphology (Fig. 2.1). However, the requirement for more than a single structural protein forming the entirety of the virus particle can also be deduced solely from considerations of the coding potential of nucleic acid molecules. A coding triplet has an M_r of

Introduction to Modern Virology, Seventh Edition. N. J. Dimmock, A. J. Easton and K. N. Leppard.
© 2016 John Wiley & Sons, Ltd. Published 2016 by John Wiley & Sons, Ltd.

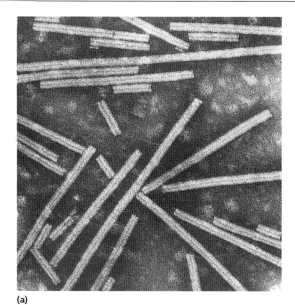

(a)

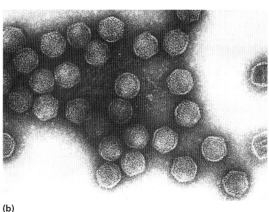

(b)

Fig. 2.1   Electron micrographs of virus particles showing their regular shape: rods and spheres. (a) Tobacco mosaic virus, (b) poliovirus.

repeated subunit. However, it is not essential that the protein coat be constructed from identical subunits, provided the combined molecular weight of the different subunits is sufficiently small in relation to the nucleic acid molecule which they protect. There is a further advantage in constructing a virus from subunits, since any misfolding of protein affects only a small part of a structural unit. Thus, provided that faulty subunits are not included in the virus particle during assembly, an error-free structure can be constructed with the minimum of wastage.

The necessary physical condition for the stability of any structure is that it is in a state of minimum free energy, so it can be assumed that the maximum number of interactions is formed between the subunits of a virus particle. Since the subunits themselves are non-symmetrical, the maximum number of interactions can be formed only if they are arranged symmetrically, and there are a limited number of ways this can be done. Shortly after their seminal work on the structure of DNA, Watson and Crick predicted on theoretical grounds that the only two ways in which asymmetric subunits could be assembled to form virus particles would generate structures with either cubic or helical symmetry. (However, an important rider to the energy status of virus particles is that at least some are suspected of being metastable, and that they can achieve a lower energy state following interaction with the receptor and other molecules during the uncoating process when infecting a cell (see Section 2.7).)

approximately 1000 but specifies a single amino acid with an average M_r of about 100. Thus a nucleic acid can at best only specify one-tenth of its mass of protein. Since viruses frequently contain more than 50% protein by mass, it is apparent that more than one protein must be present.

Obviously, less genetic material is required if a single protein molecule is used as a

2.2 The structure of filamentous viruses and nucleoproteins

One of the simplest ways of symmetrically arranging nonsymmetrical components is to place them round the circumference of a circle

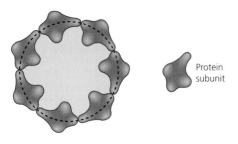

Protein
subunit

Fig. 2.2   Arrangement of identical asymmetrical components around the circumference of a circle to yield a symmetrical arrangement.

to form discs (Fig. 2.2). This gives a 2-dimensional structure. If a large number of discs are placed on top of one another, the result is a 'stacked-disc' structure. Thus a symmetrical 3-dimensional structure can be generated from a non-symmetrical component such as protein and still leave room for nucleic acid. Examination of electron micrographs of viruses reveals that some of them have a tubular structure. One such virus is the tobamovirus, tobacco mosaic virus (TMV) (Fig. 2.1). However, closer examination reveals that the TMV subunits are not arranged cylindrically, i.e. in rings, but helically. The simplest explanation for this comes from a consideration of stability. In a stacked-disc structure the subunits at different locations would not be identical in terms of the way that they interact with their neighbouring subunits. However, by arranging the subunits helically, the maximum number of bonds can be formed and each subunit will interact with its neighbours in exactly the same way as any other subunit, except, of course, for those at either end. This is called equivalence of bonding and it confers enhanced stability to the structure. Similarly, a helical arrangement offers greater stability than would be found with a cylinder which has no linking along the long axis. All filamentous viruses so far examined are helical rather than cylindrical and the insertion of the nucleic acid may be the

factor governing this arrangement (see Section 12.2). Many nucleoprotein structures inside enveloped viruses (Fig. 2.14) are constructed in the same way.

2.3 The structure is of isometric virus particles

A second way of constructing a symmetrical particle would be to arrange the smallest number of subunits possible around the vertices or faces of an object with cubic symmetry, e.g. tetrahedron, cube, octahedron, dodecahedron (constructed from 12 regular pentagons) or icosahedron (constructed from 20 equilateral triangles). Figure 2.3 shows possible arrangements for objects with triangular and square faces. Multiplying the minimum number of subunits per face by the number of faces gives the smallest number of subunits which can be arranged around such an object. The minimum number of subunits is determined by the symmetry element of the face, i.e. a square face will have four subunits, a triangular face will have three subunits, etc. Thus, for a tetrahedron the smallest number of subunits is 12, for a cube or octahedron it is 24 subunits,

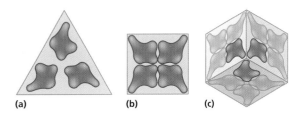

(a)   (b)   (c)

Fig. 2.3   Symmetrical arrangement of identical asymmetrical subunits by placing them on the faces of objects with cubic symmetry. (a) Asymmetrical subunits located at vertices of each triangular facet. (b) Asymmetrical subunits placed at vertices of each square facet. (c) Arrangement of asymmetrical subunits placed at each corner of a cube with faces as represented in (b).

and for a dodecahedron or icosahedron it is 60 subunits. Although it may not be immediately apparent, these represent the few ways in which an asymmetrical object (such as a protein molecule) can be placed symmetrically on the surface of an object resembling a sphere. (This can be checked by using a ball and sticking on bits of paper of the shape shown in Fig. 2.3.) Examination of electron micrographs reveals that many viruses appear spherical in outline, but actually have icosahedral symmetry rather than octahedral, tetrahedral or cuboidal symmetry. There are two possible reasons for the selection of icosahedral symmetry over the others. Firstly, since it requires a greater number of subunits to provide a sphere of the same volume, the size of the repeating subunits can be smaller, thus economizing on genetic information. Secondly, there appear to be physical constraints which prevent the tight packing of subunits required by tetrahedral and octahedral symmetry.

An unusual virus structure based on an icosahedron is seen with the geminiviruses. The particles are formed by two isometric units combined to form a *geminate* structure. An example of a geminate maize streak virus particle is shown in Fig. 29.3.

Symmetry of an icosahedron

An icosahedron is made up of 20 triangular faces, five at the top, five at the bottom and 10 around the middle, with 12 vertices (Fig. 2.4c). Each triangle is itself symmetrical and so it can be inserted in any orientation (Fig. 2.4a). An icosahedron has three axes of symmetry: fivefold, threefold and twofold (Fig. 2.4b). The simplest organization is that three capsid protein molecules form the icosahedral face. However, many viruses are icosahedral and yet achieve a much greater size than can apparently be accomplished with this simple arrangement (see Box 2.1).

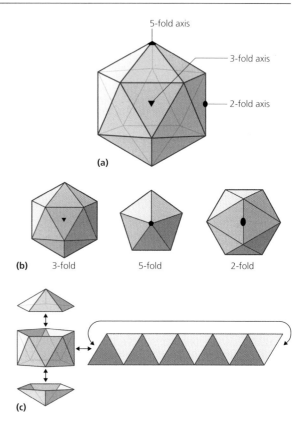

Fig. 2.4 (a) Properties of a regular icosahedron. Each triangular face is equilateral and has the same orientation whichever way it is inserted. Axes of symmetry intersect in the middle of the icosahedron. There are 12 vertices, which have fivefold symmetry, meaning that rotation of the icosahedron by one-fifth of a revolution achieves a position such that it is indistinguishable from its starting orientation; each of the 20 faces has a threefold axis of symmetry and each of the 30 edges has a twofold axis of symmetry (see (b)). The icosahedron is built up of five triangles at the top, five at the bottom and a strip of 10 around the middle (c). (Copyright 1991, from *Introduction to Protein Structure* by C. Branden & J. Tooze. Reproduced by permission of Routledge Inc., part of The Taylor & Francis Group.).

Box 2.1

The construction of more complex icosahedral virus particles

The combination of an icosahedral structure linked with evolutionary pressure to use small repeating subunits to form virus particles imposes a restriction on the achievable size of the virion. Since the size of the particle defines the maximum size of the nucleic acid that can be packed within it, it appears that viruses should not be able to package large genomes. However, many viruses have very long genomes and contain more than 60 subunits. This apparent paradox is solved as follows: if $60n$ subunits are put on the surface of a sphere, one solution is to arrange them in n sets of 60 units, but the members of one set would not be equivalently related to those in another set. If, in Fig. 2.5a, all the subunits represented by open and closed circles were identical, then those represented by closed circles are related equivalently to those represented by open circles, but open circle units do not have the same spatial arrangement of neighbours as closed circle units and so cannot be equivalently related. (The problem could be solved if the structure was built out of n different subunits, but this would require the virus to encode many more structural proteins.) In order to construct a spherical structure out of more than 60 asymmetrical subunits it is necessary to subdivide the surface of the sphere into triangular facets and organize them with icosahedral symmetry. Each triangular facet can be continuously subdivided into further triangular units in a process called triangulation (Fig. 2.5b). This is the same principle behind the construction of geodesic domes which are extremely stable structures. The device of triangulating the sphere represents the optimum design for a closed shell built of regularly bonded identical subunits. No other subdivision of a closed surface can give a comparable degree of equivalence. This is a minimum-energy structure and is the probable reason for the preponderance of icosahedral viruses.

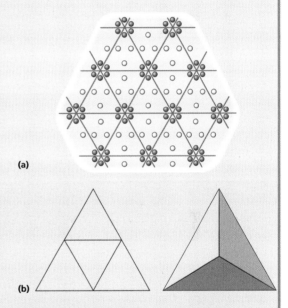

(a)

(b)

Fig 2.5 (a) Spatial arrangement of two identical sets of subunits. Note that any member of the set represented by closed circles does not have the same neighbours as a member of the other set represented by open circles. (b) Triangulation of an equilateral triangular face. The triangle on the left has been subdivided into 4 smaller equilateral triangles. The triangle on the right has been subdivided into 3 smaller equilateral triangles by distorting the plane of the original face. Each new triangle can be subdivided in turn using either process.

The triangulation of spheres – how to make bigger virus particles

It is possible to enumerate all the ways in which this subdivision of a spherical structure can be carried out. This can be demonstrated with a simple example. Starting with an icosahedron, arranging the subunits around the vertices will generate 12 groups of five subunits (Fig. 2.6a). Subdividing each triangular face into four smaller and identical equilateral triangular facets and incorporation of subunits at the vertices of those smaller triangles gives a structure containing a total of 240 subunits (Fig. 2.6b). At the vertices of each of the original icosahedron faces there will be rings of five subunits, called *pentamers* (solid circles). However, at all the other (new) vertices generated by the triangular facets will be rings of six subunits, called *hexamers* (open circles). Since some of the subunits are arranged as pentamers and others as hexamers, it should be apparent that they cannot be equivalently related; hence they are called *quasi-equivalent*, but this still represents the minimum-energy shape. Thus with the subunit being kept a constant size, greater subdivision allows the formation of larger virus particles.

T = 1: the smallest virus particle – satellite tobacco necrosis virus

In its simplest form, one subunit used in the construction of a virus particle subunit will consist of one protein. However, no independently replicating virus is known to consist of only 60 protein subunits, but satellite viruses do (Table 2.1). These encode one coat protein but depend upon coinfection with a helper virus to provide missing replicative functions. The single-stranded RNA genome of

---

Box 2.2

Calculating the number of subunits in a virus particle

The way in which each triangular face of the icosahedron can be subdivided into smaller, identical equilateral triangles is governed by the law of solid geometry. This can be calculated from the expression:

$$T = Pf^2,$$

where T, the *triangulation number*, is the number of smaller, identical equilateral triangles, P is given by the expression $h^2 + hk + k^2$. In this expression, h and k are any pair of integers without common factors, i.e. h and k cannot be divided by any whole number to give the same values.

$$f = 1, 2, 3, 4, \text{ etc.}$$

For viruses examined so far, the values of P are 1 ($h = 1$, $k = 0$), 3 ($h = 1$, $k = 1$) and 7 ($h = 1$, $k = 2$). Representative values of T are shown in Table 2.1. Once the number of triangular subdivisions is known, the total number of subunits can easily be determined since it is equal to $60T$.

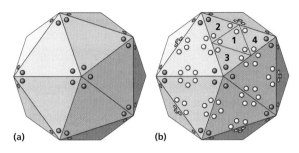

Fig. 2.6 Arrangement of 60*n* identical subunits on the surface of an icosahedron. (a) *n* = 1 and the 60 subunits are distributed such that there is one subunit at the vertices of each of the 20 triangular faces. Note that each subunit has the same arrangement of neighbours and so all the subunits are equivalently related. (b) *n* = 4. Each triangular face is divided into 4 smaller, but identical, equilateral triangular facets and a subunit is again located at each vertex. In total, there are 240 subunits. Note that each subunit, whether represented by an open or closed circle, has the identical arrangement of neighbours: see the face in which triangles 1–4 have been drawn. However, since some subunits are arranged in pentamers and others in hexamers, the members of each set are 'quasi-equivalently' related.

satellite tobacco necrosis virus is about 1000 nt. Presumably the volume of a 60-subunit structure is too small to accommodate the larger genome that is needed by an independent virus.

The virion is only 18 nm in diameter, compared the 30 nm of small independent viruses.

Detailed determination of the structure of a virus depends primarily on being able to grow crystals of purified virus, although only slightly lower resolution information can be obtained from cryoelectron microscopic examination of virus frozen in vitreous ice. The conditions required for crystallization of virus particles or proteins are not fully understood, and many will not form crystals at all. Large stable crystals are required that are then bombarded with X-rays. These are diffracted by atoms within the virion, and the image captured. Knowledge of the amino acid sequence of the proteins which comprise the particle makes it possible to determine the 3-dimensional crystal structure. X-ray analysis gives resolution to around 0.3 nm and the latest developments in cryoelectron microscopy allow resolution to around 0.4 nm. Both processes require high-powered computers that are used to make the necessary calculations and for image reconstruction.

The morphological units of virus particles seen by electron microscopy are called capsomers and *the number of these need not be the same as the number of protein subunits*. The numbers of morphological units seen will depend on the size and physical packing of the

Table 2.1

Values of capsid parameters in a number of icosahedral viruses. The value of *T* was obtained from examination of electron micrographs, thus enabling the values of *P* and *f* to be calculated.

P	*f*	*T* (= Pf^2)	No of subunits (60T)	Example
1	1	1	60	Tobacco necrosis satellite virus
3	1	3	180	Tomato bushy stunt virus, picornaviruses*
1	2	4	240	Sindbis virus
1	4	16	960	Herpesviruses
1	5	25	1500	Adenoviruses**

*In fact, picornaviruses have a pseudo T = 3 structure (see below).
**see text.

subunits and on the resolution of electron micrographs. A repeating subunit may consist of a *complex of several proteins*, such as the four structural proteins of poliovirus (see below and Section 12.3), or a *fraction of a protein*, such as the adenovirus hexon protein, half of which is considered to be a single repeating subunit.

T = 3: the molecular basis for quasi-equivalent packing of chemically identical polypeptides – tomato bushy stunt virus

Some plant virus particles achieve the $T = 3$, 180-subunit structure (Fig. 2.7) while encoding only a single virion polypeptide. They compensate for the physical asymmetry of quasi-equivalence by each polypeptide adopting one of three subtly different conformations. The virion polypeptide of tomato bushy stunt virus has three domains P, S and R (Fig. 2.8a): this is folded so that P and S are external and hinged to each other, while R is inside the virion and has a disordered structure. An arm (a) connects S to R, h connects S and P (Fig. 2.8b).

Each triangular face is made of three identical polypeptides, but these are in different conformations to accommodate the quasi-equivalent packing. For example, the C subunit has the S and P domains orientated differently from the A and B subunits (Fig. 2.8c), while the arm (a) is ordered in C and disordered in A and B (not shown). The S domains form the viral shell with tight interactions, while the P

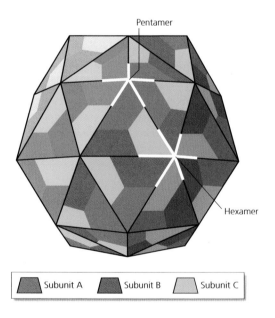

Fig. 2.7   Schematic diagram of a $T = 3$, 180-subunit virus. Each triangle is composed of three subunits, A, B and C, which are asymmetric by virtue of their relationship to other subunits (pentamers or hexamers). (Copyright 1991, from *Introduction to Protein Structure* by C. Branden & J. Tooze. Reproduced by permission of Routledge Inc., part of The Taylor & Francis Group.).

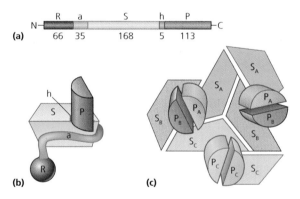

Fig. 2.8   (a) The linear arrangement of domains present in the single virion polypeptide of tomato bushy stunt virus. (b) Conformation of the polypeptide. The S domain forms the shell of the virion, while P points outwards and R is internal. (c) This shows a triangular face, composed of subunits A, B and C, and the interaction of the P domains to form dimeric projections (see text). (Copyright 1991, from *Introduction to Protein structure* by C. Branden & J. Tooze. Reproduced by permission of Routledge Inc., part of The Taylor & Francis Group.).

domains (total = 180) interact across the twofold axes of symmetry to form 90 dimeric protrusions. This virion is 33 nm in diameter and can accommodate a single-stranded RNA genome about fourfold larger than that of the satellite viruses. Thus a larger particle can be achieved without any more genetic cost.

T = 3: with icosahedra constructed of four different polypeptides – picornaviruses

Picornaviruses are made of 60 copies of each of four polypeptides: VP1, VP2, VP3 and VP4. VP4 is entirely internal. The repeating subunit of picornaviruses is the complex of VP1, VP2 and VP3. This should generate a T = 1 particle. However, despite the significant differences in amino acid sequence, the proteins adopt very similar conformations and, in geometric terms, appear as separate repeating subunits. For this reason, the assembled picornavirus particles appear to have a T = 3 structure; more accurately, this is a pseudo T = 3 (compare Figs 2.10 and 2.7).

The pentamers contain 15 polypeptides, with five molecules of VP1 forming a central vertex. These pentamers are the building-blocks in the cell from which the virion is assembled. Use of three polypeptides gives a chemically more diverse structure and may be an adaptation to cope with the immune system of animal hosts.

The cell receptor attachment site of picornavirus particles

Many virus proteins, including the VP1, VP2 and VP3 of picornaviruses, have the same type of structure – an antiparallel β barrel also known as a 'jelly-roll' (Fig. 2.9). Crystallographic, biochemical and immunological data have together identified a depression within the β barrel of VP1 which is the attachment site of picornaviruses. There are 60 attachment sites per virion. The arrangement of the β strands of VP1 is such that an annulus is formed around each fivefold axis of symmetry (Fig. 2.10). In the rhino (common cold) viruses, this is particularly deep and is called a 'canyon'. The canyon lies within the structure of the β barrel. Amino acids within the canyon are invariant, as expected from their requirement if they have to interact with the cell receptor, while amino acids on the rim of the canyon are variable. Only the latter interact with antibody. It is thought that the floor of the canyon has evolved so that it physically cannot interact with antibody. This avoids immunological pressure to accumulate mutations in order to escape from reaction with antibody, since these would at the same time render the attachment site non-functional, and hence be lethal to the virus.

An unknown structure: virus particles with 180 + 1 subunits and no jelly-roll β barrel – RNA bacteriophages

The leviviruses are 24 nm icosahedral RNA bacteriophages, and include MS2, R17 and Qβ. They encode two coat proteins. There are 180 subunits of one of these arranged with T = 3, but only a single copy of the second 'A' protein in each particle. This is the attachment protein. It is not known how the single subunit is incorporated into the particle. The main coat protein does not form a jelly-roll β barrel like those described above, but instead has five antiparallel β strands arranged like the vertical elements of battlements. Two subunits interact to form a sheet consisting of 10 antiparallel β strands.

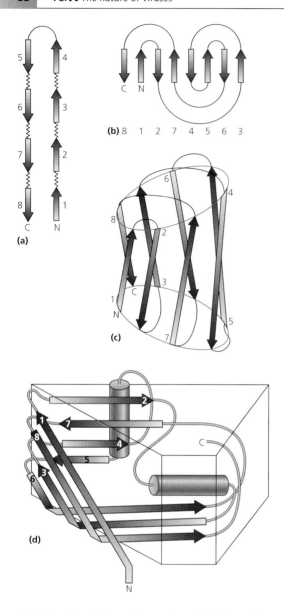

(a)

(b) 8 1 2 7 4 5 6 3

(c)

(d)

T = 25: more complex animal virus particles – adenoviruses

Careful examination of electron micrographs of adenoviruses shows that they have a $T = 25$ structure, with 240 hexons and 12 pentons, and a fibre projecting from each vertex (Fig. 2.11a, b). The fibres, the pentons and the hexons are all constructed from different proteins. This presents the problem of arranging not one, but three different proteins in a regular fashion, while adhering to the design principles outlined above. This can be achieved by arranging the pentons and the fibres at the vertices of the icosahedron and the hexons on the faces (Fig. 2.11c). However, the formula $60T$ gives the number of subunits as 1500 (Table 2.1). How is this difference resolved? The 240 hexons are composed of three identical polypeptides and each functions as *two* repeating subunits. Thus there are 1440 hexon subunits. The 12 pentons are formed from five identical polypeptides and each functions as a subunit, making 60 of these. Thus 1440 hexon subunits + 60 penton subunits make up the 1500 predicted subunits. Actually, the hexons are not all spatially equivalent, as those surrounding the vertex pentons contact five other hexons, while the others contact six hexons.

Fig. 2.9 The jelly-roll or antiparallel ß barrel: a common structure of plant and animal virion proteins. The formation of an antiparallel ß barrel from a linear polypeptide can be visualized to occur in three stages. First, a hairpin structure forms where ß strands are hydrogen-bonded to each other: 1 with 8, 2 with 7, 3 with 6, and 4 with 5, creating antiparallel ß strand pairs separated by loop regions of variable length (a). Second, these pairs become arranged side by side, so that further hydrogen bonds can be formed by newly adjacent strands, e.g. 7 with 4 (b) Third, the strand pairs wrap around an imaginary barrel forming a 3-dimensional structure. The eight ß strands are arranged in two sheets, each composed of four strands: strands 1, 8, 3 and 6 form one sheet and strands 2, 7, 4 and 5 form the second sheet (c). The dimensions of the barrel are such that each protein forms a wedge, and these wedges are subunits that are assembled into virus particles (d). ((a–c) (Copyright 1991, from *Introduction to Protein Structure* by C. Branden & J. Tooze; reproduced by permission of Routledge Inc., part of The Taylor & Francis Group. (d) From Hogle et al., 1985 *Science* 229, 1358.)

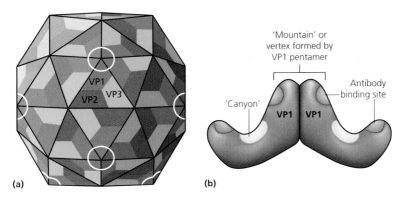

Fig. 2.10   A picornavirus particle (a) and the attachment site formed by VP1 shown as an annulus around the fivefold axis of symmetry, (from Smith et al., 1993, *Journal of Virology*, 67, 1148.) (b) A vertical section through a VP1 pentamer where the cross-section of the annulus is referred to as the 'canyon'. (From Luo et al., 1987, *Science* 235, 182).

Triple-shelled particles: capsids within capsids – rotaviruses

A different and very complex structural arrangement is found in another class of isometric viruses, the rotaviruses, the particles of which contain three distinct layers of capsids surrounding a core structure. Rotavirus encodes 12 polypeptides. Of these, six are located in the virion: three forming the inner shell, one forming the middle shell and two forming the outer shell (Table 2.2). The outer and middle shells have icosahedral symmetry with the outer shell having a T = 13 symmetry

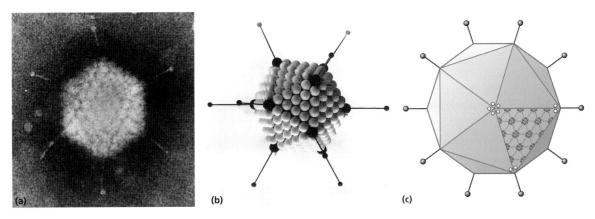

Fig. 2.11   The structure of adenoviruses. (a) A negatively stained electron micrograph of an adenovirus. (b) Model of an adenovirus to show arrangements of the capsomers. (c) Schematic diagram to show the arrangements of the subunits on one face of the icosahedron. Note the subdivision of the face into 25 smaller equilateral triangles. (Photographs courtesy of Nicholas Wrigley.)

Table 2.2

Rotavirus proteins and their location in the virion, and some of their properties. The non-structural proteins are also shown.

Location in virion or non-structural	Protein	Encoding RNA segment	Number of polypeptides per virion	Function
Inner	VP1	1	12	Virus RNA dependent RNA polymerase
Inner	VP2	2	120	Structural protein of inner capsid
Inner	VP3	3	12	Guanylyltransferase of virus RNA dependent RNA polymerase
Middle	VP6	6	780	Structural protein of middle capsid
Outer	VP4	4	120	Outer capsid spike (dimer) converted by cleavage into VP5* and VP8*
Outer	VP7	9	780	Outer capsid structural glycoprotein
Non-structural	NSP1	5		Binds several host proteins during infection
Non-structural	NSP3	7		Binds 3' end of virus mRNA for translation initiation
Non-structural	NSP2	8		ssRNA binding protein
Non-structural	NSP4	10		Virus enterotoxin
Non-structural	NSP5	11		ss RNA binding protein
Non-structural	NSP6	11		Interacts with NSP5

(13×60 subunits) and the intermediate having $T=2$ symmetry (2×60 subunits). The inner shell consists of 120 molecules of the VP2 protein which forms a stable structure to support the middle shell comprised of the VP6 protein. The core contains the genome comprising 11 segments of dsRNA and single molecules of the VP1 and VP3 proteins, which form the virion RNA polymerase, located at each of the 12 vertices of the icosahedron (Fig. 2.12). The polymerase forms an integral component of the structure at the vertices and is required for the production of virus mRNA (see Section 11.2). The precise 3-dimensional arrangement of all of the molecules in the rotavirus particle has been determined. The 11 double-stranded RNAs are thought to be tightly coiled like the DNA inside phage heads. Each molecule is thought to be associated with a polymerase complex and tethered near a vertex.

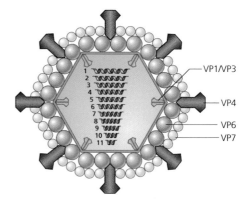

Fig. 2.12 The triple shelled structure of rotavirus showing the location of polypeptides in the virion.

2.4 Enveloped (membrane-bound) virus particles

Although they appear complex, enveloped viruses have a conventional isometric or helical structure that is surrounded by a membrane – a 4 nm-thick lipid bilayer containing proteins. The internal structure is usually referred to as a nucleocapsid to differentiate it from the capsid of viruses that lack an envelope. Examples of enveloped viruses include many of the larger animal viruses, but only a few plant and bacterial viruses. Traditionally, the enveloped viruses were distinguished from non-enveloped viruses by treatment with detergents or organic solvents, which disrupts the membrane and destroys infectivity. Thus they were sometimes referred to as 'ether-sensitive viruses'. The envelope, which is derived from host cell membranes, is obtained by the virus *budding* from cell membranes, and most contain no cell proteins (see Section 12.6). How cell proteins are excluded and why some viruses do not exclude cell proteins from their virions is not understood.

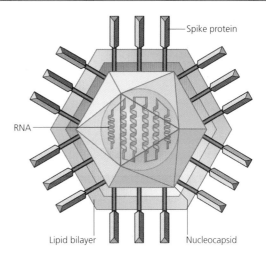

Fig. 2.13 Sindbis virus: an enveloped icosahedron. The core is $T = 3$, but the envelope is $T = 4$. See text for the explanation. (Courtesy of S. D. Fuller.)

An isometric core surrounded by an isometric envelope: Sindbis virus particles

Sindbis virus (a togavirus) has an icosahedral nucleocapsid that comprises a single protein, surrounded by an envelope from which viral spike proteins protrude. The core has $T = 3$ and 180 subunits, exactly like tomato bushy stunt virus described above. Surprisingly, the envelope also has icosahedral symmetry, but to everyone's surprise this is $T = 4$ and has 240 subunits. This apparent paradox was resolved when it was found that the two structures are complementary, so that the internal ends of the spike proteins fit exactly into depressions between the subunits of the nucleocapsid (Fig. 2.13). So far, this and its near relations are the only enveloped viruses that are known to have a geometrically symmetrical envelope.

A helical core surrounded by an approximately spherical envelope: the influenza virus particle

One of the best studied groups of enveloped viruses is the influenza viruses. The helical core is composed of the matrix (M1) protein surrounding ribonucleoprotein, itself composed of flexible rods of RNA and NP protein. This is constructed as described in Section 2.2, and arranged in a twisted hairpin structure. The genome is segmented RNA and there are eight separate core structures. Each is associated with a transcriptase complex. The core is contained within a lipid envelope that is only roughly spherical, and hence often described as pleomorphic (Fig. 2.14).

In electron micrographs (Fig. 2.14), a large number of protein spikes projecting about 13.5 nm from the viral envelope can be observed. These spikes, which have an overall length of 17.5 nm, are transmembrane glycoproteins like many of those in cell membranes. The spike layer consists solely of

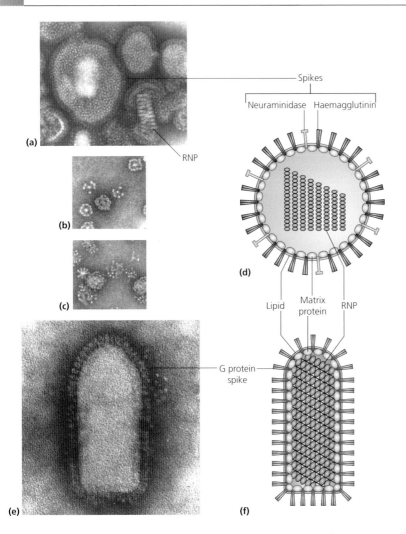

Fig. 2.14 Influenza A virus (an orthomyxovirus) and vesicular stomatitis virus (a rhabdovirus): viruses with enveloped helical structures. Although their morphology is different, these viruses are constructed in the same way. (a) Negatively stained electron micrograph of influenza A virus showing the internal helical ribonucleoprotein (RNP) core and the surface spikes. (b) Aggregates of purified neuraminidase. (c) Aggregates of purified haemagglutinin. Note the triangular shape of the spikes when viewed end-on. (d) Schematic representation of the structure of influenza virus. (e) Negatively stained electron micrograph of vesicular stomatitis virus. (f) Schematic representation of the structure of vesicular stomatitis virus. (Electron micrographs courtesy of Nicholas Wrigley and Chris Smale.)

virus-specified glycoproteins, and comprises about 800 haemagglutinin (HA) and 200 neuraminidase (NA) proteins. HA proteins are trimers and NA proteins are tetramers. NA spikes are arranged non-randomly in clusters on the virion surface. The HA protein functions in attachment and fusion-entry, and the NA protein releases infecting virus from receptors that do not lead to entry and infection. The NA also releases progeny virions which reattach to the host cell in which they were formed. Receptor analogues (e.g. oseltamivir) are antivirals that prevent virus release (see Section 27.3). The two types of spike are morphologically distinct. A few molecules of an ion channel protein called M2 are also found in the membrane. This allows the passage of protons into the core and is necessary for secondary uncoating.

A helical core surrounded by a non-spherical envelope: rhabdovirus particles

This group of viruses has a helical core consisting of just nucleoprotein, and an envelope like the influenza viruses. What is distinctive about these viruses is that they not isometric, but either bullet-shaped or bacilliform (rounded at both ends). These are unique morphologies. There are both plant and animal rhabdoviruses, and the latter are all bullet-shaped. The envelope contains a dense layer of spikes comprising just one protein, the viral attachment protein, G. The matrix protein underlies the membrane. Apart from their strikingly different overall shape, rhabdoviruses and influenza viruses have fundamentally a very similar structure (Fig. 2.14d, f). The length of the bullet appears to be controlled by the size of the RNA genome as defective-interfering rhabdoviruses, which have a massively deleted genome, form small bullet-shaped particles. Nothing is known of the structural geometry involved the formation of rhabdovirus particles.

2.5 Virus particles with head-tail morphology

The head-tail architectural principle is unique to bacterial viruses (Fig. 2.15), although many bacterial viruses have other morphologies

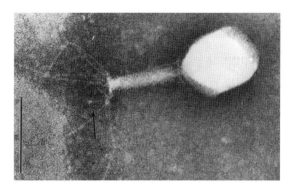

Fig. 2.15   Electron micrograph of bacteriophage T2 with six long tail fibres. Tail pins are not evident but a short fibre (indicated by the arrow) can be seen. The bar is 100 nm. (Courtesy of L. Simon.)

(Table 2.3). There is a large variation on the head-tail structural theme and bacteriophages can be subdivided into those with short tails, long non-contractile tails, and complex contractile tails. A number of other structures, such as base plates, collars, etc., may be present. Despite their complex structure, the design principles involved in head-tail phages are identical to those outlined earlier for the viruses of simpler architecture. Heads usually possess icosahedral symmetry, whereas tails usually have helical symmetry. All other structures, base plates, collars, etc., also possess a defined symmetry. The evolution of these elaborate structures may be connected with the way in which these bacterial viruses infect susceptible cells (see Section 6.5). In brief, the phage attaches to a bacterium via its tail, enzymatically lyses a hole in the cell wall and inserts its DNA, which is tightly packed into the phage head, into the cell using the tail as a conduit.

Some of the larger viruses fit none of these structures, and the rules governing their formation have not yet been elucidated. For example, the poxviruses of animals have a complex enveloped structure enclosing two lateral bodies and a biconcave core that includes

Table 2.3

Distribution of the various types of virus particle structure among families of animal, plant and bacterial viruses.

Type of particle structure	Animal viruses	Plant viruses	Bacterial viruses
Non-enveloped icosahedral	Common	Common	Uncommon: cortico-, levivi-, micro-, and tectiviruses
Non-enveloped helical	Not known	Common	Rare: ino- and rudiviruses
Enveloped icosahedral	Uncommon: arteri-, asfar-, borna-, flavi-, and togaviruses	Not known	Rare: cystoviruses
Enveloped helical	Common	Uncommon: bunya- and rhabdoviruses	Rare: lipothrixviruses
Head-tail	Not known	Not known	Common

all the enzymes required for viral mRNA synthesis, and the giant mimivirus, that infects protozoa, has a 400 nm non-enveloped, spherical particle that is surrounded by an icosahedral capsid and fibrils.

2.6 Frequency of occurrence of different virus particle morphologies

The different virus morphologies discussed above do not occur with equal frequency among animal, plant and bacterial viruses (Table 2.3). There are relatively few purely icosahedral viruses in bacteria; non-enveloped helical viruses are common and occur almost exclusively in plants; enveloped icosahedral viruses and enveloped helical viruses are common in animals and rare in plants and bacteria. Finally, head-tail virus morphology, in which an isometric head and a helical tail are joined together, is found only in bacteria. Unfortunately, there is no real explanation as to why there should be these restrictions. There also exist some very large, very complex viruses

(e.g. poxviruses of animals and mimivirus of amoebae) whose morphogenesis is beyond our current comprehension.

2.7 Principles of disassembly: virus particles are metastable

It is important to remember that all virus particles not only have to be constructed to protect the genome, but they also have to disassemble to permit the genome to enter a new target cell. This is supremely important to the virus particle as it has only one chance to do this successfully and hence have the chance to propagate its genome. One theory to explain this paradox is that the particle is metastable, i.e. it is stable but it can be triggered by interaction with the receptor, and other events to descend to a lower energy level and, in doing so, releases its genome. Not surprisingly, there are a number of fail-safe devices that tell the virus when it is safe to let go the genome. One of the simplest systems is used by enveloped animal viruses like HIV-1. This virus undergoes a succession of interactions between cell receptors and virus envelope protein binding sites, each interaction providing a molecular password that is needed to gain entry to the next level. If everything is in order, the metastable envelope protein then undergoes profound rearrangements that allow a hidden hydrophobic segment to insert into the cell membrane. This initiates fusion of the lipid bilayer of the virus with that of the cell plasma membrane, and the virus genome automatically enters the cell cytoplasm. However, if the sequence of passwords proves incorrect, the virus detaches from the cell and the process can be repeated until the correct cell is found. Mechanisms of entry are discussed in Chapter 6.

Key points

- Virus particles consist mainly of nucleic acid and protein, arranged to form regular geometric structures.
- Some viruses have particles that are surrounded by a membrane.
- Virus proteins protect the viral genome, identify the appropriate target cell, and get the genome into the target cell.
- Some viruses contain proteins with enzymatic functions that are needed for genome replication, and these enter the cell with the genome.
- Elongated virus particles have their protein subunits arranged in a helix, and are said to have helical symmetry.
- Isometric virus particles are usually icosahedrons – structures with 20 faces, with protein subunits arranged to retain the elements of icosahedral symmetry.
- Membrane-bound viruses have an internal nucleoprotein structure with helical symmetry or with isometric symmetry.

Further reading

Branden, C., Tooze, J. 1999. *Introduction to Protein Structure*, 2nd edn. Garland Publishing, New York.

Butler, P. J. G. 1984. The current picture of the structure and assembly of tobacco mosaic virus. *Journal of General Virology* 65, 253–279.

Caspar, D. L. D., Klug, A. 1962. Physical principles in the construction of regular viruses. *Cold Spring Harbor Symposium of Quantitative Biology* 27, 1–24.

Eiserling, F. A. 1979. Bacteriophage structure. In *Comprehensive Virology* (H. Fraenkel-Conrat, R. R. Wagner, Eds), Vol. 13, pp. 543–580. Plenum Press, New York.

Crowther, R. A. 2008. The Leeuwenhoek Lecture 2006. Microscopy goes cold: frozen viruses reveal their structural secrets. *Philosophical Transactions of the Royal Society of London B: Biological Sciences* 363, 2441–2451.

Johnson, J. E. 2003. Virus particle dynamics. *Advances in Protein Chemistry* 64, 197–218.

Rixon, F. J., Chiu, W. 2003. Studying large viruses. *Advances in Protein Chemistry* 64, 379–408.

Smith, T. J., Baker, T. 1999. Picornaviruses: epitopes, canyons, and pockets. *Advances in Virus Research* 53, 1–23.

Stuart, D. 1993. Virus structures. *Current Opinion in Structural Biology* 3, 167–174.

Tyler, K. L., Oldstone, M. B. A. 1998. Reoviruses I. Structure, proteins and genetics. *Current Topics in Microbiology and Immunology* 233, 1–213.

Wiley, D. C., Skehel, J. J. 1987. The structure and function of the hemagglutinin membrane glycoprotein of influenza virus. *Annual Review of Biochemistry* 56, 365–394.

Zhou, Z. H., Chiu, W. 2003. Determination of icosahedral virus structures by electron cryomicroscopy at sub-nanometer resolution. *Advances in Protein Chemistry* 64, 93–124.

For images of virus particles see the Virus Particle Explorer (VIPER) website: http://viperdb.scripps.edu

Chapter 3
Classification of viruses

Viruses are found throughout the world and infect all known organisms. Viruses may cause a range of different diseases and display a diversity of host range, morphology and genetic makeup. Bringing order to this huge diversity requires the designation of classification groups to permit study of representative viruses that can inform us about their less well-studied relatives.

Chapter 3 Outline

3.1 Classification on the basis of disease
3.2 Classification on the basis of host organism
3.3 Classification on the basis of virus particle morphology
3.4 Classification on the basis of viral nucleic acids
3.5 Classification on the basis of taxonomy
3.6 Satellites, viroids and prions

Viruses represent one of the most successful types of parasite in the world and have been isolated from representatives of every known group of organisms from the smallest single-celled bacterium to the largest mammal. Whilst in most cases the virus is specific for the host species in which it has been identified, some viruses are able to infect species from different phyla and even different kingdoms. The number of known viruses now exceeds 3000 with new viruses being discovered all the time. This very large number contains a diverse array of viruses which at first sight is very bewildering. To make the study of viruses easier and to bring order to this apparent diversity, over the years a number of different systems have been proposed to generate classification schemes which will allow us to study representative viruses rather than each individually. All of the proposed classification schemes have different strengths and weaknesses but there is now general consensus. The International Committee on Taxonomy of Viruses (ICTV) has responsibility for assignment of new viruses to specific phylogenetic groupings.

3.1 Classification on the basis of disease

The first, and most common, experience of viruses is as agents of disease and it is possible to group viruses according to the nature of the disease with which they are associated. Thus, one can discuss hepatitis viruses or viruses causing the common cold. This is attractively simple. However, this method of grouping viruses, though reflecting an important characteristic, suffers from serious deficiencies. Firstly, this approach is very anthropomorphic, focusing as it does on diseases that we recognize because they affect humans or our domestic livestock or crops. This ignores the fact that

Introduction to Modern Virology, Seventh Edition. N. J. Dimmock, A. J. Easton and K. N. Leppard.
© 2016 John Wiley & Sons, Ltd. Published 2016 by John Wiley & Sons, Ltd.

most viruses either do not cause disease or cause a disease that we do not recognize because of a lack of understanding of the host; for example, we understand little of the diseases caused by viruses of fish or amphibians. Similarly, it is possible for a single virus to cause more than one type of disease; a good example of this is varicella zoster virus, which causes chickenpox during a first infection but causes shingles when reactivated later in life. This problem is compounded when considering viruses which infect more than one host as it is common to find that they cause either no disease in one host whilst dramatically affecting the other, or that they may cause different diseases in different hosts. A classification based on disease, while it may be helpful in some settings, also fails in the important feature of being able to use the groupings to predict common fundamental features of the viruses in question. In the case of agents of hepatitis and the common cold, many different viruses with very different molecular make-ups are involved and studying just one of these tells us little, if anything, about the others.

3.2 Classification on the basis of host organism

An alternative approach has been to group viruses according to the host that they infect. This has the attraction that it emphasizes the parasitic nature of the virus–host interaction. However, there are several difficulties with this approach. This form of classification implies a fixed, unchanging link between the virus and host in question. Some viruses are very restricted in their host range, infecting only one species, such as hepatitis B virus infecting humans, so a designation based on the host is appropriate. However, others may infect a small range of hosts, such as poliovirus which can infect various primates, and the designation

here must reflect this rather than name a single species. The most serious difficulty arises with viruses which infect and replicate within very different species. This can be seen with certain viruses which can infect and replicate within both plants and insects. Similarly, the first host from which a new virus is isolated may ultimately prove not to be the most common one. Designation of a virus by the host it infects is therefore not always straightforward. Overriding all of these difficulties is the problem that, even if a number of viruses infect a single species, this characteristic does not imply any other similarities in terms of disease or genetic makeup of the various viruses.

A different level of sophistication in terms of defining the host has been used for some viruses, notably the herpesviruses. Having shown that herpesviruses are similar in a number of ways, a classification was described which defined them in terms of the nature of the host cell they infected, for example, gammaherpesviruses infecting lymphoid cells in the host animal. With the discovery of new herpesviruses which infect lymphoid cells but share characteristics with the other non-gammaherpesviruses, this definition is no longer sustainable. Consequently, studying a single virus which infects a single species or group of species, or a virus which infects a particular cell type, tells us nothing about the fundamental nature of the potentially many other viruses which also infect that host.

3.3 Classification on the basis of virus particle morphology

The structural features of virus particles and the principles which underlie these structures have been described in Chapter 2. When viruses were first visualized in the electron microscope, defining classification groups on the basis of the observed particle shape or morphology was relatively simple. A key structural feature is whether or not the virus particle has a lipid envelope and this alone can be used as a designated feature, giving enveloped and non-enveloped viruses (Section 2.4). If the virion is non-enveloped, three morphological categories are defined: isometric, filamentous and complex. Isometric viruses (Section 2.3) appear approximately spherical but are actually icosahedrons or icosadeltahedrons. Filamentous viruses (Section 2.2) have a simple, helical morphology. The complex viruses are those which do not neatly fit within the other two categories. Complex shapes for virus particles may be made up of a combination of isometric and filamentous components, such as is seen with bacteriophage T2 (Figure 2.15), or they may have a structure which does not conform to the simple geometrical rules of the majority and appear to our eye to be irregular in shape. If the virion is enveloped, a further level of classification is possible by describing the morphology of the nucleocapsid found within the membrane. Thus, there are isometric and helical nucleocapsids.

While a classification scheme based on morphology is simple and describes an unchanging feature of the virus, it suffers from several drawbacks. Primary amongst these is that knowing the shape of a virus particle does not allow us to predict anything about the biology, pathology or molecular biology of similarly-shaped viruses. Thus, two viruses with very similar morphologies may differ in all of their other fundamental characteristics. This drawback is also true even for viruses which appear to share a number of other features. For example, the polyomaviruses and the papillomaviruses were originally classified together on the basis of their very similar morphology and the similarity extended to other, deeper features of their structures including the nature and organization of their genomes. A better understanding of these

viruses at the molecular level has shown that they differ in several critical areas and they are now recognized as quite different entities. Despite these problems, a preliminary description of a virus in terms of morphology is still common.

3.4 Classification on the basis of viral nucleic acids

The nucleic acid of a virus contains all the information needed to produce new virus particles. Some of this information is used directly to make virion components and some to make accessory proteins or to provide signals that allow the virus to subvert the biosynthetic machinery of a cell and redirect it towards the production of virus. Whereas the standard form of genetic material in living systems is double-stranded DNA, viruses contain a diverse array of nucleic acid forms and compositions. The nucleic acid content of a virus has been used as a basis for classifying viruses. The key aspect of this classification scheme is that it considers the nature of the virus genome in terms of the mechanisms used to replicate the nucleic acid and transcribe mRNA encoding proteins. A detailed consideration of the nature of virus nucleic acids and the mechanisms by which they are replicated and transcribed are to be found in Chapters 7 to 11. Here we will consider only how such features can be used to generate a classification scheme.

The nature of a particular nucleic acid sample is assessed by determining its base composition, sensitivity to DNase or RNase, buoyant density, etc. Single-stranded nucleic acids are distinguished from double-stranded by the absence of a sharp increase in absorbance of ultraviolet light upon heating and the non-equivalence of the molar proportions of adenine (A) and thymine (T) (or uracil (U)) or guanine (G) and cytosine (C). From these types of analysis, it appears that viruses utilize four possible kinds of viral nucleic acid: single-stranded DNA, single-stranded RNA, double-stranded DNA, and double-stranded RNA. Each kind of genome is found in many virus families, which between them contain members that infect a diverse array of animals, plants and bacteria.

Classifying viruses – the Baltimore scheme

As considered above, viruses exhibit great diversity in terms of morphology, genome structure, mode of infection, host range, tissue tropism, disease (pathology), etc. While, as we have seen, each of these properties can be used to place viruses into groups, classifying viruses solely on the basis of one or even two of these parameters does not lead to a system where studying one virus in a particular group can be used to draw inferences about other members of the same group. Also, classification on these grounds does not give a good basis for unifying discussions of virus replication processes. To circumvent these problems, Nobel Laureate David Baltimore proposed a classification scheme which encompasses all viruses based on the nature of their genomes, and their modes of replication and gene expression. This system provides an opportunity to make inferences and predictions about the fundamental nature of all viruses within each defined group.

The original Baltimore classification scheme was based on the fundamental importance of messenger RNA (mRNA) in the replication cycle of viruses. Viruses do not contain the molecules necessary to translate mRNA and rely on the host cell to provide these. They must therefore synthesize mRNAs which are recognized by the host cell ribosomes. In the Baltimore scheme, viruses are grouped according to the mechanism of mRNA synthesis

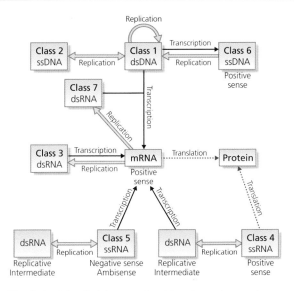

Fig. 3.1 The Baltimore classification scheme (see text for details). mRNA is designated as positive sense. Translation of protein from mRNA and positive sense RNA virus genomes is indicated with dashed red arrows. The path of production of mRNA from double-stranded templates is shown with solid black arrows. Blue arrows show the steps in the replication of the various types of genome with double-headed arrows indicating production of a double-stranded intermediate from which single stranded genomes are produced.

which they employ (Fig. 3.1). By convention, all mRNA is designated as positive (or 'plus') sense RNA. Strands of viral DNA and RNA which are complementary to the mRNA are designated as negative (or 'minus') sense and those that have the same sequence are termed 'positive' sense. Using this terminology, coupled with some additional information about the replication process, a modified classification scheme based on the original proposed by Baltimore defines seven groups of viruses, with each commonly being referred to by the nature of the virus genomes it includes:

Class 1 contains all viruses that have double-stranded (ds) DNA genomes. In this class, the designation of positive and negative sense is not meaningful since mRNAs may come from either strand. Transcription can occur using a process similar to that found in the host cells.

Class 2 contains viruses that have single-stranded (ss) DNA genomes. The DNA can be of positive or negative sense, depending on the virus being studied. For viruses in class 2, the DNA must be converted to a double-stranded form before the synthesis of mRNA can proceed.

Class 3 contains viruses that have dsRNA genomes. All known viruses of this type have segmented genomes and mRNA is only synthesized from one template strand of each segment. The process of transcription from a dsRNA genome can be envisioned as occurring using a mechanism similar to that for transcription from a dsDNA genome. However, the enzymes necessary to carry out such a process do not exist in normal, uninfected, cells. Consequently, these enzymes must be encoded by the virus genome and must be carried into the cell by the virus to initiate the infectious process.

Class 4 contains viruses with ssRNA genomes of the same (positive) sense as mRNA and which can be translated. Synthesis of a complementary strand, generating a dsRNA intermediate, precedes synthesis of mRNA. As with class 3 viruses, the RNA synthesis must be carried out using virus-encoded enzymes, although these are not carried in the virus particle.

Class 5 contains viruses that have ssRNA genomes which are complementary in base sequence to the mRNA (negative-strand RNA viruses). Synthesis of mRNA requires novel virus-encoded enzymes, and generation of new virus genomes requires the synthesis of a dsRNA intermediate, the positive sense strand of which is used as a template for replication. Viral RNA-synthesizing enzymes are carried in the virion. Some class 5 viruses use the newly-synthesized 'antigenome' RNA strand as a template for production of an mRNA and are referred to as 'ambisense' viruses.

Class 6 contains viruses that have ssRNA genomes and which generate a dsDNA intermediate as a prelude to replication, using an enzyme carried in the virion.

Class 7. More recently, it has been suggested that some viruses, termed reversiviruses, should be transferred from class 1 into a new class 7. This is based on their replication from dsDNA via a positive sense ssRNA intermediate back to dsDNA. This represents the inverse of the class 6 replication strategy, with which class 7 has many similarities.

The Baltimore scheme has both strengths and weaknesses as a tool for understanding virus properties. A particular strength is that assignment to a class is based on fundamental, unchanging characteristics of a virus. Once assigned to a class, certain predictions about the molecular processes of nucleic acid synthesis can be made, such as the requirement for novel virus-encoded enzymes. A weakness is that, whilst it brings together viruses with similarities of replication mechanism, the scheme takes no account of their biological properties. For example, bacteriophage T2 and variola virus (the cause of smallpox) are classified together in class 1 although they are totally dissimilar in both structure and biology. Similarly, the identification of a positive sense RNA genome is not sufficient to classify the virus unambiguously since viruses of classes 4 and 6 have similar genome nucleic acids.

3.5 Classification on the basis of taxonomy

The International Committee on Taxonomy of Viruses (ICTV), first founded in the late 1960s, has established a taxonomic classification scheme for viruses. This uses the familiar systematic taxonomy scheme of Order, Family, Subfamily and Genus (no Kingdoms, Phyla or Classes of virus have been described within this scheme). However, the concept of a species in the classification of viruses is complex and in many cases remains a point of ongoing debate. For viruses with RNA genomes, the concept of a species is made difficult by the absence of a proofreading function during genome replication. The result of this is that a virus exists as a member of a population where each member has a genome sequence which may be different to the others but which belongs to a collection of sequences which will combine to form a consensus for that virus. The virus is said to be a quasispecies, and there is no defined 'correct' genome sequence (this is discussed in Chapter 4).

In assigning a virus to a taxonomic group, the ICTV considers a range of characteristics. These include host range (eukaryote or prokaryote, animal, plant, etc.), morphological features of the virion (enveloped, shape of capsid or nucleocapsid, etc.) and nature of the genome nucleic acid (DNA or RNA, single stranded or double stranded, positive or negative sense, etc.). Within these parameters additional features are considered. These include such things as the length of the tail of a phage or the presence or absence of specific genes in the genomes of similar viruses, and these aspects allow allocation of subdivisions in the taxonomic designation. With the advent of rapid nucleotide sequencing of virus genomes it is now possible to derive phylogenetic trees based on the degree of conservation of specific gene sequences and this further informs the designation of phylogenetic groupings (clades). The use of phylogeny is particularly useful when considering newly-discovered viruses for which little other information may be available. For each of the individual genera defined by the ICTV a single virus has been designated as the type member. The type member is used essentially as the reference for the genus. For an example of the classification process, see Box 3.1.

It can be seen from the details used in the classification of measles virus that the system utilizes aspects of all of the other classification schemes in an attempt to clearly identify each virus group. The classification of viruses is an ongoing project as new viruses are discovered. As our understanding of specific features grows, particularly in the area of phylogenetic analysis, the taxonomic groupings are constantly reassessed.

3.6 Satellites, viroids and prions

Satellites and viroids

Satellites and viroids have features which differentiate them from the better understood 'conventional' viruses. However, they also have features which are common to several viruses and they are often referred to as subviral agents. The similarities with conventional viruses have permitted a preliminary classification of viroids into Families and Genera.

Satellites are categorized by two forms: satellite viruses and satellite nucleic acid. Neither satellite viruses nor satellite nucleic acids encode the enzymes required for their replication and require co-infection with a conventional, helper virus to provide the replicative enzymes in the same way as DI viruses (Section 8.2). However, unlike DI viruses, the sequence of satellite genomes is significantly different to that of the helper virus. Satellite viruses encode the structural protein that encapsidates them but the satellite nucleic acids encode either only non-structural proteins or no proteins at all. The satellite nucleic acids derive their structural proteins from the

Box 3.1

Example of phylogenetic assignment of measles virus using the ICTV criteria

This describes the key elements for classification. Other more detailed criteria are also used.

Order: *Mononegavirales*. Enveloped viruses containing one single-stranded negative sense RNA molecule. The genome RNA contains a number of defined features; the regulatory sequences at the 3' and 5' ends are complementary and flank the core, conserved, genes which are present in a fixed order in the genome. The nucleocapsids are helical. The ribonucleoprotein of the virus core is the 'infectious unit'. The virion contains prominent spike-like structures on the surface.

Family: *Paramyxoviridae*. The virions are usually spherical and contain spikes consisting of two or three glycoproteins arranged in homopolymers. A nonglycosylated matrix protein is associated with the inner surface of the lipid membrane. The genome RNA is between approximately 13,000 nt and 15,500 nt in length. Genome RNA does not contain a 5' cap structure or a 3' polyadenylate tail.

Subfamily: *Paramyxovirinae*. The genome contains between five and seven transcriptional regions. Phylogenetic sequence relatedness exists between the proteins, with the nucleocapsid protein, the matrix protein and the polymerase protein generally the most highly conserved. Nucleocapsids have a diameter of 18 nm and a pitch of 5.5 nm. Surface spikes are approximately 8 nm in length. The genome length is a multiple of six nucleotides. All members have haemagglutinin activity associated with the virion.

Genus: Morbillivirus. Narrow host range which also distinguishes the members from each other. Members display greater sequence relatedness in phylogenetic trees within the genus than with members of the other genera in the subfamily. All members show the same pattern of transcription and RNA editing. All members generate intracytoplasmic and intranuclear inclusion bodies which are assemblies of virus nucleocapsids. Members show cross-reactivity in some, but not all, antibody neutralisation and cross-protection tests.

Type member: measles virus.

co-infecting helper virus. In some cases, the sequences of the immediate termini of satellite genomes are similar to those of the helper virus, suggesting that these regions are involved in replication as is frequently the case for the helper virus genomes.

The presence of a satellite or satellite virus may affect the replication of the helper virus and may also increase or decrease the severity of the disease(s) caused by the helper virus. Satellite viruses and satellite nucleic acids with either DNA or RNA genomes have been identified. The RNA genomes range in size from approximately 350 nt to 1500 nt and the DNA genomes, which are either single stranded or double stranded, range from 500 nt to 1800 nt. The RNA satellite genomes have been found in linear ssRNA, circular ssRNA or dsRNA forms.

A satellite that is important because of its association with disease in humans is hepatitis delta virus (HDV). HDV is only found in association with hepatitis B virus (HBV) and is associated with enhanced pathogenicity of HBV (see Section 22.5). The HDV genome consists of a circular ssRNA molecule which is extensively base-paired and appears as a rod-like structure. HDV is unable to replicate without a helper virus (HBV) which provides the structural proteins that encapsidate the genome and allow HDV to be spread. The HDV genome is approximately 1700 nt in length and encodes two proteins. Both proteins are translated from an mRNA which is of opposite sense to the HDV genome RNA, analogous to the method of gene expression used by the class 5 viruses (Chapter 11).

Viroids are novel agents of disease in plants; their infectious material consists of a single circular ssRNA molecule with no protein component. Viroids, with genomes ranging in size from 220 to 400 nucleotides, are the smallest self-replicating pathogens known. Up to 70% of the nucleotides in the genome RNAs are base-paired, and when genomes are examined in the electron microscope they appear as rod-shaped or dumb-bell-shaped molecules (Fig. 3.2), similar to the genome of HDV. Nucleotide sequence analysis of viroid RNA has shown that no proteins are encoded by either the genome or antigenome sense RNA. It is thought that the diseases associated with viroids result from the RNA interfering with essential host cell mechanisms.

Viroids and HDV appear to be replicated by the cellular DNA-dependent RNA polymerase II which normally recognizes a DNA template. Replication of viroid RNA is described in Section 8.6.

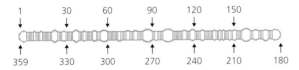

Fig. 3.2 Diagram of the circular single-stranded 359-nucleotide RNA of potato spindle tuber viroid, showing the maximized base-paired structure.

Prions

Prions were initially identified as the agents of a number of diseases characterized by slow, progressive neurological degeneration which was fatal. The diseases are associated a spongy appearance of the brain that is seen *post mortem*. Together, the diseases are termed spongiform encephalopathies and include scrapie in sheep, bovine spongiform encephalopathy (BSE; 'mad cow' disease) in cattle and the human infections of kuru and variant Creutzfeldt-Jakob disease. Prions have also been identified in the yeast *Saccharomyces cerevisiae* and the fungus *Podospora anserina*. Prions were initially thought to be viruses which replicated slowly within their hosts but, despite much work, no nucleic acid has yet been found in association with infectious material. Analysis of prions has shown that they are aberrant forms of normal cellular proteins which can induce changes in the shape of their normal homologues with catastrophic consequences for the host. The nature of these unusual agents and the diseases they cause is considered in Chapter 28.

Key points

- Viruses infect all known species.
- Viruses can be classified by a number of different methods including disease caused, host range, morphology and nature of the genome in association with the method of replication and transcription.
- Virus genomes consist of either DNA or RNA and may be single stranded or double stranded. Some viruses convert their genomes from RNA to DNA, or vice versa, for replication and transcription.
- The Baltimore system has the potential to predict general features of the replication and transcription of viruses from a knowledge of the nature of the genome.

- The best and most comprehensive classification scheme uses elements of all possible schemes, in association with phylogenetic analysis of nucleic acid sequences to generate a taxonomic system which accurately assigns viruses to a genus within an hierarchical structure.
- Satellites, satellite viruses and viroids differ from viruses in fundamental ways but satellites and satellite viruses rely on the presence of specific viruses to replicate. Satellites, satellite viruses and viroids are associated with serious disease in their hosts.

Further reading

Baltimore, D. 1971. Expression of animal virus genomes. *Bacteriological Reviews* 35, 235–241.

Diener, T. O. 2011. *The Viroids*. Springer-Verlag, New York.

Flores, R., Grubb, D., Elleuch, A., Nohales, M.Á., Delgado, S., Gago, S. 2011. Rolling-circle replication of viroids, viroid-like satellite RNAs and hepatitis delta virus: variations on a theme. *RNA Biology* 8, 200–206.

Karayiannis, P. 1998. Hepatitis D virus. *Reviews in Medical Virology* 8, 13–24.

King, A. M. Q., Adams, M.J., Carstens, E. B., Lefkowitz, E. 2011. *Virus Taxonomy: Ninth Report of the International Committee on Taxonomy of Viruses*. Elsevier Press, London.

The website of the International Committee on Taxonomy of Viruses (ICTV): http://www.ncbi.nlm.nih.gov/ICTVdb/index.htm

The GenBank sequence database (includes virus genome sequences): http://www.ncbi.nlm.nih.gov/genbank/

Chapter 4

The evolution of viruses

Like all living organisms, viruses are not unchanging entities fixed for all time but are subject to evolutionary pressures and undergo evolutionary change.

Chapter 4 Outline

4.1 Mechanisms of virus evolution
4.2 The potential for rapid evolution: mutation and quasispecies
4.3 Rapid evolution: recombination
4.4 Rapid evolution: reassortment
4.5 Evolution to find a host, and subsequent co-evolution with the host

Whatever their origins, as they undergo replication viruses are subject to evolutionary pressures, just as are all other organisms, and this has led to the extraordinary diversity of viruses that exist today. Whilst we have records indicating the interaction of viruses with humans spanning thousands of years, unfortunately the viruses themselves leave no archaeological record. Thus, we cannot compare the viruses of today with those from the long ago past to assess how they have changed over the intervening period. A recent exciting development has been the recovery from permafrost of an unusual type of giant virus called a Pandoravirus that was infecting amoeba around 30,000 years ago. This virus is remarkably similar to its relatives found today, suggesting that the changes that have occurred are not as great as might have been expected. For the more conventional viruses we have no such opportunity to study how viruses have evolved.

Coupled with the lack of a fossil record is the fact that the rapid rate of virus replication, with immense numbers of new progeny viruses produced from an infected cell, provides an opportunity for viruses to generate a vast range of mutants from which more successful variants can be selected over a time scale not achievable by other, more slowly replicating organisms. This creates a very complex landscape within which to study virus evolution. However, modern methods of nucleic acid analysis are now able to investigate the diversity of virus genomes that arise over very short timeframes and this has provided insight into the processes that drive virus evolution. An important consideration regarding the evolution of viruses is that they are obligate parasites and require a host within which to replicate. Selective pressures that alter host characteristics coupled with selective pressures brought to bear on the virus by individual hosts are important in the evolution of the virus. The various ways in which viruses interact with their hosts are discussed in Chapters 15 to 17. Evolution of any successful parasite has to ensure that the host also survives. The various possible virus–host interactions can be thought of as different ways in which viruses have solved this problem.

Introduction to Modern Virology, Seventh Edition. N. J. Dimmock, A. J. Easton and K. N. Leppard.
© 2016 John Wiley & Sons, Ltd. Published 2016 by John Wiley & Sons, Ltd.

4.1 Mechanisms of virus evolution

Two processes that contribute significantly to virus evolution are recombination and mutation. Mutation involves the well-understood processes that lead to changes in nucleotide sequences in nucleic acids (Fig. 4.1A). Classical recombination takes place infrequently between the single molecule genomes of two related DNA or RNA viruses that are present in the same cell, and generates a novel combination of genes (Fig. 4.1B). Of far greater significance is the potential for genetic exchange between related viruses with segmented genomes. Here, whole functional genes are exchanged, and this type of recombination is called *reassortment* (Fig. 4.1C). The only restriction for reassortment is the compatibility between the various individual segments making up the functional genome. Fortunately, this seems to be a real barrier to the unlimited creation of new viruses, although it is not invincible since pandemic influenza A viruses can be created in this way (see Section 4.4). When considering virus evolution it is important to appreciate that virus genomes accumulate mutations in the same way as all other nucleic acids and, where conditions enable a mutant to multiply at a rate faster than its fellows, that mutant virus will have an advantage and will predominate in subsequent generations if no additional selective pressures intervene. Each of the processes are considered here in turn.

4.2 The potential for rapid evolution: mutation and quasispecies

Evolution arising from the acquisition of mutations which confer a selective advantage is a familiar concept in all organisms, including viruses. Mutation is of particular significance to the evolution of viruses with RNA genomes (including retroviruses) as, in contrast to DNA synthesis, there is no molecular proofreading mechanism during RNA synthesis. Mutations accumulate at a rate of approximately 10^{-3} to 10^{-5} per base per genome replication event (i.e. one mutation for every 1000–100,000 newly-synthesized bases). This is equivalent to an average of 1 mutation in every new genome produced. In contrast, with DNA the figure for mutation rates is estimated to be 10^{-9} to 10^{-10} per nucleotide per cycle. In other words, while the same error processes are at work for viruses with either an RNA or a DNA genome, an RNA virus can achieve in one generation a degree of genetic variation which would take an equivalent DNA genome between 300,000 and 3,000,000 generations to achieve. However, recent analyses have suggested that for viruses with small DNA genomes the mutation rates may be higher than this, and may even approach those of the RNA viruses. Thus, following infection with an RNA virus, and possibly also a small DNA virus, a large population of progeny viruses with different genome sequences is rapidly generated rather than a homogeneous population of viruses with just one, or a small range of genome sequences. This is called a *quasispecies*. Not all of the resulting mutants will be viable but many are. Selective pressure, particularly that exerted by the defence systems of the host and any environmental pressure for survival outside of the host, will determine which of the progeny have the greatest chance of being able to infect a new host and pass on the mutated gene(s). An example of the power of a quasispecies to provide mutants from which one with a selective advantage can be selected can be seen in the extremely rapid, and concerning, appearance of virus mutants resistant to antiviral drugs almost immediately after each new drug is introduced.

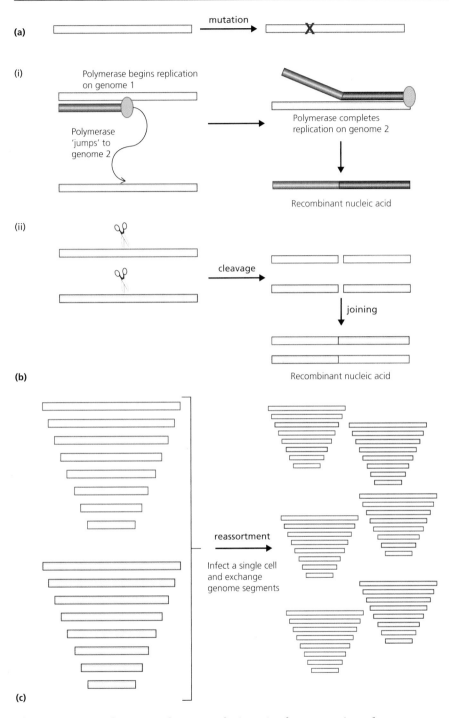

Fig. 4.1 The mechanisms of virus evolution. **A.** The generation of spontaneous mutants through one or more nucleotide transitions or transversions. **B.** (i) Recombination arising from transfer of the replication complex from one genome template to a second to generate a hybrid molecule. (ii) Recombination by cleavage and subsequent joining of nucleic acids to generate hybrid molecules. **C.** Reassortment arising from the exchange of genome segments within cells simultaneously infected with two related viruses.

The principle of the quasispecies has had several implications for the population biology of viruses. For example, it is not possible to define the genome sequence of that virus population precisely. The current concept of a quasispecies is that each contains a population of viruses which form a swarm or 'cloud' of sequences that extends outwards from a predominating sequence, with distance equating to the extent of sequence difference. Thus, if a single sequence is determined it is likely to represent the predominating one in the population from which it was obtained, but continuous sampling from that population will identify an increasing number of sequence variants that are increasingly representative of the population as a whole. The nature of the predominating sequence is determined by the success of a particular mutant under selective pressure.

Allied to the quasispecies concept is that of the 'error threshold' which suggests that the maximum mutation rate of a replication process is linked to the ability to form a viable quasispecies population i.e. if the mutation rate exceeds a specific rate, the frequency of non-viable mutations in the population is too great to permit the establishment of a stable population. This sets an upper limit on the rate of mutation that can be achieved without loss of the population and defines the limit of variation present in a quasispecies after replication.

The presence of a population of viruses in a quasispecies also leads to the principle that the members of the quasispecies can act as a cooperative consortium, with members demonstrating different levels of fitness to maintain variability in the population. In this way, a mutant virus that carries a defect can be maintained in the population, ensuring that any advantageous features it possesses are retained for the potential benefit of the population when it is subject to alternative pressures in the future. However, if the population goes through a bottleneck, such as occurs when a single virus or a very small number of individual viruses establish an infection in a new host, the variability is restricted and must be re-established following replication. It is presumed that the less fit viruses present in the quasispecies will be lost at that time. The nature of the resulting predominating sequence and the distribution of the variability in the population will be determined by the selective pressures that the viruses encounter in the new host. Thus, in principle, the viruses that arise from an infection of one individual will differ from those generated within a different host even of the same species.

The quasispecies phenomenon means that viruses, and particularly RNA viruses, are able to adapt rapidly to, and to exploit, any environment they occupy. This is seen on both a micro and a macro level, and arises because one or more members of the quasispecies will inevitably have a selective advantage over others. The micro level is infection of the individual, and virus evolution during the course of infection in the individual is seen, particularly in lifelong infections with HIV-1 (see Chapter 21) and hepatitis C virus (see Chapter 22). It is likely that, in these infections, the immune system is one of the major selective forces, and that the quasispecies phenomenon allows these viruses to persist in the face of that immune response. A demonstration of HIV-1 evolution driven by an alternative selective pressure is the increased predominance of viruses with a preference for infection of central memory T cells in the later stages of infection. The macro level of evolution is seen in viruses with a worldwide distribution. The process reflects the global distribution of people with different genetic backgrounds, especially differences in the MHC (HLA) haplotype that controls T cell immunity. This leads to the establishment of different specific viral genetic sequences, known as genotypes or clades of a virus, in different parts of the world, and a classification that is based on sequence rather than serology (e.g. HIV-1, see Chapter 21).

Interestingly, not all RNA viruses exhibit such obvious variation, even though they have the same potential for change as other RNA viruses such as HIV-1. For example, the overall antigenicity of polioviruses types 1, 2 and 3, and measles virus has not changed over a period of approximately 50 years, and the original vaccines generated decades ago still provide the same level of protection. Thus, the polioviruses and measles virus appear to be genetically stable in their key antigenic sites. However, their genomes do accumulate mutations, and minor changes in antigenicity (in individual epitopes) of the virion can be detected by probing with monoclonal antibodies. In addition, viruses from different global areas can be distinguished by their genotype. It is not understood why, for example, measles virus is antigenically stable and influenza virus is not, as in many ways these two infections are similar (both viruses are highly infectious, spread by the respiratory route, and cause acute infections). Presumably, measles virus is under a less effective or more constrained evolutionary selection pressure than influenza virus, but exactly what these pressures are is not known.

4.3 Rapid evolution: recombination

Recombination, in which there is genetic exchange between individual viruses, is a major force in virus evolution and takes place in a cell that is simultaneously infected by two viruses. The genetic exchange can occur either during the replication of the genome or independently of replication. Recombination may occur during replication of the genome when the two genomes are highly related, with regions of homology that permit the replicating enzyme to move from one genome template to another

taking with it the newly-synthesized molecule which is then extended using the second genome as a template (Fig. 4.1B(i)). Thus, in one event the daughter molecule has some of the properties of both parents. However, both parts of the resulting genome have to be sufficiently compatible for the progeny to be functional. Non-replicative recombination may occur by exchange of genetic material between two genomes involving breakage and joining together of the two or more parts (Fig. 4.1B(ii)). Both DNA and RNA genomes undergo recombination, although replication associated recombination is more common in viruses with RNA genomes. In principle, either recombination event may occur more than one time within a genome generating mosaic recombinant viruses, although multiple events are significantly less likely to occur than a single recombination event. Recombination is a major factor in the evolution of adenoviruses; the linear dsDNA genomes of many of the established human serotypes have been found to be recombinants, with parts of their sequence derived from two or more other known viruses. An interesting possibility that arises with recombination is the generation of recombinant genomes that contain duplicated or deleted sequences due to the recombination event fusing the genomes together at non-homologous regions. The generation of duplicated regions provides the possibility for one copy of a duplicated gene to mutate without detriment to the virus. This may be a mechanism by which viruses acquire new genes with new functions.

4.4 Rapid evolution: reassortment

Viruses with segmented genomes can also undergo recombination by acquiring an entire genome segment from another virus

(Fig. 4.1C). This form of recombination is known as *reassortment*, and the product is called a *reassortant virus*. Reassortment occurs at a higher frequency than is achieved by the replicative or non-replicative recombination events described above. The effect on the characteristics of the virus can be enormous; for example, an influenza virus can acquire an entirely new coat protein in a single step. With influenza A virus this process, called *antigenic shift*, is responsible for pandemic influenza in man (see Chapter 20). However, a viable reassortant influenza virus requires compatibility between all eight genome segments, and this limits the creation of new viruses.

While influenza A viruses infect many species, the natural animal reservoir of the influenza A viruses that have also been shown to infect humans comprises wild aquatic birds such as ducks, terns and shore birds (Section 20.1). The influenza A virus antigens that stimulate protective immunity in the host are the envelope glycoproteins, the haemagglutinin (HA; Fig. 6.3) and neuraminidase (NA), and immunity is mediated by virus-specific antibodies directed against these proteins. Influenza A viruses are classified formally as *types*, *subtypes* and *strains* (Box 4.1). There are 16 subtypes of haemagglutinin and 9 subtypes of neuraminidase found in nature in wild aquatic birds. In principle, this provides the possibility of generating 144 combinations of HA and NA proteins. By comparison, influenza in man is a minor component of the virus ecology, with currently only two subtypes, H1N1 and H3N2, in circulation although there have been periodic infections with other strains which have so far failed to become established in the human population.

Despite the induction of an immune response following infection, repeated infection with influenza virus is common. The repeated infections are possible because the HAs and NAs of influenza viruses also evolve continuously by mutation in a process referred to as *antigenic drift*. This means that previously-acquired immunity is rendered ineffective over time.

Box 4.1

Influenza viruses have a formal descriptive nomenclature

Influenza virus type: *A*

Subtype (example): *H3N2*

Where H is the haemagglutinin and N is the neuraminidase. There may be any permutation of H subtypes 1–16 and N subtypes 1–9.

Strain (example): *A/HongKong/1/68(H3N2)*

The strain designation indicates (in order): the type, where the strain was isolated, the isolate number, the year of isolation, the subtype.

The nomenclature of non-human strains also includes the host species, e.g. A/chicken/Rostock/1/34(H7N1)

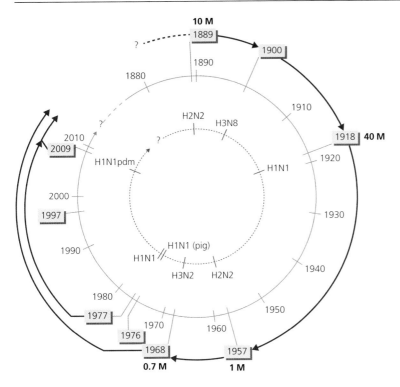

Fig. 4.2 History of antigenic shifts of influenza A viruses in humans. The outer circle denotes the year of emergence of a new subtype that is shown in the inner circle, the duration of the reign of that subtype, and when its replacement by another subtype occurred. A time scale is shown in the middle circle. Approximate worldwide mortality figures for each shift are indicated in millions. The 1900 shift did not cause a serious pandemic. In 2009, a new pandemic H1N1pdm virus appeared, supplanting the previous H1N1 subtype virus. Since that time the 2009 H1N1 virus has come to dominate whilst co-existing with the previously present H3N2 subtype. The occasional infection of people with bird or pig viruses that are not transmitted person-to-person is not shown.

The appearance of shift viruses in humans is shown chronologically in Fig. 4.2. Whilst influenza virus has been grown in the laboratory for study since 1933, retrospective information about the subtypes present before that date has been obtained by studying influenza virus-specific antibodies in human sera that had been stored in hospital freezers for other purposes. As can be seen, antigenic shift has occurred sporadically from the first recorded event of 1889 at intervals of 11, 18, 39, 11 and 41 years.

Until 1977, only one virus subtype was in circulation at any one time, and this virus was replaced completely when a new subtype was introduced. In 1977, during the reign of the H3N2 subtype, an H1N1 virus appeared that was identical to the H1N1 virus which had been in circulation in 1950. Thus this was not strictly speaking a shift, but a reintroduction. Where the 1977 virus came from is not known. It had not been infecting the human population during its 'absence' as it would have undergone 27 years of antigenic drift. It is as if it had been frozen, and some say that this is literally what happened. However, this is conjecture. From 1977 until the appearance of the new pandemic H1N1 strain in 2009 the descendants of the

1968 H3N2 and the 1977 H1N1 co-circulated and continued to undergo antigenic drift. Since 2001 the H3N2 virus has remained but the prevalence of the descendants of the 1977 H1N1 virus reduced significantly and were replaced by the descendants of the 2009 H1N1pdm strain following its pandemic emergence. The underlying mechanisms for the replacement of one influenza virus with another are not understood but they probably reflect the intricacies of the interaction of the virus and the immune system of the target host population.

Recombination by reassortment in influenza A viruses is very efficient as the genome is segmented – it comprises eight single stranded, negative-sense RNAs. The haemagglutinin and neuraminidase proteins are encoded by distinct RNA segments. When a cell is infected simultaneously with more than one strain of virus, newly-synthesized RNA segments reassort at random to the progeny (Fig. 4.3). Many, though not all, of the 2^8 (256) possible genetic permutations that can be formed between two viruses are genetically stable. Reassortment occurs readily in cell culture, in experimental animals, and in natural human infections between all type A influenza viruses, but not between type A and the genetically distinct influenza B viruses. Even after a reassortment event, more mutational adaptation is likely to be needed before the virus is able to cause a recognized pandemic. Thus it may be that a shift virus is present in the human population, undergoing mutational adaptations for a few years before it causes a pandemic.

4.5 Evolution to find a host, and subsequent co-evolution with the host

Most viruses have a restricted host range that may be limited to a single species. This presents

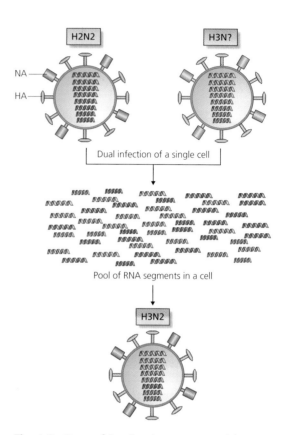

Fig. 4.3   Recombination (reassortment) between an existing human influenza A virus (H2N2) and a new virus from the wild bird reservoir (H3N?, where ? represents an unknown neuraminidase subtype; see text) that gives rise to antigenic shift. The two viruses simultaneously infect a cell in the respiratory tract, and the 8 genome segments from each parent assort independently to progeny virions. The example shows a novel progeny virion (H3N2) that comprises the RNA segment encoding the H3 avian haemagglutinin and the 7 remaining segments from the existing human virus.

a potential problem if the availability of susceptible members of the host species becomes a limiting factor. Contact between different species presents an opportunity for viruses to explore the potential for growth in a new host and viruses constantly 'test' this. This is dramatically seen with the sudden appearance of outbreaks of new diseases such as is seen with Ebola virus. Infections of humans by a virus naturally infecting another animal species, referred to as a *zoonosis* (see Section 18.5), occur rarely but can result in devastating effects if the virus evolves to become established in the human population as has been seen with HIV and influenza A virus (Chapter 20). As human activity takes people into new environments, the risk of zoonotic virus infections increases.

To successfully enter a new host species the virus must evolve to express the characteristics that allow it to infect and be transmitted within that species. Here we will consider the general principles of availability of a suitable host population in which the virus can successfully propagate, the possible consequences for the host of the appearance of a new pathogen and the subsequent development of a co-evolutionary relationship between the virus and its host.

Evolution of measles virus: the need for a minimum population level

Measles virus

- *Member of the paramyxovirus family.*
- *Infects only humans.*
- *Unusual in that every infection causes disease.*
- *Infection results in lifelong immunity.*

Given the requirement for a host within which to grow, the availability of a susceptible host as well as the selective pressures that the host brings to bear play a major role in the evolution of viruses. F. L. Black studied the occurrence of measles in island populations

Table 4.1

Correlation of the occurrence of measles on islands with the size of the island population (from Black, 1966, *Journal of Theoretical Biology* 11, 207–211).

Island group	Population (thousands)	New births per year (thousands)	Proportion of the year in which measles occurred (%)
Hawaii	550	16.7	100
Fiji	346	13.4	64
Solomon	110	4.1	32
Tonga	57	2.0	12
Cook	16	0.7	6
Nauru	3.5	0.17	5
Falkland	2.5	0.04	0

(Table 4.1), and found a good correlation between the size of the population and the number of cases of measles recorded on the island throughout the year. As the population size increased, measles cases were found for more extended periods following introduction. From this it was possible to predict that a population of at least 500,000 is required to provide sufficient susceptible individuals (i.e. new births) to maintain measles virus in the population with infection occurring throughout the year and in all subsequent years. Below that level, the virus eventually dies out, typically in less than a year, until it is reintroduced from an outside source.

On the geological time-scale, humans have appeared only recently and have only existed in population groups of over 500,000 since the river valley civilizations of the Tigris and Euphrates were established 6000 years ago. In the days of very small population groups, measles virus could not have existed in its present form. There is no evidence that it can establish persistent long-lasting infections from which it can efficiently infect new susceptibles. The virus with which we are familiar must therefore have appeared within the last few thousand years. Black speculated on the antigenic similarity of measles, canine distemper and rinderpest viruses: the latter infect dogs and cattle respectively, animals that have been commensal with humans since their nomadic days. Black suggested that these three viruses have a common ancestor which infected prehistoric dogs or cattle. The ancestral virus may have evolved to the modern measles virus when changes in the social behaviour of humans gave rise to populations large enough to maintain the infection. Once the virus was able to be maintained in this new host population, the virus would have evolved to its present form as a result of the pressures it encountered.

Evolution of myxoma virus: the potential impact on a new host species

Myxoma virus

- *Member of the poxvirus family.*
- *Natural host is the South American rabbit in which it causes only minor skin outgrowths.*
- *Also infects the European rabbit causing myxomatosis, with lesions over the head and body surface; infection is 99% lethal.*
- *Is spread by arthropod vectors that passively carry virus on their mouth parts: mosquitoes in Australia and rabbit fleas in the UK.*

Consideration of the evolution of myxoma virus demonstrates a pressure towards co-existence with the host that appears to be a common aspect of virus evolution. Myxoma virus was selected as a biological weapon in an attempt to control the populations of the European rabbit in the UK and Australia where they were responsible for serious damage to crops. The animals were extremely susceptible to infection and the effects of this intervention in nature were carefully studied with respect to the changes occurring in both the virus and the host populations and provides an object lesson in the problems of biological control.

In the first attempts to spread the disease in Australia, myxoma virus-infected rabbits were released in the wild but, despite the virulence of the virus and the presence of susceptible hosts, the virus died out. It was then realized that this was due to the scarcity of mosquito vectors, which aid transmission and whose numbers are seasonal. When infected animals were released at the peak of the mosquito season an epidemic of myxomatosis followed. Over the next two years, the virus spread 3000 miles across Australia and across the sea to Tasmania. However, during this period it became apparent that fewer rabbits were dying from the disease than at the start of the epidemic. Comparison of the virulence of the original virus with virus newly isolated from wild rabbits by inoculating standard laboratory rabbits identified two significant facts: (i) rabbits infected with new virus isolates took longer to die, and (ii) a greater number of rabbits recovered from infection. From this it was inferred that the virus had evolved to a less virulent form (Box 4.2). The explanation was simple: mutation produced virus variants which did not kill the rabbit as quickly as the parental virus. This meant that the rabbits infected with the mutant virus were available to be bitten by mosquitoes for a longer period of time than rabbits infected with the original virulent strain. Hence the mutants could be transmitted to a greater number of rabbits. In other words, there was a strong selection pressure in favour of less virulent mutants which survived in the host in a transmissible form for as long as possible.

Box 4.2

Evidence for the evolution of myxoma virus to avirulence in the European rabbit after introduction of virulent virus into Australia in 1950

After release into Australia, viruses were isolated from wild rabbits and their virulence tested by infecting laboratory rabbits. The percentage of mortality in the laboratory rabbits and the mean survival times were calculated.

Mean rabbit survival time (days)	Mortality (%)	Year of isolation			
		1950–1951	1952–1953	1955–1956	1963–1964
<13	>99	100	4	0	0
14–16	95–99		13	3	0
17–28	70–95		74	55	59
29–50	50–70		9	25	31
>50	<50		0	17	9

The second finding concerned the rabbits themselves, and the possibility that rabbits were evolving resistance to myxomatosis. To test this hypothesis, a breeding programme was set up in the laboratory. Rabbits were infected and survivors were mated and bred with other survivors. Offspring were then infected, the survivors mated and so on. Part of each litter was tested for its ability to resist infection with a standard strain of myxoma virus. The result confirmed that the survivors of each generation progressively increased in resistance. However, the genetic and immunological basis for this is not fully understood.

This work shows how a virus which is avirulent and well adapted to peaceful co-existence with its host can cause lethal infection in a new host, and how evolutionary pressures rapidly set up a balance between the virus and its new host which ensures that both continue to flourish. The latter is a phenomenon that has been recognized with many viruses after they encounter a new host and remains a stumbling-block for biological control of pests that attack animals or plants. Today, rabbits are still a serious problem to agriculture in the UK and Australia.

Co-evolution of herpesviruses and their hosts

Herpesviruses

- *Members of the herpesvirus family.*
- *Large dsDNA virus with genomes ranging in size from approximately 125,000 to 240,000 bp.*
- *As a group, they infect a very broad range of organisms but each virus is species-specific.*
- *Establish lifelong latent infections in many hosts.*

Once a virus has arisen with the necessary characteristics to successfully infect and to be efficiently transmitted between members of a new host species, the process of evolution of the virus does not stop. As indicated in the consideration of myxoma virus, the introduction of a virus into a new species frequently results in very severe disease but this usually alters due to changes in the virus and potentially also the host. Whilst it is possible to see the effect of such changes over a relatively short period for some viruses such as myxoma virus the absence of a fossil record prevents us from assessing the evolution of viruses over longer periods. However, this has been approached from a different perspective by studying the genomes of members of the herpesvirus family. A significant characteristic of the herpesviruses that infect mammalian and avian hosts is that following infection they establish a latent infection which lasts for the lifetime of the host (see Chapter 17). Typically the virus is relatively quiescent during these latent infections but periodically it re-emerges from this state to produce infectious progeny that infect other susceptible hosts. Due to the large DNA genome it is assumed that the rate of mutation in herpesviruses is likely to be considerably slower than that in RNA genome viruses. The long periods of quiescence with no, or limited, genome replication would also lead to slower rates of mutational change over time. Herpesviruses have been found in a very diverse range of hosts and analysis of the genome sequences of many viruses demonstrates that those infecting mammalian

and avian hosts show conservation of a large number of core genes, both in sequence and organization within the genome. Coupled with a number of other biological factors, this strongly suggests that the mammalian and avian herpesviruses originated from a common ancestor. Using standard phylogenetic analysis of genome sequences it is possible to draw evolutionary trees linking viruses to each other in the same way as for all other organisms. Comparison of the evolutionary trees of selected herpesviruses with trees derived from sequences of the natural hosts of the same herpesviruses shows a striking concordance (Fig 4.4).

The analysis shows that the mammalian herpesviruses have not only evolved over the same timescale as their hosts but that they have diverged from each other following the same pattern of divergence as the host. Thus, the ancestor of the mammalian herpesviruses most likely first infected the predecessor of the mammals and ungulates. Just as the original progenitor host evolved into different species that became the ones we recognize today, so too the viruses evolved from their common ancestor. The virus evolution occurred in concert with the evolution of the host, presumably following and adapting to changes in the host that affected the survival of the virus. This example of co-evolution of the

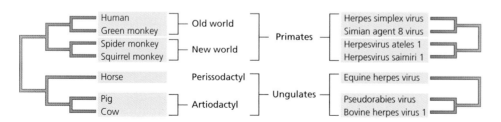

Fig. 4.4 Phylogenetic trees of primates and ungulates compared with the phylogenetic trees based on the gene encoding glycoprotein B of herpesviruses associated with the same species. The striking similarity of the evolutionary trees indicates that the hosts and their viruses have co-evolved since the introduction of the virus into the common ancestor of the primates and ungulates. (Data taken from McGeoch, D. J. and Cook, S., *J. Molecular Biology* (1994), vol. 238, pp 9–22.)

herpesviruses with their host over millennia is likely to have occurred with all viruses following introduction into a new host species and it is initiated each time this occurs. The concept of co-evolution of virus and host demonstrates the intimate relationship that the two adversaries have, with each responding to the other in a constant battle of successive rounds of attack and defence that have formed the viruses we see and experience today. As with all such events, evolution ensures that only the successful survive.

Key points

- Viruses are evolutionarily dynamic and respond to changes in the macroenvironment (the population) and microenvironment (the individual body and immune system).
- All established human viruses undergo Darwinian evolution as a result of evolutionary pressure acting on random variants arising by mutation, although the scale of change ranges from low (e.g. as in measles virus) to high (e.g. as in influenza virus).
- Every living organism has its own extensive range of viruses, and new infections of man can arise by non-human viruses crossing species, e.g. HIV, influenza A virus.
- New strains of influenza virus are introduced periodically into the human population leading to worldwide pandemics. These arise as a result of antigenic shift in which the genome segments encoding the HA and potentially also the NA proteins are exchanged with a current human virus.

- Many new infections in a host species result in high morbidity and mortality, but evolution selects hosts that are genetically more resistant to infection and less virulent progeny virus.
- Once a virus establishes a new host it will co-evolve with that host thereafter.

Questions

- Describe the processes of antigenic drift and antigenic shift in influenza virus and indicate, with reasons, which is likely to generate a more pathogenic virus.
- Using examples, discuss the proposition that viruses with RNA genomes evolve more rapidly than those with DNA genomes.

Further reading

Black, F. L. 1966. Measles endemicity in insular populations: critical community size and its evolutionary implications. *Journal of Theoretical Biology* 11, 207–211.

Domingo, E., Sheldon, J., Perales, C. 2012. Viral quasispecies evolution. *Microbiology and Molecular Biology Reviews* 76, 159–216.

Lauring, A. S., Andino, R. 2010. Quasispecies theory and the behaviour of RNA viruses. *PLoS Pathogens* 6, e1001005.

Kerr, P. J. 2012. Myxomatosis in Australia and Europe: a model for emerging infectious diseases. *Antiviral Research* 93, 387–415.

Taubenberger, J. K., Kash, J. C. 2010. Influenza virus evolution, host adaptation, and pandemic formation. *Cell Host and Microbe* 7, 440–451.

Chapter 5

Techniques for studying viruses

Viruses can either be detected directly – by identification of their nucleic acid genomes or proteins in samples – or else indirectly – by their pathological effects in cell culture or whole organisms, by their interaction with antibody or by identifying the presence of antibody that is specific for a known virus. Once isolated, viruses can be studied using a wide range of techniques. Targeted manipulation of viral genomes by reverse genetics has proved a powerful means of identifying gene function.

Chapter 5 Outline

5.1 Culturing wild virus isolates
5.2 Enumeration of viruses
5.3 Measuring infectious virus titres
5.4 Measuring physical virus titres
5.5 Detecting virus in a sample
5.6 Understanding virus replication cycles
5.7 Viral genetics and reverse genetics
5.8 Systems-level virology

The technology that underpins the study of viruses covers the full breadth of modern biology. Very many techniques of importance to virologists are equally relevant to other branches of the life sciences. The purpose of this chapter is not to attempt an exhaustive review of all these techniques but rather to consider the techniques that are specific to virology or have a particular application within this field. Other molecular and cell biology texts provide coverage of more generic methodologies.

5.1 Culturing wild virus isolates

What is our first point of scientific contact with a virus? How do we learn of its existence? Traditionally, this knowledge has come from the observation of the generation of disease (pathogenesis) in an organism that displays features suggestive of an infectious cause, followed by attempts to identify a virus (or other type of infectious agent) in samples taken from the organism. A further step is then to exclude other known pathogens as the cause of the pathogenesis.

Viruses are too small to be seen except by electron microscopy (EM) and this requires concentrations in excess of 10^{11} particles per ml, or even higher if a virus has no distinctive morphology, as well as very expensive equipment and a highly-skilled operator. Such concentrations are rarely achieved in a specimen and, even if it is possible to identify a new virus in appropriate samples directly, studying its biology will inevitably require that

Introduction to Modern Virology, Seventh Edition. N. J. Dimmock, A. J. Easton and K. N. Leppard.
© 2016 John Wiley & Sons, Ltd. Published 2016 by John Wiley & Sons, Ltd.

it be cultured in an amenable laboratory system. Such culturing is important in part because the virus will need to be amplified to produce sufficient to study, but also because most of the important knowledge about a virus will come from studying its effects in infected cells or an infected organism. For viruses of host species other than humans, a certain amount may be achievable using the intact host organism if that is amenable to laboratory work but, even here, there are serious limitations on experimentation and work in cell cultures or organ cultures is much preferable; for viruses of humans, such culture experiments are crucial because experimental infection of humans would not be ethical.

Growing and studying a virus obtained from a clinical sample in culture is not straightforward because human or animal cell cultures do not reflect very well the types of cell that viruses would infect naturally in the body (Box 5.1). Thus, virus in a primary isolate derived directly from a clinical sample will often grow poorly, if at all, in such cultures. Organ cultures may give better results, but these are limited in scale. With persistence, it is sometimes possible to persuade virus to grow from clinical samples after several blind passages (moving material on from a first infected culture to a second one without attempting to detect or count virus in the material) in a well-chosen cell type, eventually producing a stock that can be characterized. Often, the initial indication of success in this process is the appearance of cell death, known as *cytopathic effect* (cpe), in the culture (Fig. 5.1).

Box 5.1

Choosing a culture system for animal viruses

Culture system	Advantages	Limitations
Animal	Natural infection	Upkeep is expensive. Variation between individuals, even if inbred, means that large numbers are needed. Ethical considerations.
Organ culture, e.g. pieces of brain, gut, trachea	Natural infection. Differentiated cell types present. Fewer animals needed. Less variation since one animal gives many organ cultures.	Unnatural since cultures are no longer subject to homeostatic responses such as the immune system.
Cell	Can be cloned, therefore variation is minimal. Good for biochemical studies as the environment can be controlled exactly and quickly.	There are three types of cell culture: primary cells, cell lines and permanent cell lines. Primary cells are derived from an organ or tissue; they remain differentiated but survive for only a few passages. Cell lines are dedifferentiated but diploid and survive a larger number (about 50) passages before they die. Continuous cell lines are also dedifferentiated, but immortal.

 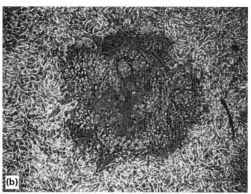

Fig. 5.1 Cytopathic effects caused by viruses in confluent cell monolayers (a layer of cells with a depth of just one cell). (a) Chick embryo cells infected by influenza A virus. In the clear central area infected cells have lysed. Some cell debris remains, and cells in the process of rounding up can be seen on the edge of the lesion. There are healthy cells around the periphery. (b) A monkey cell line infected with human respiratory syncytial virus (HRSV). HRSV does not lyse cells, but causes them to fuse together to form syncytia. A collection of syncytia forms the dark area in the centre. Individual cells are magnified to approximately 3 mm in length, and are packed close together. Note the difference in morphology between the monkey cells and the chick cells.

However, it is important to appreciate that such virus has been subject to a powerful selection pressure and is likely to differ in its genome sequence and key functions from the virus in the original clinical specimen. Thus, to study a virus almost inevitably involves changing it; this fundamental compromise underlies most virology research.

5.2 Enumeration of viruses

Perhaps the most important techniques in virology are those that allow the enumeration of viruses. Without accurate quantitation of virus to use in an experiment it is impossible to carry these out under reproducible conditions. How do we quantify the amount of virus we have in a sample or in a stock that has been grown in a laboratory?

Almost all methods to measure the amount of virus will actually determine its concentration in the sample. The virus concentration is often referred to as the *virus*

titre, expressed in virus units per ml of sample. There are two different measures of virus titre that are useful: a physical particle count, i.e. the concentration of particles present in a sample; and an infectivity count, i.e. the concentration of particles present are that are capable of completing successfully a full infectious cycle in a susceptible host system – cell or animal. You might imagine that these two numbers would be very similar for any given virus sample, but in fact the ratio of physical to infectious particles is frequently high, with values of between 10 and 100 being common in well-studied animal viruses.

Why should the physical and infectious titres of a sample differ so dramatically? In part, it reflects inefficiencies in assays of infectivity, i.e. not all particles that could potentially score as infectious in the assay actually do so on any single occasion. Reasons for this include random chance, i.e. 'does the virus happen to encounter a susceptible cell in the time allowed for this?' and also the effects of intracellular resistance mechanisms that act against an incoming infectious particle (see Chapter 13), but it is also the case that great numbers of defective particles are produced in many virus infections. Such particles normally contain an incomplete or mutated copy of the virus genetic material, such that one or more products essential to infection cannot be successfully encoded. Under conditions where a cell is only infected by a single particle, the inability to make any one of the crucial viral proteins would mean no productive infection would result.

5.3 Measuring infectious virus titres

In 1952, Renato Dulbecco (1914–2012) published his landmark paper that established the plaque assay as a means of counting viable animal viruses. Adapting a concept already established for measuring bacteriophage infectivity, he used Western Equine Encephalitis virus to produce plaques in monolayers of chick embryo fibroblasts and showed a linear relationship between virus dilution and measured plaque number.

The classic method for determining infectious virus titres is the *plaque assay*, which can be used, with appropriate hosts, to measure bacteriophage, animal virus or plant virus concentrations (Fig. 5.2). It is only applicable to those virus–host combinations where infection causes cpe. To perform the assay for a mammalian virus, a sample is serially diluted and a known volume of each dilution (often termed an *inoculum*) is then applied to infect confluent monolayers of susceptible cells. At lower dilution factors, i.e. more concentrated virus, many if not all cells in the culture will be infected; such dilutions are not informative. However, when the sample is diluted further so that the amount of infectious virus present is

sufficiently low, the infected cells will be physically isolated at discrete locations within a layer of otherwise uninfected cells. After infection, movement of virus between locations in the cell layer is prevented by overlaying the cells with nutrient medium that is solidified, usually through the addition of agar. Under these circumstances, the progeny virus produced can only infect immediately adjacent cells; after a few cycles of infection and cell death, circular zones of cell death within the cell layer, known as *plaques*, can be seen by eye after staining of cell material with crystal violet or neutral red and counted. Since each of these plaques originated from an infectious particle in the diluted inoculum applied to the cells, the titre of the stock can be calculated in *plaque-forming units (pfu) per ml* using the plaque count, the volume of inoculum used and the dilution factor of that inoculum. To obtain an accurate value, it is important to use only those plaque counts that arise from dilutions giving well-spaced plaques, but also to use an area of susceptible cells that is sufficient for between 20–100 plaques to be counted in this way. It is also crucial to perform replicate determinations and take a mean value as the titre.

Exactly the same approach is used to titre bacteriophages. Here, the host bacterial cells are infected in suspension with various dilutions of the phage and then plated in soft agar on a nutrient agar plate. The bacterial lawn, equivalent to the mammalian cell monolayers above, grows rapidly and plaques are then visible within it. For plant viruses, titration is less easy since different leaves, even on the same plant, cannot be relied upon to give reproducible titration measurements of a virus stock. However, the two halves of a single leaf can be used to give comparative titration measurements of two stocks.

An alternative method for determining infectious titre, which again relies on the ability of a virus to cause cpe, is the Tissue Culture Infectious Dose 50% (TCID$_{50}$) assay. As before,

Fig. 5.2 Plaques of viruses. (a) Plaques of a bacteriophage on a lawn of *Escherichia coli*. (b) Local lesions on a leaf of *Nicotiana* caused by tobacco mosaic virus. (c) Plaques of influenza virus on a monolayer culture of chick embryo fibroblast cells.

a virus sample is serially diluted and replicate aliquots of each dilution are applied to susceptible monolayers of cells. But rather than use a large area of cells, assays are performed using small cultures in 96 well plates. Each of the serial dilutions is applied to several equivalent cultures and spread of progeny virus from one part of the monolayer to another is allowed through the use of liquid culture medium. After a fixed time appropriate to a particular virus/cell combination such that two or three cycles of infection can occur, the cells are fixed and stained and gross cytopathic effect in each cultures is assessed to determine the dilution that gives cpe in 50% of equivalent cultures. At this dilution, the volume used to infect each culture contains, by definition, 1 $TCID_{50}$ unit of virus. The titre of the stock, in $TCID_{50}$ units/ml, can then be calculated. This is an example of an end-point dilution assay.

A more modern development is the fluorescent focus assay. The essential principles are the same as for a plaque assay but, rather than allowing several cycles of infectious spread from the cells initially infected within the assay culture, the cells are fixed during the first cycle (killed *in situ*) at a time when they would be full of viral proteins if infected. The presence of these proteins is then detected by applying an antibody to the cells that will bind to any viral proteins present (known as the primary antibody) and then the location of this antibody among the cells is determined by applying a fluorescently tagged secondary antibody whose target antigen is the constant region (Fc) of the primary antibody. The fluorescent cells are then counted by microscopy under illumination that causes the tag to fluoresce. From the fluorescent cell count, a titre can be determined in *fluorescent focus units (ffu) per ml* in the same way as for a plaque assay. It is also possible to count the infected fluorescent cells by flow cytometry. Importantly, this approach to finding out how much infectious virus is present can be applied to viruses that do not produce cpe in their host cells as well as to cytopathic viruses. The methodology of fluorescent antibody staining is more generally termed immunofluorescence and when applied at higher magnifications it has important applications in the study of virus infections.

5.4 Measuring physical virus titres

Many viruses have the ability to bind to the surface of red blood cells (RBC) and cross-link them into a clump; in this state the RBCs are said to be agglutinated, hence this virus property is described as *haemagglutination* (HA). This function is a property of proteins on the surface of the virus particles, which often are also the proteins responsible for attachment to cells during the first steps of an infection; it is for this reason, for example, that the attachment protein of influenza viruses is known as the haemagglutinin. However, HA activity in a virus is not the same as infectivity and, generally, a virus can still agglutinate RBCs when its infectivity is inactivated. For each virus, HA activity will only be observed with RBCs from specific species.

Viruses whose particles have HA activity can be quantified using this property (Fig. 5.3). Serial two-fold dilutions of the virus preparation are incubated in buffer with a fixed number of RBCs to allow agglutination to occur. The final dilution that is still capable of agglutination is taken as the end point and this amount of virus is defined as 1 HA unit. From this definition, the concentration of virus in the stock may be calculated in HA units/ml. Note that the assay volume and number of RBCs must be kept constant in order for comparisons of HA titres to be valid.

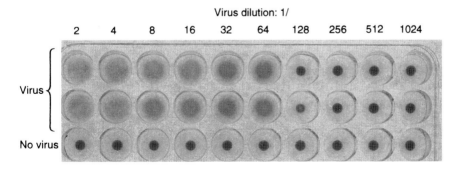

Fig. 5.3 Haemagglutination assay. Here an influenza virus is serially diluted from left to right in wells in a plastic plate. Red blood cells (RBCs) are then added to 0.5% v/v and mixed with each dilution of virus. Where there is little or no virus, RBCs settle to a button (from 1/128) indistinguishable from what is seen when no virus is added (row 3). Where sufficient virus is present (up to 1/64 dilution), the RBCs agglutinate and settle in a diffuse pattern. (Photograph by Andy Carver.)

5.5 Detecting virus in a sample

Detection of known viruses in clinical or other samples is the area of viral diagnosis. Since direct detection of virus particles is difficult, diagnostic techniques focus on the detection of specific viral components, or inferring the presence of a virus indirectly through the detection of virus-specific antibodies.

There is a vast array of techniques that are or have been used to detect the presence of virus. These vary greatly in their utility, with practical considerations being the time taken, requirement for specialized equipment and the cost per sample. These are important drivers for new developments of assays towards scalability for high-throughput with automation. This section covers in detail only those techniques that are of current or major historic importance.

It is important to realize that any assay will have a lower limit of sensitivity, below which it will give a negative result for a sample even if it is intrinsically positive. An assay also has definable specificity, which is a measure of how likely it is to produce a positive result for a sample that is genuinely negative. Generally, assay sensitivity and specificity are inversely related; as you drive up the sensitivity to reduce false-negative results, you introduce a greater risk of false-positive results. Thus unambiguous diagnosis requires a combination of tests, typically a high sensitivity test which will give some false-positives, followed by a confirmatory test of high specificity on the positive samples, which then excludes any false positive outcomes.

Detection of viral protein: ELISA

All methods for detecting specific viral proteins in samples rely on the use of specific antibodies. As described in detail in Chapter 14, antibodies are proteins produced by the immune system of higher vertebrates in response to foreign materials (antigens) which those cells encounter. Such antibodies have regions that recognize and bind specifically to the antigen that elicited them. Antibodies are secreted into the body fluids and are most easily obtained in serum, the fluid component of blood that remains after clotting. A serum that contains antibodies reactive against a particular entity is known as an antiserum.

The most widely-used method for viral protein detection is a configuration of an enzyme-linked immunosorbent assay (ELISA) that is otherwise known as a sandwich assay (Fig. 5.4). These are quick to do and the sequence of fluid additions and washes in multiwell plates can be automated with suitable equipment. In essence, the assay uses antibodies firstly to concentrate any viral antigen present in the sample and then secondly to detect the presence of that antigen in the plate well. The detection phase is two-step, involving primary antibody (step 3) and secondary antibody (step 4), which gives substantial signal amplification and hence improved sensitivity because multiple secondary antibody molecules can bind to each primary antibody. The secondary antibody is raised in a different species, using Ig from the species providing the primary antibody as antigen. It is important that the antibodies used in steps 1 and 3 come from different species, so that the secondary antibodies added in step 4 bind to the well only if antigen has been immobilized in step 2. So, for example, the capture antibody might be raised in mice, the primary detection antibody in rabbits and the secondary antibody in goats. Ultimately,

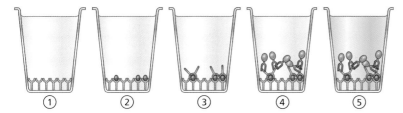

Fig. 5.4 Sandwich ELISA for antigen detection. 1: antibody A, specific for the viral antigen being tested for, is bound to wells of a 96 well plate and excess washed out. 2: the test sample is added and any specific antigen present binds to the antibody, everything else is washed away. 3: another antibody, B, also specific for the viral antigen but raised in a different species from antibody A, is added and binds only if antigen was present at step 2; excess is washed away. 4: an enzyme-conjugated secondary antibody is added, which has specificity for Ig from the species providing antibody B; it binds only if antibody B bound at step 3. 5: a chromogenic substrate for the enzyme is added and the colour intensity after a fixed time is measured by spectrophotometry using a plate reader.

the presence of bound secondary antibody is revealed by the action of the covalently coupled enzyme, which is chosen to be capable of converting a colourless substrate into a coloured product; examples include alkaline phosphatase and horseradish peroxidase. The colour intensity (absorbance) is a measure of the amount of bound enzyme, which is proportional to the amount of immobilized antigen in step 2. Hence, by assaying the samples in serial dilution, quantitative information can be obtained.

Detection of viral nucleic acid: PCR and bDNA assays

The polymerase chain reaction (PCR), which amplifies a specific DNA segment from a complex mixture of DNA molecules, was devised in 1985 by Kary Mullis. This general technique can be applied to detect viral DNA genomes in a sample among many other uses and it can be extended to detect viral RNA genomes by inclusion of a reverse transcription step that uses a retroviral enzyme (Section 9.3) to convert RNA into DNA.

The basis of PCR amplification is outlined in Fig. 5.5. The key specificity determinant in the reaction is the choice of primers, which are short synthetic DNA oligonucleotides of around 20–25 nucleotides in length. Two primers are required that base-pair specifically (hybridize) to distinct sequences in the template molecule that flank a region of defined length, typically 100–500 bp (base-pairs), such that DNA synthesis from the 3′ end of each primer will extend across this target region to the binding site of the other primer. This region is often termed the *PCR amplicon*. When using PCR to detect viral sequences, it is important to consider the variation in sequence that arises naturally in most viruses; to be confident of detecting a given virus, the primers need to target genome regions that are well conserved.

To carry out a PCR, primers and template sample are incubated with dNTPs and a thermostable DNA polymerase, usually Taq polymerase from *Thermophilus aquaticus*, in a repeated series of temperature steps that are automated in a thermocycler. After 30 cycles or so, provided the starting template material contains molecules to which the primers can bind, the reaction produces a readily detectable amount of specific product. The presence of this product at the end of the reaction is proof that

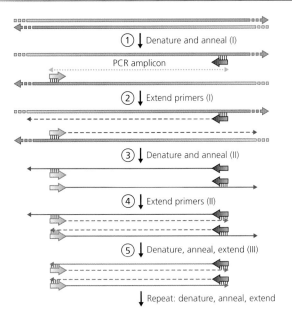

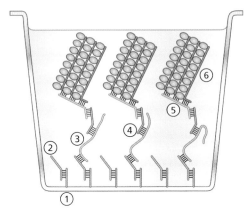

Fig. 5.5 PCR amplification of a target DNA sequence. Thick brown lines: template DNA; red lines: first cycle amplification product; purple lines: second and subsequent cycle product. Turquoise and grey block arrows represent specific left and right primers respectively. Base pairing of these primers to the template DNA defines the PCR amplicon (dotted orange line).

Fig. 5.6 Branched DNA assay. A generic capture probe (1) is bound to the wells of a microtitre plate. The well is given specificity by binding a capture extender (2). This oligonucleotide has two regions, one able to base pair with the capture probe and the other with specificity for the desired target nucleic acid (3). Bound target is detected by base pairing with a label extender (4), which also has two regions, one specific for a different region of the target and the other that can base pair to the pre-amplifier (5). The pre-amplifier oligo has repeated sequence elements and so can bind multiple copies of the amplifier oligonucleotide that in turn has multiple enzymes coupled to it (6).

the template sample contained the target being tested for. Such reactions can be very readily used to test for the presence of a particular virus, provided its genome sequence is known.

Originally, PCR products were detected by ethidium bromide staining of agarose gels in which the PCR material had been separated by electrophoresis. With DNA size markers analyzed alongside, the correct size of the observed product is further confirmation that the expected amplicon has been detected. A more modern approach is to use PCR quantitatively. By performing the reactions in a light-cycler, in which the DNA content of the reaction mix is measured repeatedly in real time during the course of the 30–40 reaction cycles, either the relative or absolute amounts of target sequence in the starting material can be

determined. Monitoring how much HIV genome there is in the blood is critical to effective and sustained therapy for HIV patients (see Section 21.6).

An alternative to PCR for the rapid and quantitative detection of viral nucleic acid sequences is a branched DNA assay (Fig. 5.6). This again relies on base-pairing to provide detection specificity, but signal amplification to give sufficient sensitivity is achieved very differently. In concept, the branched DNA assay is actually more like an ELISA. Any target nucleic acid molecules in the sample are first captured on microtitre plate wells by specific oligonucleotides and then that bound target is detected by the formation of a base-pairing network, resembling the branches of a tree, that

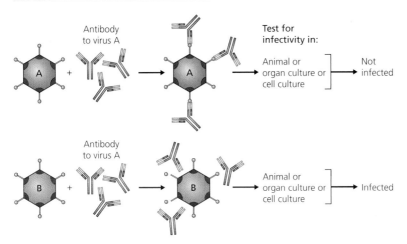

Fig. 5.7 Virus neutralization test. Virus A loses its infectivity after combining with A-specific antibody (it is neutralized). A-specific antibody does not bind to virus B, so infectivity of virus B is unaffected. The complete test also requires the reciprocal reactions.

ultimately immobilizes large numbers of enzyme molecules in the well. As with an ELISA, a positive result is revealed by the conversion of enzyme substrate to a coloured product.

Detection of virus-specific antibodies (serology, applicable only to animal viruses)

The presence in a serum sample of antibodies reactive against a virus was traditionally detected in an antibody neutralization test (Fig. 5.7). A test sample of infectious virus was incubated with the serum and the infectivity remaining was then tested on a susceptible cell culture in comparison with controls. If the infectivity was reduced by the serum, then it was deemed to contain neutralizing antibodies for that virus, those antibodies blocking infectivity in some way by their binding to the virus particle. Another property of many viruses that can be blocked by antibody and hence used

to detect the presence of specific antibody is haemagglutination (HA, Section 5.4). By pre-incubating virus and serum before adding RBCs, the ability of antibodies in the serum to reduce agglutination of constant amounts of virus and RBCs can be measured in an HA-inhibition test. If a series of serum dilutions is used, both HA-inhibition and neutralization tests can be quantified to produce serum titres of either HA-inhibiting or neutralizing antibody. HA inhibition tests are rapid, taking just an hour or so compared with an average of 3 days for a measurement of infectivity. However, they are less sensitive; approximately 10^6 plaque-forming units (PFUs) of influenza virus are needed to cause detectable agglutination.

By far the most versatile and widely applied technique for specific antibody detection is the antibody-capture ELISA. This uses exactly the same principles as the assay for viral antigen just described but reconfigured to detect antibodies (Fig. 5.8). Test antigen (either whole virus or protein subunits produced by recombinant DNA techniques) is immobilized

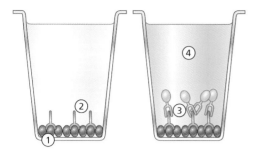

Fig. 5.8 Antibody capture ELISA. Microtitre plate wells are pre-coated with virus or viral antigen (1). Test samples, possibly containing specific antibody for that virus, are added to the wells in serial dilutions (2). After allowing time to bind, unbound Ig is washed away and any bound antibody detected by addition of a secondary antibody (3) to which an enzyme is coupled (4). A positive result is revealed by the enzyme converting its substrate to a coloured product.

on the surface of the wells, and samples that possibly contain antibody are then added in varying dilutions and allowed to bind. After washing away unbound material, any bound Ig is detected using an enzyme-conjugated secondary antibody as before.

One problem with using the presence of specific antibody as a test for the presence of a virus is that IgG remains as an aspect of immune memory long after the virus that elicited it has gone from the system. However, an antibody-capture ELISA can be specifically configured to detect only specific IgM molecules. Since IgM is only produced during the active primary response to an antigen, its presence is indicative of a current infection.

5.6 Understanding virus replication cycles

So long as virus has to be grown in whole organisms, it is difficult to move beyond a superficial understanding of its replication mechanism. However, once a suitable cell culture system has been established in which a virus can be grown successfully, that mechanism is open to detailed study. The whole suite of modern molecular cell biology techniques can be applied in the same way as for all other fields of biological research. Thus, for example, the synthesis of viral proteins can be followed by western blotting with virus-specific antibodies, or the production of viral genomes can be measured by quantitative PCR, after reverse-transcription for RNA genomes. In all cases, the central underlying approach is to monitor specific parameters over the time-course of virus growth. A typical experiment would involve the infection of a series of replicate (i.e. equivalent) cell cultures at time zero and then harvesting of individual cultures at various times thereafter (hours post-infection) to give appropriate samples for analysis. These experiments would also include equivalent mock-infected cultures, where virus is omitted but all other steps are carried out as for the infected ones. By comparing the samples from these control cultures with those obtained from the equivalent infected cultures, the effects due specifically to virus infection can be identified.

A very important development of this general approach is to compare the growth of two or more closely-related viruses by monitoring their one-step growth curves (see Section 1.3). By doing this, it is possible to link differences in observed phenotype with specific genetic differences in the viruses. Consider the simple scenario in which a wild-type virus stock is deliberately mutated to produce stocks with mutations in each of two genes A and B. Comparing the one-step growth of these three viruses reveals that the mutant in gene A produces less viral genome than the wild-type, suggesting the protein product of gene A is required for genome replication, while the mutant in gene A replicates genome normally but still produces fewer infectious progeny than the wild-type, which suggests gene B is important for the formation of virions. As discussed in the next section, it is possible to isolate mutant stocks of many viruses for use in such experiments.

An important variable in such experiments is the *multiplicity of infection* (m.o.i.), which is the ratio of infectious units of virus added to the number of cells present in the culture. To avoid the confusion caused by virus that is produced by the initially infected cells going on to infect other cells in the culture, it is normally preferable to conduct a one-step experiment in which every cell is infected at time zero (see Section 1.3). Because the interaction of virus particles with target cells is a random, stochastic process, a simple equivalence of infectious virus particles and cells is not sufficient to achieve this. In fact, it can be calculated that at m.o.i. = 1, only about 67% of cells will be infected. An m.o.i. of 10 is normally used to ensure that essentially all cells are infected; however, under such conditions many cells will be infected by multiple virus particles simultaneously. This can have consequences for the study of mutant or variant viruses where the severity of the mutant phenotype (i.e. the differences from the events seen during infection with the standard strain of that virus) is often dependent on the m.o.i. used.

5.7 Viral genetics and reverse genetics

The ability to apply genetics to the study of viruses was a crucial stepping stone towards our understanding of them. Generally, those viruses where genetic studies quickly became possible have gone on to be understood in great detail whilst those where genetics proved more difficult have lagged behind. There is an important distinction to be drawn between classical genetics and reverse genetics. The former term applies to the isolation of variant (mutant) viruses from a population – perhaps following deliberate random mutagenesis – based on their aberrant phenotype, following which the nature of the genetic lesion has to be identified to link a particular gene product with the phenotype observed. The latter refers to the directed mutation of a viral genome in a precise, predetermined way, for example to test a specific hypothesis about the function of a gene, following which virus containing that mutation must be reconstituted for study; in other words, the genotype is established first and then the phenotype is evaluated. This approach leads more directly to a genotype-phenotype link but with the significant possibility that the chosen mutation might either have no detectable phenotype, a lot of work for no reward, or might be lethal to the virus, meaning that all attempts to rescue live virus from that altered genome are doomed to failure.

Reverse genetic systems are now available for many species of animal viruses across a range of virus families (Box 5.2). For some of these families, this technology emerged very early, with infectious virus being readily recovered from cell cultures that had been transfected with cloned genome DNA. For others, particularly those RNA viruses where enzymatic functions are encapsidated with the genetic material that are essential in the initiation of infection (Chapter 8), the reconstitution of infectious virus from manipulated genome material was inherently more problematic. These problems still have not been fully resolved for the rotaviruses.

5.8 Systems-level virology

Recent technical advances have seen a push towards using systems-level analyses across many areas of biological research, including virology. This is true at all levels from virus infection of cells in a culture dish, through the

Box 5.2

Reverse-genetic systems for some animal viruses

Baltimore class	Virus family	Reverse-genetic method
Class 1	Polyomavirus	Transfected cloned genome is infectious in standard cell lines.
	Papillomavirus	Cloned genome transfected into keratinocytes gives virus in epithelial raft cultures.
	Adenovirus	Transfected cloned genome is infectious in standard cell lines.
	Herpesvirus	Transfected cloned genome is infectious in standard cell lines.
	Poxvirus	Recombination between cloned genome fragments and viral genome in transfected/infected cells.
Class 2	Parvovirus	Transfected cloned genome is infectious in standard cell lines (with helper functions for Dependovirus).
Class 3	Reovirus (exc. rotavirus)	Virus produced from transfection of plasmids encoding each genome segment as (+)sense RNA under T7 promoter in cells expressing T7 pol.
	Orbivirus	Virus produced from transfection of (+)sense RNA transcribed in vitro from plasmids encoding each genome segment under T7 promoter.
	Rotavirus	Rescue from fully cloned material not yet achieved.
Class 4	Picornavirus	Transfected genome, transcribed in vitro from cloned DNA, is infectious in standard cell lines.
Class 5	Orthomyxovirus	Virus produced from transfection of plasmids encoding each (-)sense genome segment under T7 promoter plus plasmids expressing viral NP, PA, PB1 and PB2 into cells expressing T7pol.
	Paramyxovirus	Virus produced from transfection of plasmid encoding antigenome under T7 promoter plus plasmids expressing viral N, P and L proteins (plus others in some cases) into cells expressing T7pol.
Class 6	Retrovirus	Transfected cloned genome infectious in standard cell lines.
Class 7	Hepadnavirus	Transfected cloned genome produces infectious progeny in certain liver cell lines.

interaction of virus with an individual host organism to population-level studies. These techniques are the 'omics', i.e. transcriptomics and proteomics, and high-throughput sequencing. In each case, the objective is to identify and quantify every nucleic acid or protein molecule in a sample. As with more traditional techniques, by comparing virus-infected and mock-infected samples with these techniques it is possible to identify differences that are due to infection but with the added advantage that the outcome is not biased by preconceptions; while the conventional approach only allows the discovery of differences that are anticipated and therefore are specifically tested for, system-level techniques will reveal all differences that are above the limit of detection.

High-throughput sequencing techniques have also opened up the possibility of defining the population diversity of viruses. Previously, we could say that we 'knew' the sequence of a virus whilst accepting that the sequences of individual virus particles varied considerably both between infected individuals and within individuals. But with current sequencing techniques it is possible rapidly to characterize the full depth of sequence variation within a population of virus, and then to understand how that sequence space changes with time under selective pressures.

Another major impact of modern sequencing techniques is the ability to hunt for new viruses in clinical specimens by looking for unaccounted-for sequences and then looking for sequence relatedness with known viruses. This approach has given rise to the concept of the virome, the complete description of the viruses present, for example, in a human gut or in a species of bat. Using this approach, the modern day virus-hunters are rapidly expanding the number of known viruses.

Key points

- Growth in vitro of wild virus isolates is unreliable and imposes strong selection on the virus to change.
- It is difficult to study a virus that does not grow to high titre in cultured cells or in a convenient whole organism model.
- The classic technique for quantifying infectious virus is the plaque assay.
- ELISAs, to detect specific antibody or viral antigen, and PCR assays to detect viral genome, are key diagnostic techniques.
- For routine diagnostics, speed, automation, reliability and cost are key factors.

- The role of specific viral gene products in the virus lifecycle can be dissected by making and studying specific viral mutant strains.
- Reverse genetic systems are available for many viruses to permit targeted mutational analysis.

Questions

- Explain how the specificity of antibodies can be used to achieve accurate virus detection.
- Discuss the isolation of virus mutants and their importance in developing an understanding of virus replication cycles.

Further reading

Bendinelli, M., Friedman, H. 1998. *Rapid Detection of Infectious Agents*. Plenum, New York.

Bridgen, A. 2012. *Reverse Genetics of RNA Viruses: Applications and Perspectives*. Wiley-Blackwell, Chichester.

Clementi, N. 2000. Quantitative molecular analysis of virus expression and replication. *Journal of Clinical Microbiology* 38, 2030–2036.

Crowther, J. R. 2001. *The ELISA Guide*. Humana, Totowa, NJ.

Harlow, E., Lane, D. 1999. *Using Antibodies: A Laboratory Manual*. Cold Spring Harbor Laboratory Press, Cold Spring Harbor, NY.

Mahy, B. W. J., Kangro, H. O. 1996. *Virology Methods Manual*. Academic Press, London.

Taubenberger, J. K., Layne, S. P. 2001. Diagnosis of influenza virus: coming to grips with the molecular era. *Molecular Diagnosis* 6, 291–305.

Wiedbrauk, D. L., Farkas, D. H. 1995. *Molecular Methods for Virus Detection*. Academic Press, New York.

Wild, D. 2013. *The Immunoassay Handbook*, 4th edn. Elsevier, London.

Zuckerman, A. J., Banatvala, J. E., Griffiths, P., Schoub, B., Mortimer, P. 2009. *Principles and Practice of Clinical Virology*, 6th edn. Wiley-Blackwell, Chichester.

Part II

Virus growth in cells

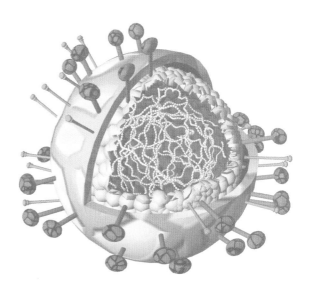

Chapter 6

The process of infection: I. Virus attachment and entry into cells

The first stage in the cellular replication cycle of a virus is the delivery of the virus genome into a suitable target cell. For animal viruses, this involves attachment to specific receptor molecules on the cell surface, which triggers internalisation in various ways. Viruses of plants and bacteria additionally need to breach the cell wall to gain entry.

Chapter 6 Outline

6.1 Infection of animal cells: the nature and importance of receptors
6.2 Infection of animal cells: enveloped viruses
6.3 Infection of animal cells: non-enveloped viruses
6.4 Infection of plants cells
6.5 Infection of bacteria
6.6 Infection of cells: post-entry events
6.7 Virus entry: cell culture and the whole organism

Attachment of a virus particle to a cell, and its entry into that cell, represent the initiating events in a virus replication cycle. For these events to occur, virus must first gain proximity to a cell within a susceptible organism. Free-living bacterial cells encounter virus in the environment through random diffusion. For plants, viruses are most commonly delivered to a new host by a vector species. In animals, viruses may be delivered to the body surface by a vector or by direct contact with particles on surfaces, or else viruses can be internalized into the body from the environment via the respiratory, gastrointestinal or genital tracts where potential target cells may then be encountered. There is no way that a virus can control these initial encounters to ensure it meets an individual of a susceptible species, so any organism will be constantly encountering viruses which cannot infect it productively. Attachment and entry into cells therefore represent the first points at which the ability of a virus to infect a host productively may be determined. This chapter considers the virus attachment and entry events that occur in animal, plant and bacterial systems.

6.1 Infection of animal cells: the nature and importance of receptors

Virus receptors are normal cell surface molecules that have been adopted by viruses to serve as

Introduction to Modern Virology, Seventh Edition. N. J. Dimmock, A. J. Easton and K. N. Leppard.
© 2016 John Wiley & Sons, Ltd. Published 2016 by John Wiley & Sons, Ltd.

docking and entry molecules for them. Binding to a receptor concentrates a virus at the cell surface to facilitate entry and prepares the particle to release its genome.

When a virus particle finds itself in proximity with a cell, something must initiate the process of infection. That event is the binding of surface components of the virus particle (*attachment proteins*) to cell surface components (*receptors*). By binding in this way, free diffusion of the particle is prevented, so increasing the chances that virus entry will happen. Binding of attachment proteins to receptors often also triggers conformational changes in the particle that are necessary for entry or genome release to occur. In these circumstances, attachment represents an irreversible commitment of the virus particle to its interaction with that cell; if entry events fail after this change, or the cell happens not to be a viable host for the virus, the particle will not be able to detach and seek another more favourable host cell.

Receptors are not provided by the cell simply to facilitate infection by viruses. Rather, they are part of the functional fabric of the cell, with essential structural or functional roles in the life of the organism. As such, even the powerful selective pressure that is imposed by their presence, making the host susceptible to a virus, is not capable of eliminating the receptor from the cell surface. Receptor molecules are usually proteins, but carbohydrates and, very occasionally, lipids are also used (Table 6.1). The virus-receptor interaction is highly specific, but multiple viruses may use the same receptor. Sometimes, related viruses share the same receptor but shared usage can occur even between members of different virus families. One example of this is the sugar *N*-acetylneuraminic acid (sialic acid) which often forms the terminal moiety of carbohydrate groups on a glycoprotein or glycolipid (Table 6.1). The unequivocal demonstration that a

molecule serves as a receptor for virus infection requires some stringent tests (Box 6.1).

The original concept of virologists regarding virus particle–host cell interactions was that each virus would use a single, definable type of molecule as its receptor, and that differences between viruses in their receptors would explain the unique patterns of infectivity that each virus displayed. However, as analyses of specific viruses have expanded, so more examples of multiple receptor usage are emerging. For example, herpes simplex virus can utilize either HVEM, a member of the tumour necrosis factor receptor family, nectins 1 & 2, which are cell adhesion molecules from the immunoglobulin superfamily, or sites in heparan sulphate as entry receptors. The ability to use alternative receptors may allow a virus to expand the range of cell types it can infect. Also, some viruses make sequential interactions first with a primary receptor and then with co-receptors as part of a multistep entry process within a single cell type. As examples of this, HIV-1 particles bind first to the primary receptor, CD4, and then to a co-receptor CXCR4 or CCR5, while human adenovirus type 5 utilizes a primary receptor CAR and then a cell surface integrin to gain entry into a target cell.

For some viruses, not all cell surface molecules that are able to bind the virus can serve as receptors to initiate infection. Molecules that bind virus in this non-productive way are known as pseudoreceptors. One example where pseudoreceptors have been shown to affect the biology of infection is the mouse virus, polyomavirus. Polyomavirus particles attach to cells via sialic acid residues on the cell surface, but the nature of the chemical linkage between the sialic acid and the next residue in the carbohydrate chain determines whether or not the interaction will lead to infection. Because the pseudoreceptor form of sialic acid is abundant, strains of virus that can bind it in addition to the true receptor remain highly concentrated near the site of production

Table 6.1
Some cell surface molecules used by animal viruses as receptors.

Molecule	Normal function	Virus	Type of receptor
Protein			
$\alpha_v\beta_x$ integrin	Adhesion to other cells via vitronectin	Adenoviruses	Co-receptor
$\alpha_v\beta_6$ integrin	Adhesion to other cells via vitronectin	Foot-and-mouth disease virus	Primary
Acetylcholine receptor	Binds the acetylcholine neurotransmitter	Rabies virus	Primary
Angiotensin converting enzyme 2	Forms active angiotensin from precursors; multiple physiologic effects	SARS coronavirus	Primary
β-adrenergic receptor	Binds the hormone adrenalin (epinephrine)	Reoviruses	Primary
CAR (Coxsackie-adenovirus receptor)	A unique protein of unknown function	Many human adenoviruses, Coxsackie B viruses	Primary
CCR5 and CXCR4	C-C chemokine and C-X-C chemokine receptors	HIV-1, HIV-2, SIV	Co-receptor
CD4	Ligand for MHC II – on T helper cells	HIV-1, HIV-2, SIV	Primary
CD46	Negative regulation of complement activity	Culture-adapted measles virus, human species B adenovirus	Primary
CD150 (SLAM)	Lymphocyte activation molecule	Measles virus	Primary
PVRL4 (nectin 4)	Epithelial cell adhesion molecule	Measles virus	Primary
CD155	Ligand for vitronectin, nectin-3 and DNAM-1	Poliovirus	Primary
CR2	Receptor for complement component C3d	Epstein–Barr virus	Primary
DC-SIGN	C-type lectin on dendritic cells that binds carbohydrate moieties	HIV-1, SIV	Primary, but transfers bound virus from a dendritic cell to a T cell
HVEM	TNF receptor family member	Herpes simplex virus	Primary
ICAM-1	Adhesion to other cells via CD54	Rhinoviruses (most but not all)	Primary
IgA receptor	Binds IgA for transport across the cell	Hepatitis B virus	Primary
MHC I	Ligand for CD8; presents peptides to T cells	Human cytomegalovirus	Primary
MHC II	Ligand for CD4; presents peptides to T cells	Lactate dehydrogenase-elevating virus	Primary

Table 6.1
(*Continued*)

Molecule	Normal function	Virus	Type of receptor
Nectin1, 2	Intercellular adhesion	Herpes simplex virus	Primary
Phosphate transporter	Transports phosphate	Some retroviruses	Primary
Virus-specific IgG bound to cells by Fc receptors	Binds to virus	Dengue virus *in vivo* and many others *in vitro*	Primary
Carbohydrate			
N-acetylneuraminic acid (sialic acid; only when terminal in the carbohydrate moiety)	Part of the carbohydrate moiety of glycoproteins and glycolipids. Gives cells much of their negative charge	Influenza virus A, B, C, paramyxoviruses, polyomavirus, encephalo-myocarditis virus, reoviruses	Primary
Heparan sulphate	Extracellular matrix glycosaminoglycan	HIV-1, herpes simplex virus, dengue virus, Sindbis virus, cytomegalovirus, adeno-associated virus, respiratory syncytial virus, foot-and mouth disease virus	Co-receptor
Lipid			
Phosphatidylserine	Constituent of lipid bilayer	Vesicular stomatitis virus	Primary

ICAM, intercellular adhesion molecule; MHC, major histocompatibility complex; HIV, human immunodeficiency virus; SIV, simian immunodeficiency virus.

Box 6.1

Evidence required to demonstrate receptor activity in a cell-surface molecule

- To be a genuine virus receptor, candidate receptor molecules:
  - Should be shown to bind to intact virus or to the known attachment protein(s) of the virus
  - Should be present on cells susceptible to infection and absent from other cells
  - Should confer infectability on non-permissive cells when expressed in those cells from a cloned cDNA.
- Alternatively, proof that a candidate receptor can actually mediate infection can come from demonstrating that a monoclonal antibody to the candidate receptor can block infection of cells by the virus.
- A molecule may fail these tests and still be a genuine receptor for the virus, since not all antibodies to a receptor will necessarily block infection and there may be factors blocking infection in a non-permissive cell other than the lack of the receptor.

and so show only limited spread and pathogenesis around the animal, whereas strains that bind only the true receptor spread more easily. The relevance to pathogenesis of a virus avoiding being trapped at the site of production by non-productive interactions is also illustrated by the importance of influenza A virus neuraminidase in virus release (see Chapter 20).

Another important feature of the interaction between virus particles and receptors is binding affinity. You might expect that tighter binding between the attachment protein and receptor would increase the chance of productive infection because it would reduce the fraction of particles that detached before irreversible commitment to entry had occurred. High affinity interactions should also promote the concentration of highly dilute incoming virus particles at the surface of a cell, also increasing the chance of infection. However, experiments in the same polyomavirus system mentioned above suggest that too high an affinity for the receptor can reduce the overall efficiency of infection in the whole organism; of two virus strains where neither could bind pseudoreceptors, the one with lower affinity for the receptor spread faster around the animal and caused more severe disease.

Finally, virus infection can itself reduce the levels of specific receptors on the cell surface. Binding to receptors will typically cause their internalization and this often results in the receptors being trafficked to lysosomes for degradation. Also, many enveloped viruses express their glycoproteins, which include the attachment functions, on the surface of infected cells prior to their inclusion into envelopes of progeny particles. By binding here with virus receptor molecules, they can again interfere with receptor trafficking to reduce receptor levels on the cell surface. A practical consequence of this is that an infected cell is often refractory to infection, at least for a time, by any other virus that uses the same receptor.

An extreme example of this is retroviruses, which typically do not cause acute cytolysis upon infection but which establish themselves in the DNA of the cells (see Chapter 9). Retroviral sequences can become incorporated into the germ line of the organism and, thereafter, the proteins from their residual envelope genes can protect progeny individuals against infection by viruses that have the same receptor tropism. This may provide a selective advantage that could explain the acquisition and retention of such sequences in evolution (see Section 31.2)

6.2 Infection of animal cells: enveloped viruses

The outermost layer of many animal virus particles is a lipid bilayer, or envelope, in which attachment proteins are embedded. These viruses deliver their genetic material and other contents into the cell by membrane fusion, either at the cell surface or after being taken into the cell in an endocytic vesicle. The lipid envelope and embedded proteins become part of the cell membrane.

The attachment proteins of enveloped viruses are transmembrane proteins that span the envelope bilayer and are usually multimeric assemblies, meaning each attachment protein is multivalent. Binding of the attachment proteins to target cell receptors brings the lipid bilayers of the virus envelope and the host membrane into close proximity. However, the charged head groups of the phospholipids in the two membranes will tend to repel each other and so membrane fusion will not occur spontaneously. Instead, these viruses must provide specific fusion functions to drive the merging of the two membranes. For some viruses, fusion occurs at the plasma membrane immediately following attachment whilst for others fusion is delayed until after the particle has been internalized by endocytosis (Fig. 6.1). Attachment and fusion functions can be provided by two different proteins in the viral envelope (e.g. members of the family *Paramyxoviridae*) while in influenza A virus and retroviruses such as HIV-1 both functions are provided within the same protein. Common principles appear to underpin fusion in all these cases: during particle formation, after folding of what is always a multimeric

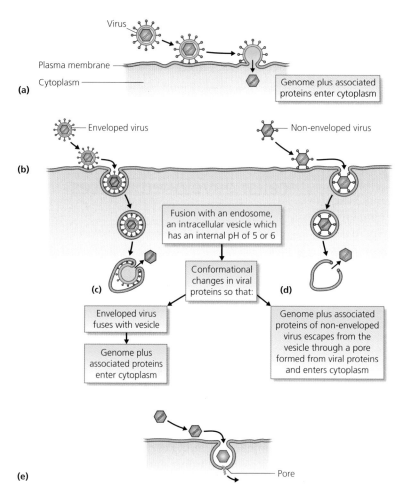

Fig. 6.1 Entry of animal virus genomes into cells. All viruses start by attaching to specific receptors on cells and then enter by different routes. (a) Entry by fusion of the lipid bilayers of an enveloped virus and the plasma membrane at neutral pH. (b) Entry by endocytosis is followed by a fusion of the vesicle with an endosome and a decrease in the pH of the endosome. This promotes conformational changes in viral proteins. For enveloped viruses this leads to fusion of the lipid bilayers of the virus and the endosome (c). For non-enveloped virus particles (d), this results in the insertion of newly-exposed hydrophobic regions of the virion into the lipid bilayer of the vesicle and the escape of the viral genome and associated proteins through this hydrophobic pore into the cytoplasm. Some non-enveloped virus types may deliver their genome into the cytoplasm through a pore that develops in the plasma membrane before endocytosis is complete (e).

precursor form of the protein into a stable conformation, site-specific cleavage destabilizes the tertiary structure to a metastable state, priming it for a conformational change to occur upon attachment that will reveal its potent fusogenic peptide domains to cause membrane fusion.

Fusion at the plasma membrane: human immunodeficiency virus

The attachment protein of HIV-1 is formed as a trimer of gp160 envelope proteins and, after cleavage, comprises three gp120 polypeptides (attachment) and three gp41 transmembrane polypeptides (fusion) that anchor the protein in the viral lipid envelope. Direct fusion of the HIV-1 envelope with the plasma membrane is initiated by the succession of virus–receptor interactions involving gp120 (Box 6.2). This triggers conformational changes in the envelope protein that result in the exposure of the hydrophobic terminal segments of the gp41

subunits from their normally concealed positions close to the virion membrane. Fusion then proceeds in a series of steps (Fig. 6.2). Firstly, some of the N-terminal hydrophobic regions insert into the lipid bilayer of the cell membrane; secondly, a small number of envelope proteins combine together to form a hydrophobic channel between the two membranes, disrupting both bilayers and allowing the disturbed lipid to flow through the hydrophobic channel and the two bilayers to come together and form a fusion pore; lastly, the pore enlarges so that the genome and associated proteins can pass through it and enter the cytoplasm and the viral and cell bilayers become united. HIV infection is considered in detail in Chapter 21.

Fusion in an endosome: influenza A virus

Similar to the HIV-1 envelope proteins, the haemagglutinin (HA) protein of influenza

---

Box 6.2

Attachment interactions of HIV-1 with its target cell

HIV-1 infects CD4$^+$ T cells or CD4$^+$ macrophages. Initially, the virus uses a cellular protein (cyclophilin A) that is incorporated into the HIV-1 particle during assembly to bind a low affinity receptor, heparan. This interaction increases the chance of contacting the primary receptor, CD4, which is a less abundant molecule. If the virus does not find a primary receptor molecule, it will dissociate from the cell completely and the process begins again. The search process consists of the virus rolling along the cell surface, repeatedly dissociating and reassociating with low affinity receptors, until it comes in contact with CD4 molecules.

The CD4–gp120 binding is a high affinity interaction but, with this association alone, the virus and cell lipid bilayers are too far apart for fusion to take place. However, binding to CD4 causes gp120 to undergo conformational changes that expose another site that is specific for one of the co-receptors, CCR-5 or CXCR4. Because these co-receptor molecules comprise 7 transmembrane regions with only small loops on the extracellular surface, binding to gp120 draws the bilayers closer together. This series of receptor interactions progressively ratchets the viral and cell bilayers into closer proximity so that fusion can take place.

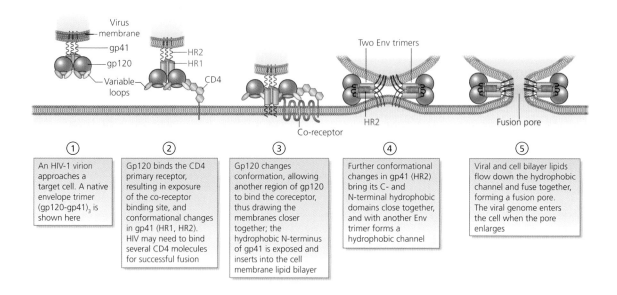

Fig. 6.2   Model for the fusion entry of the genome of an HIV-1 virion into a target cell. (Adapted from Moore, J. and Doms, R. The entry of entry inhibitors: A fusion of science and medicine. *Proceedings of the National Academy of Sciences of the USA* 100, 10598. © 2003 National Academy of Sciences, USA.)

A virus particles is formed as a homotrimer of the HA0 polypeptide and is then cleaved at a specific internal site to separate the HA1 and HA2 components, which remain associated in the trimeric protein (Fig. 6.3). This cleavage is essential for the particle to be infectious because it frees the fusion domains of the three subunits from structural constraints; the nature of the cleavage site and the availability of enzymes that can cleave it are significant determinants of pathogenicity (see Chapter 20). Attachment to receptors via the globular head of the protein (HA1) triggers endocytosis of the particle. This is a natural cell process by which small segments of plasma membrane and associated proteins are pinched off and internalized within the cell as a vesicle. It is important to realize that, although the endocytosed particle is now within the volume of the cell, it is still topologically external

to the cell – entry has not yet occurred. Subsequent fusion between the vesicle and virus envelope membranes, which then allows release of the genome into the cytoplasm, is dependent upon the internal environment of the endocytic vesicle becoming acidic (pH 5–6). This is achieved by fusion of the endocytic vesicle with an intracellular vesicle called an endosome. An endosomal membrane protein then pumps protons into the lumen of the vesicle. The low pH initiates conformational changes in the HA protein which release the hidden hydrophobic N-termini of the membrane-anchored HA2 subunits that carry fusogenic activity. Fusion of the viral envelope with the lipid bilayer of the vesicle then proceeds essentially as described above for HIV-1. Endosome acidification is also important for disassembly of the influenza A virus nucleocapsid; the M2 protein in the viral

(a) HA monomer

(b) HA trimer

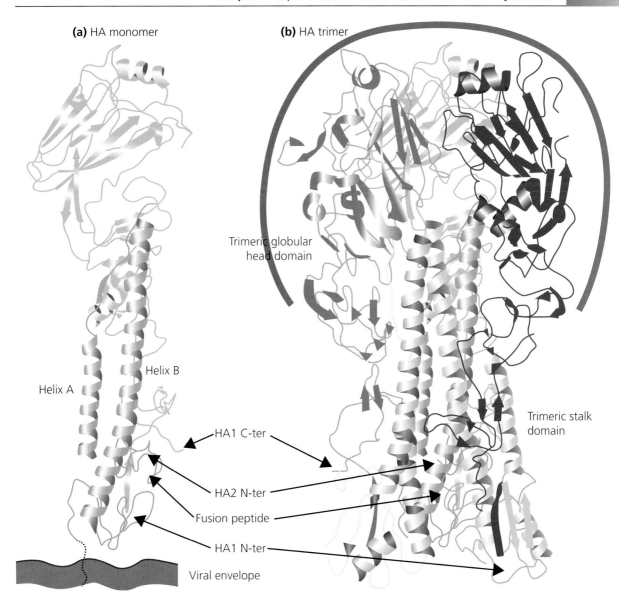

Trimeric globular head domain

Helix B

Helix A

Trimeric stalk domain

HA1 C-ter

HA2 N-ter

Fusion peptide

HA1 N-ter

Viral envelope

Fig. 6.3 Crystal structure of the influenza virus haemagglutinin (HA) from strain PR8. (a) A monomer of the HA trimer in mature form, after cleavage of the HA0 primary translation product to its HA1 and HA2 components. The distal HA1 subunit is shown in green and the membrane associated HA2 in turquoise. The C-terminal part of HA2, which crosses the viral envelope, was removed before the structure was determined. (b) The HA trimer. HA1 – green, pink, red; HA2 – turquoise, violet, yellow. The globular head of HA1 is made of a distorted jelly-roll β barrel like the capsid proteins of most of the icosahedral viruses; it bears all the antibody neutralization sites on the protein. The fusion peptide is buried in the trimeric structure. A low pH-induced conformational change causes the three parts of the globular head to swing away and helix A of HA2 to move upwards to become an extension of helix B. This carries the fusion peptide (at the N terminus of HA2) up to the top of a long helical bundle, where it can interact with the membrane of the vesicle.

envelope acts as a proton channel to acidify the particle and this is essential for infection to proceed.

6.3 Infection of animal cells: non-enveloped viruses

Virus particles that lack a lipid envelope must directly breach a cell membrane to deliver at least their genetic material, and sometimes the entire particle, into the cell. For some viruses, such entry events occur at or near the cell surface while others breach the membrane of an endocytic vesicle after being taken into the cell.

The mechanisms by which non-enveloped viruses breach the barrier of the target cell membrane to achieve entry are generally less well understood than those of enveloped viruses. However, for those examples that have been carefully studied, the same themes of induced conformational change and important hydrophobic peptide sequences emerge as key features. As for enveloped viruses, the timing of

the entry event relative to endocytosis can vary. For example, poliovirus apparently delivers its genome into the cytoplasm at or close to the cell surface whereas adenovirus achieves entry from an endosome (Fig. 6.1).

Although both these routes of non-fusion entry are equivalent in delivering the viral genome into the cytoplasm of the cell, they differ substantially in the scope for delivering viral proteins too. Entry via the endosomal route is likely to deliver substantial quantities of protein components from the infecting particle into the cell; these then have the opportunity to play functional roles in setting up the new infection. In contrast, when a genome is introduced via a pore in the membrane, it is likely that most or all of the particle proteins will remain external to the cell.

Viral genome entry at the plasma membrane: poliovirus

Poliovirus interacts with its receptor, CD155, at the cell surface leading to conformational changes in the particle (Box 6.3). The precise location of the next steps has been much

---

Box 6.3

Evidence for conformational change during poliovirus entry

- Exposing virus either to cells that display the receptor or to soluble receptor protein causes a change in the sedimentation coefficient of the particles (i.e. their size and/or shape changes) from 160S (native particles) to 135S.
- Whereas native 160S particles are resistant to protease and reasonably soluble, 135S particles are protease-sensitive and hydrophobic.
- In 135S particles, elements of the VP1 and VP4 proteins that are internal in 160S particles are now exposed on the particle surface.
- Many 135S particles detach from cells without achieving infection; these are unable to attach to further cells.
- 135S particles that remain attached are gradually converted to an 80S form that has lost the RNA genome.

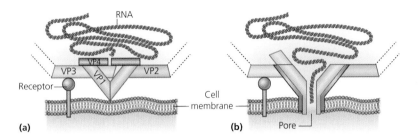

Fig. 6.4   A model for the translocation of poliovirus RNA across the cell membrane. (a) Receptor interactions bring a vertex of the virion, formed of VP1 (turquoise), close to the target cell membrane. A part of VP3 (orange) blocks the 5-fold axis of the virion while the small VP4 protein (red), which has a myristylated N-terminus, is held internally. (b) Conformational change induced by receptor binding moves the VP3 plug from the vertex and allows VP1 and the myrisyl groups on VP4 to interact with the membrane, forming a pore through which the viral RNA can enter the cell.

debated. Many experiments have suggested that entry is independent of classical endocytic pathways, leading to the idea that the virus forms a pore in the plasma membrane at the site of attachment through which the genome RNA enters the cell. The presence at the surface of infected cells of empty capsids supports this idea. However, more recent data suggest that formation of the pore actually requires the closure or near-closure of a vesicle around the virus particle first, although the genome then exits the particle (and the vesicle) while they are still very close to or attached to the plasma membrane. Poliovirus genome entry into the cell is certainly low pH-independent, meaning that there is no requirement for it to reach the endosomal compartment. The details of pore formation during poliovirus genome entry are shown in Fig. 6.4.

Virus entry via an endosome: adenovirus

Adenoviruses make their initial interaction with their receptors via the fibre proteins that protrude from the vertices of the capsid (Section 2.3). The best studied of this virus family are the human species C adenoviruses that utilize CAR as their receptor (Table 6.1). Attachment facilitates a secondary interaction between the capsid protein at the base of the fibre, known as penton base, and cell surface integrins (Fig. 6.5). Together, these interactions promote endocytosis via clathrin-coated pits, during which at least some of the fibre and penton base units are stripped from the particle. A conformational change then occurs in an internal capsid component, protein VI, which lies under the outer shell. This change is thought to expose an amphipathic (one face hydrophobic, the other hydrophilic) helical domain that can dissolve into and destabilize the vesicle membrane, permitting its rupture and the release of the residual, partially uncoated particle into the cytoplasm.

There are actually multiple mechanisms of endocytosis in eukaryotic cells. Whilst adenovirus uses the clathrin pathway to internalize itself in a vesicle, polyomaviruses such as SV40 use the caveolin pathway for this purpose. The fates of vesicles formed by these different routes are not equivalent, which perhaps explains why a virus may not be infectious if it is internalized via an interaction with the 'wrong' receptor (see Section 6.6).

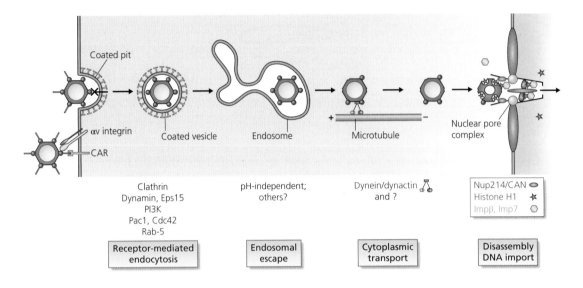

Fig. 6.5 The process of adenovirus entry into cells (see text for details). Beneath the diagram are indicated some of the host factors involved at each stage. (Reproduced with permission from Meier, O. and Greber, U. *Journal of Gene Medicine* 5, 452–462, © 2003 John Wiley & Sons.)

6.4 Infection of plant cells

> *Plants have rigid cell walls composed of cellulose. Viruses require assistance from an external trauma to damage the plant cell wall and membrane so that they can achieve entry.*

The cell wall that surrounds the plasma membrane of all plant cells prevents viruses from attaching and entering them in the ways just described for animal cells. Viruses can only reach the cytoplasm of plant cells when the tissue is damaged. Thus plants are infected either with the help of *vectors*, namely animals that feed on plants, or invading fungi, or by mechanical damage caused by the wind or passing animals, all of which allow viruses to enter directly into cells. In the laboratory, this is mimicked by the gentle application of abrasive carborundum to leaves prior to adding virus to the leaf tissue (Fig. 1.1b). This description makes the process of virus transfer to a new plant host sound random and haphazard. However, the transmission of most plant viruses is a highly specific process, requiring the participation of particular animals as vectors in each case, which feed by piercing plant tissues with their mouthparts (leafhoppers, aphids, thrips, whiteflies, mealy bugs, mites or nematodes).

Because they are introduced mechanically into the cytoplasm of a host cell, plant viruses do not appear to have specific receptors on the cell surface. Once the vector has provided the initial introduction event, how do the viruses then spread from cell to cell to spread the infection within the plant? The answer lies in the interconnection of the cytoplasm

---

Box 6.4

Evidence that plant virus genomes can move directly between cells of the plant

- Some viruses have genetic material that is directly infectious to cells (e.g. positive-sense RNA viruses).
- Infection for these viruses can be induced by directly applying genomes to plant leaf tissue, with scarification of the leaf to promote uptake.
- Infection then spreads throughout the plant.
- When the same experiment is done with cloned genomes that carry a mutation in their capsid protein gene – meaning that the genomes cannot form infectious particles – exactly the same systemic spread around the plant occurs.
- Genetic analysis shows that plant viruses encode movement proteins that form channels along the plasmodesmata connecting plant cells and facilitate virus genome transmission between cells.

---

compartments of plant cells by plasmodesmata. These allow an incoming virus genome (or its progeny) to move between cells without encapsidation and re-infection, often dependent on the virus expressing a specialized movement protein (Box 6.4).

6.5 Infection of bacteria

Bacteria, like plants, have a strong cell wall. Some viruses possess the means to breach the bacterial cell wall and inject their genome through it; others exploit specialized structures on the bacterial surface as entry portals for the viral genome. Often, most or all of the particle proteins remain outside the cell after genome entry.

Attachment of bacteriophages to the bacterial cell

Most bacteriophages (phage) attach to the bacterial cell wall, but there are some that exploit cell receptors on more specialized structures of the host cell, such as the pili, flagella or capsule. The receptor for phage λ (lambda), a tailed phage of *Escherichia coli*, is maltoporin, which is a high affinity maltose-binding/transporter protein produced by the *lam*B gene. LamB is encoded within a maltose-inducible operon, so that susceptibility of cells to this phage is dependent on the nutrient environment.

Most tailed phages attach to the bacterial cell wall, and do so by the tip of their tail. Two well-documented examples of tailed phage attachment are phage T2 and T4 of *E.coli*. These both have a complex structure, including a tail, base plate, pins and tail fibres (see Chapter 2). The initial attachment of T2 or T4 to their receptors on the bacterial surface is made by the distal ends of the long tail fibres. These fibres attach first, bend at their centre, and their distal tips contact the cell wall only some distance from the midpoint of the phage particle. After attachment, the phage particle is brought closer to the cell surface. When the base plate of the phage is about 10 nm from the cell wall, contact is made between the short pins extending from the base plate and the cell wall (Fig. 6.6).

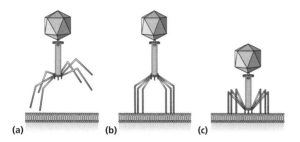

Fig. 6.6 Steps in the attachment of bacteriophage T4 to the cell wall of *Escherichia coli*. (a) Unattached phage showing tail fibres and tail pins (see Fig. 2.15). (b) Attachment of the long tail fibres. (c) The phage particle has moved closer to the cell wall and the tail pins are in contact with the wall.

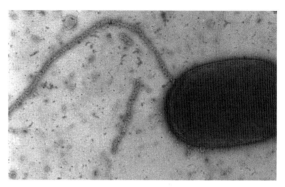

Fig. 6.7 Attachment of many spherical RNA phages to the sex pilus of *Escherichia coli*. (Courtesy of C. C. Brinton.)

Some phages, such as phage χ (chi) and PBSI, attach to flagella. The tip of the tail of these phage has a fibre that wraps around the flagellum. The phage then slides until it reaches the base of the flagellum. A further site of important cell receptors for phage is on the sex pili. Bacteria which harbour the sex factor (F) or certain colicins or drug resistance factors produce pili, and two classes of phage are known to attach to these pili. The filamentous single-stranded deoxyribonucleic acid (DNA) phage such as M13 attach to the tips of the pili, whilst many types of spherical ribonucleic acid (RNA) phage attach along the pili (Fig. 6.7).

Entry of bacteriophage genomes into bacterial cells

The experiments of Hershey and Chase, which used the tailed bacteriophage T2 (see Section 1.3), indicated that it was mainly viral nucleic acid that entered the cell during infection. The way in which this occurs is a fascinating story and the action of the phage particle in the process has been likened to that of a hypodermic syringe injecting DNA into the cell. The various steps in tailed phage genome entry

are shown in Fig. 6.8. However, most bacteriophages do not possess contractile tail sheaths and the way in which their nucleic acid enters the cell is not known. Hershey–Chase-type experiments with the filamentous DNA phages, which attach to the sex pili, suggest that both the DNA and the coat protein enter the cell and it has been postulated that, after attachment, the phage-pilus complex is retracted by the host cell to aid phage entry. However, similar experiments using RNA phage, which also attach to the sex pili, suggest the phage coat protein does not enter the cell. Instead, these particles undergo a structural change as a result of their interaction with the bacterium (perhaps similar to that seen in poliovirus, Box 6.3) that releases the attachment protein, A, from the particles. This protein is then thought to enter the cell with the RNA genome.

6.6 Infection of cells: post-entry events

For bacteriophage, entry of the genome into the cell is enough to begin the cycle of

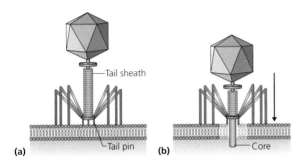

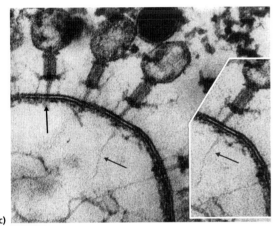

Fig. 6.8 The mechanism of entry of the phage T4 genome into the bacterial cell. The tail sheath in its extended form at the point of tail pin contact with the cell wall consists of 24 rings of small and large subunits surrounding a core (a). Following attachment, the tail sheath contracts driven by the merging of the small and large subunits to give 12 rings of 12 subunits. This shortening of the tail sheath pushes the non-contractile tail core through the outer layers of the bacterium with a twisting motion (b). This process is probably aided by the action of the lysozyme which is built into the phage tail. At the same time, contraction of the head forces the injection of the phage DNA through the tail core into the cell. 144 molecules of adenosine 5′-triphosphate (ATP) built into the tail sheath probably provide the energy for contraction. (c) Negatively stained electron micrograph of T4 attached to an *Escherichia coli* cell wall, as seen in thin section. The needle of one of the phages just penetrates through the cell wall (black arrow). Thin fibrils (red arrow) extending on the inner side of the cell wall from the distal tips of the needles are probably phage DNA.

replication. There is not much subcellular compartmentation for the virus to contend with and the naked genome is immediately exposed for gene expression. For viruses of animals and plants, the situation is more complex. If the capsid has entered the cell intact, then it must be at least partially disassembled before the processes of gene expression and genome replication can occur. This is necessary so that substrates for synthesis of new molecules, and in some cases cellular enzymes, can gain access to the viral genome. Also, the viral genomes and any necessary associated proteins must be transported to reach the part of the cell where replication is going to occur. This typically involves elements of the cytoskeleton such as actin filaments and microtubules, as in adenovirus movement from the endosome to the nuclear membrane (Fig. 6.5). For viruses that replicate in the nucleus, the genome must pass through the nuclear membrane via a nuclear pore. Finally, many viruses begin the process of reorganizing the cell environment very soon after entry in ways that will support or favour the productive replication cycle which is to follow. The events that happen under these different headings are in most cases unique to a particular virus type and so cannot be generalized, but it is important to remember that they exist.

The critical importance of post-entry events to the establishment of a productive infection means that there are lots of opportunities for a virus infection to fail at these early stages. This is one explanation for why not all virus entry events lead to an infectious outcome. Another is that different cell surface molecules undergo different pathways of internalization within the cell, leading to different fates; some may be recycled to the cell surface, others may be efficiently targeted to lysosomes for degradation. If a virus needs to reach a certain intracellular compartment in order to infect successfully, then it is important that it binds to a receptor molecule that can take it there.

6.7 Virus entry: cell culture and the whole organism

Virtually all data for the attachment and entry of animal viruses come from experiments with cells lines in vitro. Most of these cell lines are de-differentiated compared with the state of the actual target cells for infection in vivo, meaning that their composition of cell surface molecules may have changed, and they also typically divide more frequently than would the target cells of the virus in the whole organism. Thus infection of cells in the whole organism may not take place by exactly the same process as defined from studies in cell culture. In particular, a virus may use different receptors in vivo from those defined for it using a permanent cell line in culture.

Attempting to grow viruses from clinical samples in de-differentiated cell culture systems is notoriously difficult to achieve. This is simply a reflection of the differences between such cells and the true physiological targets of the virus in vivo. Putting virus from a clinical isolate into culture places a powerful selection pressure on it to adapt to this non-physiological situation. Sequence variants in the virus population that are better able to grow in the new environment than the bulk of the viruses present rapidly come to dominate the population (see Chapter 4). In principle, variation in any property encoded by the virus may become selected for in this way. One facet of such adaptation can be changes in the cell receptor specificity of the virus. For example, there are two natural receptors for measles virus, a molecule known as SLAM or CD150, that is present on the surface only of activated lymphocytes, macrophages and dendritic cells, and PVRL4 which is found on the surface of certain epithelial cells. This distribution of receptors matches the observed tropism of the virus in vivo. However, laboratory-adapted and vaccine strains of measles virus can use a ubiquitously expressed cell surface molecule, CD46, as a receptor due to the acquisition of mutations that cause specific amino acid changes in the virus attachment protein. This exemplifies the ability of many viruses to evolve rapidly under selective pressure.

Key points

- Animal viruses initially interact with a target cell by binding to specific receptor molecules (mostly proteins) on the cell surface. These have varied normal functions in the uninfected cell.
- Binding to cell receptors is an active process that causes a virus particle to undergo conformational changes which eventually lead to the viral genome entering the cell.
- The genomes of all enveloped animal viruses enter the cell cytoplasm by fusion of lipid bilayers from the virus and the cell.
- Animal viruses that are taken into the cell in a vesicle by receptor-mediated endocytosis are functionally external to the cell until they fuse with or breach the endosome membrane; the low pH environment of the vesicle can trigger these processes.
- Plant viruses enter the cytoplasm of a plant cells through a physical break in the plant cell wall caused by mechanical damage or animal vectors.
- Some bacterial (head-tail) viruses inject their genome through the cell wall into the host cell cytoplasm, but many bacterial viruses use other mechanisms.

Questions

- Discuss the varied strategies used by viruses to enter animal cells.
- Compare and contrast the mechanisms by which viruses achieve entry into mammalian cells, plant cells and bacteria.

Further reading

Arnberg, N. 2009. Adenovirus receptors: implications for tropism, treatment and targeting. *Reviews in Medical Virology* 19, 165–178.

Clapham, P. R., McKnight, A. 2002. Cell surface receptors, virus entry and tropism of primate lentiviruses. *Journal of General Virology* 83, 1809–1829.

Dhiman, N., Jacobson, R. M., Poland, G. A. 2004. Measles virus receptors: SLAM and CD46. *Reviews in Medical Virology* 14, 217–229.

Dimitrov, D. S. 2004. Virus entry: molecular mechanisms and biomedical applications. *Nature Reviews Microbiology* 2, 109–122.

Forrest, J. C., Dermody, T. S. 2003. Reovirus receptors and pathogenesis. *Journal of Virology* 77, 9109–9115.

Greber, U. F., Way, M. 2006. A superhighway to virus infection. *Cell* 124, 741–754.

Harrison, S. C. 2008. Viral membrane fusion. *Nature Structural and Molecular Biology* 15, 690–698.

Hogle, J. 2002. Poliovirus cell entry: common structural themes in viral cell entry pathways. *Annual Review of Microbiology* 56, 677–702.

Marsh, M., Helenius, A. 2006. Virus entry: Open sesame. *Cell* 124, 729–740.

Poranen, M., Daugelavicius, R., Bamford, D. H. 2002. Common principles in viral entry. *Annual Review of Microbiology* 56, 521–538.

Puntener, D., Greber, U. F. 2009. DNA tumour virus entry – From plasma membrane to the nucleus. *Seminars in Cell and Developmental Biology* 20, 631–642.

Schneider-Schaulies, J. 2000. Cellular receptors for viruses: links to tropism and pathogenesis. *Journal of General Virology* 81, 1413–1429.

Skehel, J. J., Wiley, D. C. 2000. Receptor binding and membrane fusion in virus entry: the influenza hemagglutinin. *Annual Review of Biochemistry* 69, 531–569.

Smith, A. E., Helenius, A. 2004. How viruses enter animal cells. *Science* 304, 237–242.

Stewart, P. L., Dermody, T. S., Nemerow, G. R. 2003. Structural basis of nonenveloped virus cell entry. *Advances in Protein Chemistry* 64, 455–491.

Tsai, B. 2007. Penetration of non-enveloped viruses into the cytoplasm. *Annual Review of Cell and Developmental Biology* 23, 23–43.

Chapter 7

The process of infection: IIA. The replication of viral DNA

Viruses with DNA genomes infect animals, plants and bacteria. Their genomes may be single- or double-stranded, and be either linear molecules or circles; they also vary considerably in size. Among the human viruses, genome sizes range from those of the hepadnaviruses at around 3000 base pairs (3 kbp) to the herpes and poxviruses, in the range 130–230 kbp. However, genomes can be as large as the Acanthamoeba polyphaga *mimivirus at 1180 kbp. This diversity in form and size is reflected in the differing mechanisms of genome replication of these viruses.*

Chapter 7 Outline

7.1 The universal mechanism of DNA synthesis
7.2 Replication of circular double-stranded DNA genomes:
  Polyomavirus, Papillomavirus and
    Baculovirus families
7.3 Replication of linear double-stranded DNA genomes that can form circles:
  Herpesvirus family; bacteriophage λ
7.4 Replication of linear double-stranded DNA genomes that do not circularize:
  Adenovirus and Poxvirus families
7.5 Replication of single-stranded circular DNA genomes:
  Bacteriophages φX174 and M13
7.6 Replication of linear single-stranded DNA genomes:
  Parvovirus family
7.7 Dependency versus autonomy among DNA viruses

The basic mechanism whereby a new strand of DNA is synthesized in a cell is the same, regardless of whether the DNA is of cellular or viral origin. The fact that all DNA synthesis shares these fundamental features has meant that viral DNA molecules, which can be manipulated with ease, have often been studied by biochemists attempting to unravel the mysteries of DNA replication. However, although these studies have added greatly to our understanding of this process, they have also revealed that viruses employ this basic mechanism in many different ways, in each case to suit the peculiarities of their genome structures. This chapter explores this diversity of viral DNA replication mechanisms.

Introduction to Modern Virology, Seventh Edition. N. J. Dimmock, A. J. Easton and K. N. Leppard.
© 2016 John Wiley & Sons, Ltd. Published 2016 by John Wiley & Sons, Ltd.

7.1 The universal mechanism of DNA synthesis

DNA strand polarity and the nature of DNA polymerases

The successful replication of a double-stranded DNA molecule requires that two daughter strands be synthesized. Since these must base pair with the two parental (template) strands, which are antiparallel, the daughter strands must also be antiparallel. Thus, if an enzyme complex were simply to move along a double-strand DNA template synthesizing two daughter strands, one of these strands would need to be made with $5' \rightarrow 3'$ polarity and the other with $3' \rightarrow 5'$ polarity. However, no DNA polymerase yet characterized has the capacity to synthesize DNA in the $3' \rightarrow 5'$ direction, and the biochemistry of DNA synthesis suggests that such activity is not possible.

One solution to this problem is for synthesis to proceed in the $5' \rightarrow 3'$ direction along one parental strand, the *leading* strand, and for

discontinuous $5' \rightarrow 3'$ synthesis to occur in the opposite direction along the other, or *lagging*, strand (Fig. 7.1, Box 7.1). In essence, while the enzyme complex moves forward copying the leading strand template, it reveals an increasing length of uncopied lagging strand template. Once a sufficient length of this template is available, a new daughter strand is initiated and copied from it, synthesis continuing until the polymerase comes up against the $5'$ end of the previously-synthesized lagging strand segment. This is the solution which is adopted in all organisms other than viruses, and by some viruses too, and is discussed later in the context of SV40 genome replication.

The alternative solution to the direction of synthesis problem, which is adopted by some viruses, is to separate in time the production of the two daughter strands. Once the requirement to have both strands made by the same enzyme complex is removed, then it is clearly straightforward for both the daughter strands to be made by continuous synthesis in the $5' \rightarrow 3'$ direction. An example of a virus employing this replication mechanism is adenovirus, which is discussed in Section 7.4. The problem with this mechanism, which is

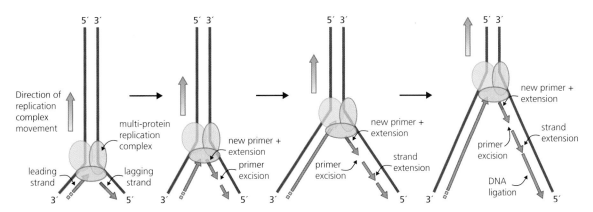

Fig. 7.1 Model for replication of double-stranded DNA through discontinuous synthesis of the lagging strand. Both strands are synthesized in a $5' \rightarrow 3'$ direction but only the leading strand is synthesized continuously, the other being formed from a series of short DNA molecules, each primed by a piece of RNA. The primers are excised and replaced by DNA before the fragments are joined by DNA ligase. Blue – parental DNA; brown – new DNA; red – RNA primer.

> **Box 7.1**
>
> **Evidence for discontinuous DNA synthesis**
>
> • When *Escherichia coli* cells infected with phage T4 are pulse-labelled with radioactive thymidine, the newly synthesized (i.e. radioactively labelled) DNA found immediately after the pulse is in small fragments of 1000–2000 nucleotides, as measured by velocity sedimentation.
> • After similarly-labelled cells are incubated for a further minute with unlabelled thymidine (a pulse-chase experiment), the labelled DNA is found in very much larger pieces.
> • The small fragments of DNA produced initially are known as Okazaki fragments after Tuneko Okazaki, the person who first published this result.

probably the reason why it has not evolved more generally, is that synthesis of one strand is necessarily delayed relative to the other and, therefore, in the interim period after the first replicating enzyme has passed through the template duplex, there are large amounts of single-stranded DNA (the second template strand) exposed. Since single-stranded DNA promotes recombination by strand invasion, such a mechanism would carry a cost of genetic instability which would be unsustainable for large genomes.

DNA strand initiation and the need for a primer

For any DNA replication mechanism, the process of new synthesis must begin somewhere. The locations on template molecules where this occurs are specific and are termed origins of replication. For replication to commence, specific proteins must bind to the origin and the two DNA strands must be separated to reveal the templates for new synthesis. These requirements lead to certain typical features in origin sequences, such as sequence symmetry to allow for specific binding of proteins and the presence of an AT-rich region for ease of strand separation.

During replication, new DNA strands must be initiated at least to begin the process and, if the discontinuous mechanism applies, repeatedly thereafter. This raises another problem since DNA polymerases are unable to start DNA chains *de novo*; all require a hydroxyl group to act as a primer from which to extend synthesis. The general solution to this problem is to invoke the action of an RNA polymerase since these do not require a primer to start synthesis. This enzyme synthesizes a short RNA primer, copying the DNA template, and this is then extended by DNA polymerase (Fig. 7.1, Box 7.2). Once they have served their purpose, the primers are excised by specific enzymes which recognize and degrade RNA that is duplexed with DNA (RNase H enzymes). The gaps so created are then filled by continuing $5' \rightarrow 3'$ DNA synthesis from the adjacent fragment and the fragments finally joined together by DNA ligase.

Why the requirement for an RNA primer?

Why has this complex process of RNA-primed DNA synthesis evolved, when it could have been avoided if DNA polymerases were able to initiate strands *de novo*? The answer probably

---

Box 7.2

Evidence for RNA-primed DNA synthesis

- Conversion of bacteriophage M13 single-stranded DNA to the double-stranded form in infected *E. coli* cells is inhibited by rifampicin, an inhibitor of *E. coli* RNA polymerase. In rifampicin-resistant *E. coli*, the inhibitor has no effect on phage DNA synthesis.
  N.B. Eukaryotes, rather than using one of their three RNA polymerases to synthesize primers, have a dedicated enzyme for this purpose known as a primase.

---

lies in a key evolutionary benefit of using DNA rather than RNA as genetic material, which is that the fidelity of DNA replication is very much greater (one error in 10^9–10^{10} base-pair replications) than that of RNA replication (one error in 10^3–10^4). The first check on fidelity of synthesis, applicable to both RNA and DNA polymerases, is the strong selection for base-pairing of an incoming nucleoside triphosphate to the template. However, DNA polymerases have an additional check, the ability of DNA polymerases to 'proofread' the DNA which they have just synthesized. Proofreading means that DNA polymerases excise any unpaired nucleotides from the 3′ end of a growing strand before adding further nucleotides. RNA polymerases do not have this activity since they have to be able to initiate synthesis of new molecules without a properly base-paired primer. Since RNA transcripts are continually turned over and have no long-term role in the organism, errors in their synthesis can be tolerated. This, however, has consequences for those viruses which have RNA genomes (see Chapter 8).

RNA primers and the 'end-replication' problem

The requirement for a primer for DNA synthesis creates a difficulty in achieving complete replication of linear molecules. If synthesis in the 5′ → 3′ direction is initiated at one end of the molecule with an RNA primer, there is no mechanism for filling the gap left when that primer is later digested away (Fig. 7.2); to fill this gap would require 3′ → 5′ synthesis and we know that this cannot occur. When the replicating fork reaches the other end of the template, a similar problem arises with the lagging strand. Without a solution, the net result would be synthesis of two daughter molecules, each with a 3′ single-stranded tail. If such molecules were to undergo further rounds of replication, smaller and smaller 3′ tailed duplexes would result.

This 'end-replication' problem is universal in biology. In eukaryotes, it is solved through the use of telomeres at the ends of each linear chromosome, which are replicated reiteratively outside of the normal replication process to maintain the chromosome length from one generation to the next. Prokaryotes, by contrast, have solved the problem by evolving circular genomes. When replicating such a molecule, the first primer can be excised and replaced by extension from the 3′ end of the fragment that is eventually synthesized adjacent to it. Some viruses employ this latter strategy to achieve replication of their genomes, whilst others have unique solutions to the 'end-replication' problem. The following sections discuss the genome replication of a series of DNA viruses to illustrate these different mechanisms.

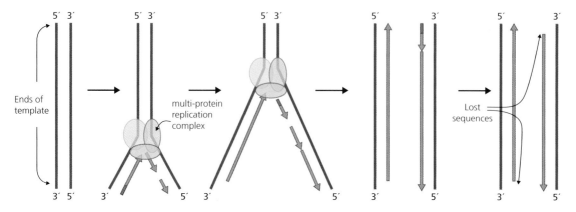

Fig. 7.2 The problem of replicating the ends of linear DNA molecules through the use of RNA primers. Once the first primer has been excised, there is no mechanism for filling the gap. Blue – parental DNA; brown – new DNA; red – RNA primer.

7.2 Replication of circular double-stranded DNA genomes

Important viruses in this category

- *The polyomavirus family, including SV40– an important model for the study of eukaryotic cell processes, and Merkel cell polyomavirus – a human tumour virus.*
- *The papillomavirus family – the causative agents of skin and genital warts, and of cervical carcinoma.*
- *The baculovirus family – insect viruses considered as agents for biocontrol of crop pests and used as laboratory tools for heterologous protein production.*

Simian virus 40 (SV40) has been studied in great detail, initially because of its tumourigenic properties (see Chapter 25) and then as an easily-analyzed model system through which to understand eukaryotic cell processes. Its genome is a circular, double-stranded DNA of around 5200 bp in length and, in its detailed

biochemical mechanism, the replication of this molecule has close parallels to the replication of eukaryotic cell genomes. The overall characteristics of SV40 replication are summarized in Fig. 7.3. The circular genome has a single origin of replication (a), at which new synthesis initiates coordinately on both strands using RNA primers. These two newly-initiated strands are then extended away from the origin in opposite directions, forming the leading strands of two diverging replication forks (c). As synthesis proceeds, the lagging strand templates are revealed at each fork (note that different strands of the template constitute the lagging strand template at the two forks), and these are then copied by discontinuous synthesis (see Section 7.1). Progress of the two forks around the template gives a structure with the appearance of the Greek letter θ ('theta'), which is therefore known as a theta-form intermediate (d). Ultimately, the two forks converge again on the opposite side of the template circle from the origin and, when they meet, both strands will have been completely copied (e). The two daughter molecules are, like the parent, closed circular duplexes. Initially, they are topologically linked in a structure

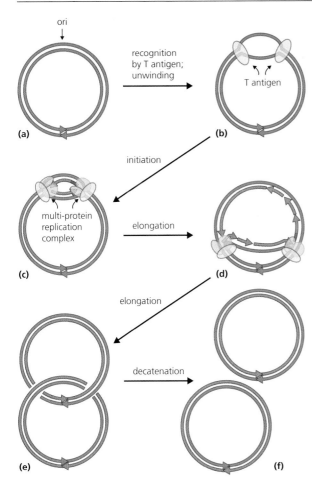

ori

recognition by T antigen; unwinding

T antigen

(a)

(b)

initiation

multi-protein replication complex

elongation

(c)

(d)

elongation

decatenation

(e)

(f)

Fig. 7.3 A general scheme for SV40 replication (see text for details). Blue – parental DNA; brown – new DNA; red – RNA primer. Arrowheads on newly-synthesised DNA fragments represent 3′ ends; the 5′ → 3′ polarity of other DNA strands is indicated by ▶ symbols.

the test tube, directed by the SV40 origin and using only purified, characterized proteins. The 10 proteins needed are listed in Table 7.1, with their activities in the replication process. It is noteworthy that only one of these proteins is encoded by the virus; all the others are provided by the cell and perform the same functions in SV40 replication as they are believed to do in host cell DNA synthesis. The one viral protein involved, large T antigen (see Chapter 10), provides the key initial recognition of the origin DNA sequence, binding specifically to it (Box 7.3). It then assembles a double hexamer around the DNA that has helicase activity, begins the separation of the DNA strands, and then recruits cellular single-strand DNA binding protein and DNA polymerase α/primase complex through protein: protein interactions, setting the scene for the initiation reaction. After initiation, T antigen continues to play a role as a helicase, unwinding the template ahead of each replication fork.

The details of how T antigen performs its roles have been the subject of extensive and detailed study. Early on, it was realized that the origin contained four pentanucleotide sequences 'GAGGC', collectively termed site 2, that were critical for T antigen binding and origin function (Fig. 10.4) but that the active helicase, which could hydrolyze ATP and melt DNA duplexes to single strands, was hexameric. The details of this structure were eventually revealed by X-ray crystallography in 2003. The basis of the transition from DNA-bound monomers of T antigen to the assembled double hexamers, and the position of the DNA relative to those hexamers, has remained uncertain – particularly whether both strands of the duplex pass through the central holes of the hexamers or only one strand, with the other one passing around the protein. It has been argued that when loaded on the DNA the hexamers are in fact spirals – akin to split-ring structures, which would provide a route for a DNA strand to

called a catenane (e); this follows necessarily from the closed circular nature of the template, one strand of which ends up in one daughter and one in the other. The final stage in SV40 replication is the separation of these circles (f).

Through work in the laboratories of Kelly, Stillman and others, the complete replication of circular DNA molecules has been achieved in

Table 7.1

Proteins involved in SV40 replication.

Protein	Function(s)	Role in SV40 replication
Large T antigen[*]	Sequence-specific DNA binding DNA helicase	Initial recognition of the viral replication origin Unwinding the DNA template duplex
RP-A	Single-stranded DNA binding	Stabilization of unwound DNA in the replication bubble at initiation, and within the replication forks
DNA polα/primase	Complex of activities synthesizing primers for initiation of new DNA strands and production of short DNA molecules	Initiation and short distance extension of the leading strands and each lagging strand fragment
DNA polδ[‡]	Extension of DNA strands; highly processive[†] in association with accessory factors	Processive[†] extension of leading strands and each lagging strand fragment
PCNA and RF-C	Accessory factors for DNA polδ	Increased processivity[†] of DNA polδ
topoisomerase I	Relieving tortional stress in duplex DNA molecules	Removes the excess supercoiling which builds up in the DNA template ahead of the replication fork due to helicase action
RNase H	Degradation of RNA within RNA:DNA hybrids	Removal of primers
DNA ligase	Joining DNA ends	Links up lagging strand fragments after primer removal and fill-in synthesis
topoisomerase II	Separating topologically linked DNA molecules	Separates the daughter duplexes at the end of the replication process.

[†]Processivity describes how far a polymerase molecule is likely to travel on a given template molecule before dropping off and having to rebind the template (or another template) molecule to recommence synthesis.
[*]Large T antigen is the only virus-specified protein – all others are encoded by the cell.
[‡]DNA polδ was originally believed to be a replicative polymerase of the host. It is now thought to function in repair synthesis of DNA, and that the virus is hijacking repair processes rather than host replicative synthesis to achieve progeny genome production.

Box 7.3

Evidence for SV40 T antigen binding to the origin of replication

- Some viruses with point mutations in the origin of replication have a 'small-plaque' phenotype, meaning they grow poorly. Selecting for revertants of such mutants gave better-growing viruses. These were found to have changes to the coding sequence of T antigen that compensated for the original origin mutation (this is termed *pseudoreversion*). This result suggests an interaction between origin DNA and T antigen.
- When SV40 DNA is fragmented with a restriction enzyme, mixed with T antigen and then the T antigen is purified back out of the mixture using a specific antibody, the DNA fragment containing the origin specifically co-purifies with T antigen while other fragments remain behind in the solution.

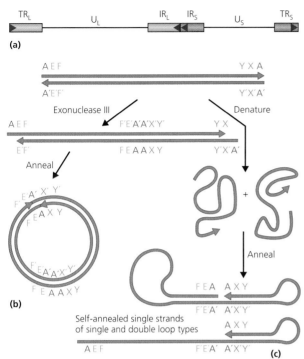

Fig. 7.4 (a) HSV1 genome organization; L = long, S = short, U = unique sequence, TR = terminally repeated sequence, IR = internally repeated sequence. Red triangles represent direct repeats (A) at the genome termini and their internal inverted copies (A′). Turquoise and purple shading represent the remainder of the inverted repeat sequences that flank the long and short unique regions respectively. The genome is not drawn to scale; U_S should represent about 8.5% and U_L about 71% of the total genome. (b,c) The HSV1 genome is depicted by two parallel lines, with terminal direct repeats shown as **A** and its complement **A′**, the repeats flanking the long unique region as **EF** and its complement **E′F′**, and the repeats flanking the short unique region as **XY** and its complement **X′Y′**. 3′ ends are represented by arrowheads. (b) Experimental demonstration of direct repeats at the HSV1 genome termini (see Box 7.4). (c) Experimental detection of inverted repeats in HSV1 DNA (see Box 7.4). The two forms of self-annealed strand that are seen are shown for the upper strand of the double-stranded genome. The same forms may be adopted by the lower strand.

escape the central hole of the enzyme during unwinding of a DNA duplex.

Papillomavirus genomes are a little larger than that of SV40, at around 8000 bp, but the mechanism of replication – studied originally in the model bovine papilloma virus, BPV1 – is very similar. The PV E1 protein functions analogously to SV40 large T antigen in the replication process, except that its sequence-specific DNA binding activity is very weak, and is only revealed in the presence of the E2 protein, which has much stronger specific DNA binding activity (see Section 10.3). The E1 and E2 proteins bind as a complex to the origin of replication of the virus. Thereafter, E1 provides DNA helicase activity whilst all other replication functions are taken from the host as for SV40.

Baculovirus genomes are very large circles, of 100–140 kbp. Their replication is not fully understood, but appears to be a combination of theta-form replication (as for SV40, above) and rolling-circle replication (see Section 7.3). The best studied virus of this type is *Autographa californica* nuclear polyhedrosis virus. The genome has multiple origins of replication and encodes most, if not all, of the proteins needed to complete its replication.

7.3 Replication of linear double-stranded DNA genomes that can form circles

Important viruses in this category

- *The herpesvirus family, including the human pathogens Epstein-Barr virus, Kaposi's sarcoma herpes virus, varicella-zoster virus and herpes simplex virus types 1 & 2.*
- *Bacteriophage λ, a classic model system for the study of control of gene expression.*

The HSV1 genome and its replication

The HSV1 genome is a linear molecule of about 153 kbp in length. This DNA has a complex structure, with several repeated elements (Fig. 7.4(a), Box 7.4). The first of these is direct repeat of about 500 bp, termed 'a' repeats, at the ends of the molecule. This is referred to as terminal redundancy, and appears to be a general feature of herpesvirus genomes. These repeats contain no genes, but are crucial in the process of genome packaging during the formation of new virus particles. The second group of repeats in the HSV1 genome are inverted copies of sequences several kb in length. These divide the DNA into two parts, each comprising a unique region flanked by a pair of inverted repeats. Each pair of repeats contains genes, for which the virus is therefore diploid. Although some other herpesvirus genomes also have this latter type of repeat organization, it is not a universal feature and does not appear to be essential for HSV1 replication. The final feature of the genome organization that is significant is the presence of single nucleotide 3′ extensions on each end; these are complementary and mediate genome circularization prior to replication, probably through the action of a protein.

The replication of HSV1 DNA is outlined in Fig. 7.5. Progeny genomes are produced as concatemers (multiple genome-equivalents covalently joined end-to-end in very long linear molecules) by rolling circle replication from circular templates (Box 7.5). Unless the incoming genome can be amplified in some way, only one such replication complex can exist in the cell, and this is not the case. This leads to the inference of a preceding circle amplification phase in HSV1 replication but there is no evidence for this. However, another herpesvirus, Epstein-Barr virus (EBV), has a well-documented mechanism for replication as a circle (similar to that used by SV40, Section 7.2) and circle amplification prior to a rolling circle phase is well documented during phage λ replication. Both these facts support the notion that HSV1 might also use such a strategy.

HSV1 rolling circle replication has been well characterized. Three origins of replication have been defined by experiment but mutants possessing only one of these sequences can still grow, indicating redundancy of function. Replication involves a considerable contribution

---

Box 7.4

Evidence for the repeat structure of the herpes simplex virus (HSV)1 genome

- DNA from virions can be treated with a strand-specific exonuclease that digests only from 3′ ends and so produces single-stranded regions at both ends of the genome. When such DNA is then incubated under conditions which allow base-pairing between complementary single-stranded sequences, it forms circles when viewed by electron microscopy (Fig. 7.4(b)). This indicates terminal redundancy, i.e. direct repeats.
- When virion DNA is denatured, to break all base pairing, and then allowed to renature in dilute solution (to promote self-annealing), single-stranded molecules with one or both termini base paired to internal sequences are seen by electron microscopy (Fig. 7.4(c)). This indicates that sequences at the genome termini are also present as inverted copies within the genome.

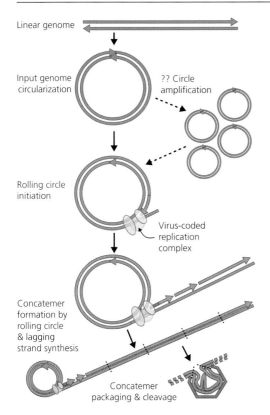

Linear genome

Input genome circularization

?? Circle amplification

Rolling circle initiation

Virus-coded replication complex

Concatemer formation by rolling circle & lagging strand synthesis

Concatemer packaging & cleavage

Fig. 7.5 A model for the replication of HSV1 DNA (see Section 7.3 for details). Blue – parental DNA; brown – new DNA; red – assembling capsid.

from viral proteins: these provide origin recognition, DNA helicase, single-strand DNA binding, primase and DNA polymerase activities. The requirement for cellular proteins is correspondingly far more limited but includes host topoisomerases. Concatemeric DNA is only cleaved during incorporation into virions to give encapsidated, unit-length, genomes (see Section 12.5).

Bacteriophage λ replication

The bacteriophage λ linear genome circularizes upon injection into the cell via its cohesive ends, which are 12 nucleotide complementary extensions on the 5′ ends of the molecule. During the early replication period, λ DNA replicates bidirectionally to generate up to 20 circular progeny copies in a single cell. Late replication is initiated by a switch to the rolling circle mechanism, which generates concatemeric linear molecules. These are then cleaved during packaging into genomes of the correct length and possessing the same cohesive ends as the linear genome which initiated infection (see Section 12.4).

Box 7.5

Evidence for rolling circle replication during HSV1 infection

- The presence of concatemeric DNA is diagnostic of rolling circle replication.
- Analysis of DNA from HSV1-infected cells by restriction digestion and Southern blotting shows that the expected fragments from the genome ends are present in low amounts compared to the remaining fragments, with a correspondingly increased amount of fragments representing covalently joined end fragments. This is evidence for abundant end-to-end joining of viral genomes, i.e. concatemeric DNA.
- Pulse-chase analysis with labelled DNA precursors shows that concatemeric DNA is produced before any labelled unit-length genomes. This discounts the possibility that the concatemers form by ligation of unit-length molecules.

7.4 Replication of linear double-stranded DNA genomes that do not circularize

Important viruses in this category

- *The adenovirus family, which includes a variety of human pathogens affecting the respiratory tract, eye or gut. These viruses are also being actively developed as vectors for gene therapy.*
- *The poxvirus family, including the eradicated human pathogen, variola virus (smallpox), and vaccinia virus which is important as a recombinant vaccine carrier.*

Adenoviruses

Adenovirus genomes are linear double-stranded DNA molecules of around 36 kbp in length. At the two ends there are inverted terminal repeat sequences of 100–150 bp that contain the origins of replication, and on each 5′ terminus there is a covalently attached, virus encoded protein known as the terminal protein (TP) (Box 7.6). The replication strategy of adenoviruses is summarized in Fig. 7.6. Understanding the details of this process was made possible through the development of cell-free replication systems for the virus by Kelly and co-workers. To initiate replication, the origins are recognized by a complex of two viral proteins: a DNA polymerase and the TP precursor (pTP). This binding is assisted by two cellular proteins, which bind specifically to sequences adjacent to the origin. These are actually transcription factors which the virus 'borrows' to assist its replication. However, it is not their transcription regulatory activity which is needed. Rather, they are used to alter the conformation of the DNA to promote binding of the pTP:pol replication complex.

Once the viral DNA polymerase–pTP complex has bound to the origin (Fig. 7.6(a)), it initiates DNA synthesis by copying from the 3′ end of the template strand, using an amino acid side chain –OH in the bound pTP as a primer (b). This priming mechanism is the key to adenovirus solving the 'end-replication' problem as there is no RNA primer complementary to viral sequence which later has to be excised and somehow replaced. However, the mechanism could be said to break the rule of universal proofreading of DNA synthesis, since initiation does not require a primer which is base-paired to the template; in some way, the pTP:DNA interaction must substitute for this requirement. Synthesis then progresses 5′ → 3′ across the length of the template, displacing the non-template strand which is bound and stabilized by a virus-coded DNA binding protein (c). This process may occur at similar times from the origins at each end of the molecule (d), in which case the template duplex falls apart when the replication complexes meet (e); replication is then completed by the polymerases moving on to the ends of their respective templates (f). When only one origin is used, the non-template strand is completely displaced without being replicated (g). Its replication is achieved via formation of a 'pan-handle' intermediate by base-pairing between sequences at its two ends (h); the short double-stranded region thus formed exactly resembles a genome end and so can serve as an origin in the way already described (i, j). Since replication of the two strands is separated in time, only one DNA strand being synthesized by a replication complex, all of the DNA can be synthesized continuously – there is no lagging strand.

Box 7.6

Evidence for adenovirus genome structure

- Denaturation of purified linear viral DNA with alkali, followed by neutralization with acid, causes the formation of unit-length single-stranded circles with short double-stranded tails that are visible by electron microscopy (below). This is indicative of inverted terminal repeats.

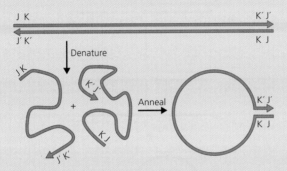

Demonstration of inverted terminal repeat sequences in adenovirus DNA. The genome is depicted by two parallel lines, with inverted terminal repeats shown as **JK** and its complement **J'K'**. 3' ends are represented by arrowheads. Denaturation and annealing of the genome allows formation of single-stranded circles with double-stranded 'pan-handle' projections.

- DNA extracted from particles by a process that includes proteolysis is linear when viewed by electron microscopy whereas when purified without the use of such enzymes, much of the DNA is circular. This result suggests that protein: protein interactions mediate linkage of the genome ends.
- A covalent phosphodiester link between a serine in TP and the 5' deoxycytidine residue in the DNA has been demonstrated biochemically.

Poxviruses

Poxvirus replication has been characterized principally through studying vaccinia virus (the smallpox vaccine virus). The vaccinia virus genome is a linear molecule of around 190 kbp that is unusual in having covalently closed ends (Fig. 7.7(a)). This means that the two pairs of nucleotides at the ends of the molecule are each linked by a standard 5' to 3' phosphodiester bond so that there are no free 5' or 3' ends on the molecule. As a result, when the DNA is completely denatured, a circular single-stranded molecule is generated. At the two ends of the genome there are inverted terminal repeats. These are much larger than those in adenovirus, extending for some 10 kbp, and they contain several genes.

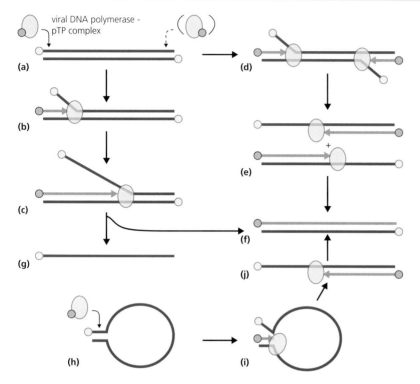

Fig. 7.6 A general scheme for adenovirus DNA replication (see Section 7.4 for details). Blue – parental DNA; brown – new DNA. Arrowheads on new DNA strands represent 3′ ends. Yellow circles represent the terminal protein (TP) attached to parental DNA 5′ ends and brown circles the terminal protein precursor molecules (pTP) which prime new DNA synthesis. The viral DNA polymerase is represented in pink.

Vaccinia virus is thought to use a novel strategy to avoid problems in replicating its genome ends. The current model for its replication is summarized in Fig. 7.7. The process occurs in the cytoplasm of infected cells using exclusively virus-coded proteins. Replication is thought to initiate through an enzyme recognizing and nicking one template strand at a specific site within the inverted repeats. These therefore constitute origins of replication. Either or both of these sites may be used simultaneously (b). DNA polymerase can then extend from the 3′ end that is produced, displacing the non-template strand, until the end of the template is reached (c). However, the polymerase does not have to stop at this point,

since the terminal repetition means that the molecule can base-pair in an alternative way which once again presents a template for extension (d). The newly-synthesized strand folds back and base pairs to itself so that the paired 3′ end is now directed back towards the centre of the molecule and synthesis can continue. As with adenovirus, exactly the same events can proceed independently at both ends of the duplex template. When the replication forks meet in the centre of the molecule, the two halves fall apart (e) and replication is completed when the polymerases run up against the base paired 5′ end and the ends are ligated (f). Notice how nicking the parental DNA to provide a primed template complex

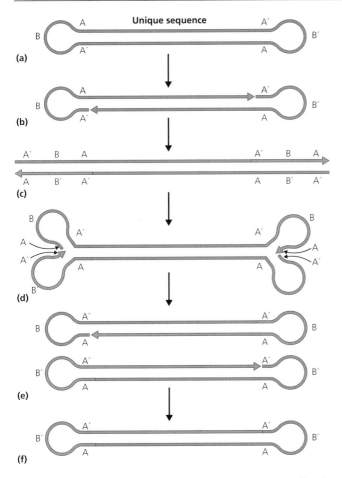

Fig. 7.7 General scheme for vaccinia virus DNA replication (see Section 7.4 for details). Blue – parental DNA; brown – new DNA. Arrowheads on new DNA strands represent 3′ ends. Complementary sequences are denoted **A, A′** etc. Note that the two alternative forms of the terminal loop sequence, **B, B′** are exchanged with each cycle of replication; individual molecules may have loops with identical or complementary sequences.

avoids the need to use RNA primers that would have to be excised later.

What happens if, on a template molecule, only one of the two origins is activated? The replication fork, as it proceeds back through the template, will not meet one coming the other way. Neither will it come up against a base-paired 5′ end. Instead, it will continue to synthesis a complementary strand from its template all the way up to and then around the covalently closed end of the genome and back along the other side. What results is a concatemer – two unit length molecules covalently joined. In fact, concatemers can be produced even when both origins are used together. When the polymerase reaches the base-paired 5′ end ahead of it, it does not have to stop. Its situation exactly resembles that at the initiation event, and just as it did then, it can continue on, displacing the non-template strand and beginning the whole cycle of events again. How then are these concatemers resolved? They can be cleaved by the same site-specific nicking enzyme which created the 3′ end on which synthesis initiated. These nicked molecules can then refold into unit length molecules which can be closed by DNA ligase to provide progeny genomes.

7.5 Replication of single-stranded circular DNA genomes

Important viruses in this category

- *Bacteriophage φX174, an early target for molecular and structural characterization.*
- *Bacteriophage M13, a crucial laboratory tool for DNA sequencing prior to the development of PCR techniques and now high-throughput methods.*

The first step in replicating a single-stranded genome is the creation of a complementary strand. Once this is achieved, replication mechanisms will resemble those for double-stranded genomes. The best-studied example of single-stranded circular DNA replication is bacteriophage φX174. The incoming single strand is converted to a double-stranded replicative form (RF) via RNA-primed DNA synthesis. This primer is synthesized by host RNA polymerase using as template a short hairpin duplex formed in the single-strand genome by intramolecular base-pairing. The parental RF is then amplified before switching to rolling circle replication for the production of a single-strand concatemer from which progeny genomes are excised and circularized.

Until quite recently, no viruses with single-stranded circular DNA genomes were known that could infect animals. However, such viruses have now been described (classified in the family *Circoviridae*) which infect pigs and various bird species, and there are also several Torque Teno (TT) viruses with single-stranded circular DNA genomes (3000–4000 nt length; family *Anelloviridae*), some of which cause inapparent infections in humans. Replication of these viruses requires a double-stranded intermediate, and circoviruses utilize a rolling circle mechanism to produce progeny genomes.

7.6 Replication of single-stranded linear DNA genomes

Important viruses in this category

- *The autonomous parvovirus B19, a human pathogen.*
- *The defective parvovirus, adeno-associated virus, which is being developed as a gene therapy vector.*

The parvovirus family comprises both autonomous and defective viruses. The autonomous parvoviruses, such as the minute virus of mice (MVM), package a negative-sense DNA strand while defective parvoviruses, such as the adeno-associated viruses (AAV), package both positive- and negative-sense DNA strands in separate virions, either one being infectious. These defective viruses are almost completely dependent on co-infection with helper virus for their replication (either adenovirus or various herpesviruses can perform this function). Parvovirus genomes contain terminal hairpins (inverted repeats). Whilst the two hairpins have distinct sequences in the autonomous viruses, they are complementary in the defective viruses (Fig. 7.8). As discussed below, this difference explains why the two types of virus differ in the polarities of DNA strand which they package.

A model for autonomous parvovirus DNA replication is presented in Fig. 7.9 with some

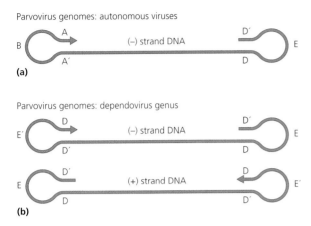

Parvovirus genomes: autonomous viruses

(a)

Parvovirus genomes: dependovirus genus

(b)

Fig. 7.8 Schematic representation of the genome structures of autonomous and defective parvoviruses. Complementary sequences are denoted **A, A**′, etc.

evidence for it in Box 7.7. The essential first step in parvovirus replication is conversion of the genome to a double-stranded form by gap-fill synthesis (Fig. 7.9 (a, b)). The terminal hairpin provides a base-paired 3′ OH terminus at which DNA elongation can be initiated without the need for an RNA primer. Once this is achieved, the replication mechanism is quite similar to that of the poxviruses. Displacement of the base-paired 5′ end allows synthesis to continue to the template strand 5′ end (c). The structure now undergoes rearrangement to form a 'rabbit-eared' structure (d). This recreates the hairpin originally present at the 5′ end of the parental genome and also forms a copy of this hairpin at the 3′ end of the complementary strand, which can serve as a primer for continuing synthesis (e). Further cycles of rearrangement and strand displacement yield a tetramer-length duplex (f), and this process can continue (g).

The concatemers created by replication can serve as intermediates from which progeny single-stranded viral DNA can be generated by displacement synthesis. A nick is introduced 5′ to a genome sequence within the concatemer by a sequence-specific nicking enzyme, NS1 in MVM and Rep68 in AAV, which remains attached to the 5′ end (Fig. 7.9 (f, h)). The 3′ OH terminus then acts as a DNA primer for displacement synthesis of DNA single strands, which are packaged concomitantly (h). After a complete genome has been displaced, excision of the progeny genome is completed by endonuclease cleavage, resulting in the release of a complete virus particle and termination of the displacement synthesis.

The defective parvovirus AAV differs from MVM in having terminal hairpin sequences that are complementary to each other. The result is that the 5′ ends of both negative- and positive-sense strands are identical and so, within the model just described for linked strand-displacement and packaging, site-specific cleavage will occur equally at the 5′ end of both positive and negative strands and therefore both will be packaged. It is still unclear why this subset of the parvoviruses shows dependence on helper functions for growth, since these required functions do not necessarily include any that are directly involved in DNA synthesis. The helper functions can be provided by various unrelated viruses and the dependence on them can also be overcome in cell culture by treating cells with genotoxic agents (chemicals which damage DNA). It is also significant that AAV is now known to be able to establish a latent infection, integrating into a specific site in the host chromosomes (see Chapter 15); this might be the preferred strategy of the virus for ensuring perpetuation of its genetic material. Thus 'helper' functions may alternatively be viewed as viral functions which alter the cell environment so as to block establishment or maintenance of latency and hence favour AAV productive replication.

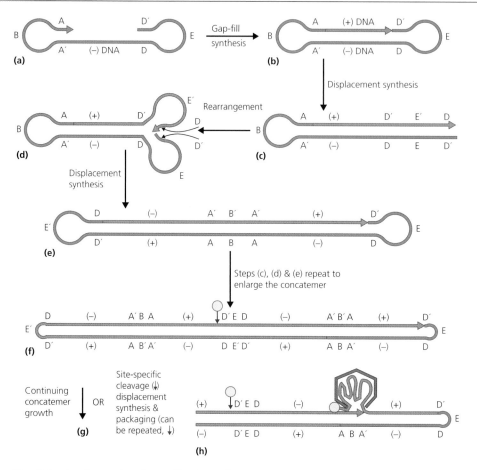

Fig. 7.9 A scheme for the replication of an autonomous parvovirus (see Section 7.6 for details). Blue – parental DNA; brown – new DNA; red – assembling capsid. Arrowheads represent 3′ ends. The red circle represents the viral site-specific nicking enzyme. Complementary sequences are denoted **A, A**′, etc.

Box 7.7

Evidence for the mechanism of parvovirus replication

- Infected cells accumulate double-stranded forms of viral DNA, a large fraction of which cannot be irreversibly denatured, suggesting that the two strands are covalently linked.
- The only virus-encoded replication activity so far defined is a site-specific endonuclease.
- Duplex DNA molecules with single-strand tails have been observed in cells infected with MVM whereas no free single-stranded DNA has been detected in such cells.
- The 5′ end of the genome, with attached viral endonuclease protein, is exposed outside the assembling virion.

7.7 Dependency versus autonomy among DNA viruses

The autonomy of viruses from their hosts as regards the replication of their DNA varies between wide limits and is a function of the size of the viral genome. At one end of the scale are large viruses, such as the poxviruses. These require little more from their host cells than an enclosed environment, protein synthesizing machinery, a supply of biosynthesis substrates and an energy source. Some may not even require this much; herpes simplex and vaccinia viruses both specify a new thymidine kinase and several other enzymes that contribute to the production of DNA precursors. At the other end of the scale are viruses, such as minute virus of mice and SV40, whose genomes can specify only a few proteins. Since some of these are needed to form the virus coat, not many genes are left to code for functions essential to replication. These viruses rely on the host not only for DNA precursors but also for polymerases, ligases, nucleases, etc. The extreme in terms of lack of autonomy is represented by viruses such as adeno-associated virus, which requires assistance for replication not only from the host but also from another virus.

Key points

- DNA as genetic material has the advantage over RNA of increased replication fidelity but this means that DNA viruses adapt and evolve more slowly than RNA viruses.
- Viruses must obey the cellular rules for DNA synthesis, which means using a primer and having to solve the 'end replication' problem on linear genomes.
- Double-stranded DNA genomes that are intrinsically circular or which can circularize

before replication avoid the 'end replication' problem. Linear genomes with covalently closed ends also avoid this problem.
- Circular double-strand DNA can replicate either via a theta-form intermediate to produce daughter molecules that exactly resemble the parent, or via a rolling circle mechanism to give concatemers that must then be resolved to unit length molecules by specific cleavage events before or during packaging.
- Adenoviruses are unique amongst animal viruses in replicating their linear genomes from the ends, using a protein primer.
- Viral single-strand DNA genomes must be converted to double-strand forms before replication can proceed as for double-stranded genome types.
- There is a roughly inverse relationship between genome size and dependence on the host for essential functions in DNA replication.
- SV40 replication is a good model for events during eukaryotic cell DNA replication.

Questions

- Compare and contrast the strategies adopted by members of the various virus families that infect mammals and belong to Baltimore class 1 for replication of their double-stranded DNA genomes.
- Discuss the basis of the 'end-replication' problem for linear double-stranded DNA in biological systems, and the various strategies employed by viruses to overcome this problem.

Further reading

Cotmore, S. F., Tattersall, P. 2006. A rolling-hairpin strategy: basic mechanisms of DNA replication in the parvoviruses. In J. Kerr et al. (Eds), *Parvoviruses*, p. 171–188. Hodder Arnold, London.

Fanning, E., Zhao, K. 2009. SV40 DNA replication: from the A gene to a nanomachine. *Virology* 384, 352–359.

Kornberg, A., Baker, T. 2005. *DNA Replication*, 2nd edn, revised. W. H. Freeman, San Francisco.

Liu, H., Naismith, J. H., Hay, R. T. 2003. Adenovirus DNA replication. *Current Topics in Microbiology and Immunology* 272, 131–164.

Méndez, J., Stillman, B. 2003. Perpetuating the double helix: molecular machines at DNA replication origins. *Bioessays* 25, 1158–1167.

Muylaert, I., Tang, K.-W., Elias, P. 2011. Replication and recombination of herpes simplex virus DNA. *Journal of Biological Chemistry* 286, 15619–15624.

Ogawa, T., Okazaki, T. 1980. Discontinuous DNA replication. *Annual Review of Biochemistry* 49, 421–457.

Chapter 8

The process of infection: IIB. Genome replication in RNA viruses

The synthesis of RNA by RNA viruses includes replication, which is defined as the production of progeny virus genomes, and transcription to produce messenger RNA (mRNA). The process of transcription for RNA viruses is described in Chapter 11 where it is discussed in terms of gene expression. This chapter focuses on the process of replication of RNA viruses.

Chapter 8 Outline

8.1 Nature and diversity of RNA virus genomes
8.2 Regulatory elements for RNA virus genome synthesis
8.3 Synthesis of the RNA genome of Baltimore class 3 viruses
8.4 Synthesis of the RNA genome of Baltimore class 4 viruses
8.5 Synthesis of the RNA genome of Baltimore class 5 viruses
8.6 Synthesis of the RNA genome of viroids and hepatitis delta virus

A very large number of viruses have been shown to contain genomes comprised of RNA (Box 8.1). The replication of RNA genomes requires the action of RNA-dependent RNA polymerases which are not encoded by the genome of the infected host cell but instead are synthesized by the virus. During the process of replication of RNA virus genomes, as for all other processes which involve synthesis of nucleic acid, the template strand is 'read' by the polymerase travelling in a $3' \rightarrow 5'$ direction with the newly synthesized material starting at the $5'$ nucleotide and progressing to the $3'$ end. Most RNA viruses can replicate in the presence of DNA synthesis inhibitors indicating that no DNA intermediate is involved. However, this is not true for the retroviruses (Baltimore class 6) and these will be considered separately (Chapter 9).

The polymerases which carry out RNA replication are either transported into the cell at the time of infection or are synthesized very soon after the infection has begun from coding information in the virus genome. Frequently, the polymerases involved in RNA replication are referred to as 'replicases' to differentiate them from the polymerases involved in transcription. However, both processes are carried out by the same enzyme exhibiting different synthetic activities at different times in the infectious cycle.

Introduction to Modern Virology, Seventh Edition. N. J. Dimmock, A. J. Easton and K. N. Leppard.
© 2016 John Wiley & Sons, Ltd. Published 2016 by John Wiley & Sons, Ltd.

Box 8.1

Evidence for RNA as the genetic material for some viruses

- The presence of uracil instead of thymidine.
- The presence of ribose instead of deoxyribose sugars.
- Buoyant density of the nucleic acid: RNA is more dense than DNA.
- Sensitivity to RNase and not DNase.
- Some purified RNA genomes, when introduced into cells yield infectious virus showing that they are sufficient to establish a complete cycle of infection.

8.1 Nature and diversity of RNA virus genomes

As demonstrated in the Baltimore classification scheme (Section 3.4), RNA genomes can be single-stranded or double-stranded and the former may be of either positive (mRNA-like) or negative sense (see Section 3.4). RNA molecules which act as virus genomes exist only as linear molecules although infectious circular RNA molecules form the genomes of a specialized type of agent, the viroid, described in Section 3.6. An unusual feature of many RNA viruses is that their genomes consist of multiple segments, analogous to chromosomes of host cells, and these viruses must ensure that at least one copy of each segment is present in the infected cell to generate a full complement of genes from which new viruses can be generated (see Section 12.7). During the replication process the RNA molecules remain linear, and covalently closed circular molecules are never observed. The RNA genomes of different viruses vary greatly in size, though they do not display the range seen with DNA virus genomes. The largest single molecule RNA virus genomes known are those of the coronaviruses which are approximately 30,000 nt, with the smallest animal virus RNA genomes being those of the picornaviruses at

approximately 7500 nt. Several bacteriophage have RNA genomes smaller than this, one of the smallest being MS2 with a single-stranded positive sense RNA genome of 3569 nucleotides. The nodaviruses which infect insects and some animals have very small genomes which consists of two RNA molecules of approximately 3000 nt and 1400 nt. All enzymes involved in RNA synthesis, whether of virus or host cell origin, are unable to proofread (i.e. to correct incorrectly inserted bases). This contrasts with DNA synthesis in which correction may occur (Section 7.1). The lack of proofreading means that the RNA genomes mutate more rapidly than DNA genomes. Estimates suggest that the error rate during RNA virus replication is approximately 10^{-3} to 10^{-5} per base per genome replication event (one mutation per 1000 to 100,000 bases per genome replication event). This has implications for the evolution of RNA viruses (see Chapter 4).

8.2 Regulatory elements for RNA virus genome synthesis

Certain features are common in the process of replication for all RNA viruses. In order to make a faithful copy of the genome, the RNA-dependent RNA polymerase must begin synthesis at the 3′ terminal nucleotide of the template strand. The 3′ terminus must therefore contain a signal to direct initiation of synthesis. As indicated by the principles which underpin the Baltimore classification scheme (Section 3.4), all RNA viruses must replicate via a dsRNA intermediate molecule. For example, a virus with a ssRNA genome must produce a dsRNA intermediate by synthesizing a full-length 'antigenome' strand. The antigenome strand will then, in turn, be used as a template to synthesize more genomes for packaging into progeny virions. As before, synthesis using an antigenome as template must begin at the 3′ terminal nucleotide if a faithful, full length copy is to be made, and the 3′ end of the antigenome must also contain a signal to direct the polymerase to begin synthesis. For most viruses the mechanism of initiation of RNA synthesis during replication is only poorly understood, if at all, but a combination of old and new analyses have identified the termini of RNA viruses as containing the regulatory elements which direct RNA synthesis. The most convincing evidence has come from the study of the genomes of defective-interfering viruses and, more recently, from reverse genetics studies on RNA viruses.

For some RNA viruses the sequences at the immediate termini of the genomes are almost completely complementary to each other. These complementary regions, which can range from approximately a dozen to over 50 nucleotides in length in different viruses, are referred to as inverted repeat sequences. For genomes with complementary termini both the genome and antigenome strands will contain nearly the same sequence at the 3′ terminus so that the initiation of synthesis with these molecules as templates can occur using the same mechanism. This is seen for many viruses such as paramyxoviruses and influenza viruses. For viruses whose genomes do not contain inverted repeat sequences, the initiation of synthesis of antigenome- and genome-sense molecules must each be controlled by different processes. In some cases, different features may be present at the 5′ and 3′ termini of an RNA genome. Examples include covalently attached proteins at the 5′ end in the case of picornaviruses, such as poliovirus, and long homopolymeric polyadenylate (polyA) tracts at the 3′ ends of many positive sense RNA genomes such as those of picornaviruses, alphaviruses, flaviviruses and coronaviruses.

The generation and amplification of defective–interfering (DI) virus RNA

A common phenomenon for viruses is the production of DI particles as the result of errors in their nucleic acid synthesis. Here we shall consider only DI RNA viruses, about which more is known. DI viruses are mutants in which the genomes have large deletions, leaving RNA which may comprise as little as 10% of the infectious genome from which they were replicated. DI viruses are unable to reproduce themselves without the assistance of the infectious virus from which they were derived (i.e. they are defective). The portions of the genome which are retained in DI viruses necessarily contain all of the elements essential for replication, and packaging, of the genome. Analysis of DI genomes therefore provided the first evidence to show the location of replication sequences and their subsequent analysis, coupled with modern molecular biological techniques, has assisted in identifying key *cis*-acting elements in virus replication. Propagation of DI virus is optimal at a high multiplicity of infection when all cells contain both a DI genome and an infectious virus genome. DI genomes depress (or interfere with) the yield of infectious progeny, by competing for a limited amount of one or more product(s) synthesized only by the infectious parent, referred to as the helper virus. Many DI viruses contain genomes which are deleted to such an extent that they synthesize no proteins and some have no open reading frame. Because they depend upon parental virus to provide those missing proteins, DI and parental viruses are composed of identical constituents, apart from their RNAs. Thus it is usually difficult to separate one from the other. A notable exception is the DI particle of rhabdoviruses such as vesicular stomatitis virus (VSV) whose particle length is proportional to that of the genome it encloses. When centrifuged these short particles remain at the top of sucrose velocity gradients and are thus called T particles to distinguish them from infectious B particles, which sediment to the bottom. Some biological implications of DI particles are discussed in Section 15.2.

A clue leading to one hypothesis explaining how DI RNAs are generated came from electron microscopic examination of genomic and DI RNAs from single-stranded RNA viruses. Both were found to be circularized by hydrogen bonding between short complementary sequences at the termini, forming structures called 'panhandles' or 'stems' (Fig. 8.1a). The deletion that results in DI RNA may arise when a polymerase molecule detaches from the template RNA strand and reattaches either at a different point of the genome (Fig. 8.1a) or to the newly-synthesized, incomplete, strand (Fig. 8.1b). Thus the polymerase begins to replicate faithfully but fails to copy the entire genome. Most VSV DI viruses are of the latter type and lack the 3′ end of the standard virus genome, having instead a faithful copy of the 5′ end and a complementary copy of the 5′ end at the two termini. There are no DI viruses known which lack the 5′ terminus. In other DI viruses parts of the genome can be duplicated, often

several times over, during subsequent replication events making complex structures which bear little resemblance to the standard genome from which they were derived. It is not known whether the event(s) that occur to produce the DI RNA genome take place during synthesis of the antigenome, as indicated in Fig. 8.1, or during synthesis of genome RNA, or both. The three classes of DI genome are summarized in Fig. 8.2. A key point for the DI RNA viruses is that their genomes always contain the same termini as those of the parent virus, or consist of inverted repeats of sense and antisense copies of the 5′ end of the normal genome while the remainder of the genome can be substantially deleted without impairing the ability of the DI genome to replicate. This is taken to indicate that the genome termini are essential and contain the regulatory elements to direct genome synthesis.

Usually, interference only occurs between the DI virus and its parent. This is because DI virus lacks replicative enzymes and requires those synthesized by infectious virus. Specificity resides in the enzymes, which only replicate molecules carrying certain unique nucleotide sequences. Intuitively, it can be seen that in a given amount of time an enzyme will be able to make more copies of the smaller DI RNA. Thus, as time progresses, the concentration of DI RNAs increases relative to the parental RNA in an amplification step. However, this does not completely explain all aspects, as some large DI RNAs interfere more efficiently than smaller ones. Such DI RNAs seem to have evolved a polymerase recognition sequence which has a higher affinity for the enzyme than that of the infectious parent and hence confers a replicative advantage. This may indicate that, while the essential minimal sequences for directing genome synthesis are located at the termini, sequences located elsewhere may also play an enhancing role. The only other sequence that all DI genomes must

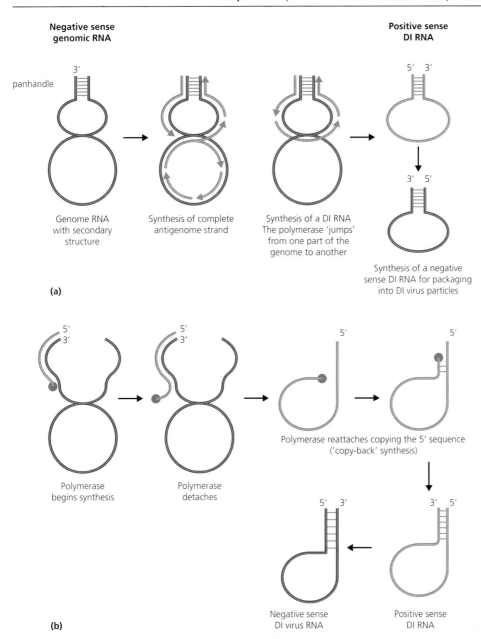

Fig. 8.1 Hypothetical schemes to explain the generation of DI RNAs having sequences identical with both the 5′ and 3′ regions of the genome (a) and with the 5′ region only (b).

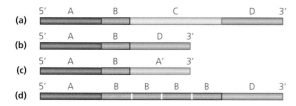

Fig. 8.2 Comparison of the genome organizations of normal and DI virus. (a) A representation of a normal virus genome. The coloured segments, labelled A to D, represent different regions of the genome (not to scale). (b) The genome of a 5′,3′ DI virus in which a large portion of the genome (region C) has been deleted. (c) The genome of a 5′ DI virus. Large portions of the genome (regions C and D) are not present and the turquoise block (region A′) represents the complement of the normal 5′ end sequence of region A. (d) A complex DI virus genome containing multiple copies of a short portion (region B) of the normal genome with other sequences (region C) deleted.

retain is a packaging or encapsidation sequence since without this they cannot be recognized by virion proteins and form viral particles.

Reverse genetics of RNA viruses

Over recent years, one of the most exciting developments has been the generation of reverse genetic systems for a wide range of RNA viruses, particularly those with negative sense ssRNA genomes (Box 8.2). The absence of a DNA intermediate in the replication cycle of RNA viruses had limited research as there are no tools to manipulate RNA molecules in the same way that there are for DNA. However, for many RNA virus systems it is now possible to generate DNA copies of virus genomes and convert these back into RNA, and ultimately into infectious virus particles. The DNA can be manipulated in a variety of ways, including the generation of deletions or specific mutations which are then mirrored in the synthetic replicas of the virus genomes. Analysis of these mutated genomes has shown, for representatives of all RNA virus families, that the immediate termini of the genome, or each genome segment, contain the elements which are essential to direct RNA synthesis for replication. This is true whether the virus genome contains inverted repeat sequences or different, unique sequences at the termini of the genome. Mutation of the sequences at the termini has serious deleterious consequences for the ability of the RNA molecules to be replicated by virus proteins.

Box 8.2

Reverse genetics systems for viruses with RNA genomes

Reverse genetics systems have been described for several viruses in Baltimore classes 3, 4 and 5 including:

Class 3 viruses – birnaviruses, orbiviruses
Class 4 viruses – picornaviruses, alphaviruses, coronaviruses
Class 5 viruses – rhabdoviruses, filoviruses, paramyxoviruses, bunyaviruses, orthomyxoviruses

8.3 Synthesis of the RNA genome of Baltimore class 3 viruses

Class 3 viruses contain multiple segments of dsRNA, the fewest number being 2 segments in the partiti- and birnaviruses and the greatest number being 12 segments in the coltiviruses. In reoviruses the genome consists of 10 segments and in the rotaviruses the genome consists of 11 segments. Each genome segment replicates independently of the others. By analogy with DNA replication, this dsRNA could replicate by a *semi-conservative* mechanism such that the complementary strands of the parental RNA duplex are displaced into separate progeny genomes, or alternatively the parental genome could be conserved or degraded. In fact, dsRNA genomes are replicated conservatively. This has been best studied for reoviruses but is also the case for rotaviruses (Box 8.3).

Following initiation of infection, several proteins in the reovirus particles are removed by protease digestion during the uncoating process to form a subviral particle which is found in the cytoplasm where replication takes place. The dsRNA genome is retained within the subviral particle and does not leave it during the infectious cycle. This observation indicated that replication of the dsRNA could not occur in the normal semi-conservative way as seen for DNA as the two parental genome strands remain together. The only virus nucleic acid found in the cytoplasm outside the subviral particles is mRNA which is generated by transcription using the particle-associated RNA-dependent RNA polymerase as described in Chapter 11. Since both strands of the genome RNA are retained in the subviral particle, it was clear that the single-stranded mRNA transcripts must be the sole carriers of genetic information from parent to progeny. This means that only one strand of each of the 10 genome segments is used as template and the newly-synthesized RNA must then be replicated to form a new dsRNA genome segment.

Newly-synthesized negative sense RNA is only found as part of a dsRNA molecule. The dsRNA segments are never found free in infected cells but are always associated with an immature virus particle. Each particle must contain a single copy of each of the 10 reovirus genome segments. The mechanism by which a virion specifically packages one of each of the 10 RNA segments is not yet understood.

8.4 Synthesis of the RNA genome of Baltimore class 4 viruses

Box 8.3

Evidence for conservative replication of the reovirus genome

In 1971, Michael Schonberg and colleagues provided evidence that the reovirus genome was replicated conservatively.

Infected cells were labelled with ^{3}H-uridine for 30 min at various times during the replication phase of the reovirus infectious cycle. Total dsRNA was then isolated and hybridized to excess unlabelled positive sense RNA which had been prepared in vitro. Three results could be obtained, depending on the mode of replication:

i. If replication was semi-conservative, then the dsRNA should be equally radioactively labelled in both strands. Hybridization to an excess of unlabelled positive sense strands would occur with only 50% of the label.

ii. If the positive sense strand was used as template, then hybridization of the radioactively labelled negative strand would result in 100% of the label forming a hybrid.

iii. If the negative sense strand was used as template then radioactively labelled positive sense RNA would be produced and hybridization would generate dsRNA which did not contain any radioactivity.

The result showed that when isotope was added in a short pulse at an early stage of the infection all of the label was associated with the negative strand and thus replication utilizes the positive sense mRNA as template. In cells exposed continuously to ^{3}H-uridine, the label was found to be divided between both strands. Varying the time of the short pulse of radioactivity showed that there is a lag between the synthesis of the positive template strand, and the onset of replication of the negative sense strand. Presumably this allows the mRNA to be translated to produce the polymerase before being used as a template in replication. When cells are labelled for long periods at later times the label is distributed between positive and negative sense RNA as the newly-synthesized positive sense mRNA is used as template to make new negative sense strands.

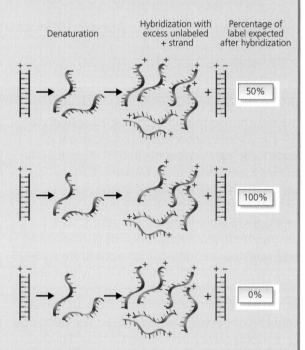

Diagrammatic representation of the experiment that demonstrates that the reovirus genome is replicated conservatively.
Radioactively labelled RNA strands are represented by red lines and unlabelled strands by blue lines.

The coronaviruses which include important veterinary viruses and the SARS virus.
Rubella virus, the causative agent of German measles.
Hepatitis C virus.
Yellow fever virus.

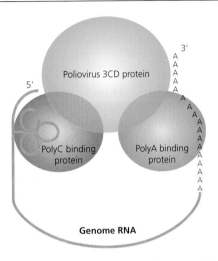

Fig. 8.3 Circularization of picornavirus genome RNA in the replication complex.

While there are many differences in the details of the replication cycles of class 4 viruses in terms of gene expression (Sections 11.3–11.5) and assembly, the process by which their positive sense ssRNA genomes are replicated is very similar. A great deal of information is available about the mechanism of picornavirus genome replication and this will be described here in detail. In principle, the process used by picornaviruses is applicable to all class 4 viruses, with the generation of the same type of intermediate molecules in vivo and in vitro. A similar process is also applicable to the production of the coronavirus subgenomic mRNAs described in Section 11.5.

Following infection, the picornavirus genome RNA is the only nucleic acid which enters the infected cell so it must act as a template for both translation and transcription, just as the reovirus mRNAs do after their synthesis in the infected cell. The picornavirus genome RNA has a 3' polyA tail and a small 22 amino acid protein called VPg covalently attached to the 5' end. Only picornavirus genomes have a VPg protein attached to the RNA. Translation is necessary as a first step in the infectious cycle to produce the polymerase enzyme which will synthesize the new genomes. Aspects of translation of picornavirus genome RNA are discussed in Section 11.3. Replication takes place on smooth cytoplasmic membranes in a replication complex and, as indicated earlier, must involve the generation of a dsRNA intermediate structure.

Initiation of replication occurs at the 3'-end of the positive sense, genome/mRNA polyA tail,

the presence of which is essential for replication to occur. An interesting feature of picornavirus replication is that the genome RNA is held in a circular form due to the action of both virus and host proteins (Fig. 8.3). Cells contain a polyA binding protein which plays a role in translation of mRNA and this protein binds to the 3' end of the picornavirus genome RNA. A second protein which has an affinity for polyC tracts in RNA binds to a region of the virus genome RNA near the 5' end. This region, which is rich in pyrimidine residues, forms a complex 3-dimensional structure referred to as a cloverleaf. The third protein in the complex is the picornavirus 3CD protein, the precursor of the polymerase, which is activated by proteolytic cleavage. By associating with this RNA:protein complex, the polymerase is located near the 3' end of the genome where replication will begin.

It is not known whether other class 4 virus genomes are circularized in a similar way as a necessary requirement for replication but the involvement of host cell proteins in replication of RNA virus genomes is common. Several regions of the poliovirus genome form complex secondary structures which are essential for replication. Some of these structures are located far from the genome termini and they may interact with virus and/or host proteins or with other regions of the genome RNA to form higher order structures. Such structures, essential for replication, are seen in the genome RNAs of most class 4 viruses. The process of initiation of picornavirus RNA replication is not yet understood, nor is it known how the VPg protein becomes attached to the 5′ end of the newly-synthesized RNA. One possibility is that the attachment of VPg and initiation of RNA synthesis are coupled. If so, this presumably happens on initiation of both positive and negative strand synthesis, since both strands have VPg at their 5′-ends. Once RNA synthesis begins, the polymerase proceeds along the entire length of the template RNA. The replication process requires concurrent protein synthesis and addition of protein synthesis inhibitors, even when replication has started, inhibits any further rounds of replication.

The suggested mode of replication is that, initially, the positive sense genome ssRNA is used as a template to generate negative sense, antigenome RNA with VPg at the 5′-end. Once the polymerase complex has moved along the template the 3′ end will become available for a further round of replication even though the preceding complex has not yet completed copying the template. In fact, multiple initiations may occur on the template before the first replication complex has completed its work and it is estimated that up to five functional replication complexes may be associated with a picornavirus template at any one time (Fig. 8.4). The newly-synthesized RNA is in turn used as template for production of new copies of the genome RNA, with VPg attached to the 5′ end, in a similar way. Since

much more positive sense genome ssRNA than negative sense ssRNA is produced, the synthesis process must be biased, or asymmetric. Unlike the negative sense RNA, the completed positive sense RNA is released from the replication complex for packaging into virions, use as template for further replication, or for translation (with concomitant loss of VPg).

Analysis of virus-specific RNA isolated from picornavirus-infected cells shows that the replication complex contains an RNA molecule which is partially ssRNA and partially dsRNA. This is called the replicative intermediate (RI). Negative sense RNA is only ever found in the cell in association with positive sense RNA in the RI. In the RI the nature of the association of the positive and negative strands is not known, but it is likely that they are not fully base-paired in a dsRNA structure but are only loosely linked and the two strands may only associate at the region where synthesis is occurring. After purification of the RI, deproteination and ribonuclease (RNase) treatment generates a pseudo dsRNA the same length as the virus genome and with nicks in the newly-synthesized strand of RNA. This structure, termed the replicative form (RF), is produced by removing the ssRNA tails found in the RI (Fig. 8.4). A similar dsRNA RF complex is also found in poliovirus infected cells when treated with inhibitors of host cell RNA polymerase but it is not known whether this complex is involved in the replication of the virus RNA or is an artefact of the drug treatment.

For the class 4 viruses, other than the picornaviruses, which lack a VPg on the genome, RNA synthesis is nonetheless initiated at the 3′ end of both the genome and antigenome RNAs. In all cases, RI structures are produced suggesting that the basic principles of the replication process are the same, despite the absence of VPg.

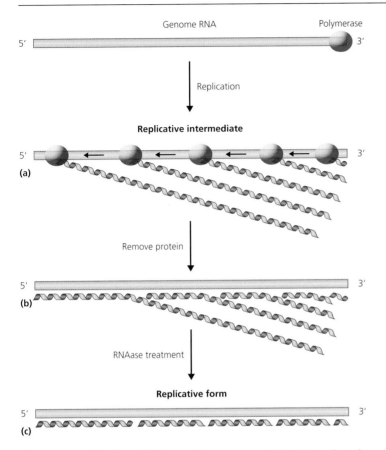

Fig. 8.4 Replication of picornavirus genome RNA. Note that the replication takes place using a template which is held in a circular form by the action of three proteins but is represented here as linear for simplicity. (a) Proposed structure for a molecule of replicative intermediate (RI) before deproteinization. (b) RI after deproteinization. (c) The effect of treating deproteinized RI with RNase to produce a replicative form (RF) RNA.

8.5 Synthesis of the RNA genome of Baltimore class 5 viruses

Important viruses in this category

Class 5 viruses contain a negative sense ssRNA genome.
Measles virus, a paramyxovirus, responsible for many child deaths worldwide.

Influenza virus, an orthomyxovirus, responsible for severe respiratory disease. Worldwide pandemics have killed many millions of people.
Rabies virus, a rhabdovirus.
Ebola and Marburg viruses, filoviruses.

Class 5 viruses are of two types, those with genomes consisting of a single molecule and those with segmented genomes. The former are grouped together in a taxonomic order, the

Mononegavirales, indicating that they have a single, negative sense ssRNA molecule as genome. The Mononegavirales include the paramyxo-, rhabdo- filo- and bornavirus families. The viruses with segmented genomes include the orthomyxo-, arena- and bunyaviruses. The details of many aspects of RNA replication of class 5 viruses are not yet known but the general principles appear to be common to all, irrespective of the number of segments which make up the genome. An important feature for the replication of all class 5 virus genome RNAs is the presence of complementary sequences at the termini as described above (Section 8.1).

Since the negative sense genome RNA cannot be translated into protein to produce the required polymerases, as is seen with positive stranded RNA genomes, class 5 virus RNA synthesis can only occur using RNA-dependent RNA polymerase synthesized in the previous cycle of infection and present in the virus particle that is introduced into the infected cell at the time of infection. Polymerase activity can be detected for some class 5 viruses in vitro after partial disruption of the virus with detergent in the presence of the four ribonucleoside triphosphates and appropriate ions. Viruses 'activated' in this way in the absence of whole cells synthesize RNA at a linear rate for at least 2 hours. In general, the in vitro systems yield only mRNA, not positive sense antigenome RNA. However, the virus-associated polymerase is responsible for both replication and transcription in infected cells.

Replication of the RNA genome of non-segmented class 5 viruses: rhabdoviruses

Most information is available about the process of replication of rhabdoviruses, especially vesicular stomatitis virus (VSV), and they have been used to generate a model of replication applicable to many other class 5 viruses with either a single or multiple genome segments, including all Mononegavirales and the bunya- and arenaviruses. Recent studies have confirmed that the general principles of the model derived for VSV are true for all members of the order Mononegavirales.

The ssRNA genome (and antigenome produced during replication) of VSV is always found closely associated with three virus proteins in a helical complex. The most abundant protein is the nucleoprotein (NP), with smaller amounts of a phosphoprotein (P), and only a few molecules of a large (L) protein. The L protein is the catalytic component of the replication complex responsible for carrying out RNA synthesis but the NP and P proteins are essential for its activity. The complex of NP, P and L proteins together with genomic RNA, referred to as the nucleocapsid complex, also carries out transcription to produce mRNA (see Section 11.9). It is not known what causes the complex to transcribe mRNA at some times and to replicate the genome at others. The relative amounts of the three proteins appear to be critical for the function of the nucleocapsid. In the paramyxoviruses, the NP protein is known as the N protein, which is functionally analogous.

The replication of class 5 viruses requires continuous protein synthesis, and addition of inhibitors of protein synthesis results in an immediate cessation of replication. Class 4 viruses are the same in this respect. Consequently, virus mRNA and protein synthesis must occur prior to the onset of replication. As for all nucleic acid synthesis, replication begins at the 3′ end of the negative sense RNA template where the polymerase binds to a specific sequence and continues to the 5′ end, generating a positive sense, antigenome RNA. Due to the complementarity of the terminal sequences, a very similar polymerase-binding sequence is present at the

3' end of the antigenome where the polymerase begins synthesis to generate new genome RNA molecules. Circular nucleocapsids are often observed within infected cells suggesting that the complementary termini interact to form a panhandle structure, similar to that shown in Figure 8.1. The interaction of the termini may occur either by direct base pairing or in association with the nucleocapsid proteins. This circularization, which is more commonly seen with the RNAs from viruses with segmented genomes, may be a necessary step in the replication process, though this is not yet clear.

The molecular structure of the RI generated during replication of rhabdoviruses is similar to that of picornaviruses. The RI is closely associated with the three replication proteins and antigenome RNA is only found in the RI. Purification of the RI and treatment with RNase generates a double-stranded RF RNA analogous to that generated from the picornavirus RI (Fig. 8.4).

The antigenome RNA is used as template to produce negative sense ssRNA for progeny virus genomes. The negative sense RNA is found as a nucleocapsid structure which can be used in further rounds of replication prior to incorporation into virus particles. Since considerably more negative sense RNA is produced than positive sense RNA, the replication process must be asymmetric to favour production of one strand over the other. During replication the polymerase must ignore the signals for termination of mRNA synthesis which are recognized during transcription.

Replication of the RNA genome of segmented class 5 viruses: orthomyxoviruses

As with the Mononegavirales, the genome RNA of class 5 viruses that is comprised of multiple segments also forms helical nucleocapsid

structures. These often appear as circles, probably due to the formation of panhandle structures held together by the nucleocapsid proteins. The major protein of the influenza virus nucleocapsid is the nucleoprotein (NP). The NP protein interacts directly with the genome RNA, binding to the sugar phosphate backbone and leaving the nucleotide bases exposed on the surface of the structure. The location of the other nucleocapsid proteins PA, PB1 and PB2 (which are also involved in transcription, Section 11.7), is less clear, but they are thought to be associated with the nucleotide bases on the outside of the helix.

Each of the virus genome segments replicates independently. The positive sense RNA which is generated from the genome template is probably initiated *de novo*. This contrasts with the use of a primer derived from the host cell mRNA during transcription of mRNA (Section 11.7). The reason for the different activities of the nucleocapsid complex during replication and transcription is not known. The replication complex is dependent on the continued synthesis of at least one viral protein. This is similar to the situation for the other RNA viruses. The newly-synthesized positive sense antigenome RNAs, which are present in nucleocapsid complexes, are used as templates for the production of negative sense ssRNAs, which in turn can be used for further rounds of replication prior to formation of new progeny virions. The acquisition of genome segments during influenza virus assembly is discussed in Section 12.7.

During replication the synthesis of RNA does not stop in response to the transcription termination and polyadenylation signals in the genome segments that are used in mRNA synthesis (Fig. 11.8), but continues to the end of the template molecule. The production of the positive sense antigenome RNA is thought to occur by way of an anti-termination event, i.e. in the absence of a cap structure on the RNA being synthesized, NP protein acts in an unknown way to prevent termination at the polyadenylation signal, so that synthesis continues to the end of the template. This may be achieved by direct interaction of the NP with the RNA and the PB and PA proteins in the nucleocapsid structure. This complex may be different depending on whether or not a cap has been used to initiate RNA synthesis. However, the details of this process are not yet fully understood.

8.6 Synthesis of the RNA genome of viroids and hepatitis delta virus

Replication of viroids

The covalently closed ssRNA genomes of plant viroids do not encode any proteins and so there is no positive sense mRNA made. By convention, the strand of RNA which is most abundant in the infected cell is termed positive sense though this has no practical meaning. When the naked RNA enters the cell it must be replicated by the proteins already present in the host plant. The enzyme most likely to be responsible for replicating the genome RNA is a host cell DNA-dependent RNA polymerase. For the viroids that replicate within the nucleus of the host cell (members of the family *Pospiviroidae*) it is likely to be the host RNA polymerase II that is used and for the viroids that replicate within chloroplasts (members of the family *Avsunviroidae*) it is likely to be a nuclear encoded polymerase that is transported to the chloroplast. It is not known how these enzymes are made to function on an RNA template but this may be due, at least in part, to the extensive base-pairing structure of the genome RNA (Section 3.6).

The replication of the genome RNA for members of both viroid families begins by adopting a rolling circle mechanism as described

for circular DNA (Section 7.5) but the subsequent events differ for the two viroid families. The RNA polymerase begins replication at a precise point on the viroid genome RNA but the nature of the initiation event is not known. For the *Pospiviroidae* the replication process generates a linear concatemeric RNA, of opposite sense to the genome, which is used as a template to synthesize linear genome sense concatemeric RNA from which genome length RNA molecules are excised. The excision relies on the action of a host encoded RNase III-like enzyme. For the *Avsunviroidae* the concatemeric RNA produced from the initial circular molecule is cleaved as it is synthesized to form genome length antigenome RNA which is circularized. This is then used as a template for production of genome sense concatemeric RNA which is also cleaved into genome lengths as it is being synthesized. The cleavage of the antigenome and genome sense concatemeric *Avsunviroidae* RNA occurs by the action of an unusual RNA sequence within the newly-synthesized molecule, called a ribozyme. These ribozyme sequences adopt a complex 3-dimensional structure which has an enzymatic function that autocatalytically cleaves the RNA at a specific site, generating genome length linear ssRNA molecules. For both viroid families the newly-generated genome length molecules are finally converted into circles by an unknown process.

Hepatitis delta virus

Hepatitis delta virus (HDV) has a circular RNA genome of 1.7 kb. The genome is extensively base-paired and has a rod-like appearance in the electron microscope. However, unlike the plant viroids, HDV encodes two proteins, the large and small delta antigens. These are produced from the same mRNA but the larger protein results from translation of a proportion of the mRNA molecules which is modified by a host cell enzyme to allow translation beyond the stop codon that normally terminates protein synthesis to generate the small delta antigen. This means that the two proteins share amino acid sequence but the large delta antigen is extended by 19 amino acids at the carboxy terminus. HDV can only spread as a virus having been replicated in cells also infected with hepatitis B virus. The process of HDV genome replication follows a similar process as described for the *Avsunviroidae* viroids. The host cell RNA polymerase replicates the HDV RNA by a rolling circle mechanism and the linear genomes which are produced by cleavage of the concatemer by the HDV ribozyme sequence are circularized by an unknown process. The small delta antigen has been implicated in the initiation of replication but this remains an area of debate and research. The genome is packaged into particles using hepatitis B virus structural proteins together with both delta antigens. The resultant particle is antigenically indistinguishable from those of hepatitis B virus.

Key points

- Viruses with RNA genomes use novel, virus-encoded enzymes to synthesize their RNA.
- Class 3 viruses have segmented genomes and each segment replicates independently using a polymerase contained within the virion which is taken into the cell. Replication occurs within virus-like structures in the cytoplasm of the infected cell.
- The genome of class 4 viruses is translated immediately after infection to synthesize the RNA polymerase and other proteins.
- The picornavirus genome is held in a circular form essential for replication by a combination of host and virus proteins.
- Class 5 virus RNA genomes have complementary sequences at the termini and these sequences are critical for replication.

- As with class 3 viruses, class 5 viruses carry an RNA polymerase in the virion which enters the infected cell.
- The replication of RNA molecules, either whole genomes or individual genome segments, follows the same process of formation of a replication intermediate (RI) which consists of partially ssRNA and partially dsRNA in a complex with the replication proteins.
- Viroids use the host cell DNA-dependent RNA polymerase to replicate their RNA genomes in a rolling circle process, followed by cleavage and subsequent circularization of the new genome RNA.

Questions

- Compare and contrast the mechanisms of replication of the RNA genomes of viruses belonging to Baltimore classes 4 and 5.
- Discuss the different roles of reovirus mRNA in translation and as the templates for genome replication.

Further reading

Conzelmann, K.-K. 1998. Nonsegmented negative-stranded RNA viruses: genetics and manipulation of viral genomes. *Annual Review of Genetics* 32, 123–162.

Curran, J., Kolakofsky, D. 1999. Replication of paramyxoviruses. *Advances in Virus Research* 54, 403–422.

Dimmock, N. J. 1991. The biological significance of defective interfering viruses. *Reviews in Medical Virology* 1, 165–176.

Flores, R., Gas, M-E., Molina-Serrano, D., Nohales, M-A., Carbonell, A., Gago, S., De la Peña, M., Daròs, J. A. 2009. Viroid replication: rolling-circles, enzymes and ribozymes. *Viruses* 1, 317–334.

Lai, M. M. C., Cavanagh, D. 1997. The molecular biology of coronaviruses. *Advances in Virus Research* 48, 1–100.

Marriott, A. C., Dimmock, N. J. 2010. Defective interfering viruses and their potential as antiviral agents. *Reviews in Medical Virology* 20, 51–62.

Marriott, A.C., Easton, A. J. 2000. Paramyxoviruses. In, reverse genetics of RNA viruses. *Advances in Virus Research* 53, 312–340.

Portela, A., Digard, P. 2002. The influenza virus nucleoprotein: a multifunctional RNA-binding protein pivotal to virus replication. *Journal of General Virology* 83, 723–734.

Taylor, J. M. 1992. The structure and replication of hepatitis delta virus. *Annual Review of Microbiology* 42, 253–276.

Taylor, J. M. 2003. Replication of human hepatitis delta virus: recent developments. *Trends in Microbiology* 11, 185–190.

Chapter 9

The process of infection: IIC. The replication of RNA viruses with a DNA intermediate and vice versa

Some viruses switch their genetic material between RNA and DNA forms during their infectious cycles. The idea of DNA synthesis from an RNA template was once regarded as heresy to the doctrine of information flow from DNA to RNA to protein. However, it now has an established place in molecular biology. Indeed, many mammalian genome DNA sequences (some pseudogenes, many highly repetitive sequences and certain types of transposable elements) are known to have been created in this way.

Chapter 9 Outline

9.1 The retrovirus replication cycle
9.2 Discovery of reverse transcription
9.3 Retroviral reverse transcriptase
9.4 Mechanism of retroviral reverse transcription
9.5 Integration of retroviral DNA into cell DNA
9.6 Production of retrovirus progeny genomes
9.7 Spumaviruses: retrovirus with unusual features
9.8 The hepadnavirus replication cycle
9.9 Mechanism of hepadnavirus reverse transcription
9.10 Comparing reverse transcribing viruses

This chapter discusses the replication of the retroviruses and the hepadnaviruses, two important virus families that have their genetic information in both RNA and DNA forms at different stages of their lifecycles. The form of nucleic acid packaged into particles differs between the viruses, being RNA in most retroviruses and DNA in hepadnaviruses. The process by which DNA is copied from an RNA template is known as reverse transcription. This step is essential in the replication of both virus families but, on its own, does not achieve genome amplification. Instead, the increase in genome number needed for progeny particle formation only comes when RNA copies are transcribed from the DNA.

Introduction to Modern Virology, Seventh Edition. N. J. Dimmock, A. J. Easton and K. N. Leppard.
© 2016 John Wiley & Sons, Ltd. Published 2016 by John Wiley & Sons, Ltd.

9.1 The retrovirus replication cycle

Important viruses in the retrovirus family

- *Human immunodeficiency virus types 1 and 2, members of the lentivirus genus in this family, are the cause of the global AIDS pandemic (see Chapter 21).*
- *Human T-cell lymphotropic virus type 1 is associated with a neuromuscular condition, tropical spastic paraparesis, and adult T-cell leukaemia (see Section 25.7).*
- *Animal retroviruses have been important model systems in studies to understand the events that occur during cancer development (see Section 25.6).*
- *Retroviruses, including HIV, are important as gene therapy vectors–carriers to get therapeutic DNA into cells (see Section 30.3).*

The various stages in the cycle of retrovirus replication are considered in detail in the following sections of this chapter. However, it is useful to see the bigger picture of how these stages fit together before trying to understand them in detail. All retroviral particles contain two identical single-stranded genome RNA molecules, typically 8000 to 10000 nucleotides in length, that are associated with one another

(see Box 9.1). These RNAs have the same sense as mRNA and also have the characteristic features of eukaryotic mRNA, with a 5′ cap structure and a 3′ poly-adenylate (poly-A) tail. Despite these features, the genome RNA is never translated after the particle enters a cell. Instead, it is used as a template for the synthesis of a double-stranded DNA molecule. This event, known as reverse transcription, occurs in the cytoplasm within the incoming virus particle. The DNA then moves to the nucleus where it is integrated into the host genome. Only then can progeny genomes be produced by transcription of the DNA to give mRNA molecules. These can then be translated to give protein and/or packaged into progeny particles that then leave the cell, so completing the cycle.

9.2 Discovery of reverse transcription

Reverse transcriptases allow DNA copies to be created from RNA molecules. This process is crucial to the replication of important human pathogens, has shaped the structure of large parts of our own genome, and is key to the study of gene function in the laboratory.

Box 9.1

Evidence for dimerization of retroviral genomes

- Native genomes sediment with a size of 70S by ultracentrifugation whereas the size after denaturation is 35S.
- Electron microscopy analysis of several retroviral genomes shows pairs of similar RNA molecules linked together close to one end.
- Mutation analysis maps a dimerization function to one or more stem-loop structures near to the 5′ end of the genome RNA.

Box 9.2

Evidence for a DNA intermediate in retroviral replication

- Infection can be prevented by inhibitors of DNA synthesis added during the first 8–12 hours after exposure of the cells to the virus, but not later.
- Formation of virions is sensitive to actinomycin D, an inhibitor of host RNA polymerase II, which uses DNA as a template for RNA synthesis.
- Infection of cells by Rous sarcoma virus* (a virus of birds) confers stably inheritable changes to the cells' appearance and growth properties (see Sections 25.1 and 25.6). The details of these changes are virus-strain specific, indicating heritability of viral genetic information in the cells, a DNA property.

*Rous sarcoma virus infection is not cytolytic so the cells do not die from the infection.

The hypothesis of a DNA intermediate in retroviral replication was developed by Howard Temin in the 1960s (see Box 9.2). His 'provirus' theory postulated the transfer of the information of the infecting retroviral RNA to a DNA copy (the provirus) which then served as a template for the synthesis of progeny viral RNA. It is now clear that this theory is correct.

Temin's theory required the presence in infected cells of an RNA-dependent DNA polymerase or 'reverse transcriptase' but, at the time, no enzyme had been found that could do this. If such an enzyme existed, then the retrovirus must either induce a cell to make it or else carry the enzyme into the cell within its virion. A search was begun for such an enzyme in retrovirus particles and David Baltimore and Howard Temin each independently reported its isolation in 1970, work for which they were subsequently awarded the Nobel prize.

Reverse transcriptase has since become a cornerstone of all molecular biology investigations because it provides the means to produce complementary DNA (cDNA) from mRNA in the laboratory and so allows cDNA cloning. It has also become clear that reverse transcription has played a big part in the shaping of the genomes of complex organisms such as ourselves. Within our DNA there are many intronless pseudogenes (non-functional genes) and also lots of repetitive elements, each of which has the hallmarks of a reverse-transcribed and integrated sequence.

9.3 Retroviral reverse transcriptase

The reverse transcriptase protein (RT) provides three enzymatic activities: (i) reverse transcriptase – synthesis of DNA from an RNA template; (ii) DNA polymerase – synthesis of DNA from a DNA template; and (iii) RNaseH activity – digestion of the RNA strand from an RNA:DNA hybrid to leave single-stranded DNA. Note that an enzyme with RNaseH activity is also required to excise the primers during DNA replication (see Sections 7.1 and 7.2). For more detail about the RT protein, see Fig. 9.1 and Box 9.3.

The RT enzyme, like all other DNA polymerases, requires a primer from which to initiate DNA synthesis (see Section 7.1). The primer used to reverse transcribe the viral RNA is a host tRNA that is carried in the

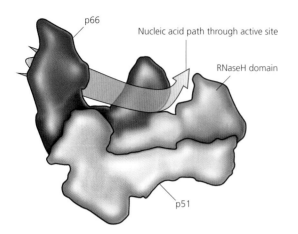

p66

Nucleic acid path through active site

RNaseH domain

p51

Fig. 9.1 Retroviral heterodimeric reverse transcriptase. The enzyme from human immunodeficiency virus is formed of p66 and p51 subunits. P66 has five structural domains, coloured blue, red, green, yellow and orange in order from the N-terminus. The last of these is the RNaseH domain, not present in p51. The remaining four domains found in p51 are coloured in pale versions of the equivalent p66 domains. The structural arrangement of these domains in p51 is distinct from their organization in p66, where they form the active site for DNA synthesis. Drawn from data presented in Ding et al. (1998). *Journal of Molecular Biology*, 284, 1095–1114.

virus particle in association with the viral genome. The 3′ end of the tRNA is base paired to the genomic RNA near to its 5′ terminus (Fig. 9.2). Each retrovirus contains a specific type of tRNA; for example, Rous sarcoma virus has tryptophan tRNA and Moloney murine leukaemia virus has proline tRNA. These specific tRNAs are probably selected by their sequence complementarity with the genomic RNA at the tRNA binding site.

To perform its function in the virus life cycle, RT must enter a cell in the virus particle together with the genomic RNA. Mutant retroviral particles that do not contain active RT cannot establish proviral DNA in a cell even if the cell has been engineered to contain RT already. This indicates that reverse transcription occurs without full uncoating of the genome and explains why the genome, although having all the features of mRNA, is never translated; host ribosomes do not have access to it.

Box 9.3

The reverse transcriptase (RT) protein

- The active form of the RT enzyme is a dimer formed of two related polypeptides (i.e. a heterodimer) but the details of its composition vary between retroviruses.
- RT from avian retroviruses is composed of α (60000 M_r) and β (90000 M_r) subunits. The β subunit comprises the α polypeptide with another enzyme, integrase, linked to its C-terminus (for the production of retroviral proteins, see Section 10.9).
- RT from the human retrovirus, HIV, is fully cleaved from integrase and instead the subunits of the heterodimer differ by the presence (p66) or absence (p51) of the RNaseH domain of RT. Although the protein domains needed to form the reverse transcriptase/polymerase active site are present in both subunits, there is no catalytic activity in p51 because it adopts a very different structure from p66 in the heterodimeric enzyme (Fig. 9.1).

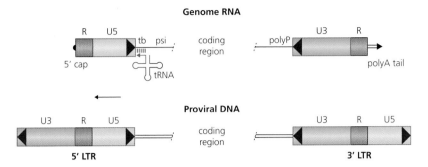

Fig. 9.2   Comparison of the structures of retrovirus genome RNA (top) and the proviral DNA created from it by reverse transcription. U5 and U3 are unique sequences at the 5′ and 3′ ends of virion RNA; R is a directly repeated sequence at the RNA termini. Short inverted repeat sequences are represented as ◀, ▶. tb is the binding site for a transfer RNA and polyP is a polypurine region, both significant in reverse transcription (see Section 9.4 and Fig. 9.3). Long terminal repeats (LTRs) comprise duplications of the sequences U3, R, U5. psi is the specific packaging signal for RNA genomes. Not to scale.

9.4 Mechanism of retroviral reverse transcription

Comparing the structures of the genome RNA and proviral DNA

If you compare the sequences of a viral genomic RNA and a proviral DNA that is copied from it, the two molecules are not precisely co-linear (Fig. 9.2). The directly repeated R sequences, found at the 5′ and 3′ ends of the genome RNA adjacent to the cap and polyA tail, are internal in the proviral DNA molecule. In other words, additional sequence has been added outside each R sequence in the double-strand DNA provirus as compared with the genomic RNA. Where do these additional sequences come from? A search of the genome RNA sequence shows that they are present, but are internal to the R repeats. The sequence U_5, which is copied to the 3′ end of the provirus, lies just inside the 5′ end of the genome, adjacent to R. Conversely U_3, which is copied to the 5′ end of the provirus, lies originally just in from the 3′ end of the

genome, again adjacent to R. Therefore, the process of reverse transcription has to duplicate sequences from one end of the genome and place the copy at the other end. The result is long directly repeated sequences at each end of the provirus, comprising the elements U_3, R, and U_5; these are known as the long terminal repeats or LTRs. LTR formation is essential for retroviral gene expression from the provirus (see Section 10.8). The proviral DNA has one further significant feature. At the outer ends of the LTRs are short *inverted* repeats that derive from sequences originally present at the internal ends of the U_5 and U_3 sequences in the genome. These are important in retroviral integration (see Section 9.5).

The model for reverse transcription to form proviral DNA

Figure 9.3 presents a model for proviral DNA synthesis, some evidence for which is highlighted in Box 9.4. The tRNA that is base-paired to the genome provides the primer for synthesizing the DNA negative-sense strand ((−) DNA). It is positioned exactly adjacent to

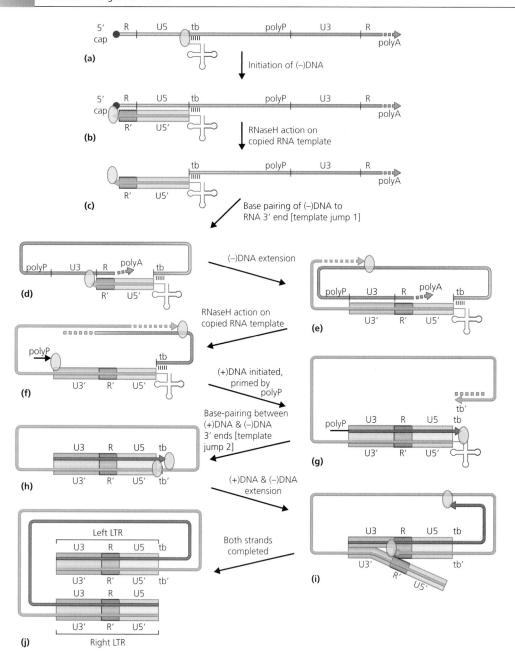

Fig. 9.3 A scheme for the synthesis of retroviral linear proviral DNA by reverse transcriptase (see Section 9.4 for details). Brown lines: RNA; orange lines: (−)DNA; red lines: (+)DNA; green ellipses: reverse transcriptase enzyme. Arrowheads represent 3′ ends. The colour coding of the U3, R and U5 elements of the LTRs is carried forward from Fig. 9.2 to illustrate how the LTRs are created. Abbreviations and other symbols as for Fig. 9.2.

Box 9.4

Evidence for the current model of retroviral reverse transcription

- Using detergent-lysed virions in DNA synthesis reactions in vitro, a major product is a short fragment corresponding to a copy of the 5′ end of the genome from the tRNA binding site (Fig. 9.3b). This is termed negative strong-stop DNA.
- In infected cells, a discrete DNA intermediate is detected that corresponds to the (+)DNA fragment in Fig. 9.3f. This is termed positive strong-stop DNA.

the U_5 sequence (a). This positioning effectively defines U_5 as it is the sequence between this point and R sequence that becomes duplicated during DNA synthesis. Synthesis initiated at the tRNA extends to the 5′ end of the template (b), forming an RNA:DNA hybrid duplex. This duplex is a substrate for the RNaseH activity of RT (see Section 9.3), which degrades the template RNA that has been copied (c). As a result of this RNaseH action, the newly-synthesized DNA sequence (R′) is now free to base pair with another R sequence (d). This could be at the 3′ end either of the same RNA molecule (as shown) or of the second genome copy in the particle. This template switch is often referred to as the first 'jump'; once it has occurred, synthesis of the (−)DNA can then proceed along the body of the genomic RNA (e) with the RNaseH activity of RT continuing to degrade the template RNA as it goes (f).

A key question is how synthesis of the positive sense strand of the provirus ((+)DNA) is begun, as this too needs a primer. A comparison of genome and provirus sequences predicts that (+)DNA initiation must occur immediately 5′ to the U_3 sequence, as this is the sequence that will form the 5′ end of the provirus. Adjacent to U_3 in the genomic RNA, all retroviruses have a conserved purine-rich region (polyP in Fig. 9.2). During degradation of the genome by RNaseH following (−)DNA synthesis, polyP is relatively resistant to

degradation. It remains base-paired to the (−) DNA for long enough to provide the required primer for (+)DNA synthesis (f) but is eventually removed by RNaseH. Synthesis proceeds rightwards from this primer using the newly-synthesized (−)DNA as a template (g). Some retroviruses, such as HIV, additionally prime (+)DNA synthesis from other RNaseH-resistant genome oligonucleotides and so produce a (+)DNA strand that is fragmented. These pieces are presumably joined together later by host DNA repair enzymes.

The 5′ end of the (−)DNA strand is still attached to the primer tRNA. RT uses the 3′ segment of this, which had been paired with the genome originally and therefore has the exact complementary sequence, as template for further (+)DNA synthesis (g). In the meantime, synthesis of the (−)DNA continues to the end of the available template RNA (g), which will be the sequence which originally bound the tRNA primer. Both of these RNA sequences have been spared from RNaseH degradation up to this point as they have been paired as an RNA: RNA hybrid, which is not a substrate for this enzyme, but once they have been reverse-transcribed they too are degraded. This exposes single-stranded DNA sequences at the 3′ ends of the (+)DNA and (−)DNA that are complementary (tb and tb′). These then base pair (h), allowing RT to make the second template jump that occurs during reverse transcription. This jump gives the molecules of RT that are synthesizing (+)DNA and (−)DNA strands the templates they need to complete the synthesis of a double-stranded DNA provirus with LTRs at each end (i, j).

The existence of template jumps during reverse transcription suggested a possible explanation for the presence of two genome RNA copies in the particle. Perhaps the RT could not jump between ends of the same molecule but had to jump from the 5′ end of one genome copy to the 3′ end of the other. However, the experimental evidence is that proviral DNAs can be formed solely by intramolecular jumps, as shown here. It is probable, though, that the possibility of making intermolecular jumps minimizes the effects of genome damage on virus viability and so confers an evolutionary advantage. It would, of course, have been easier to describe reverse transcription if the tRNA primer binding site was at the 3′ end of the genome! However, this would have made the provirus an exact copy of the genome and so it would not have had LTRs. As discussed in Section 9.6, these are essential to the generation of progeny viral RNA genomes. This simple mechanism would also leave unsolved the 'end-replication' problem (see Section 7.1).

The action of RNaseH during reverse transcription means that the process can only generate one proviral DNA molecule from each genome RNA, i.e. it is a conversion rather than amplification step. The cycle of retroviral genome replication is only completed when multiple RNA copies are transcribed from the proviral DNA later in the infectious cycle (see Section 9.6). Nonetheless, reverse transcription is a crucial first step in the infection of a cell by a retrovirus – without it, the infection cannot progress any further. Thus it is not surprising that this process is an important target for drugs designed to treat infection by the human retrovirus, HIV. However, RT lacks any proofreading activity and so retroviral sequences undergo high levels of mutation with every cycle of infection. This rapid evolution of sequence is a major problem in the treatment of HIV infection, because drug resistance can evolve very rapidly (see Section 21.6). The resulting antigenic variation is also one of the reasons why an effective vaccine against this virus has not so far been developed.

9.5 Integration of retroviral DNA into cell DNA

Proviral DNA is produced in the cytoplasm within the partly uncoated virus particle. This complex of DNA with residual virion proteins, including the virus-coded integrase enzyme that entered the cell in the particle, is termed the pre-integration complex or PIC. It migrates to the nucleus where the DNA is integrated into cellular DNA. Successful integration requires the short inverted repeat sequences at the proviral termini (Fig. 9.2) and the integrase enzyme. As with RT, this enzyme must enter the cell in the infecting particle to be able to act; it cannot be added later.

There are three steps in the integration process (Fig. 9.4, Box 9.5). First, two bases are removed from both 3′ ends of the linear proviral DNA molecule by the action of the viral integrase (b). Second, the 3′ ends are annealed to sites a few (four to six) bases apart in the host genome (c). These sites are then cleaved by the integrase with the ligation of the proviral 3′ ends to the genomic DNA 5′ ends (d). This part of the integration reaction requires no input of energy from ATP, etc., because the energy of the cleaved bonds is used to create the new ones (i.e. it is an exchange reaction), and thus is reversible. Finally, gaps and any mismatched bases at the newly-created junctions are repaired by host DNA repair functions (e); this renders the integration irreversible.

Infected cells typically contain 1 to 20 copies of integrated proviral DNA. There are no

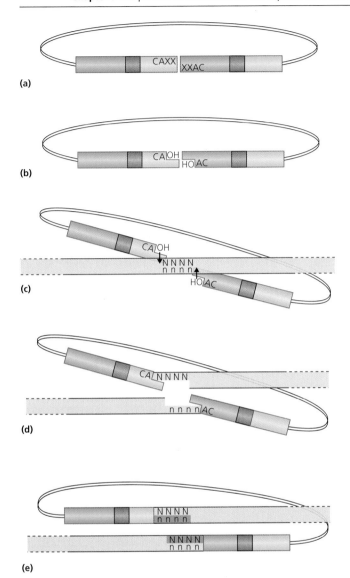

Fig. 9.4 Integration of retroviral DNA into the host genome. For details see Section 9.5. Turquoise, purple and orange: the U3, R & U5 elements of the retroviral LTRs; solid black lines: retroviral DNA; pink: host DNA random integration target; brown: DNA repair synthesis.

specific target sequences for integration in the host genome, although there is some preference for relatively open regions of chromatin, i.e. regions with active genes. Different classes of retrovirus also show preferences for integration, e.g. murine oncoviruses target active promoter regions rather than the bodies of genes. For

most retroviruses, integration only occurs in cells which are moving through the cell cycle. This is thought to reflect an inability of the PIC to cross the intact nuclear membrane; breakdown of the membrane at mitosis therefore allows the PIC to access the host chromosomes. However, HIV does not suffer

---

Box 9.5

Evidence for the current model of retroviral integration

- Integrated proviruses always lack terminal nucleotides as compared with unintegrated DNA and the integrated DNA is always flanked by a short duplication of cellular sequence.
- Appropriate integration intermediates have been characterized from in vitro integration reactions.

---

this restriction; proteins within its PIC can mediate nuclear uptake and hence allow infection of quiescent cells.

While most retroviral integration events during an infection will occur in somatic cells, at various points in evolutionary history there have clearly been retroviral integrations into the germ line of humans and other species. These events have led to the establishment of endogenous retroviruses that are inherited just like standard genetic loci. Most of these loci have suffered mutations that prevent the production of virus from them, but some are still capable of producing virus-like particles under appropriate conditions (see Section 31.2).

9.6 Production of retrovirus progeny genomes

New retroviral RNA genomes are transcribed from integrated proviral DNA. In all respects, this process resembles the production of cellular mRNA. Cellular DNA-dependent RNA polymerase II (RNA pol II) transcribes the provirus and the primary transcript is capped and polyadenylated by host cell enzymes. Clearly, many RNA copies can be transcribed from a single provirus, and it is this, rather than reverse transcription, that gives genome amplification during the replication cycle.

The progeny genomes must resemble exactly the parental genome that produced the

provirus, otherwise the virus will not have reproduced itself successfully. To achieve this, the transcription start site must be exactly at the 5′ end of the R element within the left-hand LTR. However, RNA pol II promoters lie upstream of the transcription start point. If the provirus were simply a copy of the genome, there would be no viral sequences upstream of this start point to provide the promoter and so genome RNA synthesis would depend on fortuitous integration of the provirus adjacent to a host promoter. This is why the creation of the LTRs during reverse transcription is so important. With the creation of the left LTR, a virus-coded sequence, U_3, is placed upstream of the required start site at R and this provides the necessary promoter elements for RNA pol II. Equally, at the other end of the genome, the polyA addition site must be fixed exactly at the 3′ end of the R element within the right-hand LTR. Since sequences both upstream and downstream of a polyadenylation site are important in determining its position, it is again essential that proviral sequences extend beyond the intended genome 3′ end; these sequences are provided by the U_5 element within the right-hand LTR. Thus, the construction of the LTRs by sequence duplication during reverse transcription is crucial to retroviral replication. By virtue of this process, a retroviral genome manages both to encode, and to be encoded by, an RNA pol II transcription unit.

9.7 Spumaviruses: retrovirus with unusual features

The spumaviruses of the retrovirus family (also known as the foamy viruses) came to be studied in detail rather later than other retroviruses. Most work has been done on a virus initially known as human foamy virus. However, this virus is actually a chimpanzee virus which crossed into man as a dead-end zoonotic infection and is now termed simian foamy virus. Unlike the standard retrovirus replication cycle, reverse transcription in spumaviruses at least begins (and may even be completed) within assembling progeny particles before they are released from a cell. In other words, DNA synthesis occurs at the end of the replication cycle rather than at the beginning. Thus the genetic material within at least a proportion of spumavirus particles is DNA rather than RNA. In this sense, the spumaviruses somewhat resemble the hepadnaviruses, which are also reverse-transcribing viruses with DNA in their particles (see Section 9.8). Other details relating to gene expression (see Section 10.9) also suggest that the spumaviruses have similarity to both standard retroviruses and hepadnaviruses. These differences from standard retroviruses mean that the spumaviruses are now regarded as a subfamily of the retroviruses, the *Spumavirinae*, rather than a genus in that family.

9.8 The hepadnavirus replication cycle

> *Important viruses in this family*
>
> - *Human hepatitis B virus creates a large global burden of chronic liver disease and is the cause of many cases of hepatocellular carcinoma (see Chapter 22).*

Human hepatitis B virus (HBV) particles contain a partially double-stranded circular DNA genome. This comprises two linear DNA strands that form a circle through base pairing (Fig. 9.5 (a)). It is possible to designate positive and negative strands because all the genes are arranged in the same direction (see Section 10.9). The (−)DNA strand is covalently linked to the virus-coded P protein at its 5′ end and extends the full circumference of the circle and beyond. The (+)DNA strand overlaps the 5′–3′ junction of the (−)DNA and acts as a cohesive end to circularize the genome; it is normally incomplete.

When an HBV particle enters the cell, the genome is transported to the nucleus where it is completed to give an intact double-stranded circle. This is then transcribed to give a variety of mRNA species (see Section 10.9), including one type that can be packaged into progeny capsids as an alternative to being translated. It is the transcription of this pregenome RNA that represents the genome amplification stage in the hepadnavirus lifecycle. Once in the capsid, the pregenome RNA serves as a template for reverse transcription, giving rise to the partially double-stranded DNA that is found in particles after they have left the cell.

Not all the DNA-containing progeny particles leave the cell. They can also re-infect the nucleus of the cell that produced them. Since the infection itself does not kill the cell, this process leads to an amplification of the number of copies of viral DNA in the nucleus that are available for viral gene expression and progeny production.

9.9 Mechanism of hepadnavirus reverse transcription

Molecular analysis of HBV replication was relatively slow in coming as the virus is difficult

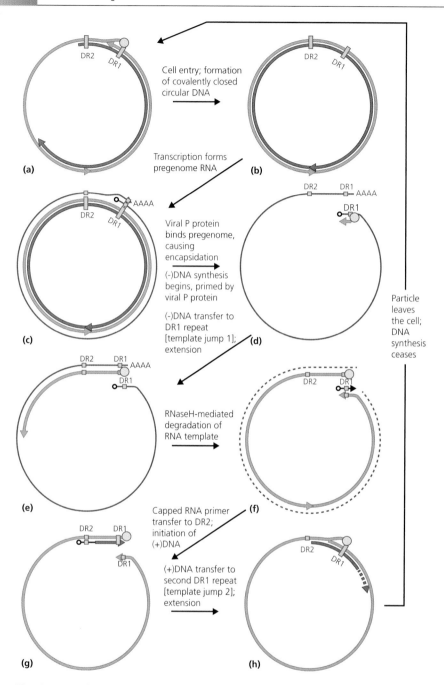

Fig. 9.5 Replication of the hepadnavirus genome by reverse transcription (see Section 9.9 for details). Green lines – RNA; red lines – (+)DNA; orange lines – (−)DNA; black circle – RNA cap structure; AAA – RNA polyA tail. DR1, DR2 represent two copies of a short, directly repeated sequence in the genome. The viral P protein (terminal protein and reverse transcriptase) attached to the genome 5′ end is shown as a red circle. Filled arrowheads represent 3′ ends. Other arrows indicate the polarity (5′ → 3′) of nucleic acid strands.

Box 9.6

Evidence for the current model of hepadnavirus replication

- Formation of covalently closed viral DNA circles precedes the appearance of viral RNA.
- There is a substantial excess of (−)DNA over (+)DNA in infected cells.
- Synthesis of (−)DNA is insensitive to actinomycin D and some of this DNA can be found in RNA:DNA hybrid molecules.
- Radioactive precursor incorporation into viral DNA is associated with immature particles in the cytoplasm.

---

to grow in culture. However, a scheme for replication has been derived from studies involving HBV and its relatives, woodchuck and duck hepatitis viruses (Fig. 9.5), some evidence for which is summarized in Box 9.6.

After the HBV particle enters the cell, the DNA genome is transported to the nucleus, where the attached P protein is removed and both strands are completed and ligated by host DNA repair systems to give a covalently closed circle, CCC (b). The (−)DNA strand of the CCC then provides a template for transcription by host RNA pol II. Note that, in contrast to the retroviral provirus, there is no requirement for the hepadnavirus CCC to integrate into the host genome for it to be transcribed. Transcription produces various mRNAs (see Section 10.9) which are exported to the cytoplasm. The longest class of mRNA (c), which has a terminal repetition because it extends over more than the full circumference of the circular template, encodes the P protein and also serves as the pregenome, i.e. the template for genome DNA synthesis. P protein is multifunctional, with terminal protein, reverse transcriptase/DNA polymerase and RNaseH domains. As soon as it is synthesized, the P protein interacts with a sequence, known as ε, close to the 5′ end of the RNA molecule which encoded it, and directs the packaging of this RNA by core protein. Once

this has occurred, creation of the DNA genome can begin.

The terminal protein domain of P protein serves as a primer for synthesis of (−)DNA by the reverse transcriptase domain of P. This explains why P is found attached to the 5′ end of this genome strand in the particle. DNA synthesis begins at the ε sequence in the pregenome RNA 5′ end (d). The short (−)DNA fragment produced then moves, with the associated P protein, to base-pair with the second copy of its template sequence at the RNA 3′ end. By doing this, the polymerase makes the first template jump of hepadnavirus reverse transcription. DNA synthesis then continues (e), with RNaseH activity degrading the template as it does so (f). A specific positive sense RNA oligonucleotide from the very 5′ end of the RNA, containing the repeat sequence DR1, is spared this degradation (f). It transfers to base-pair with the second repeat, DR2, near the 5′ end of the new negative strand, where it primes synthesis that extends to the very 5′ end of its template (g). To continue synthesis, a second template jump is then needed. The newly-synthesized (+)DNA copy of DR2 switches its base-pairing to the second copy of DR1 at the other end of the (−)DNA template. Synthesis of (+)DNA can then continue onwards into the body of the genome (h). It is very unusual for this positive strand to be completed before the particle exits the cell, depriving the particle of substrates for DNA synthesis and so terminating further strand extension. Hence, the double-stranded genomes that are seen in virus preparations are normally incomplete.

9.10 Comparing reverse transcribing viruses

Use of a reverse transcriptase is not restricted to the animal retro- and hepadnaviruses. The caulimoviruses are the only truly double-stranded DNA virus family in the plant kingdom. Investigation of the representative virus, cauliflower mosaic virus (CaMV), has shown that it is a reverse-transcribing virus, with properties intermediate between retroviruses and hepadnaviruses. Like HBV, the CaMV genome is a double-stranded DNA circle, with a complete but gapped negative strand and an incomplete positive strand. However, like retroviruses, negative strand DNA synthesis is primed by a host cell tRNA which base-pairs to the template RNA close to its 5′ end.

The retroviruses, hepadnaviruses and caulimoviruses are in most senses completely unrelated; their protein coding strategies are different and only the retroviruses carry a specific integration function. However, some molecular aspects of their replication show considerable similarity. All three viruses have reverse transcription mechanisms which involve shifts or jumps of the extending polymerase from the 5′ end to the 3′ end of a template molecule, mediated through sequences repeated at the two ends of the template. Also, in each type of virus, host RNA polymerase II is used to produce RNA which serves as either genome or pregenome. It is the timing of this event in the viral lifecycle that varies, leading to the difference observed in the nature of the nucleic acid in mature virions.

This variation can be seen even within a virus family, as the spumavirus subfamily of the retroviruses illustrates. Thus, in essence, the replication cycles of all these viruses are temporal permutations of the same set of events.

Key points

- Reverse transcriptase enzymes can use RNA as template to generate new DNA strands, and thus are able to reverse the classical flow of genetic information.
- Retroviruses and hepadnaviruses use reverse transcription as an obligatory step in their replication cycles and encode reverse transcriptases.
- Reverse transcription provides genome conversion for these viruses, not replicative amplification. This latter event is provided by RNA synthesis using the DNA created by reverse transcription as a template.
- Most retroviruses use reverse transcription at the beginning of their replication cycle, immediately after entry into a cell, whereas hepadnaviruses use this process at the end of the cycle, within maturing virions.
- Integration of retroviral DNA (proviral DNA) is essential to the virus lifecycle and is catalyzed by a virus-coded integrase carried in the particle. In contrast, hepadnavirus DNA integration is not required and there is no specific mechanism provided for this to occur.

Questions

- Explain the molecular events that allow retroviruses to utilize the machinery of a eukaryotic host cell for mRNA production, despite the viruses having positive sense single-stranded RNA genomes.

- Compare and contrast the mechanisms of genome replication employed by hepadnaviruses and retroviruses.

Further reading

Basu, V. P., Song, M., Gao, L., Rigby, S. T., Hanson, M. N., Bambara, R. A. 2008. Strand transfer events during HIV-1 reverse transcription. *Virus Research* 134, 19–38.

Delelis, O., Lehmann-Che, J., Saib, A. 2004. Foamy viruses – a world apart. *Current Opinion in Microbiology* 7, 400–406.

Delelis, O., Carayon, K., Saif, A., Deprez, E., Mouscadet, J. F. 2008. Integrase and integration: biochemical activities of HIV-1 integrase. *Retrovirology* 5, DOI: 10.1186/1742-4690-5-114.

Desfarges, S., Ciuffi, A. 2010. Retroviral integration site selection. *Viruses-Basel* 2, 111–130.

Goff, S. P. 2007. Host factors exploited by retroviruses. *Nature Reviews: Microbiology* 5, 253–263.

Herschhorn, A., Hizi, A. 2010. Retroviral reverse transcriptases. *Cellular and Molecular Life Sciences* 67, 2717–2747.

Lewinski, M. K., Bushman, F. D. 2005. Retroviral DNA integration – Mechanism and consequences. *Advances in Genetics* 55, 147–181.

Li, X. A., Krishnan, L., Cherepanov, P., Engelman, A. 2011. Structural biology of retroviral integration. *Virology* 411, 194–205.

Lindemann, D., Rethwilm, A. 2011. Foamy virus biology and its application for vector development. *Viruses-Basel* 3, 561–585.

Linial, M. L. 1999. Foamy viruses are unconventional retroviruses. *Journal of Virology* 73, 1747–1755.

Nassal, M. 2008. Hepatitis B viruses: reverse transcription a different way. *Virus Research* 134, 235–249.

Paillart, J. C., Shehu-Xhilaga, M., Marquet, R., Mak, J. 2004. Dimerization of retroviral RNA genomes: an inseparable pair. *Nature Reviews Microbiology* 2, 461–472.

Schultz, S. J., Champoux, J. J. 2008. RNase H activity: structure, specificity and function in reverse transcription. *Virus Research* 134, 86–103.

Wilhelm, M., Wilhelm, F. X. 2001. Reverse transcription of retroviruses and LTR retrotransposons. *Cellular and Molecular Life Sciences* 58, 1246–1262.

Chapter 10

The process of infection: IIIA. Gene expression in DNA viruses and reverse-transcribing viruses

The process of gene expression by the various DNA viruses closely parallels that of their host organisms, with many using host enzymes for transcription and translation. Most of the viruses impose a temporal phasing on their gene expression. Initially, their focus is on producing proteins that either are required for genome replication or that modify the host environment to make it more favourable for the virus to grow. Later, the emphasis switches to the production of large amounts of the proteins needed to form new virions.

To be expressed, the genomes of DNA viruses must be transcribed to form positive sense mRNA, and this must then be translated into polypeptides. For viruses of eukaryotes, further steps intervene: RNAs are usually capped and polyadenylated, and may also need to be spliced, to give functional mRNA. If it has been produced in the nucleus, the mRNA must be moved to the cytoplasm to allow translation to occur. Each stage in the pathway of gene expression, from transcription of RNA through to post-translational modification of protein, represents a potential point of control and DNA viruses exploit these possibilities in various ways so that each one achieves an organized program of gene expression.

For most DNA viruses, gene expression is phased, with early genes being expressed before DNA synthesis begins and late genes only being activated after this event. In some cases, further temporal divisions are also apparent. These patterns of expression have been characterized in considerable detail and, as well as furthering our understanding of virus growth cycles, such studies have increased greatly our understanding of the molecular biology of eukaryotic cells. This chapter and the

Introduction to Modern Virology, Seventh Edition. N. J. Dimmock, A. J. Easton and K. N. Leppard.
© 2016 John Wiley & Sons, Ltd. Published 2016 by John Wiley & Sons, Ltd.

Box 10.1

Evidence for detailed patterns of viral transcription

- The viral mRNA species that come from each part of the genome are revealed by probing Northern blots of infected cell RNA with labelled probes from each genome region.
- The co-linearity of mRNA and genome can be measured by R-loop mapping in the electron microscope.
- Precise mRNA 5′ and 3′ end positions on the genome, also the positions of splice donor and acceptor sites, are determined by S1 nuclease mapping or RNase protection analysis. Both assays use the encoding DNA (or its derivative) as a probe and map the point of discontinuity between the mRNA and the probe.
- Primer-extension analysis can also be used to map RNA 5′ ends.
- Low abundance splice variant mRNAs are detected by reverse-transcriptase PCR and sequencing of the product to reveal the junction sequences, or by new 'deep-sequencing' approaches used to characterize total transcriptomes.
- Promoter elements in the genome capable of directing transcription are detected and mapped using reporter gene assays (see Box 10.5).

---

accompanying one on RNA virus gene expression (Chapter 11) demonstrate these points by examining gene expression and its control in a variety of virus systems.

10.1 The DNA viruses and retroviruses: Baltimore classes 1, 2, 6 and 7

All DNA viruses synthesize their mRNA by transcription from a double-stranded DNA molecule; those from Baltimore class 1 can express their genes directly while classes 2 and 7 have to convert their genomes to double-stranded form or complete the second strand before transcription begins. The retroviruses of class 6 can also be considered here since they must create a dsDNA version of their genome RNA before transcription can take place. For each of the viruses featured in this chapter, detailed information is available about the

specific mRNAs produced, although this information is only included here for the simpler viruses. Evidence for these viral transcription patterns has come from applying a common set of techniques (Box 10.1). More recently, a number of these viruses have been shown to encode microRNAs, which can regulate viral or cellular gene expression, in addition to mRNAs.

Among those viruses that infect eukaryotes, all except the pox-, irido- and asfarviruses, carry out transcription in the cell nucleus. Thus, for the majority of DNA viruses, the cell's own transcription and RNA modification machinery is available and the viruses make use of this to produce their mRNA. These viruses are therefore good model systems for studying the synthesis of cellular mRNA and several significant milestones in understanding eukaryotic gene expression have come from this work. For example, transcription enhancers and transcription factors which regulate RNA polymerase II (RNA pol II) activity were first identified during studies of SV40 gene expression, and splicing was discovered by analyzing adenovirus mRNA.

The amount of genetic information varies by more than 100-fold between different DNA virus families, as considered in Chapter 7. Some have severe restrictions on their coding capacity and depend greatly on their host for essential replicative functions, despite having evolved to maximize their use of the available genetic information, whereas others have had the opportunity to evolve or acquire many functions that are not essential for virus multiplication. The larger viruses are less dependent upon the cell and can, for example, multiply when cells are not in S phase (synthesizing cellular DNA), whereas the smallest viruses are unable to do so. These larger viruses can also use inhibition of host macromolecular synthesis as a strategy to focus all of the host's resources towards their own replication. Several of the viruses have the capacity to induce the host cell to enter S phase to promote their replication. Such functions may not be crucial in rapidly-growing experimental cell cultures, but are probably essential during natural infections when the virus infects cells that are not dividing. As discussed in Chapter 25, several of the animal DNA viruses are capable of transforming cells. Transformation can result when the virus functions that regulate cell division become permanently expressed in cells that have not died as a result of virus infection.

10.2 Polyomaviruses

The polyomaviruses, e.g. SV40, use host mechanisms to achieve temporally phased expression from their two transcription units. The SV40 enhancer was the first DNA element of this type to be recognized and production of the two SV40 early proteins was the first characterized example of differential splicing. It is now clear that both are widespread in host gene expression. The SV40 early protein, large T antigen, is a classic example of a multifunctional protein.

Polyomavirus (PyV, of mice) and SV40 (of monkeys) exemplify the polyomavirus family. There are also several human viruses in the family, including BK, JC, KI, WU, trichodysplasia spinulosa-associated polyomavirus (TSV) and Merkel cell polyomavirus. SV40 and PyV have double-stranded circular genomes, of around 5.3 kbp in length, which encode early and late genes on opposite strands of the template, each occupying about half the genome (Fig. 10.1). The genes are transcribed by RNA pol II to produce a single primary transcript in each case; these are then differentially spliced to form multiple mRNA species. Differential splicing means that there is more than one acceptable splicing pattern for the primary transcript. Which pattern is used on a specific RNA molecule will depend on the relative affinity of

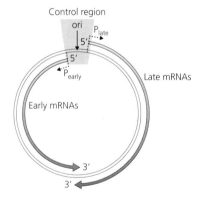

Fig. 10.1 Map of the genome of SV40 virus showing the positions of the early and late transcription units. P denotes promoters of transcription and ori denotes the origin of replication. The mRNAs are shown in detail in Figs 10.2 and 10.3, and the control region in Fig. 10.4.

the splicing machinery for the alternative splice sites available on the molecule. Altering the balance of use of alternative splice sites is a potential method for regulation of gene expression during infection (see adenoviruses and influenza viruses, Sections 10.4 and 11.7 respectively) although it does not appear to operate in the polyomaviruses.

The two mRNAs from the SV40 early region encode a major early protein, the large T (for tumour-specific) antigen, and a second protein called small t (Fig. 10.2a); the other human viruses are similar. Each protein is made from a discrete spliced mRNA (Box 10.2). The PyV early region codes for three proteins, called large T, middle T and small t antigen, using a similar strategy to SV40 (Fig. 10.2b). The late mRNAs produced by each virus also encode three virion proteins, VP1, VP2 and VP3, by differential RNA splicing (Fig. 10.3). VP1 is the major capsid protein, with VP2 and VP3 being minor components. There is no viral protein produced to complex with the DNA in the virion; histones H2A, H2B, H3 and H4 from the

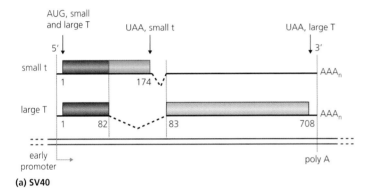

(a) SV40

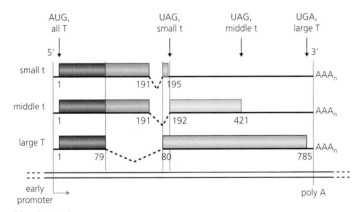

(b) polyomavirus

Fig. 10.2 Early gene expression by (a) SV40 and (b) polyoma virus. Differential splicing of the primary transcript in each case produces mRNAs with distinct coding capacities (see text and Box 10.2 for details). Brown lines denote DNA and blue lines, RNA; dotted lines denote introns and thin lines untranslated RNA regions. Proteins are represented in linear form, with colour coding to indicate the sequences that are common between two or more proteins and numbers indicating amino acid residues.

Box 10.2

Expression of the SV40 early proteins

- The longer mRNA present early in the infectious cycle is translated to give the 174 amino acid small t protein; a short intron, which is removed during processing of the primary transcript for this mRNA, lies downstream of the small t antigen open reading frame (ORF).
- The shorter early mRNA is formed by removal of a larger intron, which includes the C-terminal half of the small t antigen ORF. This alternative splicing event fuses together the first 82 codons of the small t antigen ORF in frame with 626 codons from a much longer ORF located downstream. Translation from this mRNA generates the 708 amino acid large T antigen.
- Although the 626 amino acid ORF is also present in the small t mRNA, it has no initiation codon and so is not accessed by ribosomes.
- SV40 large T antigen is an excellent example of a multifunctional protein, with roles in DNA replication, viral gene expression and modulation of host functions. These are mediated through a combination of intrinsic enzyme activity and the ability to interact with viral DNA and an array of host cell proteins.

host cell are used for this purpose. The Agno protein (whose gene gained its name because, when it was defined, it had no known protein product) functions very late in the virus life cycle; its role is not yet well understood.

The SV40 early and late promoters lie close together, flanking the origin of replication, in a region of the genome known as the control region (Fig. 10.4, Box 10.3). The early promoter is activated by host cell transcription factors and

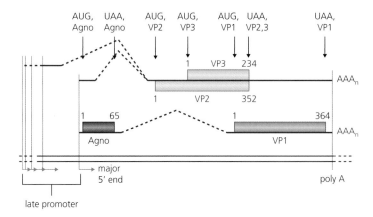

Fig. 10.3 Late gene expression by SV40 virus. The late mRNAs have heterogeneous 5′ ends because the late promoter lacks a TATA box, which normally serves to direct initiation to a specific location. Legend details as for Fig. 10.2. Polyomavirus is very similar but the virion protein (VP1) coding region does not overlap with that of VP2/VP3.

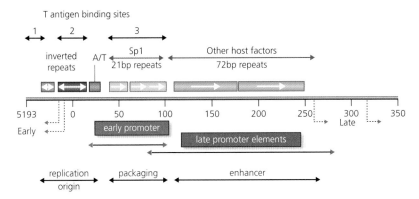

Fig. 10.4 The control region from laboratory strains of SV40. A number of sequence features are evident in the DNA and these correspond to functional units as indicated. Sites 1, 2 and 3 are three DNA regions, each containing multiple binding sites for large T antigen.

by its product, large T antigen, when this is present at low concentrations. However, in larger amounts, large T antigen inhibits early gene expression by binding to sites 1, 2 and 3 in its own promoter, contributing to the transition from early to late gene expression; binding to site 2 is also required to initiate DNA replication (see Section 7.2). The early promoter is controlled by an enhancer, the first of this type of control element to be described. The late promoter is activated by DNA replication and by

T antigen, which acts here through its effects on host cell transcription factors. As well as its functions in DNA replication and in controlling viral gene expression, large T antigen (together with small t) is able to alter host gene expression and hence activate cell cycle progression, driving cells into S phase. These effects on cell cycle regulation are considered in Section 25.3.

A further layer of regulation of SV40 early mRNA levels is provided by a virus-encoded

---

Box 10.3

Features of the SV40 control region

- The early promoter includes a TATA box which recruits general transcription factor TFIID, and binding sites for the host cell transcription factor Sp1 within the 21 bp repeats.
- The late promoter lacks a TATA box and hence there are start points for transcription spread over some distance on the genome.
- The enhancer comprises the 72 bp repeats plus some adjacent sequences, but only one of the two repeat copies is needed for activity. Wild strains of SV40, in contrast to laboratory-adapted isolates, only contain one copy of this repeat element.
- The inverted repeat at the centre of the origin of replication, with its four binding sites for T antigen, is essential for DNA replication.
- Progeny genome packaging is initiated by an interaction between capsid proteins and Sp1 bound to the 21 bp repeats.

microRNA. MicroRNAs are short (~21bp) and are produced by cleavage of short hairpin RNA from precursor RNAs that are transcribed by conventional RNA polymerases. Such RNAs are a relatively recently discovered feature of eukaryotic gene expression and are now known to be crucial regulators of gene expression networks. They act on mRNAs to which they have base-pairing complementarity, causing translational arrest or mRNA cleavage and degradation. The 3' untranslated region (UTR) of the SV40 late mRNA encodes a microRNA that, because of the overlap of early and late genes (Fig. 10.1), has perfect complementarity with the 3'UTR of SV40 early mRNA. The microRNA mediates cleavage of SV40 early mRNA to further downregulate T antigen expression during the late phase of infection.

10.3 Papillomaviruses

The papillomaviruses all infect epithelial surfaces in the body. They show phasing of gene expression not only through the time course of virus reproduction but also spatial phasing of expression through the different layers of the epithelium. The mechanisms of gene expression are, like those of SV40, similar to those of the host organism. Some human papilloviruses are significant causes of certain cancers.

The first papillomavirus to be studied in detail was bovine papillomavirus (BPV) type 1. More recently, improved study methods have allowed the focus to shift to the various human papillomaviruses (HPV) that are of disease interest. Details of the gene expression of these viruses were much harder to work out than those of the polyomaviruses because of the difficulty of growing them in cell culture systems (Box 10.4). A simplified transcription map of HPV18, as an example of the human papillomaviruses, is shown in Fig. 10.5. As with the polyomaviruses, all transcription occurs in the nucleus mediated by host RNA pol II; however, a striking difference from SV40 is that on the slightly longer genome (~8 kbp) the genes are all organized in the same orientation, i.e. only one genome strand is transcribed. The HPVs encode probably 9 proteins and mRNA for these is produced from two major promoter regions. In HPV18 these are P_{55} (early) and P_{811} (late); the positions vary slightly between HPV types. Alternative splicing and polyadenylation are used to give multiple mRNAs from each

Box 10.4

Systems for studying papillomavirus gene expression

- Transfection of cloned DNA into permanent cell lines has revealed the mRNAs expressed in the early phase of gene expression.
- Analysis of RNA from wart tissue identified novel mRNAs made only during productive infection.
- In situ hybridization to sections through a wart showed that so-called early mRNAs are produced in different regions of the lesion from late mRNAs.
- The full productive lifecycle of papillomavirus can now be achieved, on a small scale, in epithelial raft cultures (cells grown on a solid support at an air-medium interface that form a pseudo-epithelium).

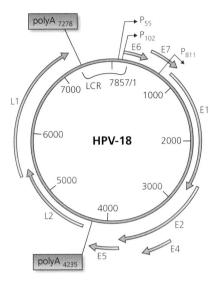

Fig. 10.5 The genome organization of human papillomavirus type 18. The curved arrows E1, L2, etc. represent open reading frames. Reading frames were originally designated E or L based on the observation that a 69% fragment of the BPV1 genome was sufficient for cell transformation *in vitro* and was hence defined as the early region by analogy with SV40. However, within the early region, the E4 gene is now known to be expressed primarily within the late temporal class of mRNA and E1 and E2 are expressed in both phases. Primary transcripts can undergo differential splicing to produce multiple mRNAs (not shown) that in some cases fuse parts of reading frames together. The region lacking open reading frames is known as the long control region (LCR). P_{55}, etc.: RNA polymerase II promoters; poly A: sites of mRNA polyadenylation.

primary transcript; for example, the late transcripts are spliced to encode three different proteins (E4, L1 and L2).

Papillomavirus gene expression is regulated by the full-length E2 protein, E2TA, which dimerizes to form a DNA-binding transcription factor; this same protein also facilitates E1 protein binding to DNA during replication (see Chapter 7). There are E2-dependent enhancer elements in the long control regions and

associated with the viral promoters (Box 10.5). As well as E2TA, some papillomaviruses also encode truncated forms of E2 that contain only the C-terminal half of the E2TA protein. As originally shown for BPV1, such truncated forms can form homo- and heterodimers with each other or E2TA; these retain specific DNA binding activity but lack the transcription activation function of the E2TA dimer. Thus, the production of truncated forms of E2 can modulate the level of E2-mediated transactivation. HPV E2 is an important regulator of gene expression from the viral early promoter, particularly controlling E6 and E7 protein levels. Under normal conditions in vivo, HPV E2 appears initially to stimulate early promoter activity but then promotes production of late mRNA at the expense of early gene expression, so down-regulating E6 and E7 expression.

The biology of papillomavirus gene expression is particularly fascinating. These viruses cause warts, i.e. benign growths in epithelia, and expression of their early and late genes is separated in time and space. In infected epithelia, early genes are expressed in the dividing cells of the basal cell layer and, as the cells divide, the viral DNA is maintained at an approximately constant copy number per cell by limited replication. The E6 and E7 proteins act to alter cell cycle control within these cells using mechanisms similar to those shown by SV40 large T antigen (see Section 25.3). The late events (accelerated DNA replication and late protein synthesis) only begin once an infected cell has left this layer and is committed to terminal differentiation. Thus the pattern of expression of the virus genes is determined by the differentiation state of the host cell. The factors which turn on this so-called vegetative phase in the growth cycle are not known. The human papillomaviruses are also of interest for their involvement in the development of certain human cancers (see Chapter 25).

Box 10.5

Evidence that the BPV E2 protein is a sequence-specific transcription factor

Reporter gene assays test for promoter or enhancer activity in pieces of DNA, by linking them in a plasmid with the coding sequence of a protein that is easy to assay for, and measuring the amount of protein produced when the plasmid is introduced into cells.

- Using the long control region (LCR) DNA of BPV as an enhancer to increase expression of the reporter gene chloramphenicol acetyl transferase (CAT) from a very weak promoter, significant CAT expression was only found when the cells also contained BPV DNA, i.e. the LCR could act as an enhancer provided one or more proteins from BPV was present.
- Using a series of plasmids in place of the complete BPV DNA, each designed to express just one of the BPV proteins, the E2 gene alone was found to cause upregulation of CAT expression.
- The response of the CAT reporter construct to E2 protein was shown to depend on a specific sequence motif in the LCR by creating a series of LCR fragments containing different sequence deletions and testing each one for enhancer activity in the assay.
- Purified E2 protein produced in bacteria was found to bind specifically to the sequences from the LCR already shown to be crucial for enhancer activity.

10.4 Adenoviruses

Studies of adenoviruses have told us a great deal about the workings of mammalian cells. The first description of RNA splicing was the recognition that the 5′ region of mRNA from the adenovirus major late transcription unit was not collinear with the genome. Detailed study of the E1A proteins has been crucial to shaping our understanding of how transcription can be controlled in eukaryotes. Mechanisms of adenovirus gene expression resemble those of the host, but differential RNA processing is taken to an extreme, with large numbers of distinct mRNAs and, hence, proteins arising from most of the transcription units.

The >50 distinct human adenoviruses are divided into six subgroups or species A–F. Their gene expression has been best studied in the closely-related species C serotypes 2 and 5. In these viruses, transcription occurs from both strands of the 36 kbp linear genome. All mRNA is produced by host RNA polymerase II; there are also two short RNAs (the VA RNAs) produced by RNA pol III. Gene expression shows temporal regulation (Fig. 10.6a) and the terms early (pre-DNA synthesis) and late (post-DNA synthesis) were used originally to classify the genes as E or L. However, it is now appreciated that expression of the E1A gene commences before the other early genes, and so should be classified as immediate-early, and that the small IVa2 and IX protein genes as well as two proteins from the 'late' L4 region form an intermediate class with roles in initiating full late gene expression. The observation which defines E1A as 'immediate-early' is that, unlike the other early genes, it is transcribed in an infected cell when protein synthesis is blocked

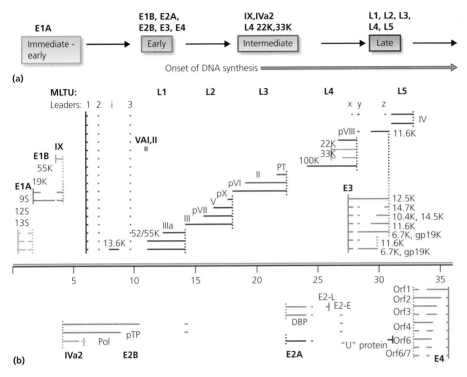

Fig. 10.6 Gene expression by adenovirus type 5. (a) The phases of gene expression. The numbers E1A, L1, etc. refer to regions of the viral genome from which transcription takes place. (b) A transcription map of the adenovirus 5 genome. The genome is represented at the centre of the diagram as a line scale, numbered in kilobase pairs from the conventional left end, with rightwards transcription shown above and leftwards transcription below. Genes or gene regions are named in bold type. Promoters of RNA polymerase II transcription are shown as solid vertical lines and polyadenylation sites as broken vertical lines. VAI and VAII are short RNA polymerase III transcripts. Individual mRNA species are shown as solid lines, colour-coded according to the temporal phase of their expression (panel a), with introns indicated as gaps. The protein(s) translated from each mRNA is indicated above or adjacent to the RNA sequence encoding it. Structural proteins are shown by roman numerals (major proteins: II - hexon; III – penton base; IV - fibre; pVII – core; see Chapter 12). PT – 23K virion proteinase; DBP – 72 K DNA binding protein; pTP – terminal protein precursor; Pol – DNA polymerase. (Modified and reproduced with permission from Leppard K. N. (1998), *Seminars in Virology* **8**, 301-307. Copyright © Academic Press.)

by a chemical inhibitor. This result also shows that E1A proteins are needed for expression of the remaining viral early genes. Controlling promoter activity is just one of the ways in which the pattern of gene expression is modulated during adenovirus infection. Some of this complexity is summarized in Box 10.6, although a detailed discussion is beyond the

scope of this book. Overall, these mechanisms achieve production of each of the viral proteins at the time, and in the amount, it is required.

The full gene expression map of adenovirus 5 is complex and can appear daunting (Fig. 10.6b). However, the key principle is that a relatively modest number of transcriptional promoters drive expression of a much larger

Box 10.6

Control of adenovirus gene expression at multiple levels

- Transcription is regulated so it takes place in an ordered sequence. The E1A proteins, especially the 13S mRNA product (Fig. 25.3), activate expression of the remaining genes, but significant transcription from the intermediate and late genes awaits the onset of DNA synthesis. Host transcription factors are also important.
- RNA processing is regulated so that the relative proportions of the possible mRNA products of a gene change as the infection proceeds. For example, when the major late gene is first transcribed, RNA processing produces L1 52/55 K mRNA almost exclusively while, later on, mRNA is produced from all of the regions L1–L5 and the predominant L1 mRNA is that which encodes the IIIa protein. L4 22 K & 33 K intermediate proteins are needed for this change in late RNA processing.
- Movement of mRNA within the cell is regulated so that viral mRNA reaches the cytoplasm in preference to cellular mRNA. Early proteins from E1B and E4 are required to achieve this.
- Translation is switched from a cap-dependent to a cap-independent mode during the late phase of infection through inactivation of cap recognition. This shuts off host protein synthesis whilst permitting continued viral late protein production since the late viral mRNAs, although capped, can recruit ribosomes in a cap-independent manner due to features of their 5' untranslated regions. These events depend on the L4 100 K protein.

number of mRNAs (which then encode different proteins) through differential splicing and polyadenylation. In other words, several distinct open reading frames are arranged downstream of a single promoter and, after transcription occurs, splicing is used to bring a downstream reading frame up to a 5' proximal position in the RNA, so that it can be translated (according to the normal rules of translation in eukaryotes, only the 5' proximal open reading frame in a mRNA is expressed). Picking out some of these features in Fig. 10.6b in detail, each of the five early (E) regions has its own promoter. Two of them (E2 and E3) have two alternative polyadenylation sites A and B, leading to two families of mRNA in each case, while transcription from the major late transcription unit (MLTU) produces five families of mRNA (L1–L5), as a result of polyadenylation at any one of five possible sites.

At the same time, alternative splicing within each mRNA family produces multiple mRNAs, each ending at the same polyadenylation site. Except for E1A, where the principal products are closely related in sequence (see Section 25.3), these related mRNAs generally encode unrelated proteins.

The multiple proteins encoded by an adenovirus gene region tend to have related functions. The E2 region, for example, encodes the three viral proteins which are directly involved in DNA replication (see Section 7.4), while splice and polyA site choice within the five families of MLTU mRNA allows synthesis of at least 14 different proteins, most of which are involved in forming progeny particles. The E3 region is the most variable when different adenoviruses are compared. It encodes proteins involved in avoiding aspects of the host immune response.

RNA processing from the MLTU is particularly wasteful of the cell's resources; every late mRNA, each one 1–5 kb long, is made from a precursor up to 28 kb in length. There is no possibility of making multiple mRNAs from the same precursor molecule as all the mRNAs include the same three RNA segments from the precursor's 5′ end; these are spliced together to form the so-called tripartite leader sequence. Thus, a lot of newly-synthesized viral RNA, both from the major late gene and other genes, is removed as intron sequence and degraded. However, the advantage of this system is that the virus can achieve coordinated expression of functionally-related proteins using a minimum of genome space to direct transcription.

10.5 Herpesviruses

Herpesviruses, including the eight known human viruses, all establish lifelong latent infections in specific tissues. To achieve this, they have to have two alternative programmes of gene expression, for lytic infection and latency, both of which employ host mechanisms for mRNA production but with very little splicing involved in the lytic programme. Whilst the productive or lytic programme is probably quite similar for all herpesviruses, the latent programmes are completely different. The fact that reactivation of herpesvirus infections, such as the periodic appearance of cold sores due to herpes simplex virus type 1, is a significant clinical problem means it is important to try to understand how these programmes of gene expression are controlled. The process of latency is considered in Chapter 17.

Herpes simplex virus types 1 and 2

Herpesvirus genomes are very much larger than those of adenoviruses and polyomaviruses, and show considerable diversity in length (130 kbp–230 kbp) and, hence, gene content. However, there are conserved blocks of genes that are common among the herpesviruses and which are probably key genes in virus reproduction. Herpesvirus gene expression has been studied principally using herpes simplex virus type 1 (HSV1), which productively infects epithelial cells in vivo and shows a productive, cytolytic infection of cell lines in culture. Gene expression during HSV1 lytic infection is considered here while gene expression in relation to latency, which is very different, is considered in Chapter 17.

HSV1 has around 70 genes interspersed on both strands of its 153 kbp genome. An immediate contrast with the smaller DNA viruses is that transcription of most HSV1 genes is controlled separately by specific promoter elements and there is very little use of splicing during mRNA production. A further contrast is that about half of the genes can apparently be mutated without affecting the growth of the virus in cell culture. These dispensable genes are presumed to be important in vivo.

As with the other animal DNA viruses, HSV1 gene expression is arranged into temporal phases. These are generally termed α, β and γ corresponding to immediate early, early and late phases. α, β and γ genes are not grouped together but are found scattered throughout the genome. The α genes of HSV1 produce various regulators of gene expression, the β genes produce proteins required for DNA replication and the γ genes produce mainly structural proteins for progeny particle formation. Although expression of the γ genes accelerates with the onset of replication, it is in fact more-or-less independent of this process. Thus, when DNA synthesis is inhibited, both early and late transcripts are detected in the nucleus.

The different phases of HSV1 gene expression are linked through the production of proteins in one phase which activate the next (Fig. 10.7). Thus, in the same way that adenovirus E1A proteins activate other viral

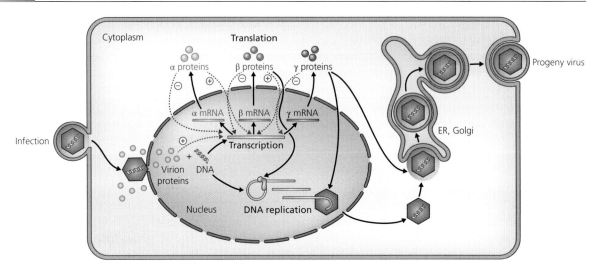

Fig 10.7   Gene expression in herpes simplex virus type 1 infection. The three successive phases of gene expression are denoted α, β, γ corresponding to immediate-early, early and late (see text). Solid arrows indicate the flow of material. Dashed arrows indicate regulatory effects on gene expression; + indicates activation; - indicates repression.

genes, so an HSV1 α gene product, known variously as ICP4 or IE175, turns on transcription of the β genes. One or more β gene product(s) then inhibit expression of α genes and induce γ gene expression, proteins from which inhibit expression of the β genes. A particularly interesting feature of this cascade of regulated HSV1 gene expression is that it is

circular. A protein produced during the late phase of infection (VP16, a γ gene product) is packaged into the particle, in the layer between the capsid and the envelope that is known as the tegument, and serves as an activator of α gene expression in the next round of infection (Box 10.7). Overall, this pattern of gene expression, with a series of genes controlled by

---

Box 10.7

Action of the tegument protein VP16 in activating herpes simplex virus α genes

- VP16 has a potent transcriptional activation function when recruited to a promoter.
- Two host proteins, Oct1 and HCF, are required for VP16 to upregulate α gene expression.
- HCF associates with VP16 and mediates its transit into the nucleus.
- Oct1 is a DNA binding transcription factor that binds to α gene promoters adjacent to a VP16 binding site. VP16/HCF complex can only bind DNA in the presence of Oct1.
- Why the virus has evolved this indirect mechanism for activation by VP16 is uncertain; it may relate to the switch into and out of latency.

activators and inhibitors, ensures that the relevant proteins are present when required and so enhances the efficiency of the virus replication cycle.

The tegument of HSV1 also carries at least one other active protein, UL41 (also known as vhs, standing for virion host shutoff). This protein is a ribonuclease which degrades cytoplasmic mRNA that is being translated. It is not specific for host mRNA, but serves to decrease the half-life of all mRNA in the cell, so favouring the expression of protein from genes that are being actively transcribed, i.e. viral genes. Later in the infection, another more specific host shut-off function is expressed by the virus while the large amounts of UL41 that the ongoing infection produces are kept inactive via binding to VP16, so allowing late viral mRNA to accumulate.

Epstein-Barr virus

Like HSV1, Epstein-Barr virus (EBV) productively infects epithelial cells. However, it establishes latency in B lymphocytes rather than sensory neurons (see Sections 15.3 and 17.4). The molecular events of the productive phase of the EBV life cycle are poorly characterized, because productive infection is hard to achieve in cell culture; however, they are believed to resemble those defined for HSV1. In contrast, the pattern of EBV gene expression during latency is well defined.

10.6 Poxviruses

Poxviruses replicate in the cytoplasm and are independent of the host for mRNA production. They use host ribosomes for protein synthesis.

Unlike the other families of DNA viruses of eukaryotes discussed in this chapter, poxviruses multiply in the cytoplasm. The molecular details of their gene expression have been studied using vaccinia virus. Its linear double-stranded genome (190 kbp) encodes a large number of proteins and these allow the virus to replicate with a considerable degree of autonomy from its host. It appears that viral gene expression is totally independent of the host nucleus since infection can proceed in experimentally enucleated cells. This cytoplasmic lifecycle requires that the infecting virion carries the enzymes needed for mRNA synthesis, including a DNA-dependent RNA polymerase and enzymes that cap, methylate and polyadenylate the resulting mRNAs. Splicing activities are not known to be used by these viruses.

Using the enzymes from the infecting particle, viral transcription begins in cytoplasmic replication 'factories'. Only a subset of viral genes (the early genes) is transcribed initially. Early gene products include proteins needed for replication and to activate the intermediate genes, which then activate late genes. Thus, infection leads to a phased sequence of gene expression punctuated by viral DNA synthesis, much as is seen for the other DNA viruses. Independence from the cell's own transcription machinery allows the virus the opportunity to shut the cell nucleus down, so that all the metabolic resources of the cell are devoted to the virus. Thus, unlike the other families of DNA viruses considered here, all of which can establish chronic, persistent or latent infections in vivo, poxvirus infection is exclusively cytolytic. Poxviruses make a large array of proteins involved in modulating the host immune response to infection.

10.7 Parvoviruses

Adeno-associated virus (AAV) is the best characterized parvovirus. Its linear single-stranded genome, once converted to double-stranded form (see Section 7.6) contains three

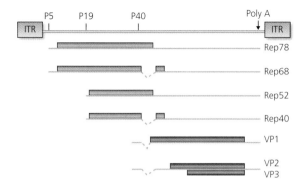

Fig. 10.8 Adeno-associated virus gene expression. The 4.6 kb genome is represented as a line scale, with mRNAs beneath it with $5' \rightarrow 3'$ polarity from left to right. ITR – inverted terminal repeat (see Section 7.6). Encoded proteins are shown as coloured boxes: blue – early; red - late. See Section 10.7 for further details.

promoters, each of which produces at least two related mRNAs by differential splicing (Fig. 10.8). The proteins produced by expression from P_5 and P_{19} (the Rep proteins) are all sequence-related, and can be regarded as early proteins because they activate the third promoter, P_{40}. The unique functions of the Rep proteins have not yet been fully defined but include the site-specific DNA cleavage activity which is essential to genome replication (see Section 7.6). The P_{40} mRNAs code for the structural proteins, VP1, 2 and 3, from which new virions are assembled. These three proteins are all sequence-related, the major capsid protein being VP3. This protein is produced by internal initiation at an AUG codon within the shortest mRNA. Read-through to this initiation point occurs in the majority of translation events because the alternative initiation event, specifying VP2, uses a non-canonical ACG codon. As noted in Section 7.6, AAV requires a helper virus for growth under most circumstances and can establish latency when such help is not available. However, although

transcriptional activators from the helper (such as adenovirus E1A, Section 10.4) can upregulate AAV gene expression and other proteins may facilitate gene expression at a post-transcriptional level, there does not appear to be an obligatory requirement for any helper function to achieve AAV gene expression (see Section 7.6).

10.8 Retroviruses

Retroviruses are RNA viruses but, having converted their genomes into a DNA form, the mechanisms of their mRNA production exactly resemble those of the host and, hence, many DNA viruses. The principal proteins, common to all retroviruses, are made as polyproteins and cleaved later into their functional components. Translation of the polyprotein that includes the crucial enzymes for reverse transcription and integration of the proviral DNA requires the ribosome to make a programmed 'error' that causes it to avoid a stop codon.

Retroviral gene expression occurs exclusively from the provirus, a DNA copy of the viral RNA genome (typically 8–10 kbp), which is normally integrated into the host chromosome (see Chapter 9). The promoter for transcription lies in the upstream LTR of the provirus, and this is regulated by a variety of cell transcription factors. In more complex retroviruses, such as human immunodeficiency virus, there are additional virus-coded regulators of gene expression (see Chapter 21). The mechanism of retrovirus mRNA production exactly mirrors the process for any host cell gene, i.e. it involves host RNA pol II, capping, splicing and polyadenylation functions. Subsequently, mRNAs are transported to the cytoplasm where they are translated. Unlike the true DNA viruses, simple retroviruses have no temporal

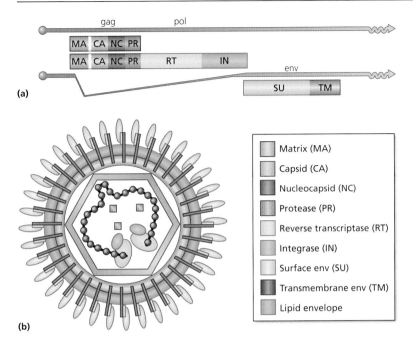

(a)

(b)

Fig. 10.9 (a) The mRNAs and proteins synthesized in avian leukosis virus-infected cells. In non-avian retroviruses, the protease, PR, is encoded as part of Pol and so is only present in the Gag-Pol fusion protein (see text for details). (b) The location of retroviral proteins within the mature virion. The role of the protease in the maturation of retrovirus particles is considered in Section 12.8. Note that the complexes of SU and TM in the envelope are actually trimers of each protein.

phasing to their gene expression. However, the additional functions of complex retroviruses do allow them to control their gene expression in this way.

The mRNAs and protein-coding strategy for a typical simple retrovirus, avian leukosis virus (ALV) are shown in Figure 10.9a. The two ALV mRNAs, produced from a start site in the 5′ LTR, share the same 5′ and 3′ end sequences. One is equivalent to the full-length genome and the other has a section removed by splicing. The spliced mRNA has *env* as its 5′ proximal open reading frame and is translated to produce the Env polyprotein. This is subsequently cleaved by a host protease into the associated surface (SU) and transmembrane (TM) components of the mature trimeric envelope protein during its

passage through the endoplasmic reticulum/ Golgi body. Spliced mRNAs are also used by complex retroviruses for expression of their extra genes which are located between *pol* and *env* and/or between *env* and the 3′ LTR (see Section 21.2).

The full-length ALV mRNA encodes both *gag* and *pol* gene products. This requires a special mechanism since translation in eukaryotes is normally initiated at the first AUG in the message, scanning from the 5′ end, and such a mechanism would produce exclusively Gag protein. In fact, no free Pol protein is produced by ALV and the protein is only synthesized fused to the C-terminus of Gag. The *pol* gene is in the −1 reading frame with respect to *gag*. Ribosomes translating *gag* avoid its termination

Box 10.8

Evidence for expression of a Gag-Pol fusion protein from avian retroviruses by ribosomal frameshifting

- The *pol* gene was known to be expressed as a C terminal fusion with the Gag protein, through characterization at the protein level, before the mechanism for achieving this was established.
- The idea that a spliced mRNA was produced that encoded the fusion protein was disproved by showing that RNA transcribed from Gag-Pol DNA in vitro (where no splicing machinery is present) and then translated in vitro produced both Gag and Gag-Pol proteins in a ratio similar to that seen in vivo. This result shows that the fusion arises as an option at the translation stage – i.e. ribosome frameshifting.
- A sequence common to the region of overlap between Gag and Pol reading frames in a number of retroviruses was shown to be necessary for Gag-Pol fusion protein expression by making point mutations in the sequence. This is the 'slippery sequence'.
- The amino acid sequence of the fusion protein exactly matched that expected for a -1 frameshift at the mapped sequence.
- Deletion/mutation analysis of the region downstream of the slippery sequence defined an RNA secondary structure element (the pseudoknot) as being required for frame-shifting at the slippery sequence.

codon in about 5% of translation events by slipping backwards one nucleotide, an event known as a ribosome 'frame-shift' (Box 10.8). Once this has occurred, the translation termination codon for the *gag* gene becomes invisible to the ribosome because it is no longer in the reading frame being used, so translation continues into the *pol* gene, producing a Gag-Pol fusion protein. The extremely high frequency of ribosome 'error' at this position at the 3′ end of *gag* is caused by a specialized sequence context. This comprises a 'slippery sequence', on which the frameshift event actually occurs, and a pseudoknot downstream, which causes the ribosome to pause on the slippery sequence and so promotes frameshifting. As a result, lower levels of *pol* gene expression are achieved than of *gag*, reflecting the relative amounts in which these proteins are needed to form progeny particles. This mechanism closely resembles that operating in coronaviruses (see Section 11.5). Other retroviruses face the same problem in achieving expression of their *pol* genes, but they find a variety of solutions. In the murine equivalent of ALV, known as MLV, the *gag* and *pol* genes are in the same frame separated only by a stop codon. This is misread by a specific loaded tRNA, producing a Gag-Pol fusion, again with about 5% efficiency.

For all retroviruses, the *gag* and *pol* genes code for polyproteins that together contain the proteins needed to build the internal parts of the virion (Fig. 10.9b). The protein encoded at the *gag-pol* boundary is a specific protease (PR), which is required to process the polyproteins in a post-assembly maturation event (see Section 12.8). In ALV, PR is coded at the C-terminus of *gag*, whereas in MLV it is at the N-terminus of *pol*. A further variant mechanism for *pol* expression then arises in other retroviruses, such as human T-cell lymphotropic virus

(HTLV), where the protease is encoded in a reading frame distinct from both *gag* upstream and *pol* downstream. One frame shift takes the ribosome from the Gag reading frame into the PR reading frame and a second shift takes it on into the Pol reading frame; three different polyproteins, Gag, Gag-PR and Gag-PR-Pol, are therefore produced. Finally, in the least well-characterized retrovirus genus, the spumaviruses, there is evidence for a separate spliced mRNA being used to encode Pol, so that no Gag-Pol fusion protein is made.

The spumaviruses actually represent an interesting intermediate state between standard retroviruses and the hepadnaviruses (see Section 10.9), based on features of both their genome replication (see Section 9.7) and their gene expression. As well as expressing Pol protein independently of Gag (in the same way as hepadnavirus P and C proteins are separate), they also make use of a second promoter to express two proteins from reading frames 3′ to *env*. Although other complex retroviruses have additional genes in this same position, they generally use splicing to express them from the same LTR promoter as is used for all gene expression in other retroviruses.

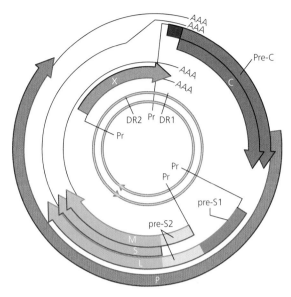

Fig. 10.10   Hepatitis B virus genome map. The closed circular DNA is transcribed from four promoters (Pr); arrowheads indicate 3′ ends. For details of protein expression, see text. Pre-S1, pre-S2 are the designations given to the upstream portions of the S reading frame which, when added to the N-terminus of S, give rise to the longer L and M surface proteins. DR1, DR2 are repeated sequences shown to orient the map (see Fig. 9.5).

10.9 Hepadnaviruses

Hepadnaviruses use host mechanisms for gene expression. They manage to make a lot of proteins from a very small genome by a combination of multiple promoters, overlapping coding regions and the use of alternative translation start codons on an mRNA.

The hepadnaviruses have very small circular genomes, e.g. hepatitis B virus, 3.2 kbp. The particular feature of note regarding the gene expression of this virus is the density at which information is compressed onto the genome,

with protein encoded by every part of the DNA and two unrelated proteins being encoded in alternative reading frames over a significant part of its length (Fig. 10.10). There are four RNA pol II promoters which produce pregenomic and three classes of subgenomic mRNA respectively. None of these mRNAs is spliced and all end at the same polyadenylation site. Transcription is regulated by cell-type specific factors and this is part of the basis for the tropism of this virus for the liver.

mRNAs from the four promoters encode seven proteins (Fig. 10.10). All except the X and pre-C proteins are structural components of the virion. Pre-C is secreted from the cell (this is known as 'e antigen') and probably serves to modulate the immune response to the virus, whereas X is a transcriptional activator that affects a wide range of viral and cellular promoters. The mRNAs transcribed from the genomic and S promoters show heterogeneous 5′ ends. This means that some molecules include the translation start sites for the pre-C and M proteins respectively, while the remainder do not. Translation from these latter mRNAs then begins at downstream AUGs, encoding the C (core) and S (surface) proteins. Synthesis of the P protein presents a problem since there is no mRNA in which its reading frame is proximal to the 5′ end. Although the position of the P reading frame, overlapping the C-terminus of the core protein sequence, is reminiscent of the retroviral arrangement of *gag* and *pol* genes, no C-P fusion protein is produced. Instead, a modified ribosome scanning mechanism is believed to allow a subset of ribosomes, which load onto the mRNA at its 5′ end, to initiate translation at the start codon for the P protein. Again, this achieves a lower level of P expression relative to C, reflecting the differing requirements for these proteins. In fact, translation of P only occurs once per mRNA since, once made, it binds to the RNA and commits that molecule to encapsidation as pregenome (Section 9.9).

10.10 DNA bacteriophages

As with the DNA viruses of eukaryotes, bacteriophage with DNA genomes show increasing sophistication of control of gene expression as the genome size increases, with temporal phasing of production of the proteins required early and late in the infection. Bacteriophage gene expression is controlled mainly at the level of transcription, since RNA processing is not a feature of prokaryotic gene expression and the short half-life of prokaryotic mRNA makes transcriptional control particularly effective. Regulated transcription of

phage genes can occur either through modification to the specificity of the host RNA polymerase, as occurs in infections by the *Escherichia coli* bacteriophage T4, or through the provision of a new phage-specified RNA polymerase, as in *E. coli* bacteriophage T7. The unique promoter specificities of the T7 RNA polymerase and similar enzymes from phages SP6 and T3 make them valuable tools for RNA transcription from cloned genes in the test-tube. One of the best studied examples of control of gene expression in a prokaryotic virus is bacteriophage λ, some information on which is provided in Chapter 17, focusing on lysogeny of this phage.

Key points

- Most DNA viruses of animals replicate in the nucleus and use host systems for mRNA production and protein synthesis.
- Those that replicate in the cytoplasm must provide their own enzymes for mRNA production.
- Retroviruses can be considered as DNA viruses when it comes to gene expression.
- Gene expression for all but some of the smallest DNA viruses is temporally phased, with different genes expressed at different stages of the infection.
- Where phasing occurs, proteins made early in an infection are required for DNA synthesis and to modify the host environment while proteins made later are directly involved in progeny particle formation or facilitate particle assembly.
- The polyoma, papilloma, adeno, parvo and retroviruses all make extensive use of RNA splicing in the expression of viral genes.
- Herpesviruses make little use of RNA splicing for lytic gene expression but extensive use of splicing during latency.
- Poxviruses and other cytoplasmic DNA viruses do not use RNA splicing.

Questions

- Discuss the strategies employed by animal DNA viruses to maximize the coding capacity of their genomes for the production of viral proteins.
- Using appropriate examples, explain how many viruses achieve a temporally regulated pattern of gene expression over the course of an infection.

Further reading

Berk, A. J. 2005. Recent lessons in gene expression, cell cycle control, and cell biology from adenovirus. *Oncogene* 24, 7673–7685.

Flint, J., Shenk, T. 1997. Viral transactivating proteins. *Annual Reviews of Genetics* 31, 177–212.

Johansson, C., Schwartz, S. 2013. Regulation of human papillomavirus gene expression by splicing and polyadenylation. *Nature Reviews: Microbiology* 11, 239–251.

White, M. K., Safak, M., Khalili, K. 2009. Regulation of gene expression in primate polyomaviruses. *Journal of Virology* 83, 10846–10856.

Wysocka, J., Herr, W. 2003. The herpes simplex virus VP16-induced complex: the makings of a regulatory switch. *Trends in Biochemical;1; Sciences* 28, 294–304.

Gene expression of retroviruses and hepadnaviruses is considered in the further reading suggested in Chapter 9.

Some aspects of gene expression for various viruses covered in this chapter are included in the further reading suggested in Chapter 25.

Chapter 11

The process of infection: IIIB. Gene expression and its regulation in RNA viruses

The presence of RNA as genome means that the process of gene expression by RNA viruses differs from that of the host and of DNA viruses. However, once an RNA genome has been transcribed to form positive sense mRNA, this is then translated into proteins using host cell ribosomes just as for DNA viruses. As for all systems, including those of the normal cell, the virus proteins produced by translation may be subject to modification that activates or deactivates them and the steady-state levels of the virus proteins are also affected by different rates of degradation; for instance, it is known that the virus-specified polymerase activity is unstable.

RNA viruses control their gene expression using a number of processes.

All RNA viruses synthesize their mRNA from a template consisting of a negative sense RNA molecule; those from Baltimore class 3 use the negative sense RNA of the double-stranded genome as template while class 4 viruses must replicate their genomes to generate a replicative intermediate (RI; Section 8.4) which contains the negative sense RNA template, before transcription can occur. Class 5 viruses can use the genome RNA directly as a template for transcription. Excepting viroids (Section 8.6), host cell enzymes cannot transcribe an RNA template, so all RNA viruses must encode their own RNA-dependent RNA polymerases. Viruses of classes 3 and 5 carry the polymerase as an essential, internal component of the virus particle and introduce it into the infected cell so that transcription can begin immediately (Box 11.1).

Class 4 virus particles do not carry a polymerase; instead, the enzyme is generated by translation of the positive sense RNA

Introduction to Modern Virology, Seventh Edition. N. J. Dimmock, A. J. Easton and K. N. Leppard.
© 2016 John Wiley & Sons, Ltd. Published 2016 by John Wiley & Sons, Ltd.

Box 11.1

Evidence that viruses of Baltimore classes 3 and 5 carry a polymerase in the virus particle

- Treatment of cells with specific inhibitors of cellular RNA polymerases, such as actinomycin D, prior to and throughout infection with class 3 viruses does not inhibit virus RNA synthesis. Class 5 viruses which replicate in the cytoplasm are not affected by actinomycin D, though those replicating in the nucleus may be affected.
- Treatment of cells with inhibitors of translation, such as the drug cycloheximide, prior to and throughout virus infection, prevents synthesis of both host and virus proteins. However, protein synthesis inhibitors do not prevent class 3 and class 5 virus mRNA synthesis.
- For some viruses, purified virions can transcribe RNA *in vitro*.
  Taken together, these data indicate that the polymerase responsible for class 3 and class 5 virus transcription is not a cellular enzyme. Since no protein synthesis is required for virus mRNA synthesis to occur, the enzyme must be introduced in an active form by the virus at the time of infection.

---

genome. The enzymes which carry out transcription of RNA viruses are frequently referred to as 'transcriptases' to differentiate the process of mRNA formation from replication. However, as indicated in Chapter 8, both transcription and replication are carried out by the same enzyme for each virus, though the enzymes may be modified in different ways to function in the two separate processes. Messenger RNAs are synthesized and translated in a 5′ → 3′ direction. Consequently, their 5′ termini are of some interest, particularly with regard to their ability to regulate gene expression.

11.1 The RNA viruses: Baltimore classes 3, 4 and 5

Class 3 viruses have a dsRNA genome.
Class 4 viruses have a positive sense ssRNA genome.

Class 5 viruses have a negative sense ssRNA genome.

In general, temporal control of gene expression in RNA viruses, if present, is not usually as pronounced as it is for many DNA viruses. Commonly, the difference in gene expression is a quantitative one, with all genes expressed simultaneously but some more abundant than others at different times, and where there is true temporal regulation, such as with the alphaviruses, the early genes are also expressed at late times during infection. Unlike the DNA viruses, and despite the necessity of using the host cell ribosomes for translation, many RNA viruses produce mRNA which is either not capped, or not polyadenylated, or neither capped nor polyadenylated. The lack of these structures, particularly the 5′ mRNA cap that is normally crucial for translation initiation, requires the virus to use novel mechanisms to ensure translation of its mRNA.

While most RNA viruses replicate in the cytoplasm of the host cell, some do so in the nucleus. Viruses cannot deviate from this 'choice'. Since, unlike the DNA viruses, RNA viruses do not use the host cell DNA-dependent RNA polymerase, there are other reasons why some RNA viruses replicate in the nucleus, which are explained below. Despite the fundamental difference in the nature of the template nucleic acid, control of gene expression by RNA viruses frequently shows similarities with that seen in DNA viruses.

11.2 Reoviruses

Transcriptional regulation of gene expression

The genome of mammalian reoviruses, the type member of the family of reoviruses, is double-stranded RNA. While much of our understanding of the basic features of gene expression of this family of viruses has come from the study of mammalian reoviruses, in recent years most attention has turned to the study of rotaviruses due to their disease-causing capacity and medical and agricultural importance and we will consider the rotaviruses here. Electrophoretic analysis of RNA extracted from rotavirus particles shows that the dsRNA genome consists of 11 different segments of RNA ranging in size from approximately 3300 bp to approximately 660 bp. Nucleotide sequence analysis of all 11 segments has confirmed that they are unique and so could not have arisen by random fragmentation of the genome. Biochemical studies and analysis of virions suggest that each virus particle contains one copy of each segment. Each segment is perfectly base-paired, with no overlapping single-stranded regions at either end of the molecule.

By analogy with the mammalian reoviruses, it was anticipated that each rotavirus genome

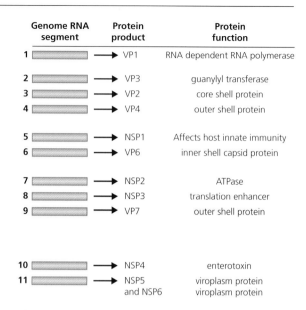

Genome RNA segment	Protein product	Protein function
1	VP1	RNA dependent RNA polymerase
2	VP3	guanylyl transferase
3	VP2	core shell protein
4	VP4	outer shell protein
5	NSP1	Affects host innate immunity
6	VP6	inner shell capsid protein
7	NSP2	ATPase
8	NSP3	translation enhancer
9	VP7	outer shell protein
10	NSP4	enterotoxin
11	NSP5 and NSP6	viroplasm protein viroplasm protein

Fig. 11.1 Genome organization and coding assignment of rotavirus. Post-translational modification is needed in many cases to produce the functional protein. Segment 11 also encodes an additional protein.

segment would direct synthesis of a single gene product. This was confirmed for all but one genome segment that encodes NSP5 and NSP6. Additional proteins were shown to be generated by post-translational modification of the precursors generated from several segments. The rotavirus genome organization, with the proteins encoded by each segment, is summarized in Fig. 11.1.

The mechanism by which viruses with dsRNA genomes initiate the expression of their genome presented a problem to early investigators since it was known that double-stranded RNA could not function as mRNA. The first virus mRNAs in the cell therefore must result either from transcription of the parental genome or by separation of the two strands. However, attempts to demonstrate strand separation were unsuccessful, and all cellular polymerases tested were unable to transcribe the dsRNA genome. This was not unexpected as all mammalian reovirus family members replicate exclusively in the cytoplasm, while host cell RNA polymerases are confined to the nucleus.

A dsRNA-dependent RNA polymerase has now been identified as an integral part of the virion particle for all class 3 viruses. The polymerase is inactive in intact virions, but becomes active during the uncoating of the virion in the infected cell. The activation process is not the result of a modification of the inactive enzyme but is due to removal of a suppression imposed by the structure of the intact virion as it loses structural proteins during the uncoating process. The polymerase is probably a multi-subunit complex intimately associated with the structure of the particle core, since it has proved impossible to separate the activity from the virus particles.

Rotavirus transcription uses only one strand of RNA as template, and the mRNA produced is *exactly* the same length as the genome RNA. dsRNA added to actively transcribing preparations is not recognized by the polymerase, suggesting that the genome is transcribed while still within the virus core. Newly-synthesized mRNA leaves the cores of the virions which initiate the infection (the 'parental' cores) by way of channels through the external protein spikes. This process has been visualized by electron microscopy for reovirus and has been clarified using cryoelectron microscopic reconstructions of rotavirus particles (Fig. 11.2).

In prokaryotes, and viruses of prokaryotes, the 5′ end of many mRNAs is a triphosphorylated purine corresponding to the residue that initiated transcription. In contrast, most eukaryotic cellular and viral mRNAs, as well as native nucleic acid from some RNA viruses of eukaryotes, are modified at the 5′ end by the addition of a so-called 'cap' structure, the general structural features of which are shown in Fig. 11.3. The terminal 7-methylguanine and the penultimate nucleotide (the first nucleotide

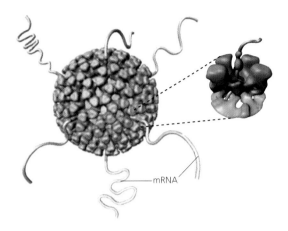

Fig. 11.2   Cryoelectron microscopic reconstruction of a rotavirus particle from an infected cell showing mRNA leaving through structures at the vertices of the icosahedral virion. (Taken from Pesavento et al., 2006, *Current Topics in Microbiology and Immunology* 309, 189–219. Reproduced with permission.)

copied from the genome template) are joined by their 5′ hydroxyl groups through a triphosphate bridge. This 5′–5′ linkage is inverted relative to the normal 3′–5′ phosphodiester bonds in the remainder of the polynucleotide chain and is formed post-transcriptionally.

The 'cap' protects the RNA at its 5′ terminus from attack by phosphatases and nucleases and promotes mRNA function at the level of initiation of translation. In eukaryotic cells, translation of mRNA is 'cap-dependent', i.e. only RNA molecules with a 5′ cap are accepted by ribosomes for translation. The functional ribosome contains a number of initiation factors and one of these, eIF4E, binds to the cap structure and brings mRNA into the translation complex in association with other proteins, including the factors eIF4GI and eIF4GII. If the cap structure is not present the RNA is not recognized as a potential mRNA. The normal cellular capping enzymes are found only in the nucleus but the rotavirus particle contains

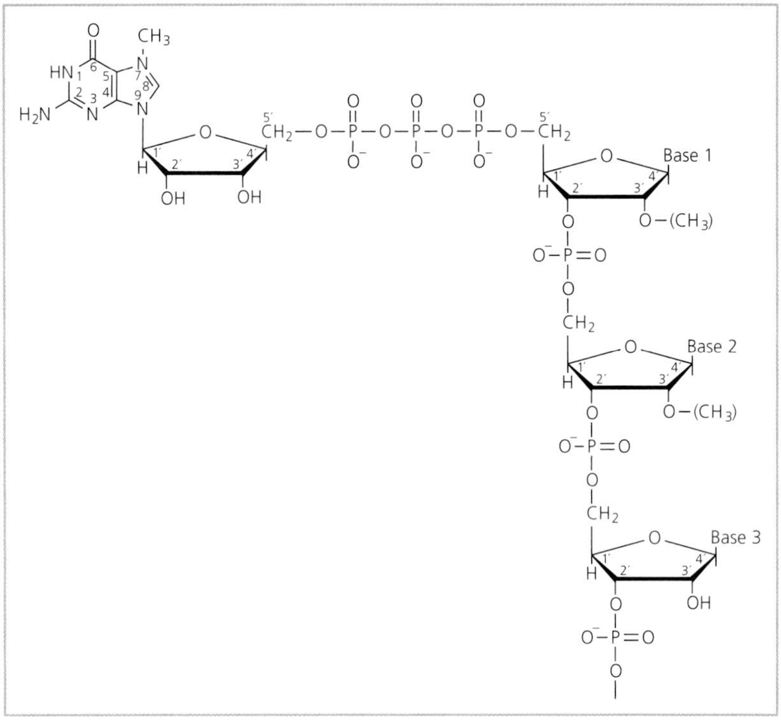

Fig. 11.3 Structure of a capped RNA molecule. Note the 5′–5′ phosphodiester linkage. Bases 1, 2 and 3 can be any of the four nucleic acid bases. If the 2′ positions of the ribose of bases 1 and 2 are not methylated the structure is referred to as a cap 0 structure, if only the 2′-O-methyl group on the ribose of base 1 is present it is called a cap 1 structure, and if a 2′-O-methyl group is present on both bases 1 and 2 it is a cap 2 structure.

several viral methylase activities which carry out this reaction. Accordingly, rotavirus mRNAs, made either from purified virions or during infection, are modified at their 5′ ends by the addition of a cap structure; however, they do not contain the 3′ poly(A) tail which is another feature of normal cellular mRNAs that is needed for efficient translation.

In contrast to the otherwise similar rotaviruses, in mammalian reoviruses the capping enzyme that is synthesized during infection is not active in the newly-generated immature particles and, as a result, the virus mRNAs they synthesize do not contain a cap structure at the 5′ end. Such mRNAs would compete poorly with host mRNA for the translation apparatus so reoviruses adapt the translation apparatus from being cap-dependent to become cap-independent; this favours production of virus proteins. One of the reovirus proteins specifically inactivates the host cap-binding protein complex which forms part of the functional ribosome. This has the result that the affected ribosomes can now translate only uncapped mRNA. (Box 11.2).

Following replication of the rotavirus genome (see Chapter 8), new virus cores containing one of each of the 11 genome segments are generated in the cytoplasm and these also synthesize mRNA from all segments

Box 11.2

Evidence that the translation process in reovirus infected cells is altered from cap-dependent to cap-independent

During translation the initiation factor eIF4E binds to the 5′ cap structure found on cellular mRNAs. The molecule 7-methyl guanosine triphosphate (m7G(5′)ppp) is an analogue of the cap structure and binds to eIF4E, preventing cap-dependent translation by competition. Using cellular lysates from uninfected and reovirus-infected cells to translate capped or uncapped mRNA in the presence or absence of m7G(5′)ppp, the requirement for a cap during translation can be assessed. The data below summarize this experiment.

Source of extract	mRNA	m7G(5′)ppp present	Translation seen
Uninfected cells	None	No	No
	None	Yes	No
	Capped	No	Yes
	Capped	Yes	No
	Uncapped	No	No
	Uncapped	Yes	No
Infected cells	None	No	No
	None	Yes	No
	Capped	No	No
	Capped	Yes	No
	Uncapped	No	Yes
	Uncapped	Yes	Yes

These data show that in reovirus infected cells translation is cap-independent. When the lysates are prepared at various times after infection it can be seen that there is a gradual change from cap-dependence to cap-independence as the infection proceeds.

throughout the infectious cycle. The genome segments are not transcribed with equal efficiencies and, as a result, the mRNAs produced fall into three distinct abundance classes. The first class of mRNAs, represented by those transcribed from genome segments 1, 2, 6, 8 and 11 in bovine rotavirus, are produced and accumulate at a steady rate throughout infection. The mRNAs from bovine rotavirus genome segments 3, 4, 5, 9 and 10 are in the second class and are synthesized slowly at early times during infection and accumulate more rapidly towards the end of infection. Finally, mRNA from genome segment 7 accumulates rapidly in the early phase of the infection and reaches a plateau that is maintained for the remainder of the infection. The fact that these abundance differences are seen demonstrates that the genome segments are transcribed independently of each other. The mechanisms that control these different rates and patterns of transcription are not understood.

Translational regulation of gene expression

Although rotavirus mRNAs lack a 3′ poly(A) tail that is found in almost all cellular mRNAs and which is required for their efficient translation, they ensure that their mRNAs are translated more efficiently than cellular mRNAs in the infected cell. The rotavirus NSP3 protein binds to a short nucleotide sequence, UGACC, found at the 3′ end of all of the virus mRNAs, and also interacts with the host cell translation initiation factor eIF4GI complex that binds to the 5′ end cap of mRNAs prior to initiation of translation. This interaction replaces the one that normally occurs between poly(A) binding protein and eIF4GI and ensures that, like cell mRNA, the virus mRNAs are circularized for translation. The stronger affinity of NSP3 for eIF4GI actually prevents binding by the poly(A) binding protein and, as a result, cellular mRNAs are denied access to the translation machinery. Additionally, the NSP3 protein is involved in a poorly-understood process that localizes the poly(A) binding protein to the nucleus of the cell where it prevents export of cellular mRNA. In this way, the virus mRNAs are able to compete very efficiently with host mRNAs for the available protein synthesis capacity.

From the earliest sequence analyses of virus genomes it has been known that open reading frames (ORFs) may overlap each other such that the same RNA sequence encodes two or more distinct proteins, as seen in Section 10.2 for SV40 and polyomavirus. Similar features are seen with some RNA genome viruses. An exception to the one segment–one protein rule for rotavirus is seen with segment 11. Segment 11 contains a large ORF with an AUG translation initiation codon located near the 5′ end of the mRNA. This ORF directs the synthesis of the non-structural protein NSP5 of 197 amino acids. Further from the 5′ end is a second ORF which begins with another AUG codon and which is in a different reading frame

Fig. 11.4 Arrangement of the overlapping open reading frames of the rotavirus gene 11 mRNA. The relative locations and protein products of the two ORFs are indicated. The numbers refer to the position of the first or last nucleotide of the relevant codon within the mRNA, numbered from the 5′ end.

(Fig. 11.4). Occasionally, ribosomes which have failed to initiate translation at the first AUG do so at the second AUG, in a process called leaky ribosomal scanning, producing a protein of 92 amino acids called NSP6. Because NSP5 and NSP6 are encoded from different reading frames they are unique and do not share any amino acid sequences. The relative level of the two proteins is determined by the efficiency of initiation of translation at the AUG initiation codons of the two ORFs, with the 5′ proximal AUG being utilized most frequently. This arrangement ensures that less of the protein encoded by the second ORF, NSP6, is produced.

While the 11 rotavirus mRNAs are present in approximately equimolar amounts late in the infectious cycle after the differences in kinetics of expression have evened out, the proteins are nevertheless made in greatly differing amounts, from very low levels of polymerase to high levels of other, major structural proteins. This means that the various mRNAs must be translated with different efficiencies. How is this achieved? The sequence surrounding the AUG initiation codon for cellular mRNAs is important in determining the efficiency of translation. The consensus sequence, identified by Marilyn Kozak, which is most favoured for initiation of translation is G/AnnAUG(G), where n is any nucleotide, and mRNAs bearing sequences most closely matching this are most readily recognized by ribosomes. In the case of segment 11, the Kozak rule appears to apply in as much as the sequence surrounding the first AUG which initiates the NSP5 protein is not optimal. This counteracts the preference for ribosomes to initiate translation at the 5′ proximal AUG codon in mRNA, allowing some ribosomes to move to the second AUG to begin translation of the second ORF and produce the NSP6 protein, although the levels of production still do not equate to those of NSP5. However, for the other rotavirus mRNAs, while the sequences around the AUG initiation codons will play a role in determining efficiency of translation, this cannot explain the degree of variation in protein accumulation that is seen. It has been proposed that the different mRNAs are likely to have different affinities for limiting factors in the cell which are necessary for initiation of translation and that their ability to bind to these factors is the key determinant of how frequently each rotavirus mRNA is translated.

11.3 Picornaviruses

The picornavirus genome is a positive sense ssRNA molecule of approximately 7500 nt containing a single large ORF. It has a polyA tract at the 3′ end, resembling that seen in host cell mRNA, but the 5′ end has a small covalently attached protein, VPg, rather than a cap structure. The first event in infection following uncoating of the picornavirus genome is translation using host cell ribosomes. The presence of VPg and the absence of a cap means that the picornavirus genome RNA cannot be translated directly by host ribosomes as it cannot interact with the standard cap-binding protein which forms part of the translation initiation complex. To overcome this barrier to gene expression, picornavirus genomes contain a region of RNA, 5′ of the AUG initiation codon for the ORF, which adopts a specific 3-dimensional conformation that directs ribosomes to initiate translation; this event therefore occurs independent of a cap structure and internal to the mRNA. This region is

referred to as the internal ribosome entry site (IRES).

Since the virus does not depend on cap recognition for synthesis of its proteins, it has evolved functions that block cap-dependent translation, so diverting translation capacity to its own cap-independent protein synthesis. The host translation system is altered due to proteolytic cleavage and consequent inactivation of both eIF4GI and eIF4GII. This cleavage prevents the interaction of these factors with the cap binding protein eIF4E. The result of this inactivation is that, although an initiation complex can still form, it does not do so on capped mRNA, translation becomes cap-independent and the cell is unable to translate its own mRNA. As a result of these virus-induced changes to the translation machinery (together with virus-encoded cleavage inactivation of the host cell TATA-binding protein in the nucleus which arrests host cell polymerase II transcription) host cell protein synthesis is inhibited and only virus-specific synthesis takes place. Production of more mRNA after this initial round of translation is by replication of the genome as described in Section 8.4.

Control of gene expression by post-translational cleavage

The entire genome of picornaviruses is translated as a huge polyprotein approximately 2200 amino acids in length, which is cleaved in a series of ordered steps to form smaller functional proteins (Fig. 11.5), both structural and non-structural. Cleavage starts while the polyprotein is still being synthesized and is carried out by virus-encoded proteases that are active within the polyprotein but which also cleave themselves from the growing polypeptide chain. The best-known is protease 3C which is produced by autocatalytic excision when P3 achieves the required conformation. Only by inhibiting protease activity or altering the cleavage sites through the incorporation of

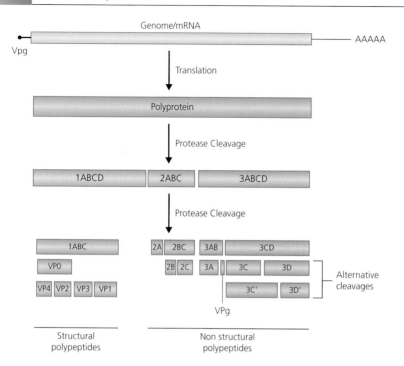

Fig. 11.5 Representation of the cleavages of the poliovirus polyprotein and the smaller products to yield mature viral proteins. The 3C′ and 3D′ proteins are produced by cleavage of 3CD at an alternative site to that producing 3C and 3D.

amino acid analogues that render the protease functions inactive can the intact polyprotein be isolated. Cleavage can also be demonstrated by pulse–chase analysis.

In theory, all picornavirus proteins should be present in equimolar proportions as they are all derived from the same precursor molecule, but this is not found in practice. The rates of cleavage of the precursors vary markedly and this allows a great degree of control over the amounts of each protein produced.

11.4 Alphaviruses

Synthesis of a subgenomic mRNA

The genome of alphaviruses consists of one positive sense ssRNA molecule of approximately 11,400 nt with a sedimentation coefficient of

42S. In contrast to the situation with picornaviruses, in vitro translation of the positive sense genomic RNA of alphaviruses, such as Semliki Forest virus, generates non-structural proteins including the polymerase, but not virion proteins. In virus-infected cells two types of positive sense RNA are produced by the action of the polymerase: one of 42S identical to the genome RNA and one with a sedimentation coefficient of 26S. Both of these positive sense RNA molecules have a cap at the 5′ end, a polyA tail at the 3′ end and function as mRNAs. The 26S RNA is approximately one-third (3800 nt) the length of the genomic RNA and represents the 3′ terminal portion of the genomic RNA; it is translated into the structural proteins (Fig. 11.6a). Thus the gene expression of these viruses has two phases, with initial production of the viral RNA polymerase followed later by expression of the virion proteins.

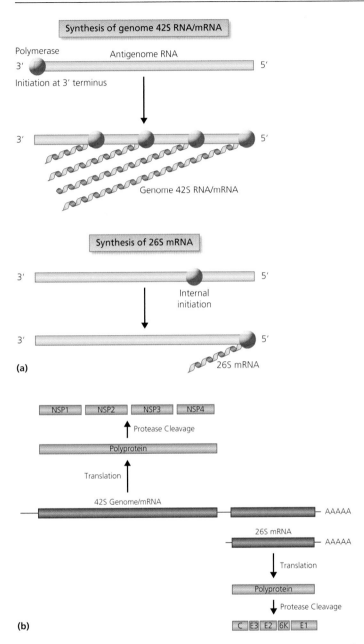

Fig. 11.6 Alphaviruses synthesize two mRNAs in the infected cell. The larger mRNA also acts as the genome; the smaller, 26S mRNA has the same sequence as the 3′ end of the genome. (a) The mechanism for production of the two mRNAs is shown. (b) The proteins encoded by each mRNA and the pattern of proteolytic cleavage that results in the generation of functional proteins.

The primary products of translation of both 42S and 26S mRNA are processed by proteolytic cleavage to produce functional proteins in a manner similar to that described for picornaviruses. Pulse-chase experiments and tryptic peptide maps show clearly that the small functional proteins are derived from larger precursor molecules (Fig. 11.6b). Since the sequence of the 26S mRNA is contained within the genomic RNA, and the latter does not direct the synthesis of the structural proteins, this infers that there is an internal initiation site for the synthesis of virion proteins within the genomic RNA which cannot be accessed by ribosomes.

The only negative sense virus RNA found in infected cells sediments at 42S and this must be the template from which both the 42S and 26S positive sense RNA are transcribed. Since these two mRNAs are co-terminal at the 3′ end, the virus polymerase must be able to initiate transcription at both the 3′ terminus of the negative sense template and also at a point within the template approximately one-third of the distance from the 5′ end (Fig. 11.6). It is not known how transcription is controlled but the levels of 26S mRNA are very high, indicating that the internal initiation of transcription is efficient and this ensures that the structural proteins are abundant. Since the 26S mRNA can only be produced following replication of the genome, this is a late mRNA and control of gene expression in this system is determined at the level of transcription. Synthesis of the 26S mRNA appears to be an adaptation for making large amounts of structural proteins without a similarly high level of non-structural proteins.

11.5 Coronaviruses

Ribosomal frameshifting

Coronaviruses have the largest RNA genomes described to date with some reaching almost 30,000 nt. The single, linear, positive sense ssRNA molecule has a 5′ cap and a 3′ polyA tail. The first event after uncoating is translation of the genome RNA to produce the virus polymerase. This enzyme generates double-stranded RNA using the genomic plus-strand as template, and then transcribes positive sense RNA from a negative sense antigenome RNA template in an RI replication complex (Section 8.4).

The first translation event directs the synthesis of two proteins from the genome RNA. The smallest, the 1a protein, is translated from the first ORF which represents only a small proportion of the coding capacity of the molecule. A second, larger protein is also produced from the same RNA, though in smaller amounts, and this is thought to contain the enzymatic component of the polymerase. The way in which the two proteins function together is not known but the manner in which they are synthesized is understood and is strikingly similar to the synthesis of certain retrovirus *gag-pol* fusion proteins (Section 10.8). Nucleotide sequence analysis of the coronaviruses identified an ORF near the 5′ end of the genome RNA capable of synthesizing the 1a protein but no ORF long enough to encode the larger protein. Instead, a second ORF, called the 1b ORF, lacking an AUG initiation codon and overlapping with the 3′ end of the 1a ORF is present. The 1b ORF is in the -1 reading frame with respect to the 1a ORF. The large protein is generated by a proportion of ribosomes which initiate translation in the 1a ORF, switching reading frames in the region of the overlap of the 1a and 1b ORFs while continuing uninterrupted protein synthesis to generate a fusion 1a-1b protein. This is known as ribosomal frame-shifting. The amount of the fusion protein produced is determined by the frequency of the frameshifting event. This is an example of translational regulation of gene expression.

The frameshifting event in coronaviruses is determined by the presence of two structural features in the genomic RNA. The first is a 'slippery' sequence in the region of overlap between the ORFs at which the frameshift occurs and the second is a 3-dimensional structure called a pseudoknot in which the RNA is folded into a tight conformation. These two features combine to effect the frameshift.

Functionally monocistronic subgenomic mRNAs

Coronaviruses carry the alphavirus strategy of producing a subgenomic mRNA to extremes, producing not just two but several mRNAs. For mouse hepatitis virus (MHV), six mRNAs in addition to the genome are found in infected cells. The size of these mRNAs added together exceeds that of the total genome. This observation was explained by sequence analysis which showed that the sequences present in a small mRNA were also present in all the larger mRNAs; all of the mRNAs shared a common 3' end with each other and the genome RNA. This is described as a nested set of mRNAs. In addition, each mRNA has a short sequence at the 5'-end which is identical for each (Fig. 11.7). Each mRNA has a unique first AUG initiation codon allowing translation of a unique protein. Although the largest (genome) mRNA contains all of the coding sequences for all of the proteins, the first ORF is used preferentially, in accordance with the normal process of eukaryote translation. For some MHV mRNAs, more than one protein is produced by each individual mRNA though the second protein is present in low levels, as with certain rotavirus and paramyxovirus mRNAs. The mRNAs are present in different quantities with respect to each other, some abundant, some less so, but the ratios of each do not alter during the infectious cycle. There is therefore no temporal control of gene expression.

The subgenomic mRNAs cannot be produced using the cell's splicing enzymes since these are located in the nucleus and coronaviruses replicate in the cytoplasm. The mechanism by which they are generated is novel and some uncertainty remains. Whilst it is possible that the negative sense antigenome RNA is used as the template for synthesis of mRNA, most evidence points to the genome RNA as the primary template. The polymerase molecule begins transcription at the 3' end of the genome RNA, producing a negative sense RNA, and at specific points the polymerase terminates and the complex of protein and newly-synthesized negative sense RNA dissociates, at least partially, from the template. The nascent RNA is then taken to a point near the 5' end of the genome RNA and synthesis continues to the end. This produces a series of negative sense RNA molecules which contain sequences complementary to the 3' end of the genome and to a small region from the 5' end. The transfer mechanism is not understood and may be related to the secondary structure of the template RNA. Certainly the outcome is reminiscent of the production of DI RNA during RNA virus replication for which one explanation is the secondary structure of the template (Section 8.2). The frequency with which the polymerase terminates within the template determines the abundance of each RNA. The negative sense RNA molecules are then used as templates to make faithful positive sense copies which act as mRNAs (Fig. 11.7). In effect, each mRNA is the product of the replication of a subgenomic negative sense RNA, produced using a replication intermediate as described for the ssRNA virus genome replication process (Sections 8.4 and 8.5). Each mRNA is then transcribed from a unique subgenomic negative sense RNA and then translated into the appropriate protein. Coronavirus gene expression is therefore controlled primarily at the level of transcription.

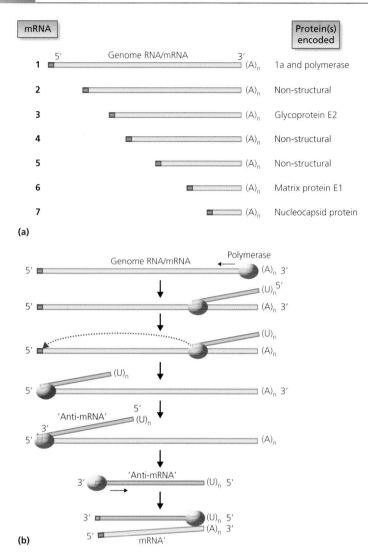

(a)

(b)

Fig. 11.7 Control of gene expression in the coronaviruses mouse hepatitis virus. (a) The products of transcription of mouse hepatitis virus and the proteins encoded by each mRNA. The common sequence found at the 5′ end of the genome RNA and all mRNAs is shown in red. (b) Negative sense RNAs are produced by a novel mechanism in which the virus-encoded RNA polymerase copies the 3′ end of the genome RNA template and completes synthesis by 'jumping' to finish copying the immediate 5′ end. The negative sense RNAs are then used to synthesize individual mRNAs. Each mRNA contains the same 5′ end sequence.

11.6 Negative sense RNA viruses with segmented genomes

The genomes of viruses belonging to the Baltimore class 5 vary in the number of ssRNA segments of which they comprise (Box 11.3), but gene expression is predominantly, but not exclusively, from monocistronic mRNAs. For viruses with segmented genomes each segment must be transcribed into mRNA offering the opportunity for transcriptional control of gene expression. Normal cells do not contain enzymes capable of generating mRNA from an RNA template and all viruses with non-segmented negative sense RNA genomes synthesize mRNA using virus-encoded transcription machinery which is carried into the cell with the infecting genome. Almost all of the negative sense RNA viruses replicate in the cytoplasm. The exceptions are the orthomyxoviruses and Borna disease virus.

Box 11.3

Baltimore class 5 viruses have a variety of numbers of genome segments which are packaged into the virus particle

Family	No. of segments/genome
Orthomyxoviridae	8
Bunyaviridae	3
Arenaviridae	2
Rhabdoviridae	1
Paramyxoviridae	1
Bornaviridae	1
Filoviridae	1

11.7 Orthomyxoviruses

Transcriptional regulation of gene expression

Of the orthomyxoviruses, most is known about the type A influenza viruses of humans and other animals. An unusual feature of influenza virus replication is that it occurs in the nucleus of the infected cell. Immediately after infection, the uncoated virus particle is transported to the nucleus, where viral mRNA is synthesized. The mRNA is exported from the nucleus to the cytoplasm for translation and later in infection certain of the newly synthesized proteins migrate from their site of synthesis in the cytoplasm into the nucleus. Since passage of molecules across the nuclear membrane is a highly selective process, this constitutes a level of control which is unique to eukaryotic cells.

Influenza viruses can only replicate in cells with a functional cellular DNA-dependent RNA pol II, though this enzyme cannot transcribe the virus genome. Two types of positive sense RNA are generated in infected cells – mRNA and the template for replication (antigenome RNA). These two types of RNA can be differentiated from each other by a number of criteria (Box 11.4).

The production of mRNA from the genome initiating the infection, termed primary transcription, uses the virion-associated polymerase and it is this step which requires active pol II. Each of the eight genomic RNA segments are used as templates for transcription of a monocistronic mRNA using a virus-encoded polymerase complex. Each genomic RNA segment contains its own signals to initiate transcription. The first 13 bases at the 5′ end of every segment are identical, and similarly the last 12 bases at the 3′ end of every segment are identical, but different from bases at the 5′ end. The signal for initiation of transcription of the genome segments is contained in the 3′

Box 11.4

Differences between the influenza virus mRNA and antigenome positive sense RNA

mRNA	Antigenome RNA
Shorter than the template genome segment	Exact copy of the template genome segment
Contains 3′ poly(A) tail	No 3′ poly(A) tail
Contains 5′ cap	No 5′ cap
Synthesis does not require protein synthesis (insensitive to protein synthesis inhibitors)	Synthesis requires continuous virus protein synthesis (sensitive to protein synthesis inhibitors)

The method of influenza virus transcription is unusual and explains the reliance on host pol II. The ribonucleoprotein (RNP) complex which carries out the transcription process has no capping or methylation activities associated with it, but the virus mRNAs have a cap 1 structure at the 5′ end (Fig. 11.3). Protein PB2, a component of the RNP complex, binds to the capped 5′ end region of newly-synthesized host cell mRNAs. The host mRNA is then cleaved 10 to 13 bases from the cap, preferably after an A residue though occasionally after a G. This small segment of RNA is then used as a primer for influenza virus mRNA synthesis in a mechanism termed 'cap-snatching'. The proteins PB1 and PA of the RNP complex initiate transcription and extend the primer, respectively. Transcription is terminated at a specific point on the template at a homopolymer run of uracil residues 17 to 22 residues from the template 5′ end. These are copied into A in the mRNA and the virus polymerase continues to add A residues, in a reiterative fashion, to generate a poly(A) tail before the mRNA dissociates from the template. In contrast, synthesis of the antigenome positive sense RNA does not use a primer and

terminal sequences. The features of the three types of virus RNA genome, antigenome and mRNA that are found in influenza virus infected cells are shown in Fig. 11.8.

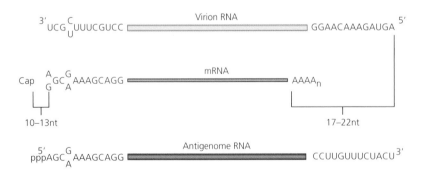

Fig. 11.8 The three primary types of virus RNA found in influenza virus type A infected cells showing their key differentiating features. All eight genome segments have these three forms. The genome RNA has conserved terminal sequences which are present in complementary form in the antigenome, which is an exact replica of the genome segment from which it was copied. The mRNA contains additional sequences at the 5′ end not present in the genome template, which derive from a captured host mRNA, and the mRNA is terminated before the end of the template is reached at which time a polyA tail is added by reiterative transcription (Box 11.4).

generates an exact copy of the template RNA. Influenza virus transcription is shown in Fig. 11.9.

Initially, similar amounts of each influenza virus mRNA are produced (as are the proteins),

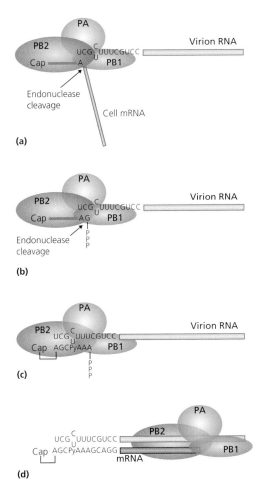

Fig. 11.9 The process of transcription from influenza virus genome segments. The three virus proteins, PB2, PB1 and PA, act together (a) to remove cap structures together with 10–13 nucleotides from the 5′ end of host cell mRNA (PB2) and (b) to use these as primers to initiate transcription (PB1) followed by (c) extension of the newly-synthesized RNA (PA) to produce (d) an mRNA molecule.

but within an hour of infection beginning, the levels of mRNAs encoding the nucleoprotein (NP) and NS1 (non-structural protein 1) increase greatly with respect to the others. Later, after genome replication has started, the levels of mRNAs encoding the haemagglutinin (HA), neuraminidase (NA) and matrix (M1) proteins predominate. Thus, although all of the genes are expressed throughout infection, there is some temporal control of transcription determined by the relative rates of transcription of some segments.

Control of gene expression by mRNA splicing

Superimposed on the basic pattern of transcription of a single mRNA from each influenza A virus genome segment is a system which permits the synthesis of two additional proteins from influenza virus segments 7 and 8 by the process of RNA splicing. This mechanism is available because the virus RNA is transcribed in the nucleus. Transcription of each of these segments produces an mRNA which encodes a specific protein: M1 and NS1, respectively. However, splicing can remove an internal section from a proportion of these primary transcripts (Fig. 11.10). In both cases, the splicing events delete a substantial portion of the first ORF, leaving the AUG initiation codon in place. The result is that an alternative ORF, in a different reading frame and not previously accessed by ribosomes, is fused to the first few codons of the M1 or NS1 ORFs. The newly-generated ORFs direct the synthesis of novel proteins, called M2 and NS2. This is similar to the situation for the polyomaviruses (Section 9.2). The levels of expression of NS2 and M2 are determined by the frequency of the post-transcriptional splicing event. An additional, rare mRNA is generated from the segment 7 primary transcript by an alternative splicing process. The splice removes the first

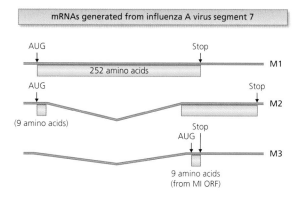

mRNAs generated from influenza A virus segment 7

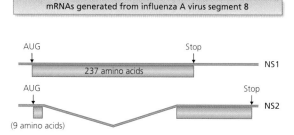

Fig. 11.10 Splicing in influenza virus RNA: synthesis of mRNAs encoding M1 and M2 proteins from segment 7 and NS1 and NS2 proteins from segment 8 of influenza virus type A. The shaded areas represent the coding regions. The reading frames of the unspliced mRNAs are shown but only the first ORFs, encoding M1 and NS1 from segments 7 and 8, respectively, are used in these mRNAs. Note that the two major products from each pair of mRNAs share a short common amino-terminal amino acid sequence but differ in the majority of the sequence. Although the mRNA is present, the M3 protein has not been detected in infected cells.

AUG codon in the primary transcript which is used to synthesize the M1 and M2 proteins. The first ORF in the alternatively spliced segment 7 mRNA contains only 9 codons. The putative protein produced by translation of this mRNA is called M3 but it has not been detected in infected cells and its role in the infectious cycle is not known.

Translational regulation of gene expression

Some human and animal influenza A viruses encode an additional protein from the mRNA transcribed from genome segment 2 that principally encodes PB1. The protein, called PB1-F2, is encoded in a short second ORF which is contained within the PB1 ORF but is in a different reading frame similar to the situation seen in the rotavirus segment 11 mRNA and some paramyxovirus mRNAs. The PB1-F2 proteins from various influenza viruses differ in size between strains, ranging from 57 to 90 amino acids in length. In many strains, this protein interacts with the host immune system at many points and is involved in the development of pathogenicity in the animal or human host. Several additional proteins are also generated from segment 3 using leaky ribosomal scanning to access alternative reading frames and one, called PA-X is produced as a result of a ribosomal frame shift, similar to that seen in coronaviruses (Section 11.5), to produce a protein containing the amino terminal 191 amino acids of the PA protein fused to 61 carboxy terminal amino acids generated from an alternative reading frame.

Post-translational protein cleavage

While some RNA viruses, such as the picornaviruses, encode proteases to produce functional proteins, most do not encode enzymes to specifically cleave their proteins and it might be expected that post-translational cleavage would not be involved in their gene expression. However, post-translational cleavage is frequently the final maturation event for virus surface proteins; for example, after its synthesis, the influenza virus HA glycoprotein undergoes a single protease cleavage. The HA protein is synthesized as a precursor HA0 which is cleaved to give two products. The larger of these is HA1, which is derived from the carboxy terminus of HA0, while the smaller HA2 comes from its amino terminus. Uncleaved HA0 molecules occur in cells that are deficient in protease activity but these are still incorporated into virus particles which are morphologically normal and can agglutinate red blood cells. However, such virus particles are non-infectious until the HA protein is cleaved. The cleaved HA changes conformation and a hydrophobic domain, referred to as the fusion peptide, is now free at the N terminus of HA2 (see Section 6.2). The cleavage of the HA0 precursor is carried out by host cell proteases and the ability of the HA0 to be cleaved is a determinant which contributes to the severity of the disease seen in the host. The restriction of such enzymes to specific tissues such as the respiratory tract determines the site of productive virus replication. If HA is cleaved by a more ubiquitous protease such as is seen in avian influenza virus infection in birds, the virus can spread systemically. Some viruses, such as Sendai virus, require cleavage activation of proteins which can occur after the virus has left the host cell rather than inside the cell as for influenza virus.

11.8 Arenaviruses

Ambisense coding strategy

Although arenaviruses appear similar to conventional segmented negative-strand viruses, albeit with just two (L and S) genomic RNAs, they have an unusual gene organization. A virus-encoded polymerase carried in the particle transcribes only the 3′ part of the genome RNAs into a capped, polyadenylated subgenomic mRNA. The polymerase also adds the cap to the newly-synthesized mRNA during the synthesis process. In the case of the S RNA, the mRNA produced encodes the nucleocapsid

(N) protein, while for the L RNA it encodes the large polymerase protein. Following replication, the 3′ end of the antigenome copy of each RNA segment is also used as a template for transcription. This generates a subgenomic mRNA from the S antigenome RNA which encodes a protein called GPC, the precursor to the virus structural glycoproteins. A protein called Z is encoded by the mRNA copied from the L antigenome RNA. Since the GPC and Z mRNAs can only be synthesized after replication they are, by definition, late mRNAs. This mechanism means that both the genome and antigenome act as negative sense RNA templates for transcription and therefore are termed 'ambisense' RNA (Fig. 11.11). Some bunyaviruses such as Punta Toro virus also use an ambisense strategy for expression from their S genome segment.

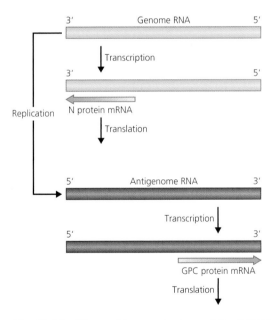

Fig. 11.11 The ambisense strategy of the S RNA segment of the arenavirus, lymphocytic choriomeningitis virus.

11.9 Negative sense RNA viruses with non-segmented, single stranded genomes: rhabdoviruses and paramyxoviruses

Transcriptional regulation of gene expression

Of the rhabdoviruses, most is known about vesicular stomatitis virus (VSV). The proposed model for VSV transcription is used as a paradigm for all other viruses with non-segmented negative sense RNA genomes, though the precise molecular details of each virus in terms of regulatory sequences, etc. differ. Transcription of the VSV genome yields five separate monocistronic mRNAs and gene expression is controlled at the level of transcription, with the abundance of the mRNAs determining the abundance of the proteins. A similar pattern of transcription is believed to occur with the paramyxoviruses though these may encode 6, 7 or 10 mRNAs, depending on the virus.

The ribonucleoprotein (RNP) complex which carries out transcription (and replication: Section 8.5) of VSV consists of the genome RNA in association with large amounts of the virus nucleoprotein (N), lesser amounts of the virus phosphoprotein (P) and a few molecules of a large (L) protein. The L protein is thought to contain the catalytic sites for RNA synthesis, and also capping of the mRNA, and is often referred to as the polymerase. The RNP complex can only initiate transcription at the 3' end of the genomic template RNA. For VSV, transcription begins with the production of a small, 49 nucleotide uncapped, non-polyadenylated RNA which is an exact copy of the 3' terminus. This is the leader RNA for which no function is known. Each of the remaining five transcription units on the genome RNA is flanked by consensus sequences which direct the polymerase to firstly initiate and subsequently terminate transcription. The initiation sequences can only be recognized by polymerases travelling from the 3' end; the polymerase cannot bind directly to the internal transcription initiation sites. As the polymerase moves along the genome it meets a transcription initiation signal and begins mRNA synthesis.

A cap is added very quickly to the growing RNA molecule and this is carried out by a component of the RNP complex since no capping enzymes are found in the cytoplasm where VSV transcription and replication occur (though there are some rhabdoviruses, notably some infecting plants, which replicate in the nucleus, so in these cases capping may use host processes). At the consensus transcription termination signal, which includes a homopolymer uracil tract, the polymerase adds a poly(A) tail by a reiterative stuttering process as described for influenza virus. At this point, the mRNA dissociates from the template and is removed for translation. The polymerase then does one of two things. A proportion, possibly 50%, of the polymerase molecules also dissociate from the template and, having done so, can only rebind at the 3' terminus where the transcription process begins again with the most 3' proximal gene on the genome. The remaining polymerases move along the genome without transcribing until they encounter the next consensus transcription initiation signal where they begin to transcribe the next region of the genome. At the end of this transcription unit the polymerase exercises the same two options of either dissociation, or translocation followed by reinitiation. The result of this process is that the mRNAs are synthesized sequentially in decreasing proportions as the polymerase moves towards the 5' terminus of the genome (Fig. 11.12). The relative abundances of the proteins reflect those of the mRNAs. A similar strategy is used by paramyxoviruses such as Sendai virus.

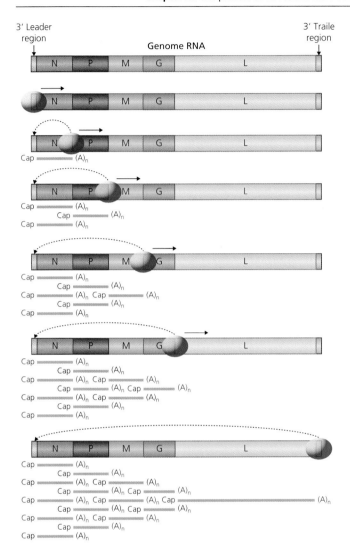

Fig. 11.12   Diagrammatic representation of the sequential transcription process of the rhabdovirus vesicular stomatitis virus (VSV). Transcription occurs in a sequential, start-stop process described in the text. Regions transcribed as mRNA are separated by non-transcribed regions, the lengths of which are virus-specific. As the polymerase complex moves towards the 5′ end of the template progressively fewer transcripts are produced.

Translational regulation of gene expression

The mRNA encoding the phosphoprotein (P protein) of several, but not all, paramyxoviruses contains two ORFs. The 5′ proximal AUG codon directs translation of a large ORF encoding the P protein (Fig. 11.13). The second ORF, initiated by the next AUG, using the process of leaky ribosomal scanning, directs the synthesis of a protein called C. The amino acid sequences of the P and C proteins are completely different, similar to the situation for the rotavirus

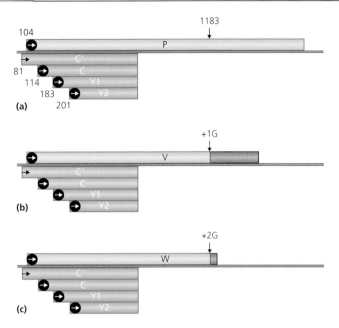

Fig. 11.13 Arrangement of the overlapping open reading frames of the Sendai virus P gene mRNA and the various potential protein products. (a) The relative locations and protein products of the ORFs are indicated. The vertical arrow shows the insertion site for non-templated G residues at nucleotide 1183 in the mRNA. The altered reading frames leading to the novel carboxy termini of (b) the V and (c) W proteins are indicated in purple. Initiation of translation at AUG (⊖)codons and the ACG (→) for the C′ protein are indicated. The numbers refer to the nucleotide positions of the first nucleotide of the relevant codon.

NSP5 and NSP6 proteins (Section 11.2). Analysis of the proteins produced by translation of the Sendai virus P mRNA indicates that several other proteins are also generated. Initiation of translation at the second and third AUG codons in the C protein ORF generates proteins called Y1 and Y2, respectively (Fig. 11.13a). Since these proteins are translated from the same reading frame as that used for the C protein, the Y1 and Y2 proteins are identical in sequence to the carboxy-terminus of the C protein. However, the differences at the amino termini of the three proteins mean that they are likely to have different functions in the virus replication cycle. This is similar to the papillomavirus E2 proteins described in Section 10.3.

The Sendai virus P mRNA also encodes a protein which is initiated at a codon ACG upstream of the first AUG codon that is used to initiate synthesis of the P protein. Initiation of translation at the ACG uses the same reading frame as that for the C protein and the protein produced, called the C′ protein, contains additional amino acids at the amino terminus when compared to the C protein. As with the Y1 and Y2 proteins, this difference may be sufficient to give the C′ protein a unique function. Use of a codon other than AUG for initiation of translation is rare but is seen in certain retroviruses and parvoviruses.

Non-templated insertion of nucleotides during transcription

In addition to the alternative translation initiation products of the Sendai virus P gene, nucleotide sequence analysis of mRNAs transcribed from this gene identified a novel strategy for gene expression which is also used by many other paramyxoviruses. During transcription of any gene it is important that the polymerase makes a faithful copy of the template to ensure fidelity of the protein sequence. However, RNA synthetic enzymes have no proofreading capacity so any errors cannot be corrected. When the Sendai virus P gene is being transcribed, most of the mRNA produced is a faithful copy of the template. However, in approximately 30% of molecules an additional G nucleotide is inserted at a precise point, nucleotide 1183 in the mRNA, which lies in the ORF encoding the P protein. The mRNA with the additional G residue directs the synthesis of a novel protein, termed V protein, which has the same amino terminal sequence as the P protein but a unique carboxyl terminus (Fig. 11.13b). Strangely, some paramyxoviruses encode the V protein from the mRNA that is faithfully copied from the genome template and the P protein from the mRNA with the additional base(s). The function of the V protein is to interfere with the host immune response to infection, thus giving the virus an increased possibility of a successful infection (Section 13.3). Furthermore, in approximately 10% of transcription events, two G residues are inserted at the same location (Fig. 11.13c). Translation of the resulting mRNA generates yet another protein, the W protein, which has the same amino terminal sequence as the P and V proteins but a unique carboxy terminus. The frequencies of these non-templated insertion events determine the relative abundance of the novel proteins.

Post-translational protein cleavage

In paramyxoviruses two glycoproteins, the attachment (G or HN) and fusion (F) proteins, are inserted into the lipid bilayer of the cell. The F protein is inserted with the carboxyl terminus on the external surface of the cell and, eventually, of the virus (a type I membrane protein) while the amino terminus of the attachment protein is external (a type II membrane protein). The F protein is synthesized as a precursor, F_0, which has an amino-terminal hydrophobic signal sequence cleaved off by a host cell protease during insertion into the cell membrane. A further cleavage event, also carried out by a host cell protease but this time in the Golgi network, is required for the F protein to become functional. The two components of the F protein, called F_1 and F_2, that are generated by this cleavage activation event are covalently attached to each other by disulphide bonds. This is analogous to the cleavage of the influenza virus HA protein (see Section 11.7). The attachment protein of paramyxoviruses is not cleaved.

Key points

- Some RNA viruses may regulate their gene expression at the level of transcription (mRNA abundance, non-templated nucleotide insertion).
- Some RNA viruses may regulate their gene expression at the level of post-transcription (RNA splicing).
- Some RNA viruses may regulate their gene expression at the level of translation (alternative open reading frames, alternative translation initiation codons, ribosomal frameshifting).
- Some RNA viruses may regulate their gene expression at the post-translational level by protein cleavage.
- Viruses may alter normal cellular process to gain an advantage in expression of their genes such as the alteration of translation on ribosomes from cap-dependent to cap-independent seen picornaviruses.
- While all of these options are available in principle, each virus utilizes only a selection

to suit its own circumstances and requirements.

Questions

- Compare and contrast the strategies adopted by alphaviruses and coronaviruses for generation of mRNA from a single-stranded positive sense RNA genome.
- Compare and contrast the mechanisms used by coronaviruses and rhabdoviruses to control the relative levels of their various mRNAs.

Further reading

Cheung, T. K. W., Poon, L. L. 2007. Biology of influenza A virus. *Annals of the New York Academy of Sciences* 1102, 1–25.

Conzelmann, K.-K. 1998. Nonsegmented negative-stranded RNA viruses: genetics and manipulation of viral genomes. *Annual Review of Genetics* 32, 123–162.

Curran, J., Kolakofsky, D. 1999. Replication of paramyxoviruses. *Advances in Virus Research* 54, 403–422.

Curran, J., Latorre, P., Kolakofsky, D. 1998. Translational gymnastics on the Sendai virus P/C mRNA. *Seminars in Virology* 8, 351–357.

Emonet, S. E., Urata, S., de la Torre, J. C. 2011. Arenavirus reverse genetics: new approaches for the investigation of arenavirus biology and development of antiviral strategies. *Virology* 411, 416–425.

Jayaram, H., Estes, M. K., Prasad, B. V. V. 2004. Emerging themes in rotavirus cell entry, genome organization, transcription and replication. *Virus Research* 101, 67–81.

Kormelink, R., de Haan, P., Meurs, C., Peters, D., Goldbach, R. 1992. The nucleotide-sequence of the M RNA segment of tomato spotted wilt virus, a bunyavirus with 2 ambisense RNA segments. *Journal of General Virology* 74, 790–790.

Neumann, G., Brownlee, G. G., Fodor, E., Kawaoka, Y. 2004. Orthomyxovirus replication, transcription, and polyadenylation. *Current Topics in Microbiology and Immunology* 283, 121–143.

Sawicki, S. G., Sawicki, D. L., Siddell, S. G. 2007. A contemporary view of coronavirus transcription. *Virology* 81, 20–29.

Strauss, J. H., Strauss, E. G. 1994. The alphaviruses: gene-expression, replication, and evolution. *Microbiological Reviews* 58, 491–562.

Chapter 12

The process of infection: IV. The assembly of viruses

Assembly of virus particles is a highly ordered process which requires the bringing together of various particle components, including the genome nucleic acid, in a controlled sequence of events.

Chapter 12 Outline

12.1 Self-assembly from mature virion components
12.2 Assembly of viruses with a helical structure
12.3 Assembly of viruses with an isometric structure
12.4 Assembly of complex viruses
12.5 Sequence-dependent and -independent packaging of virus DNA in virus particles
12.6 The assembly of enveloped viruses
12.7 Segmented virus genomes: the acquisition of multiple nucleic acid molecules
12.8 Maturation of virus particles

In infected cells, virus-encoded proteins and nucleic acid are synthesized separately and must be brought together to produce infectious progeny virus particles (virions) in a process referred to as assembly or morphogenesis. For some viruses, there is an additional aspect when one or more steps in assembly takes place in the nucleus and proteins synthesized in the cytoplasm must be moved to that location. Little information about assembly is available for most viruses. Despite the diversity of virus structure (see Chapter 2), there appears to be only three ways in which virus assembly occurs. Firstly, the various components may spontaneously combine to form particles in a process of self-assembly. In principle, this is similar to crystallization, in which the final product represents a minimum achievable energy state. Secondly, the assembly process may require specific, virus-encoded and/or host cell proteins which do not ultimately form a structural part of the virus. These are referred to as scaffolding proteins. Finally, a particle may be assembled from precursor proteins which are then modified, usually by proteolytic cleavage, to form the infectious virion. In the last two cases it is not possible to dissociate and then reassemble a mature infectious particle from its constituent parts. The presence of a lipid envelope in some viruses, surrounding a nucleocapsid core, introduces an additional aspect to assembly of an infectious particle since this must be acquired separately from the main assembly process.

Introduction of the virus genome into the progeny virus particle is a critical step in the assembly of viruses. There are two processes by which this can be achieved. Reconstitution and

Introduction to Modern Virology, Seventh Edition. N. J. Dimmock, A. J. Easton and K. N. Leppard.
© 2016 John Wiley & Sons, Ltd. Published 2016 by John Wiley & Sons, Ltd.

other studies have shown that, for some viruses, the genome which forms part of the infectious particle plays an integral role in assembly, acting as an initiating factor with the particle components assembling around the nucleic acid. For other viruses, the particle, or a precursor of the particle, is formed first and the nucleic acid is then introduced at a late stage in the assembly process. The first process obligatorily involves the interaction of virus-encoded structural proteins with specific nucleotide sequences in the genome, referred to as packaging signals. While the second process may also require packaging signals to be present in the genome this is not always the case.

The structure of virus particles, features of which are considered in Chapter 2, determines, to some extent, the process of assembly by which they are generated. Recent advances in the determination of the 3-dimensional structures of virus particles by cryoelectron microscopy and X-ray crystallography has shed more light on the precise nature of interactions between virion components. This has led to suggestions of how some of these components may come together during the assembly process.

12.1 Self-assembly from mature virion components

Absolute proof of virion self-assembly requires that purified viral nucleic acid and purified structural proteins, but no other proteins, are able to combine in vitro to generate particles which resemble the original virus in shape, size and stability, and which are infectious. Given that the components for such experiments must come from particles or infected cells, a critical step in demonstrating assembly of such components in vitro is the process used to effect their disassembly. Disassembly should not irretrievably alter the subunits to allow them to

retain the ability to reassociate in a specific manner to assemble the virus particle. Ideally, the disassembled constituent monomers of the virion should not be irreversibly denatured by the disassembly process. This has been demonstrated for very few viruses to date.

For many viruses, it is not possible to demonstrate spontaneous self-assembly even though it is suspected that it is an integral aspect of the process. For example, the assembly of a virus may be spontaneous but acquisition of infectivity then requires a maturation event which modifies one of the structural proteins after assembly has occurred. Following dissociation of purified infectious particles, the modified protein may not be able to interact spontaneously with the other components to reform the virion. Similarly, where a virus is enclosed in a lipid envelope, the disassembly process is likely to destroy this structure so an infectious particle cannot be regenerated.

12.2 Assembly of viruses with a helical structure

The best-studied example of self-assembly is the in vitro reconstitution of tobacco mosaic virus (TMV) which has been used as a paradigm for the assembly of many other viruses with helical symmetry.

Assembly of TMV

The TMV particle consists of a single molecule of positive sense ssRNA, embedded in a framework of small, identical protein molecules (A protein), arranged in a right-handed helix (Chapter 2), with each protein binding to 3 nucleotides of RNA. TMV can be disassembled to yield protein and RNA components, which can be reassembled in vitro to yield active virus. However, the isolated protein, free from any

RNA, can also self-assemble into a helical structure, indicating that bonding between the subunits is a specific property of the protein. Whilst the most likely model for the assembly of the virus would be for the protein molecules to arrange themselves like steps in a spiral staircase, enclosing the RNA as a corkscrew-like thread, research has indicated that the assembly of TMV is a much more complicated process, in which the genome RNA plays an essential role.

In solution, TMV A protein forms several distinct kinds of complex, depending on the environmental conditions, particularly ionic strength and pH (Fig. 12.1). The complexes differ in the number of individual proteins which make them up. Of these complexes, the disc structure is considered the most important, since it is the dominant one found under physiological conditions. Each disc consists of two rings with 17 subunits per ring. This is close to the 16.34 protein subunits per turn of the helix seen in the virus particle, so the bonding between the subunits in these discs is probably very similar to that seen in the virus. Whilst it is tempting to suggest that the discs could simply align to form a helix which would have a

slightly different packing arrangement for the A protein subunits, this does not occur. Instead, the key components in the spontaneous assembly of TMV are 'lock washer' structures. These are very short helical structures, slightly more than two turns of a helix in dimension (Fig. 12.1). The lock washers are produced from the discs by subtle changes in the conformation of the A protein subunit (Box 12.1). However, while lock washers interact to form helices like those in the virus particle this is a very slow process, indicating that there is another component in the system which catalyzes the assembly process. This essential catalyst is the TMV genome RNA.

When A protein subunits and small aggregates of dimers and trimers, etc. are mixed with TMV RNA, assembly is slow and formation of virus particles requires about 6 hours. However, when discs are mixed with RNA under the same conditions, assembly is rapid and mature virus forms within minutes. Addition of small aggregates as well as discs to the RNA does not increase the rate of assembly. A possible model for the assembly of TMV is shown in Fig. 12.2 (but note that an alternative model is presented later). In this model, the interaction of the genome RNA with a disc neutralizes the charges on the adjacent carboxyl group in the A protein subunits. This causes conversion of the disc into a lock washer, trapping the RNA in the groove between successive turns of the helix (see Box 12.1). Following conversion to the lock-washer form, a second disc can join, undergoing the same structural conversion to a lock washer as the helix extends. The genome RNA is therefore a catalyst for disc-to-lock washer conversion and hence for the rapid assembly of a helix, and becomes contained within the helical structure. Subsequent analysis has shown that TMV genome RNA contains a specific region near the 3' end, called the packaging site, at which the first disc binds and is converted into lock-washer form, and from where the helix is extended in both directions,

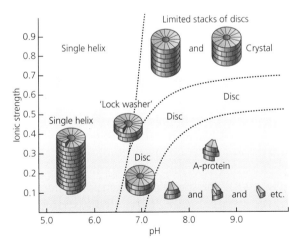

Fig. 12.1 Effect of pH and ionic strength on the formation of different assemblies of TMV A protein.

Box 12.1

Evidence that the lock washer structure is the intermediate in TMV assembly from A protein subunits to virus particle

Analysis of electron micrographs of the discs formed by A protein in vitro found that the structure at the edges of the subunits is slightly different to that seen in the virus particle. The two-ring disc structure presents a slightly different region to the underside of any further disc adding to it than would be the case in the assembled virus. This very small difference in structure appears to prevent the addition of further discs, so no virus-like helices are formed from disc:disc association. When the pH of the solution containing the discs is reduced, association progressively occurs, with the appearance of short rods made up of imperfectly meshed sections of two helical turns. After many hours, these short rods combine to give the regular virus-like structure (Fig. 12.1). This is thought to be due to the lower pH reducing the repulsion between the carboxylic acid side chains of two adjacent amino acids in the A protein subunits leading to a conversion of the disc into a 'lock-washer'. It is the lock washers which then associate to form the helix. However, this takes many hours, indicating that the process is not as efficient as assembly seen in vivo and suggesting that another component may be involved.

though at different rates in each direction (Box 12.2). Computer-assisted secondary structure prediction of the packaging site suggests strongly that it exhibits a hairpin configuration (Fig. 12.3).

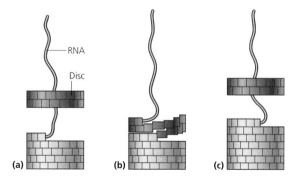

(a) (b) (c)

Fig. 12.2 Simple model for assembly of TMV. Discs of A protein convert to 'lock washer' form upon interaction with RNA, allowing their association with the growing particle. See text for details.

While the model described above is compatible with the available data, it requires that discs thread on to the end of the RNA which seems conceptually difficult. Hence, an alternative explanation had been offered. This is the 'travelling loop' model which suggests that the hairpin structure at the packaging site in the TMV genome RNA inserts itself through the central hole of the A protein disc into the groove between the rings of subunits. The nucleotides in the double-stranded stem then unpair and more of the RNA is bound within the groove. As a consequence of this interaction, the disc becomes converted to a lock-washer structure trapping the RNA (Fig. 12.4). The special configuration generated by the insertion of the RNA into the central hole of the initiating disc could subsequently be repeated during the addition of further discs on top of the growing helix; the loop could be extended by drawing more of the longer tail of the RNA up through the central hole of the

Box 12.2

Evidence that TMV genome RNA contains a specific packaging site

When purified TMV RNA is mixed with limited amounts of viral A protein in the form of discs, the RNA is incompletely encapsidated. Treatment of the incomplete structure with nucleases to destroy RNA not protected by the A protein identified a unique region of the RNA which is resistant to digestion and hence is preferentially covered by A protein. The protected RNA consists of a mixture of fragments up to 500 nucleotides long. The shortest fragments define a core about 100 residues long common to all the fragments. Analysis of the larger fragments show that they are not equally extended in both directions from the packaging site. Rather, the helix is extended more rapidly in one direction than the other. These data are interpreted as showing that assembly is initiated at a unique internal packaging site on the RNA, and that the helix is extended bidirectionally but is more rapid in the 3′ → 5′ direction with extension 5′ → 3′ being much slower. This difference in rate is acceptable as the packaging site is close to the 3′ end of TMV RNA.

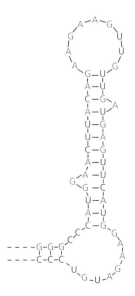

Fig. 12.3 The packaging site of TMV RNA. The loop probably binds to the first protein disc to begin assembly. The fact that guanine is present in every third position in the loop and adjacent stem may be important in this respect.

growing virus particle. Hence, the particle could elongate by a mechanism similar to initiation of packaging, only now instead of the specific packaging loop there would be a 'travelling loop' of RNA at the main growing end of the virus particle (Fig. 12.4). This loop would insert itself into the central hole of the next incoming disc, causing conversion to the lock-washer form and continuing the growth of the virus particle.

The 'travelling loop' model shown in Fig. 12.4 overcomes the disc-threading problem of the model in Fig. 12.2 as far as growth in the 5′ direction is concerned, as incoming discs would add directly on to the growing protein rod through recognition of the travelling loop in the RNA. Discs would still have to be threaded on to the 3′ end of the RNA and thus elongation in this direction would be much slower, as has been observed experimentally. One prediction of the 'travelling loop' model is that both the 5′ and 3′ tails of the RNA should protrude from one end of partially-assembled TMV particles. Electron micrographs of such structures have been observed. Currently it is not known which model is correct.

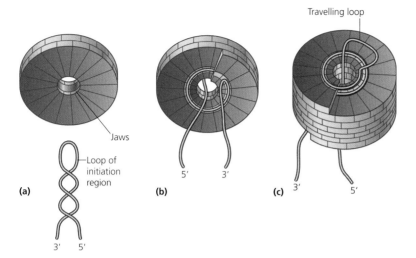

Travelling loop

Jaws

Loop of
initiation
region

(a)

(b) 5′ 3′

(c) 3′ 5′

3′ 5′

Fig. 12.4 The 'travelling loop' model for TMV assembly. The process begins with the insertion of the hairpin loop of the packaging region of TMV RNA into the central hole of the first protein disc (a). The loop of the packaging sequence intercalates between the two layers of subunits and binds around the first turn of the disc, catalyzing the structural conversion to the lock washer (b) and trapping the RNA. As a result of the mode of initiation, the longer RNA tail is doubled back through the central hole of the rod (c), forming a travelling loop to which additional discs are added rapidly through a repetition of these steps.

12.3 Assembly of viruses with an isometric structure

As indicated in Chapter 2, the particles of all isometric viruses are icosahedral with 20 identical faces. The process of triangulation describes the mechanism used by isometric viruses to increase the size of their capsid with a concomitant increase in the number of capsomere subunits. As indicated above, reconstitution experiments have had only limited success and the most detailed study so far reported on the self-assembly and reconstitution of an isometric virus is that of cowpea chlorotic mottle virus (CCMV; a plant bromovirus of Baltimore class 4). The formation of infectious CCMV particles from a stoichiometric mixture of initially separated CCMV RNA and protein affords proof of the ability of this virus to self-assemble.

Assembly of picornaviruses

Over recent years, the 3-dimensional structures of a number of viruses have been determined by X-ray crystallography. Many of these have been picornaviruses, including poliovirus, which have icosahedral particles of approximately 30 nm in diameter. Poliovirus particles consist of 60 copies of each of the four structural proteins VP1-4. The proteins associate together in a complex and these complexes are arranged in groups of three on each of the 20 faces of the icosahedron (Fig. 2.7). Much information is available about the sequence of events in the assembly of poliovirus, which illustrates how an icosahedral virus particle can be generated.

The entire genome of poliovirus is translated as a single giant polypeptide, which is cleaved as translation proceeds into smaller polypeptides (see Section 11.3). The first cleavage generates a

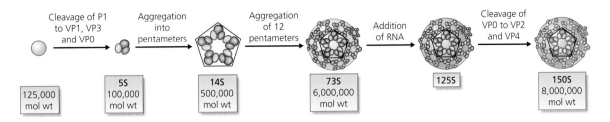

Fig. 12.5   Summary of the steps involved in the assembly of poliovirus.

polypeptide called P1 or 1ABCD (Fig. 11.5), which is the precursor to all four virion proteins. Synthesis of 1ABCD is directed by the 5′ end of the genome and it is synthesized completely before being cleaved, suggesting that either folding is necessary for cleavage or that the virus 2A protease requires a period of time to act. Subsequent cleavages of 1ABCD give rise to proteins called VP0, VP1 and VP3. Poliovirus assembly (summarized in Fig. 12.5) begins from these three proteins, which associate with each other in infected cells to produce a complex with a sedimentation coefficient of 5 S. Five of the 5 S complexes then come together to form a 14 S pentamer complex. Twelve of the 14 S complexes in turn assemble together to form an empty 73 S capsid. At this point, the genome positive sense ssRNA is added to the particle. The mechanism by which the genome is recognized and then inserted into the capsid is not known. The VPg protein covalently attached to the 5′ end of the genome RNA may be involved in its acquisition but this cannot be the only factor since the antigenome RNA produced during replication (see Section 8.4) also has VPg at the 5′ end and it is not encapsidated. Consequently, it is believed that a packaging signal sequence must be present in the genome RNA. After the RNA is encapsidated, VP0 is cleaved to yield the virion proteins VP2 and VP4 (see Section 12.8), causing an alteration in the sedimentation coefficient of the particle. VP4 is located inside the particle

with the other three proteins on the particle surface. Because of this late cleavage event, dissociation and reassembly to form infectious picornavirus particles in vitro is not possible as only the intact VP0 can form the structures necessary for the initial steps in the assembly process.

Assembly of adenoviruses

Adenovirus particles range in size from 70 to 90 nm in diameter, depending on the strain being studied, and appear icosahedral in the electron microscope (Fig. 2.11). The particle is more complex than that of the picornaviruses, containing at least 10 proteins. The 252 capsomeres which form the external surface consist of 240 protein multimers arranged such that they have sixfold symmetry (called the hexon capsomeres) and the remaining 12 arranged at the vertices of the icosahedron with fivefold symmetry (penton proteins). Each penton capsomere contains a fibre structure consisting of a trimer of fibre proteins projecting from the penton base. Thus the adenovirus particle is not a perfect icosahedron but is an icosadeltahedron. While many details of the assembly of adenoviruses remain to be established, those available indicate that, unlike the picornavirus assembly process, the individual components of the adenovirus particle are assembled independently of each other and are brought together in a directed fashion (Fig. 12.6).

The later stages of assembly of adenovirus take place in the nucleus of the infected cell where the virus genome is replicated, so the proteins which form the virion must be translocated there from their site of synthesis in the cytoplasm. The proteins which form the

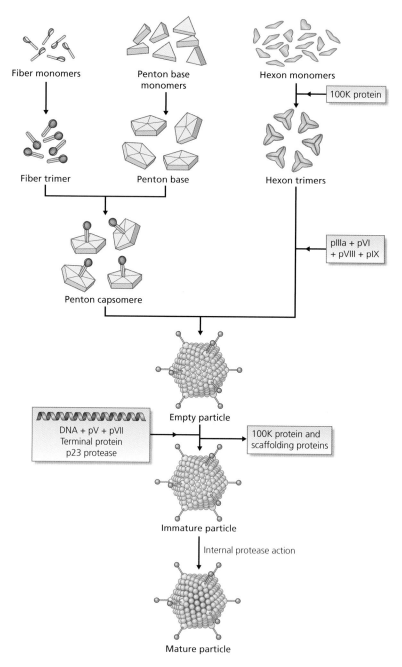

Fiber monomers

Penton base monomers

Hexon monomers

100K protein

Fiber trimer

Penton base

Hexon trimers

pIIIa + pVI + pVIII + pIX

Penton capsomere

Empty particle

DNA + pV + pVII
Terminal protein
p23 protease

100K protein and scaffolding proteins

Immature particle

Internal protease action

Mature particle

Fig. 12.6   Summary of the steps involved in the assembly of adenovirus.

fibre, the base of the penton capsomere and the hexon capsomere are synthesized independently of each other. The fibre and hexon proteins come together to form independent trimer intermediates, while the penton monomers form a pentameric penton base. The fibre trimers and penton base then associate to give a penton capsomere. The formation of the hexon trimer unit and its import into the nucleus require the presence of an additional adenovirus protein with an M_r of 100,000 which interacts directly with the hexon proteins. The M_r 100,000 protein is not found in the mature virus particle and is termed a 'scaffolding' protein.

The remaining steps in the adenovirus assembly process have been inferred from detection and analysis of putative intermediate structures. A key aspect is the formation of immature virus particles which do not contain the genome dsDNA or core proteins but which do contain at least three proteins which are not found in the infectious particle. These three scaffolding proteins may be removed, in part, by proteolytic degradation. The hexon capsomeres come together in a nonamer complex and, subsequently, 20 of these complexes interact to produce an icosahedral cage-like lattice. This structure then acquires the virus DNA, core proteins and remaining structural components, with concomitant loss of the scaffolding proteins. It is not known whether the DNA and the core proteins enter the immature particle together in a complex or in rapid succession. Finally, a virus-coded protease in the particle cleaves various components to create an infectious virion.

Adenovirus DNA contains a protein covalently attached to the termini (Section 7.4) but this is not involved in packaging of the genome into particles. Analysis of DNA packaged into adenovirus DI particles and the generation of deletion mutants has identified a region essential for packaging. Approximately 400 bp at one end of the DNA, adjacent to the E1A gene, must be present for the DNA to enter the immature particle. The conclusion that this is a packaging signal responsible for the specific acquisition of DNA by the particle is supported by evidence that the genome enters the particle in a polar fashion, one end first, and that this polarity is lost when the sequence is duplicated at the other end of the DNA. The packaging sequence is recognized by a combination of structural (IVa2) and non-structural proteins to drive its insertion and compaction within the immature particle. The IVa2 protein remains associated with the assembled particle, marking the vertex through which the DNA was inserted.

12.4 Assembly of complex viruses

The assembly process for animal viruses with particles which are neither helical nor isometric is not well understood. From our understanding of assembly, it is likely that the processes for these viruses are also highly organized. In the case of phage, such as phage λ, which has a complex structure consisting of an isometric head attached to a helical tail, the process has been well characterized using mutants which are unable to proceed beyond certain points in assembly. The isometric head structure of phage λ is similar to that of adenovirus and also acquires the DNA genome after capsid assembly. A head-like structure, called procapsid I, is produced with the assistance of scaffolding proteins. The 'entrance' to the procapsid is formed by the presence of a portal protein complex (Fig. 12.7). The scaffolding proteins are removed, causing a change in shape to produce a procapsid II structure. A protein within the phage procapsid II head recognizes a *cos* sequence in the concatemeric DNA that is the source of progeny genomes and cleaves at that point to produce the single-

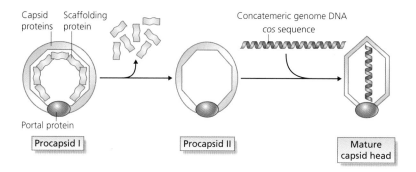

Fig. 12.7   Diagram of the morphological changes in the head structure of bacteriophage λ during assembly. The procapsid I structure loses the scaffolding proteins to produce a procapsid II structure which in turn acquires the phage genome DNA to form a mature head.

stranded overhang described in Section 7.3. The DNA is then brought into the empty head until it is full, followed by a second cleavage at the next *cos* site to produce a mature capsid (Fig. 12.7).

The phage λ tail and tail fibre components assemble independently of each other (Box 12.3). The tail takes the form of a helical structure which includes the head-tail connector, the core to which it is attached and

---

Box 12.3

Evidence that the components of tailed phage are assembled separately before combining to produce progeny particles

The process of assembly of bacteriophage lambda and T4 are very similar, with the head and tail components being assembled separately before coming together. Several conditional lethal mutants of T4 are available which are unable to produce infectious particles when bacteria are infected under non-permissive conditions. Electron microscopic examination of extracts from these bacteria revealed the presence of structures readily recognizable as components of phage particles. Whilst some of the mutants were defective in one or other of the genes which encoded the structural proteins, many of these mutants were not. This indicated that additional, non-structural gene products were involved in the assembly process. The ability of the mutants to synthesize some recognizable structural components of the phage T4 particle, such as the head, tail or tail fibres also showed that these are assembled independently of each other, similar to the situation for adenovirus capsid components (Section 12.3).

A significant advance came with the discovery that the morphogenesis of phage T4 from partially-assembled components could be made to occur in vitro. In one experiment, purified fibreless phage isolated from cells infected with a tail fibre mutant were mixed with an extract from cells infected with a mutant which could not synthesize heads. Infectivity rapidly increased by several orders of magnitude, indicating that the 'headless' mutant extract was acting as a tail fibre donor, whilst the other extract supplied the heads.

the surrounding sheath. For both the tail and tail fibres, scaffolding proteins are involved in assembly. The tail and tail fibres then spontaneously come together, with the possible involvement of one or more scaffolding proteins. Following the independent assembly of the heads and tails, the two components come together in an unknown way. It is possible that one end of the phage DNA protrudes through a vertex, disturbing the fivefold symmetry, and that this promotes tail addition. Indeed, protrusion of the DNA a short way into the tail may be a necessary structural feature for successful injection following contact of the phage with a susceptible bacterium (see Section 6.5).

12.5 Sequence-dependent and -independent packaging of virus DNA in virus particles

The adenovirus assembly process indicates clearly that packaging of genome DNA is sequence-dependent. Many other viruses also rely on the presence of specific sequences to select the virus DNA or RNA though the processes differ in detail. In the case of herpes simplex virus, the assembly process is coupled to the excision of the genome from the concatemeric replication intermediate (Section 7.3).

The herpes simplex virus particle is covered by a lipid envelope which is acquired as the final stage in assembly in a process described in Section 12.6, but the internal component of the particle, which is an isometric nucleocapsid, is formed first in a process probably similar to that described for adenovirus. The herpes simplex virus assembly process generates an immature form of the nucleocapsid lacking the virus DNA. As indicated in Section 7.3, the HSV genome contains direct repeats (the 'a' sequence) at the ends of the DNA. Each 'a' sequence contains within it two short regions, called *pac*1 and *pac*2, which are required for packaging. During assembly, the immature nucleocapsid recognizes the *pac* sequences within a DNA concatemer and cleaves the DNA to generate the precise genome termini whilst at the same time inserting it into the interior of the particle. The presence of the 'a' sequence alone cannot be sufficient to specify cleavage and ensure that the entire genome is taken into the particle since an additional copy of the 'A' sequence is located within the genome at the boundary of the internal long and short repeat elements (IR_L and IR_S; Fig. 7.4). The most likely explanation is that the length of the DNA also plays a role in assembly, with constraints on the minimum and maximum permissible lengths of the DNA accepted into a particle.

A similar process occurs with bacteriophage λ which also generates concatemeric DNA from a circular intermediate. The introduction of the correct length of DNA within the phage head ensures that the next *cos* sequence in the concatemer is readily accessible to the cleavage enzyme and the DNA is cut to leave an overhang. If two *cos* sequences are brought close together by a deletion of the genome, no *cos* sequence is available to the enzyme when the phage head is full and particle assembly aborts. Similarly, if an insertion occurs to increase the distance between *cos* sequences, the head becomes full before the next cleavage site is reached and again the assembly process aborts. In this way, phage λ regulates the size of the genomic DNA which can be packaged and ensures that all of the genes are present. However, the phage λ packaging system tolerates some flexibility in genome length which permits the virus to carry additional DNA derived from the host during the process of specialized transduction (see Section 17.2).

For a long time, it was considered essential that dsDNA animal viruses had controls in the assembly of particles to ensure that only virus DNA could be packaged. For those viruses which have a packaging signal, this is the only control present and the rest of the sequence is irrelevant. An example is SV40 which shows specific packaging via the six GC boxes in its control region. These sequences bind to the host transcription factor SP1 which then interacts with viral capsid proteins to initiate packaging around the genome. However, it has been shown that polyomavirus can, under certain conditions, package foreign DNA. This indicates that the control mechanisms for some viruses may not be as precise as was originally thought and it may be possible to exploit this in the development of new ways to transfer DNA into cells for gene therapy. True sequence-independent packaging of DNA is seen in those phage, such as phage P1, which exhibit generalized transduction.

12.6 The assembly of enveloped viruses

A large number of viruses, particularly viruses infecting animals, have a lipid envelope as an integral part of their structure. These include herpes-, filo-, retro-, orthomyxo-, paramyxo-, corona-, arena-, pox- and iridoviruses, the tomato spotted wilt virus group and rhabdoviruses of plants and animals. For each virus, the component held within the envelope, the nucleocapsid, is of a predetermined morphology which may be helical, isometric or of a more complex nature. For most enveloped viruses, the nucleocapsid is formed in its entirety prior to acquisition of the lipid envelope.

Assembly of helical nucleocapsids

The assembly process of TMV is usually considered to be a model for the generation of all helical virus structures. However, the structures of nucleocapsids of enveloped viruses differ from this in many ways. This is particularly true of the negative sense ssRNA viruses (Baltimore class 5) such as filo-, paramyxo-, rhabdo- and orthomyxoviruses. For members of some of these virus groups, the RNA genome in the nucleocapsid is protected from degradation by nucleases in vitro while for the others it is not. This indicates that the structures must be arranged differently. For all Baltimore class 5 viruses except the orthomyxoviruses, the basic structure of the nucleocapsid consists of the genome RNA encapsidated by a nucleoprotein, called N or NP, in association with smaller numbers of a phosphoprotein and a few molecules of a large protein. This complex carries out the processes of replication (Chapter 8) and transcription (Chapter 11). For orthomyxoviruses, the nucleoprotein associates with three proteins, PA, PB1 and PB2 (Section 11.7), which carry out RNA synthesis. While some of the steps in the assembly process for nucleocapsids of specific viruses are understood, no precise details are available to describe all aspects.

Analysis of genomes of negative sense ssRNA DI viruses (Section 8.2) and the generation of synthetic genomes for reverse genetics studies have shown that the termini of the genomes are essential for the encapsidation process. It is thought that specific sequences are located in the terminal regions of the genome (and antigenome) which initiate the generation of nucleocapsid structures, and ensure that virus RNA is packaged. Expression of the measles virus nucleoprotein in bacteria, in the absence of virus genome RNA, results in the production of short nucleocapsid-like RNA:protein complexes, indicating that the nucleoprotein has the inherent ability to form nucleocapsids around any RNA. Presumably the signal in the genome termini gives a significant advantage to the genome in competition with any other RNA for binding nucleoprotein, sufficient to ensure virtually exclusive packaging of viral RNA. For the paramyxovirus Sendai virus, the phosphoprotein which is part of the nucleocapsid associates strongly with the nucleoprotein prior to the interaction with RNA. This association prevents the nucleoprotein from interacting non-specifically with RNA and gives specificity to the encapsidation process. This may be a general feature of many class 5 viruses.

For several paramyxoviruses, such as Sendai virus and measles virus, the genome RNA must always have a number of nucleotides that is divisible by 6. This 'Rule of Six' is interpreted to be the result of the nucleoprotein binding to groups of six nucleotides in the RNA. If the genome does not conform to the Rule of Six, unencapsidated nucleotides will be present at one, or both, termini and this will prevent replication and, hence, propagation of the genome. However, for most viruses there is no such strict length requirement for the genome. Recently, structural analyses using synthetic ring-like nucleocapsid complexes have provided some insight into the organization of the complex. In particular, the structure of the nucleocapsid complex of respiratory syncytial virus, a paramyxovirus that does not adhere to the Rule of Six, has shown that the genome RNA is associated with a 'groove' formed by the virus N protein with alternating groups of three and four nucleotide bases exposed or buried within the groove, respectively. Similarly, the genome RNA of the rhabdovirus vesicular stomatitis is retained within a groove generated by the N protein, though in this case the RNA is held inside the coiled nucleocapsid complex rather than on the external surface as seen for respiratory syncytial virus. The retention of the genome RNA in a groove may be a common feature of negative strand RNA virus nucleocapsid complexes though the precise orientation of

the nucleotide bases is likely to differ for different viruses.

A significant difference between the structure of the TMV capsid and the nucleocapsid of some negative sense ssRNA viruses lies in the location of the genome RNA. In the TMV particle, the RNA is located entirely within the capsid. For influenza virus, respiratory syncytial virus and vesicular stomatitis virus, the RNA is wound around the outside of the nucleocapsid complex with the nucleotide bases exposed. This leaves the nucleotides available to be used as templates during replication and transcription. This feature is not common to all of the class 5 viruses but the difference between this structure and that of the TMV capsid reflects their different roles; in TMV, the capsid serves to protect the genome RNA as the particle does not contain the extra, lipid layer found in the class 5 viruses.

Assembly of isometric nucleocapsids

In the absence of detailed information, it is assumed that the assembly of isometric nucleocapsids is similar to the processes described for non-enveloped viruses. Thus, the various components are assembled either around the virus genome in response to a packaging signal, as is the case for the togaviruses, or an immature particle is formed and the genome is added subsequently, as in herpesviruses (see Section 12.5).

Acquisition of the lipid envelope

Cells contain a large quantity of lipid bilayer membranes. These membranes define boundaries between cellular compartments, such as the nucleus and cytoplasm, as well as making up the external surface of the cell itself. The majority of enveloped viruses acquire their

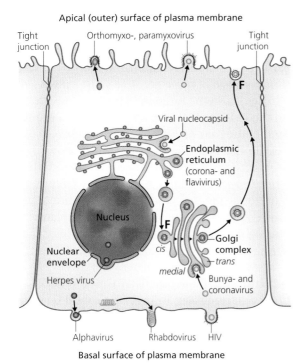

Apical (outer) surface of plasma membrane

Tight junction

Orthomyxo-, paramyxovirus

Tight junction

Viral nucleocapsid

Endoplasmic reticulum (corona- and flavivirus)

Nucleus

F

Nuclear envelope

Golgi complex

cis

medial

trans

Herpes virus

Bunya- and coronavirus

Alphavirus   Rhabdovirus   HIV

Basal surface of plasma membrane

Fig. 12.8 Sites of maturation of various enveloped viruses. F, fusion of a vesicle with a membrane.

envelope by budding from the plasma membrane of the infected cell (Fig. 12.8). Four events leading to budding have been identified. Firstly, the nucleocapsids form in, or are moved into, the cytoplasm. Secondly, viral glycoproteins, which are transmembrane proteins, accumulate in patches of cellular membrane. Thirdly, the cytoplasmic tails of the glycoproteins which protrude from the membrane interact with the nucleocapsid. This interaction may be direct, or indirect via an intermediary matrix protein which becomes aligned along the inner surface of the modified patch of membrane. Once these interactions have taken place the virion is formed by budding. A simple way of envisioning the

process is that the glycoproteins in the membrane progressively interact with the nucleocapsid, or the matrix protein and nucleocapsid where appropriate. As the number of interactions increases, the membrane with the glycoproteins inserted in it is 'pulled' around the nucleocapsid. When the nucleocapsid is completely enclosed, the lipid bilayer pinches off to release the virus particle. As a result of this, viruses which bud from the plasma membrane are automatically released when the budding process is complete. An unusual feature of the budding process is that host membrane proteins are excluded entirely from viral envelopes except for the retroviruses. The mechanism by which this exclusion is achieved is not fully understood but it is likely that the virus glycoproteins are inserted into membranes in lipid rafts which are specific areas enriched for cholesterol and sphingolipid where host proteins are not readily found. It might also be that the linkage of the glycoprotein tails to the internal matrix/capsid structures of the virus dictates a number of viral glycoproteins in the membrane that is sufficient to physically exclude any host proteins. Even for retrovirus envelopes, the non-exclusion is not random; the envelopes contain specifically-selected host proteins.

Some viruses, such as corona- and bunyaviruses bud into the endoplasmic reticulum (ER), acquiring cytoplasmic membranes as their envelopes, and are then released to the exterior via the Golgi complex. First, the virus particle buds into the ER, acquiring an envelope in a process similar to that described for budding with the plasma membrane. The enveloped virus then exits the ER within a vesicle, following the normal mechanism and route for trafficking of material from the ER to the Golgi complex. The vesicle containing the enveloped virus moves to and fuses with the *cis* Golgi complex. The virus then moves through the Golgi and exits in another vesicle which buds off from the concave *trans* Golgi network surface. This vesicle is transported to the plasma membrane where it fuses to release the enveloped particle to the exterior of the cell. The virus envelope does not fuse with any of the vesicle membranes, so the original lipid bilayer is retained throughout this process.

Herpesviruses assemble in the cell nucleus, with many of the viral proteins synthesized in the cytoplasm being transported into the nucleus. After entry of the genomic DNA into the nucleocapsid, the virus must acquire an envelope. The source from, and process by, which the envelope is obtained is still not completely clear. The most likely process is that the virus buds through the inner nuclear membrane and thus becomes enveloped. This envelope then fuses with the outer nuclear membrane, releasing the nucleocapsid into the cytoplasm. The nucleocapsid then undergoes a second budding event into cytoplasmic vesicles so acquiring the envelope, and embedded viral glycoproteins, that will stay with them. The vesicles progress through the normal Golgi system and fuse with the cell membrane, releasing their contents including the enveloped HSV virions outside the cell (see Fig. 10.7). This is similar to the release of the bunya- and coronaviruses.

For other viruses, the process by which the lipid envelope is acquired by the nucleocapsid is unclear. Very little is known about the assembly of lipid-containing phage and the iridoviruses, but it appears that the envelope is not incorporated by a budding process. In this respect, they resemble poxviruses, whose morphogenesis has been studied extensively by electron microscopy of infected cells. In thin sections, poxvirus particles initially appear as crescent-shaped objects within specific areas of cytoplasm, called 'factories', and even at this stage they appear to contain the trilaminar membrane which forms the envelope. The crescents are then completed into spherical structures. DNA is added, and then the external surface undergoes a number of modifications to yield mature virions. The origin of the lipid in the virus is not known since it does not appear to be physically connected to any pre-existing cell membrane.

One complication with envelope acquisition is that many differentiated cells in vivo are polarized, meaning that they carry out different functions with their outer (apical) surface and their inner (basolateral) surface. For instance, cells lining kidney tubules are responsible for regulating Na^+ ion concentration. They transport Na^+ ions via their basolateral surface from the endothelial cells lining blood vessels to the cytoplasm and then expel them from the apical surface into urine. Some cell lines, such as Madin–Darby canine kidney (MDCK) cells, retain this property in culture, but this can only be demonstrated when they form confluent monolayers and tight junctions between cells. The latter serve to separate and define properties of the apical and basolateral surfaces. These properties reside in cellular proteins which have migrated directionally to one or other surface. The lipid composition of a polarized cell is also distributed asymmetrically between apical and basal surfaces. Directional migration and insertion into only one of the cell surfaces also occurs with some viral proteins. For instance, if a cell is dually infected with a rhabdovirus and an orthomyxovirus, electron microscopy shows the rhabdovirus budding from the basal surface, which is in contact with the substratum, and the orthomyxovirus from the apical surface (Fig. 12.8). This property is determined by a molecular signal on the viral envelope glycoproteins which directs them to one surface or the other; tight junctions are needed or proteins diffuse laterally and viruses can then bud from either surface. This aspect of virus assembly has implications for viral pathogenesis (see Section 19.7).

12.7 Segmented virus genomes: the acquisition of multiple nucleic acid molecules

Several virus families use two or more separate and distinct nucleic acid molecules as genomes. While retroviruses package two RNA molecules which are necessary for genome replication (Section 9.4), these RNA molecules are both identical and are packaged simultaneously into the same particle as they are coupled together by proteins. In the case of other viruses, the multiple segments contain different nucleic acid sequences and each encodes proteins unique to that nucleic acid. Examples of virus families containing segmented genomes are shown in Table 12.1.

For viruses with multiple unique genome segments to be fully infectious all of the segments must be packaged into virions and introduced into the infected cell. The viruses achieve this in one of two ways. In some cases, such as the comoviruses of plants, the segments are packaged into separate virion particles. The particles are identical in terms of structural components and only differ in the nucleic acid they carry. For this to occur, the genome segments must each contain the same packaging signals recognized by the virion proteins during assembly. Hence, the two segments effectively compete with each other

Table 12.1
Examples of viruses with segmented genomes.

Virus family	Virus name	Nature of nucleic acid	Number of genome segments
Arenaviridae	Lassa virus	Negative ssRNA	2
Birnaviridae	birnavirus	dsRNA	2
Comoviridae	cowpea mosaic virus	Positive ssRNA	2
Geminiviridae (subgroup B)	cassava mosaic virus	circular ssDNA	2
Nodaviridae	Nodamura virus	Positive ssRNA	2
Bromoviridae	brome mosaic virus	Positive ssRNA	3*
Bunyaviridae	Bunyamwera virus	Negative ssRNA	3
*Nanoviridae***	Banana bunchy top virus	Circular positive ssDNA	6
	Faba bean necrotic yellows virus	Circular positive ssDNA	8
Orthomyxoviridae	influenza A and B viruses	Negative ssRNA	8
	influenza B virus	Negative ssRNA	7
	Thogoto virus	Negative ssRNA	6
Orthoreoviridae	reovirus	dsRNA	10
	rotavirus	dsRNA	11
	Colorado tick fever virus	dsRNA	12

Bromoviridae also frequently package an additional mRNA in a separate particle.
**some members of the *Nanoviridae* family contain up to 11 circular ssDNA molecules packaged into virions but several of these may be satellite molecules and do not encode virus proteins.

for packaging and each virion packages only one nucleic acid molecule. A consequence of this for the virus is that a productive infection which generates new viruses can only occur if the cell is infected with at least one virion of each type. If only one particle initiates infection no progeny can be produced, as a subset of the virus genes will be absent.

The other mechanism for packaging a segmented genome is that each virion contains at least one molecule of all of the different genome segments. In this situation a productive infection is generated by a single infectious event. This then raises questions about how the virus ensures that the virion receives a copy of each genome segment. In principle, this may be achieved if the virion randomly selects genome segments for packaging and accumulates the correct number irrespective of which segments they are. However, simple consideration of the statistical probability of acquiring a full set of the correct segments shows that this would be unacceptably inefficient. For example, for influenza A and B viruses which have eight segments, if the virion randomly packages exactly eight segments the probability of a virion acquiring one of each of the different segments necessary to produce an infectious particle is 16,777,216:1 against; this is obviously not a viable strategy. Whilst it is theoretically possible that a virion may package more than eight segments, so improving the likelihood of acquiring at least one copy of each unique segment, this does not appear to be the case. For influenza virus, the problem is solved by each segment having a different packaging signal. It is proposed that the genome segments associate in arrays, each containing one molecule of each of the eight genome segments incorporated into individual nucleocapsid complexes. These arrays of eight nucleocapsid complexes are held together by the interaction of virus proteins and act as one structure, budding through the cell plasma membrane to generate a virion containing the necessary genome RNA. Many

features of this process remain to be determined, including what process ensures that each array contains only one of each of the necessary nucleocapsid complexes. For other viruses, the process by which the selection of specific genome segments is achieved remains a mystery.

12.8 Maturation of virus particles

For most animal and bacterial viruses, formation of infectious virions requires the cleavage of precursor protein molecules into functional proteins. These cleavages may occur before or after the precursors have been assimilated into the virus particle. Examples of cleavage of a precursor prior to assembly into the virion are seen with the glycoproteins of orthomyxo- and paramyxoviruses, as well as many others. For the orthomyxoviruses, like influenza virus, the haemagglutinin (HA) protein is synthesized as an inactive precursor (HA0). Depending on the host and the specific type of influenza virus, cleavage of the HA0 protein into HA1 and HA2 is carried out either by a host protease immediately prior to insertion into the plasma membrane of the infected cell, or by an extracellular protease after the virus has been released from the cell. The HA protein is acquired by the virus during budding when the envelope is added. The cleavage of HA protein is essential as uncleaved protein cannot fuse with the host cell membrane to initiate the next round of infection. A similar situation is seen with the fusion proteins of the paramyxoviruses. For Sendai virus and the economically important avian paramyxovirus, Newcastle disease virus, the protease responsible for the cleavage of the fusion protein is restricted only to cells of the respiratory tract and this distribution prevents the viruses replicating elsewhere in the body. It

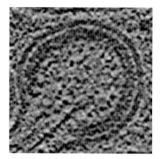  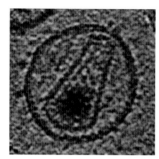

Fig. 12.9   Morphogenetic changes seen in HIV particles following budding from the surface of the host cell. (Left) Budding of an HIV capsid. (Middle) Immature HIV particle. (Right) Mature HIV virion with characteristic cone-shaped core. (Taken from Summers et al., *Journal of Molecular Biology*, 2011, 410, 491–500. Reproduced with permission.)

is for this reason that they cause only respiratory disease in their host.

An example of a maturation cleavage occurring after particle formation is seen with poliovirus where the polypeptide VP0, which is assembled into a particle (Section 12.3), is cleaved subsequently to form VP2 and VP4. Without this cleavage the particle is not infectious. A more dramatic example of morphogenetic alterations in a virus particle following assembly is seen with HIV. The HIV virion consists of a nucleocapsid surrounded by a matrix layer underlying a lipid envelope. The HIV nucleocapsid in newly-budded particles does not show the distinctive blunt conical structure that is seen with infectious, mature particles. Electron microscopic examination of particles at various times after budding has shown that the core of the particle undergoes considerable alteration in structure before achieving the final, infectious form. This process is shown in Figure 12.9. Its basis is the action of a virion protease which cleaves the assembled Gag precursor proteins (Section 10.8) into their functional products, the matrix, capsid and nucleocapsid proteins, after which the morphology of these particle components can mature. The HIV envelope contains two

glycoproteins, gp41 and gp120, which are cleaved from a precursor by a cellular protease similar to the influenza virus HA protein and the paramyxovirus fusion proteins. This cleavage is also essential for infectivity of the progeny particle.

Key points

- Virus assembly occurs spontaneously within the infected cell.
- Assembly is a highly ordered process involving intermediates.
- Assembly may involve scaffolding proteins that are essential components of the process but which do not form part of the final particle.
- The genome may form an integral part of the particle and be a component in the assembly process or it may be added to preformed immature virus particles.
- Genome acquisition may be sequence-specific, involving packaging signals or, rarely, it may be sequence-independent.
- Viruses with segmented genomes may package the segments into separate virions or specifically package a copy of each segment into a single virion.

- Viruses acquire the lipid envelope from one of the several membrane structures in the cell.

Questions

- Compare the processes of virion assembly used by bacteriophage lambda and adenovirus.
- Discuss the role of packaging signals in the process of acquisition of virus genomes during particle assembly.

Further reading

Aksyuk, A. A., Rossmann, M. G. 2011. *Bacteriophage Assembly. Viruses* 3, 172–203.

Butler, P. J. G. 1984. The current picture of the structure and assembly of tobacco mosaic virus. *Journal of General Virology* 65, 253–279.

Earnshaw, W. C., Casjens, S. R. 1980. DNA packaging by the double-stranded DNA bacteriophages. *Cell* 21, 319–331.

Henrik Garoff, H., Sjöberg, M., Cheng, R. H. 2004. Budding of alphaviruses. *Virus Research* 106, 103–116.

Homa, F. L., Brown, J. C. 1997. Capsid assembly and DNA packaging in herpes simplex virus. *Reviews in Medical Virology* 7, 107–122.

Hunter, E. 1994. *Macromolecular interactions in the assembly of HIV and other retroviruses Seminars in Virology* 5, 71–83.

Hutchinson, E. C., von Kirchbach, J. C., Gog, J. R., Digard, P. 2010. Genome packaging in influenza A virus. *Journal of General Virology* 91, 313–328.

Jones, I. M., Morikawa, Y. 1998. The molecular basis of HIV capsid assembly. *Reviews in Medical Virology* 8, 87–95.

Mettenleiter, T. C. 2002. Herpesvirus assembly and egress. *Journal of Virology* 76, 1537–1547.

Rossman, J. S., Lamb, R. A. 2011. Influenza virus assembly and budding. *Virology* 411, 229–236.

Part III

Virus interactions with the whole organism

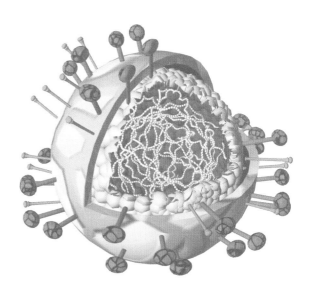

Chapter 13

Innate and intrinsic immunity

Viruses replicate very rapidly and would quickly overwhelm a host organism if it were undefended. The adaptive immune response is restricted to animals and takes several days to gather momentum. In that period, processes of innate and intrinsic immunity slow down and contain a virus so that the host can gain ascendancy over it. In plants, immunity mechanisms with parallels to these systems represent the major line of defence against pathogens.

Chapter 13 Outline

13.1 Innate immune responses in vertebrates – discovery of interferon

13.2 Induction of type 1 interferon responses

13.3 Virus countermeasures to innate immunity

13.4 TRIM proteins and immunity

13.5 Intrinsic resistance to viruses in vertebrates

13.6 Innate and intrinsic immunity and the outcome of infection

13.7 RNAi is an important antiviral mechanism in invertebrates and plants

13.8 Detecting and signalling infection in invertebrates and plants

13.9 Virus resistance mechanisms in Bacteria and Archaea

When a virus first enters a susceptible host, that organism is vulnerable to potentially catastrophic damage from the virus's uncontrolled replication and consequent damage to host tissues and functions. It is not surprising, therefore, that natural selection has led to the evolution of a variety of defence mechanisms in cellular organisms, the purpose of which is to slow down and ultimately to eliminate the virus, or other pathogens, from the system. Traditionally, these defence mechanism have been categorized as either innate or adaptive – the former being non-specific in nature whilst the latter develop specifically to counter the particular pathogen that is present. Adaptive immunity is considered separately in Chapter 14; however, it is important to remember that there are links between these two types of response. More recently, a third category of immune function has been recognized – intrinsic immunity – encompassing functions that are present in cells all the time rather than being induced by infection. In the face of this network of host defences, viruses have been far from passive bystanders. They have evolved functions that act against these defences, to blunt their effects and so permit successful virus replication nonetheless.

Introduction to Modern Virology, Seventh Edition. N. J. Dimmock, A. J. Easton and K. N. Leppard.
© 2016 John Wiley & Sons, Ltd. Published 2016 by John Wiley & Sons, Ltd.

13.1 Innate immune responses in vertebrates – discovery of interferon

Classical innate responses in vertebrates are mediated by interferon (IFN), a secreted cytokine that was the first element of the network of innate immunity to be identified. Its discovery by Isaacs and Lindenmann in 1957 was a watershed in virology. They described a substance secreted from cells treated with heat-inactivated influenza A virus that could, if applied to uninfected cells, drastically inhibit (interfere with) subsequent infection of those cells by a number of different viruses, not just

the influenza virus used initially to induce it (Box 13.1). It is now known that only vertebrate cells produce IFN; plants and insects do not. Subsequent to its discovery, IFN has been divided into three types, all of which are polypeptides encoded in the genome. Type 1 IFN comprises IFNα and IFNβ, with most mammals having multiple IFNα genes and a single IFNβ gene. Type II IFN is IFNγ, a cytokine produced by specific cell types that functions within the adaptive immune response (Chapter 14), while Type III IFN – the relatively recently-discovered IFNλ – plays a similar role to Type I IFN but may be particularly important in the response to certain pathogens.

Box 13.1

The discovery of interferon by Isaacs and Lindenmann (*Proc Roy Soc London series B, vol 147, 258–267 and 268–273, 1957*):

- Pieces of fertile hens' egg chorio-allantoic membrane were incubated in culture at 37 °C for 24 h with either buffer alone or buffer containing heat-inactivated influenza A virus.
- These membrane pieces were then washed and placed in fresh buffer containing a low concentration of live influenza A virus at 37 °C to test the ability of the membranes to support virus growth, which was measured by HA assay (Section 5.4) after 48 h incubation.
- Virus titre was reduced by 100-fold by this pre-incubation with heat-killed virus.
- The interference effect required time and metabolic activity to develop as pre-incubation with heat-killed virus for just 15 min followed by 24 h at 37 °C gave interference when the membranes were subsequently incubated with live virus whereas the same brief pre-incubation followed by 24 h at 2 °C did not.
- The membranes actively secreted an interfering activity into the medium as culture fluid from membranes that had been briefly incubated with heat-killed virus and then washed before the 24 h pre-incubation was itself able to inhibit virus infection when added to fresh membrane cultures. This activity was termed 'interferon'.
- The biological activity of interferon could be titrated in dilution assays.
- Interferon was active against (a) strains of influenza A virus unrelated to the 1935 Melbourne virus used to generate the interferon, and (b) unrelated viruses such as Sendai, Newcastle disease and vaccinia.

13.2 Induction of type 1 interferon responses

Binding of pathogen components by a series of molecular detectors in cells triggers the type 1 interferon response, setting in motion a series of events leading to a gross change in gene expression within the cell. The first purpose of this is to produce an environment that is more hostile to pathogen replication, and the second is to signal to neighbouring cells that they might be at risk, so they too initiate production of that hostile environment.

Sensing a pathogen

Innate immune responses begin at the level of the individual infected cell. To mount an innate response a cell must become aware that a pathogen is present that may be about to, or is beginning to, initiate an infection. This detection is carried out by several classes of receptor, known collectively as pattern recognition receptors or PRRs, which act as pathogen sensors. Collectively, the structures recognized by PRRs are known as pathogen-associated molecular patterns, or PAMPs. These comprise various features of pathogen molecular architecture that are either unique to pathogens or to the infected cell (i.e. not found in the uninfected host) or at least are not normally found in that compartment of the host cell where the PRR is located, such as free DNA in the cytoplasm.

Normal cell activity does not produce double-stranded RNA (dsRNA) whilst, in contrast, it is very much a part of the lifecycle of many viruses; it is thus a key hallmark of the infected state and an important PAMP. Some viruses, such as the rotaviruses, have dsRNA genomes. Many other viruses have single-stranded RNA genomes, where replication/

transcription requires the synthesis of molecules of opposite polarity which therefore have the potential for base-pairing to form dsRNA. Even the viruses with DNA genomes have genes densely packed on them, often with opposite strands overtly coding for different products, in which case mRNA molecules will be made that can base-pair. Other viral PAMPs include DNA in which CpG dinucleotides are unmethylated, this feature distinguishing it from host DNA, and uncapped cytoplasmic RNA 5′ ends.

Several classes of PRR have been identified to date, including: the Toll-like receptors (TLRs) which are membrane proteins present in either the plasma membrane or the endocytic compartment; the RIG-I-like receptors (RLRs) which are cytoplasmic RNA helicases; the DDX/DHX receptors which are RNA/DNA helicases; and the NOD-like receptors (NLRs), also cytoplasmic. Several other DNA sensors are also located in the cytoplasm, though these PRRs have not been classified by structure so far. The activities of PRRs that are important in virus infection are summarized in Box 13.2. Generally, RLRs are thought to be most important in the response to RNA viruses whilst the cytoplasmic DNA sensors and TLR9 respond to DNA viruses. There is also a mechanism using RNA pol III whereby cytoplasmic DNA can be transcribed to give RNA molecules that can then engage RLRs such as RIG-I.

Vertebrate TLRs are structural homologues of Toll, a receptor expressed in insects, and have ligand binding domains formed of leucine-rich repeats, similar to PRRs in plants (Section 13.8). There is species variation in the nature and number of TLRs, suggesting that evolution may be shaping the PRR repertoire to counter different types of pathogen that are relevant for the host organism. It is important also to remember that not all cell types express every PRR so the ability of different cells to mount innate responses to particular forms of

Box 13.2

Pattern recognition receptors that respond to virus infection[*]

Receptor[†]	Pathogen pattern recognized	Examples of viruses affected[‡]
TLR2 (PM)	Envelope fusion proteins	Class 1: CMV, HSV, VZV Class 5: LCMV, measles, VSV
TLR3 (endo)	dsRNA	Class 1: EBV Class 3: Reovirus Class 4: EMCV, West Nile virus Class 5: RSV
TLR4 (PM)	Envelope fusion proteins	Class 5: Ebola virus, RSV Class 6: MMTV
TLR7/8 (endo)	GU-rich ssRNA	Class 4: Coxsackie B, Sendai virus Class 5: influenza A virus, VSV Class 6: HIV1
TLR9 (endo)	DNA with unmethylated CpG	Class 1: CMV, HAdV, HSV
RIG-I (RLR; cyto)	RNA with 5′ triphosphate; short dsRNA	Class 1: EBV Class 3: Reovirus Class 4: flaviviruses Class 5: orthomyxo, paramyxo and rhabdoviruses
Mda5 (RLR; cyto)	Long dsRNA	Class 1: vaccinia Class 5: PIV5 (was SV5), other paramyxoviruses
IFI16 (cyto)	dsDNA	Class 1: HSV, KSHV
AIM2 (cyto)	dsDNA	Class 1: vaccinia
DDX41 (cyto)	dsDNA	Class 1: HAdV, HSV
DDX1/3/21 DHX9/36 (cyto)	dsRNA	Class 3: Reovirus Class 5: influenza A virus, VSV
DHX9/36 (cyto)	CpG dsDNA	Class 1: HSV
PolIII (cyto)	AT-rich dsDNA	Class 1: EBV, HAdV, HSV
DAI (cyto)	dsDNA	Class 1: HSV
cGAS (cyto)	dsDNA	Class 1: HSV; vaccinia

[*] The fact that a virus is not listed for a given receptor does not mean that the receptor is necessarily irrelevant for that virus.

[†] Receptor abbreviations: cGAS – cyclic GMP-AMP synthetase; cyto – cytoplasmic; DDX/DHX – DEAD/DEAH box helicase; endo – endocytic compartment; PM – plasma membrane; PolIII –RNA polymerase III; RLR – RIG-I-like receptor; TLR – Toll-like receptor.

[‡] Virus abbreviations: CMV – cytomegalovirus; EBV – Epstein Barr virus; EMCV – encephalomyocarditis virus; HAdV – human adenovirus; HIV1 – human immunodeficiency virus 1; HSV – herpes simplex virus; KSHV – Kaposi's sarcoma herpesvirus; LCMV – lymphocytic choriomeningitis virus; MMTV – mouse mammary tumour virus; PIV5 – parainfluenzavirus 5; RSV – respiratory syncytial virus; SV5 – simian virus 5; VSV – vesicular stomatitis virus; VZV – varicella-zoster virus.

pathogen varies. Phagocytic cells typically express the fullest range of PRRs.

Receiving the signal – how PRRs tell a cell about infection

Whenever a PRR is engaged by its cognate PAMP, it initiates signalling cascades in the cell that lead to altered gene expression (Fig. 13.1). These cascades differ in detail between PRRs but, in summary, PAMP binding by a PRR triggers association of a PRR with its adaptor protein, via which interactions take place to activate signalling cascades. Some PPRs share the same adaptor but each adaptor commissions a different signalling cascade. These cascades lead to activation of enzymes that are variously protein kinases or ubiquitin ligases. Ultimately, two key transcription factors are activated: IRF3 and NFκB. IRF3 dimerizes upon phosphorylation and enters the nucleus, while destruction of the NFκB inhibitor, IκB, by the proteasome following IκB phosphorylation

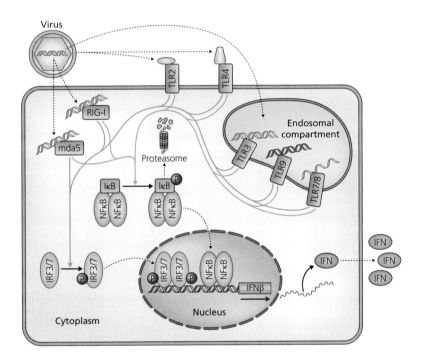

Fig. 13.1   Detection of virus by pattern recognition receptors (PRR). Infection is sensed by a combination of plasma membrane, endosomal and cytoplasmic PRRs of which the most significant for virus infection are TLR2, 3, 4, 7/8, 9, RIG-I and mda5. Cytoplasmic DNA receptors (not shown) are also important for DNA virus recognition. Each virus will be recognized by one or more PRR. Activated PRRs recruit specific adaptor molecules and, via signalling cascades, cause the phosphorylation of IRF3 or IRF7 (blue arrows) and IκB (orange arrows). Phospho-IRF3/7 forms homo- or hetero-dimers which bind target promoters in the nucleus, including the one controlling IFNβ gene expression. Phospho-IκB is a target for ubiquitin-mediated degradation, releasing active NFκB to bind target promoters. IRF3/7 and NFκB, with ATF/c-jun (not shown) activate IFNβ gene expression. Movement of factors is indicated by black dotted arrows. Note that TLR7/8 and 9 activate IRF7, not IRF3, and operate principally in plasmacytoid dendritic cells; most cell types do not express IRF7 prior to stimulation.

releases NFκB to enter the nucleus also. IRF3 and NFκB together activate the IFNβ promoter, causing the cell to secrete IFNβ, while NFκB is also involved in activating a variety of pro-inflammatory genes. A further factor, IRF7, can also be activated by these networks to activate IFN promoters but since IRF7 is normally undetectable in most cell types until induced by IFN, its role is principally in amplifying the IFN response rather than initiating it.

Establishing an antiviral state – protecting uninfected cells

The production of IFNβ in response to PAMP sensing is the means by which the presence of virus is transmitted to other neighbouring cells. IFNβ is a secreted cytokine that can act via IFN α/β receptors either on the producer cell or on other cells in which no PRR activation has occurred. In both cases, IFNβ engagement of the receptor activates the JAK/STAT signalling pathway which leads the transcription of IFN-stimulated genes (ISGs; Fig. 13.2). In transcriptomics studies, several hundred genes have been found to respond to IFN in this way. This pathway also activates expression of IFNα, which is secreted to give massive amplification of the initial response. Several ISG products, such as IRF7, STAT1, dsRNA-activated protein kinase (PKR) and 2′–5′oligoadenylate synthase play known roles in the signalling or effector phases of the IFN response while others are assumed to play some direct or indirect role in the response based on their induction profile. Some ISG products, such as MxA, are also present at lower levels in the absence of IFN and are now being characterized as having intrinsic antiviral activity (Section 13.5).

In addition to being an important viral PAMP, dsRNA is also an activator of various antiviral effectors that are induced by IFN. Examples include 2′,5′ oligoA synthetase and

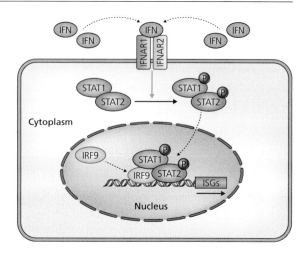

Fig. 13.2 Induction of interferon-stimulated genes (ISG) by interferon. IFNα or IFNβ binds its receptor in the plasma membrane, activating receptor-associated tyrosine kinases that phosphorylate a STAT1/STAT2 heterodimer. This binds to IRF9 in the nucleus to form a complex known as ISGF3, which binds to the promoters of a large number of ISGs and stimulates their transcription.

PKR. When bound to dsRNA of more than a certain length, PKR phosphorylates the translation initiation factor eIF2α, which is normally involved in recycling the initiation factor eIF2 to participate in further cycles of translation initiation. Phosphorylated eIF2α is inactive, meaning that this recycling of an essential protein no longer occurs and translation initiation is blocked. It is important to note that this inhibition is not virus-specific; all translation in the cell is affected. The viability of the cell is thus compromised as the price for stopping the infectious agent, hence the importance of the two layers of regulation; PKR is first induced by IFN, the presence of which indicates infection is ongoing in the immediate vicinity, and then activated by dsRNA, which indicates the cell itself is infected.

13.3 Virus countermeasures to innate immunity

Most viruses encode factors that limit the host's interferon responses. Strains that lack these functions are typically attenuated, i.e. are less or non-pathogenic, indicating the importance of interferon responses to the outcome of infection.

The ability of innate immunity to limit virus replication and spread in a host has selected for the emergence of encoded functions in most or all viruses that inhibit these host responses. Typically, such inhibition is not complete, but is sufficient to blunt the impact of the response on virus replication. Viral functions are known that target all phases of the IFN response from disrupting PAMP recognition through to blocking the activity of specific antiviral effectors that are expressed from ISGs (Box 13.3).

One of the first of these viral functions to be characterized in detail was the adenovirus VA1 RNA, which blocks the activation of PKR and hence prevents the arrest of translation. It does this by providing lengths of dsRNA within its

---

Box 13.3

Examples of viral functions that inhibit innate immunity.

Virus Classification	Virus and viral function	Host target and effect
Class 1: adenovirus	Human adenovirus 5: VA1 RNA	Protein kinase R; blocks its activation by dsRNA
	Human adenovirus 5: E4 Orf3	Promyelocytic leukemia protein; other PML body components; permits virus growth in IFN-treated cells
	Human adenovirus 5: E1A	Inhibits IFN transcription
Class 1: herpesvirus	Herpes simplex type 1: ICP0 Cytomegalovirus: IE72	As for adenovirus E4 Orf3
Class 1: poxvirus	Vaccinia virus: E3L, K3L	Block RNaseL, PKR activity
	Vaccinia virus: B18R	Binds type I IFN; acts as decoy receptor
	Vaccinia virus: A46R	Binds the essential adaptors for TLR activation; blocks TLR signalling
Class 3: rotavirus	Group A rotavirus: NSP1	In various isolates, targets RIG-I, IRF3, IRF7 and/or βTrCP for degradation
Class 4: picornavirus	Picornaviruses: 3C protease	Cleaves RLRs and/or their essential activating adaptors
	Enterovirus: 2A protease	Inhibits IFN downstream signalling
Class 4: hepacivirus	Hepatitis C: NS3/4A	Cleaves the essential adaptor for RIG-I/ mda5 activation, MAVS
Class 5: paramyxovirus	Paramyxoviruses: V proteins	Blocks mda5 activation and causes STAT1 degradation
	Pneumoviruses: NS1, NS2	Block activation of IRF3
Class 5: orthomyxovirus	Influenza A virus: NS1	Binds dsRNA and PKR; blocks IFN induction at multiple levels
Class 5: filovirus	Ebolavirus: VP35	Binds dsRNA

Box 13.4

Evidence for the action of paramyxovirus V proteins in inhibiting interferon responses

- Induction of type 1 IFN-responsive genes by IFN treatment of human diploid fibroblasts is inhibited by SV5 infection.
- STAT1, a key transcription factor for activation of type 1 IFN-response genes, is degraded in SV5-infected cells and this requires the proteasome.
- A vector expressing only the SV5 V protein mimics both these effects.
- Several paramyxovirus V proteins bind to mda5, an RLR-type PRR that is activated by intracellular dsRNA.
- These V proteins also inhibit the activation of the IFNβ promoter by dsRNA.

  Didcock et al. (1999) J Virol 73, 9928-33; Andrejeva et al. (2004) Proc Natl Acad Sci USA 101, 17264-69.

---

structure that are just sufficient to bind PKR but not enough to activate it; VA1 therefore acts as a decoy to prevent PKR binding to any longer dsRNAs that would activate the enzyme. More recently, the V proteins of paramyxoviruses were found to block both the induction of IFNβ and the subsequent induction of interferon stimulated genes (Box 13.4). The effect of the V proteins on STAT1 exemplifies a strategy found widely among viral proteins that disrupt host functions, which is the targeting of specific host proteins for proteasomal degradation. They achieve this by forming a novel substrate-recognition subunit for host ubiquitin ligase enzymes, redirecting them to modify new targets for poly-ubiquitination, a post-translational modification tag which is then used to direct their degradation. Several viruses express proteins that disrupt the structure of promyelocytic leukemia (PML) bodies in the nucleus. Components of these bodies are important for interferon responses, though the mechanism is not defined, and also have intrinsic activity against some viruses.

In all cases, the molecules mediating viral inhibition of host responses have to make specific interactions with host molecular targets in order to block their activity and, because there is sequence divergence between orthologous molecules in different species, these interactions are frequently found to be host species-specific. In other words, the countermeasures that a particular virus deploys against host innate immunity will work in some species and not others. A virus that has adapted to grow in host species X which finds itself within host species Y will typically not be able to block innate responses in the new species and so will be highly attenuated or even incapable of replicating in that new host. Thus, one of the key properties a virus must acquire when making a zoonotic transition into a new host species is an ability to moderate the innate and intrinsic responses that it encounters. It follows that the ability of a host to counter the replication of an invading virus is a key determinant of virus host range and pathogenicity.

13.4 TRIM proteins and immunity

Mammals express a large family of proteins that share a tripartite motif (hence the name TRIM) of conserved protein sequence elements in their N-terminal 360 amino acids. Although most of these are constitutively expressed, many family members are transcriptionally upregulated by type 1 IFN, which implicates their products in antiviral responses. A few of these TRIM proteins have been shown to have intrinsic antiviral activity, such as the interaction of TRIM5α with HIV that limits uncoating, whilst many others play important roles in regulating the signalling networks that lead to the IFN response and at least one, TRIM21, functions as an effector of adaptive immunity (Box 13.5). Thus, the idea has emerged that this whole gene family may be involved in countering infection in various ways. Certainly, the TRIM family

appears to be a rapidly-evolving group of genes within the mammals, which fits with the notion that they are under positive selective pressure from various pathogens. Equally, there are some members of this family whose functions, as defined so far, have nothing to do with infection. These individual TRIM proteins may have acquired functions outside of their original purpose as antiviral proteins or, alternatively, what was a primordial gene with a function in normal development or cell function has, through gene duplication and diversification, been utilized more recently to protect against infectious threats.

How can members of the TRIM gene family have such diverse functions? Many of the TRIM family members have been shown to have ubiquitin E3 ligase activity, each targeted towards different specific proteins and, given the essential role of the TRIM motif in that activity, it is suspected that all may share this property. E3 ligases regulate target protein

---

Box 13.5

Examples of TRIM protein involvement in immunity to viruses

TRIM protein	Function(s) in intrinsic or innate immunity
TRIM5α	Intrinsic inhibitor of HIV1 uncoating and other retroviruses; acts as PRR
TRIM8	Potentiates STAT1 activation in IFN response
TRIM11	Intrinsic inhibitor of HIV1 transcription
TRIM15	Intrinsic inhibitor of HIV1 assembly
TRIM19 (PML)	RNA virus resistance factor; DNA virus resistance factor; regulates IFN response
TRIM21	Intracellular Ig receptor which targets antigen complexes for proteasomal degradation; modulates IRF3/5/7/8 activation
TRIM22	Intrinsic inhibitor of HIV1 transcription/assembly
TRIM23	Promotes activation of IRF3 and NFκB
TRIM25	Activates RIG-I signalling in IFN response
TRIM27	Inhibits IRF3/7 and NFκB activation during PRR signalling
TRIM28 (TIF1β)	Intrinsic inhibitor of retroviral gene expression
TRIM30α	Inhibits NFκB activation during PRR signalling
TRIM32	Intrinsic inhibitor of HIV1 transcription; promotes activation of NFκB
TRIM56	Activates signalling from cytoplasmic DNA sensors

function, interaction, location and/or stability via the covalent post-translational attachment of either ubiquitin or similar molecules such as SUMO and ISG15; these effects include the targeting of proteins for degradation by the proteasome by K48 poly-ubiquitin addition. This means that the TRIM family as whole could, by targeting suitable proteins, influence any pathway in the cell.

13.5 Intrinsic resistance to viruses in vertebrates

Cells do not have to wait for the induction of interferon responses to begin defending themselves against an infecting virus. An increasing number of factors are being discovered that are present in normal healthy cells and which serve to inhibit infection by certain viruses. These are known as intrinsic resistance factors. Cells can also deploy autophagy to destroy incoming viruses and, if necessary, can activate apoptosis to protect the rest of the organism from infection.

Virus-specific resistance factors

The term 'intrinsic immunity' has come into use to describe the features of an uninfected host that are already in place before infection and which block or limit virus infection. This is in contrast to IFN and its effects, which are induced by infection. However, the distinction between innate and intrinsic immunity is not yet clearly drawn and is likely to be somewhat artificial since several known intrinsic immunity factors are upregulated by interferon. Often intrinsic immunity factors have a substantial measure of virus specificity, so rather than there being just one or a few intrinsic resistance factors that are globally effective against all viruses, there are instead multiple factors, each of which acts on just one or a few viruses.

Although some evidence had been found for the existence of intrinsic immunity factors through more directed studies, a step-change occurred with the application of systems-level approaches to the problem. By screening the complete mRNA population from a host for the ability to reduce virus replication in a chosen cell line, it was possible to take a global view of the impact of the host cell environment on the outcome of the infection. Exactly the same approach has also revealed many host components that are unexpectedly advantageous to the virus in terms of the productivity of the infection. Clearly, the identity of both negative and positive factors is virus-dependent. From the rate at which knowledge of these factors has emerged in recent years, and the relatively restricted range of viruses that has so far been examined in detail, it must be very likely that many more such factors remain to be discovered.

Some of the best characterized intrinsic immunity factors are listed in Box 13.6, with their known targets. To these should be added TRIM5α (Box 13.5), which is the best understood of the TRIM family members active against HIV. The mechanism of action of factors that limit HIV replication are considered in Chapter 21. In many cases, the importance of these intrinsic immunity factors is underscored by the fact that the viruses targeted have evolved functions that oppose their actions. As with viral functions opposing innate immunity, the interactions of these functions with intrinsic immunity factors are also important determinants of cross-species transmission and pathogenicity.

Autophagy and apoptosis as intrinsic defences

Autophagy is a mechanism for the recycling of redundant or damaged cell content, including

Box 13.6

Important non-TRIM viral intrinsic resistance factors

Resistance factor	Virus targeted	Action
APOBEC3G	HIV	Causes genome hypermutation
IFITM3	Influenza A	Interferes with attachment or entry
MxA	Influenza A; other viruses	Binds nucleoprotein complexes
Tetherin	HIV, other retroviruses; paramyxo, filo, rhabdo and arenaviruses; KSHV	Reduces particle release

organelles such as mitochondria. It involves the envelopment of the material within a newly-formed autophagic membrane, followed by fusion of that autophagosome with a lysosome, which exposes the autophagosome content to degradative enzymes. The same mechanism can also be directed towards the destruction of bacterial and other pathogens that invade a cell.

One of the triggers for autophagy which is relevant to viral infection is the exposure to the cytoplasm of lipids that are normally concealed on the inner face of endocytic vesicles, i.e. the side of the lipid bilayer that is topologically equivalent to the outside of the plasma membrane. Many viruses first enter a cell within endocytic vesicles (Chapter 6). Whilst enveloped viruses escape the vesicle by fusion, which does not breach the integrity of the vesicle membrane, the escape of non-enveloped viruses such as adenovirus can trigger assembly of an autophagic membrane. Depending on the speed of movement of the residual virus away from the damaged vesicle, this autophagic response could reduce the proportion of initial infection events that actually lead to establishment of the viral genome within the cell.

Apoptosis is, in effect, an inbuilt self-destruct program within cells. A key feature of apoptosis, that distinguishes it from necrotic cell death (lysis; See Section 15.7), is that the dying cell remains intact and its contents stay within the plasma membrane. This prevents release of cell contents which are inflammatory and which have to be tidied up by scavenger cells, particularly those of the immune system. During apoptosis a cell undergoes profound internal changes which include fragmentation of its chromosomal DNA. These processes follow a clear, well-regulated pattern. Ultimately, the cell rounds up and is disposed of by being engulfed by a phagocyte, within which it is hydrolyzed. Apoptosis serves to regulate cell number during development, familiar manifestations being the separation of webbed digits in the human embryo to give fingers and toes, the loss of the tail of the amphibian tadpole, and the removal of self-reactive T cells during the development of the immune system. It is also a potent force in intrinsic antiviral defence.

Apoptosis can be triggered either by appropriate external signal molecules binding to cell surface receptors (extrinsic pathway) or within a cell (intrinsic pathway). The pathway can be activated if the survival of a cell might pose a threat to the longer-term health of the organism as a whole. Virus infection at the cell level clearly represents such a threat, so it is unsurprising that the intracellular events that occur during virus infection, some commissioned directly by the virus – e.g. the production of viral proteins or the shut-down of host gene expression – and others arising indirectly from the host's response to the presence of the virus, frequently activate pro-apoptotic signalling. If unchecked, such signals result in the shut-down of the cell and the ordered destruction of its contents occurring with a time-span of a few hours. This can limit the potential productivity of the infection. As a result, apoptosis apparently imposes a selective pressure on viruses since many have evolved functions that block apoptotic signalling pathways. A key target of these viral functions is the cellular stress/damage regulator p53 which, when activated, can drive the transcription of pro-apoptotic genes. Some of these viral functions are considered further in Section 25.3 in the context of carcinogenesis driven by virus infection. Some viruses also inhibit the extrinsic pathway of induction, e.g. disruption of signalling from cell surface receptors by adenovirus E3 proteins. Virus-induced apoptosis can be an important factor in pathogenesis.

13.6 Innate and intrinsic immunity and the outcome of infection

Viral pathogenesis is a major topic that is covered in Part IV of this text. However, it is important to appreciate here that the nature and extent of the innate or intrinsic immune response is a crucial determinant of viral pathogenicity in most, if not all, infections in vertebrates. If the host response is too weak, then virus replication will spread widely and potentially cause severe disease before there is time for an adaptive response to control it. Equally, a strong innate response will itself cause signs and symptoms that are frequently associated with virus infection. The strength of

the innate response will be determined in part by the number of PRRs that are triggered by the virus, but also by the success of the virus in inhibiting these innate host responses via inhibitory functions which they express upon infection. Virus mutants that lack these functions are normally attenuated.

Although innate immunity has evolved to limit virus replication, the superior rate of virus evolution over that of any eukaryotic host means that the virus will always be a step ahead in the evolutionary arms race. Thus the adaptive immune response, where a highly-specific response even to a totally new pathogen can be generated outside of the constraints of the slow rate of host evolution, is necessary to achieve elimination of a virus from the system. The detailed workings of the adaptive response are considered in Chapter 14. However, it is important to note that adaptive responses are stimulated by the preceding innate response, which increases the expression of cytokines to attract immune cells to the site of infection and enhances the presentation of antigen to those cells by up-regulating the expression in the tissue of components such as MHC Class 1 molecules.

13.7 RNAi is an important antiviral mechanism in invertebrates and plants

It is now well understood that virtually all eukaryotes possess a microRNA (miRNA) system. These systems act on short hairpin regions found within RNAs transcribed in the nucleus. The hairpins are cleaved out, exported to the cytoplasm as dsRNA and then further trimmed by a Dicer enzyme before one of the strands becomes associated with a so-called silencing complex (RISC) that includes an Argonaut protein. RISC then binds to cytoplasmic mRNA to which the short RNA has base-pairing complementarity and either cleaves that mRNA (inactivating it) or else represses its translation. This mechanism has proved to be a very significant means by which organisms regulate their gene expression.

In invertebrates and plants, it is now clear that a very closely-related RNAi system is a crucial feature of their immunity against viruses (Fig. 13.3). Cytoplasmic viruses give rise to

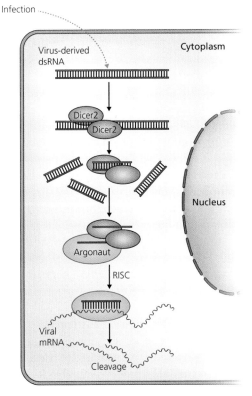

Fig. 13.3 RNAi pathway in insect cells. Cytoplasmic dsRNA arising during virus infection is recognized by Dicer2, leading to cleavage of the dsRNA into dsRNA oligonucleotides. After associating with Argonaut protein, the duplex is unwound and one of the two strands is selected as the guide strand and loaded into the RISC (RNA silencing) complex. RISC associates with viral mRNA by base-pairing with the guide strand, leading to cleavage of the mRNA by Argonaut and consequent inactivation of the mRNA.

RNAs that are cleaved by Dicer-like (DCL) activities which, in these organisms, include enzymes that are specialized for recognizing long dsRNA molecules. DCL action therefore yields lots of vsiRNAs, potentially derived from sites across the full viral genome length, that are then loaded into RISC to target viral RNA and inhibit infection. If the viral genome is dsRNA or is formed of ssRNA but with regions of secondary structure, it may be a direct target for DCL action upon entry into the cytoplasm. Alternatively, transcription and replication from either RNA or DNA genomes will produce self-complementary RNAs that can base-pair to provide a substrate for DCL activity. In practice, whether particular dsRNAs or regions of secondary structure are actually substrates for DCL will depend on whether they are protected by bound proteins, and by their compartmentation within the cytoplasm.

An important feature of the RNAi system, in plants and nematodes at least, is amplification. Once DCLs have acted upon a viral RNA molecule to produce primary vsiRNA, these can be copied by a resident RNA-dependent RNA polymerase to greatly expand the pool of vsiRNA available for incorporation into RISC. This feature also allows the immunity to a virus that results from vsiRNA action to spread systemically as secondary vsiRNAs are released from producer cells, travel to other parts of the organism and are then taken up again. In every cell that takes up vsiRNA, re-amplification can occur. Even though insects (*Drosophila*) appear not to have this amplification system, they also achieve systemic spread of a vsiRNA response by some means. In some organisms, such as the nematode *C. elegans*, the vsiRNA system is even capable of delivering an immune memory across generations.

Given the importance of RNAi as an antiviral mechanism in these organisms, you would expect viruses that infect them to have evolved mechanisms to moderate its effects. Certainly, many of the plant and insect viruses that have been studied do indeed encode proteins that interfere specifically with DCL or Argonaut function. Just as with viral evasion of innate immunity in vertebrates, these plant/insect viral functions appear to be important pathogenicity determinants. It has also been suggested that arboviruses (viruses that are carried in insects and are transmitted into vertebrate hosts) have a different interaction with RNAi systems than viruses which are acute insect pathogens and that this is linked to their long-term carriage in the insects without significant pathogenesis.

13.8 Detecting and signalling infection in invertebrates and plants

Innate immunity is an evolutionarily ancient form of antiviral defence. There are considerable similarities between the mechanisms of pathogen sensing and response between vertebrates, invertebrates and plants. These include the architecture of transmembrane pattern receptors and the key signalling pathways.

Although RNAi is the major immune system active in insects against virus infection, they also have systems for pathogen sensing and response that have clear molecular and functional parallels with the interferon response pathway in vertebrates. The primary characterization of these pathways has been in the context of responses to bacterial or fungal pathogens, but they also function in an antiviral context. Two receptors, including Toll which has structural similarity to the TLRs of vertebrates and so gave them their name, are activated by ligands produced as a consequence of bacterial infection. Each then activates a signalling cascade that leads to activation of

NFκB orthologues and changes in gene expression. How virus infection activates these pathways is not certain, but deficiencies in the pathways can enhance virus replication which suggests they are important.

Insects also have a Jak/STAT pathway that has similarities with the signalling pathway from interferon receptors in vertebrates (Section 13.2). This pathway is known to be activated by two polypeptide ligands binding to cell-surface receptors. One of these ligands is expressed in response to bacterial infection while the other, *vago*, is expressed in response to signalling from activated Dicer-2, which is the insect enzyme that is active in the RNAi response. Once the Jak/STAT pathway is activated, a phosphorylated STAT dimer enters the nucleus to drive transcription of response genes that encode antiviral products. Since other proteins involved in the RNAi pathway are not needed for this effect of Dicer-2, the response is independent of the RNAi response itself. Instead, Dicer-2 appears to be acting as a pattern-recognition receptor (PRR), responding to a pathogen signal (dsRNA) to elicit a downstream response. Importantly, since *vago* is secreted, this effect can spread to neighbouring cells. This can be seen as analogous to the effect of secreted interferon in vertebrates.

Plants also have receptors that trigger responses to pathogens; as with insects, these have been largely characterized in the context of bacterial or fungal pathogens but also have relevance to virus infection. Initially, the presence of a pathogen is recognized by cell surface PRRs binding conserved features of the pathogen (*PAMP-triggered immunity*, PTI) whilst in the second phase, intracellular effectors released by the pathogen to counteract that initial host PTI response are themselves recognized by cytoplasmic receptors (*effector-triggered immunity*, ETI). Activation of ETI often leads to a hypersensitive response, causing a zone of necrosis in the leaf tissue but, importantly, it also leads to protection of the surrounding tissue and other tissues of the plant against infection.

Both classes of receptor involved in this pathway are characterized by leucine-rich repeats that give binding specificity for diverse pathogen features; in the cell surface PRRs this domain is extracellular and is linked to a cytoplasmic protein kinase domain which is activated by dimerization and reciprocal phosphorylation to initiate a signalling cascade. The intracellular receptors are homologous to the NLR class of intracellular receptor in vertebrates. The cell surface PRRs are functionally analogous to the TLRs of vertebrates, though structurally related, but the gene family is vastly more diverse. In the much-studied plant *Arabidopsis*, there are more than 600 PRR genes, as compared with 10–12 TLRs in mammals.

13.9 Virus resistance mechanisms in bacteria and archaea

Prokaryotes can store a genetic memory of infections by copying short segments of pathogen genomes into their own. Transcripts from these segments can cause the destruction of incoming DNA with matching sequence.

Unlike the relationship between viruses and large, multicellular eukaryote hosts, the potential rate of evolutionary adaptation in prokaryotic hosts is a lot closer to that of the bacteriophage that infect them. Until recently, the prokaryotes were thought not to have any form of active phage resistance mechanism but rather that they survived both by rapid adaptation, e.g. to lose surface features used by the phage to attach and enter the cell, and by

co-existing, e.g. when a bacterium becomes a phage lysogen and carries the phage genome inserted into its own, whilst at the same time operating potent DNA restriction/modification systems that minimize the likelihood of successful invasion by any foreign DNA. However, with the discovery of the CRISPR system (Fig. 13.4), it is now clear that this is not the case; just like animals and plants, the Bacteria and Archae have a means to repel infection and even to pass on a memory of that infection to daughter cells.

CRISPR stands for 'clustered regularly interspaced short palindromic repeat', a phrase which describes the key genomic feature that encodes this resistance mechanism. In essence, CRISPR loci function as an expanding library of short DNA sequences that the cell (or its ancestors) has encountered. Each locus – and a cell may have several loci – comprises a series of short elements, each one having the form of a spacer (S) located between repeated (R) sequences that can be represented as 'R-S-R'; the spacers in these elements represent the sequence memory that is held in the library. A new 'memory' is formed when a foreign DNA molecule, such as a phage genome, enters the cell; sequence features in the invader are recognized and cleaved by Cas (CRISPR-associated) proteins to release 30 bp fragments that are then inserted as spacers within new R-S-Rs that are formed at the 5′ end of the CRISPR locus, which therefore grows by one R-S-R. As a result, there can be diversity at the level of the individual in the length of CRISPRs and in the precise collection of sequences carried within these loci among different isolates of a species.

The collection of spacer elements in a CRISPR locus is used to protect against future infection by any phage whose sequence matches one or more of the spacers. Each CRISPR locus is routinely transcribed to produce a long RNA in which the repeat units in the R-S-R elements form hairpin structures

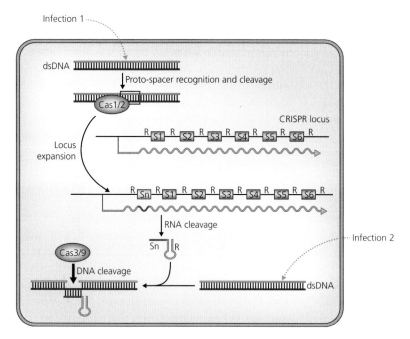

Fig. 13.4 The CRISPR system in Bacteria and Archaea. Foreign DNA is recognized and processed by Cas proteins to yield short DNA oligonucleotides that are then integrated as spacers between repeat elements in CRISPR loci in the genome. CRISPR RNA transcripts fold to give hairpin structures derived from each repeat element transcript; these are recognized in processing the RNA to yield molecules that can base-pair with target DNA should it be encountered again on subsequent infection. Base-pairing leads to cleavage and inactivation of the target DNA by the action of further Cas proteins. The precise details of these mechanisms vary between CRISPR systems.

that flank and separate the transcripts of each spacer. This RNA is processed by specific nucleases to yield short CRISPR RNAs that can base-pair with their target in the phage genome, an event which leads ultimately to specific cleavage of that genome and, hence, inactivation of its infectivity. It is even possible that some CRISPR systems can store sequences that are encountered as RNA, and hence be active against RNA phages, though the details of how this works are not yet clear.

Key points

- Vertebrates possess a type 1 interferon response, which is induced within an infected cell but spreads to other cells to establish a general antiviral state.
- The interferon response is initiated when molecular features of a pathogen are recognized by one or more pattern recognition receptors.
- The interferon effect is mediated through the induction of transcription of a large set of genes.
- Invertebrates and plants have pattern recognition receptors and signalling pathways that have similarity to those in vertebrates.
- Invertebrates and plants use RNAi, a system that deploys short RNAs to target mRNA for cleavage, as a key means of defense against virus infection.

- Bacteria and Archaea are protected by CRISPR systems, which preserve a nucleic acid 'memory' of past infections and use this as a defense based on nucleic acid base-pairing and cleavage.
- Intrinsic defense factors, always present and specific against certain viruses, are being increasingly recognized in vertebrate systems.

Questions

- Explain the activation of the interferon response by virus infection and how viruses counteract this.
- Compare and contrast pathogen sensing in vertebrates, invertebrates and plants.
- How do we know that innate and intrinsic immunity are important antiviral mechanisms?

References

Cullen, B. R. (ed.) 2013. *Intrinsic Immunity. Current Topics in Microbiology and Immunology*, vol. 371, Springer, Heidelberg, New York.

Gürtler, C., Bowie, A. G. 2013. Innate immune detection of microbial nucleic acids. *Trends in Microbiology* 21, 413–420.

Kawai, T., Akira, S. 2011. Toll-like receptors and their crosstalk with other innate receptors in infection and immunity. *Immunity* 34, 637–650.

Kingsolver, M. G., Huang, Z., Hardy, R. W. 2013. Insect antiviral innate immunity: pathways, effectors, and connections. *Journal of Molecular Biology* 425, 4921–4936.

Mandadi, K. K., Scholthof, K.-B. G. 2013. Plant immune responses against viruses: how does a virus cause disease? *The Plant Cell* 25, 1489–1505.

McNab, F. W., Rajsbaum, R., Stoye, J. P., O'Garra, A. 2011. Tripartite-motif proteins and innate immune regulation. *Current Opinion in Immunology* 23, 46–56.

Randall, R. E., Goodbourn, S. 2008. Interferons and viruses: an interplay between induction, signalling, antiviral responses and virus countermeasures. *Journal of General Virology* 89, 1–47.

Randow, F., MacMicking, J. D., James, L. C. 2013. Cellular self-defense: how cell-autonomous immunity protects against pathogens. *Science* 340, 701–706.

Sorek, R., Lawrence, C.M., Wiedenheft, B. 2013. CRISPR-mediated adaptive immune systems in Bacteria and Archaea. *Annual Review of Biochemistry* 82, 237–266.

Chapter 14

The adaptive immune response

- The immune system is a complex interacting mixture of cells and soluble components that protects us from infections and neoplasms.
- The adaptive immune system is activated in response to an infection and is stimulated by the preceding innate immune response.
- The adaptive immune system brings to bear a range of different components to fight the current infection and establish a memory to provide protection from future infections by the same agent.
- The adaptive immune system has evolved to kill infected cells and this lethal capability has to be tightly controlled to ensure that it is only deployed against foreign cells or host cells invaded by pathogens.
- The adaptive immune system establishes a memory of past battles to allow it to respond more efficiently when re-encountering the same pathogen.
- All micro-organisms have evolved mechanisms to combat or evade the adaptive immune system. How these work depends on the micro-organism's survival strategy.
- The immune system is a finely balanced compromise – if set too broadly it attacks harmless foreign matter that we eat or inhale, and causes bystander damage through friendly fire (e.g. allergy, hypersensitivity), or even autoimmunity; if set too narrowly then some micro-organisms may go unrecognized.

Chapter 14 Outline

14.1 General features of the adaptive immune system
14.2 Cell-mediated immunity
14.3 Antibody-mediated humoral immunity
14.4 Virus evasion of adaptive immunity
14.5 Age and adaptive immunity
14.6 Interaction between the innate and adaptive immune systems

The mammalian immune system contains two elements: innate and adaptive. Following infection with a virus, the innate immune response is the first line of defence. Innate immunity detects molecular signatures of invading pathogens and activates a series of processes that, if possible, act to prevent establishment of a successful infection. If that is unsuccessful, the innate immune system will attempt to limit the scope of an infection. Aspects of the innate immune response are discussed in Chapter 13. The innate and adaptive immune responses differ in two key ways: specificity and longevity. Innate immunity provides a general response that is not tailored to the pathogen but recognizes very general features of molecules associated with certain types of pathogen, such as bacterial or viral glycoproteins. In contrast, the adaptive immune response relies on exquisite specificity

Introduction to Modern Virology, Seventh Edition. N. J. Dimmock, A. J. Easton and K. N. Leppard.
© 2016 John Wiley & Sons, Ltd. Published 2016 by John Wiley & Sons, Ltd.

against the pathogen by recognizing short peptides as coming from a foreign organism. Also, the innate immune response is very short-lived and is typically induced and completed within a matter of days, while the adaptive immune response begins equally rapidly and ultimately generates a memory that is retained throughout life. Despite the differences between the two immune systems they are inextricably linked, and the innate immune response plays important roles in the initial stages of the adaptive immune response by attracting to the site of infection the cells that will initiate an adaptive immune response and by enhancing the presentation of virus antigens that are targeted by that response.

An important aspect to bear in mind is that, just as with the innate immune system, viruses have evolved mechanisms to counter, subvert and evade adaptive immune responses and the outcome of the infection depends on the balance achieved between the armaments available to the host and the virus. Study of the adaptive immune system is a discipline in itself and in this chapter we will consider only a limited range of aspects that relate to virus infection of humans and other mammals. For a more comprehensive consideration, reference to a specialized text is recommended.

14.1 General features of the adaptive immune system

The adaptive immune system consists of two elements: the cell-mediated immune system and the humoral immune system. These are differentiated by the nature of the cells that mediate the two types of response and the mechanism of action following the initiation of the responses. The cells involved in the adaptive immune system are highly mobile and most are present in the blood and lymphatic circulatory systems that radiate throughout the body, with smaller numbers resident within specific tissues in the body. Two major types of cells are involved in the adaptive immune response, *T and B lymphocytes*, which act as sentinels constantly surveying the body for evidence of infection. The T lymphocytes are involved in the cell-mediated adaptive immune response and B lymphocytes are the effectors of the humoral immune response. Other cell types also play additional roles. Whilst able to function independently, the two arms of the adaptive response interact with each other at several levels and also interact with the innate immune system as described in Section 14.6.

Once the T and B cells locate a signal indicating the presence of an infection they are unable to react immediately and have to increase in number by division to produce large numbers of genetically identical daughter cells, in a process called *clonal expansion*. They must also differentiate from an inactive non-dividing state into an active cell that targets the foreign material introduced by the virus. Adaptive immunity is specific for foreign molecules, typically proteins that, in the situation being discussed here, are synthesized during infection or form part of the invading virus particle. The molecules recognized by the T and B cells are referred to as *antigens* and these may comprise complete virus-encoded proteins or fragments of these proteins. Individual T and B cells recognize only small regions of these antigens, referred to as *epitopes*, and one protein may contain several epitopes, each recognized by different individual T and B cells. A glossary of common immunological terms is given in Box 14.1 and these will be used throughout this chapter.

The adaptive immune system can also be considered as consisting of the systemic immune system (which looks after the body as a whole) and the mucosal immune system (which looks after mucosal surfaces). The latter comprise the surfaces of the respiratory tract,

Box 14.1

An alphabetical glossary of essential immunological terms

Antibody: a soluble, epitope-specific protein synthesized and secreted by a B effector cell.

Antigen: a molecule that reacts, though its epitopes, with receptors on B or T cells.

Antigen presentation: display on the cell surface of a peptide complexed with an MHC class I or MHC class II protein that is capable of reacting with a receptor on T cells.

Antigen processing: digestion of self and non-self proteins to peptides in normal cells that then complex with MHC class I or II proteins.

B cell receptor (BCR): integral plasma membrane protein of B cells that recognizes an epitope (any type of molecule); antibody made by that B cell has the same epitope specificity as the BCR.

B effector cell: short-lived cell that makes antibody; also known as a plasma cell, an activated B cell, and an antibody-forming cell.

B and T lymphocytes: the main components of adaptive immunity; inactive cells, each of which has an epitope receptor of one specificity.

CD4: an integral membrane protein on the surface of T cells that recognizes MHC class II proteins; diagnostic of helper T cells; T cells have *either* CD4 or CD8.

CD8: an integral membrane protein on the surface of T cells that recognizes MHC class I proteins; T cells have *either* CD8 or CD4.

Cell-mediated immunity (CMI): immunity mediated by various T cell responses.

Chemokines: a superfamily of small (ca. $10,000$ M_r) soluble protein mediators and communicators involved in a variety of immune, inflammatory and other processes; involved in the development of dendritic, B and T cells, and lymphoid cell trafficking.

Cytokines: a superfamily of soluble protein mediators and communicators released from cells by specific stimuli; e.g. when activated lymphocytes bind antigen (these are also known as lymphokines); includes the interleukins.

Cytotoxic T lymphocyte: activated T cells that can kill target cells by lysing them.

Epitope: part of a molecule that binds the paratope of a specific T or B cell receptor.

Humoral immunity: antibody-mediated immunity.

Immunogen: a molecule that stimulates adaptive immunity to its epitopes.

Interleukin: any member of a family of cytokines (IL-1, IL-2, etc.).

MHC (major histocompatibility) protein (in man usually referred to as human leukocyte antigen (HLA)); MHC class I: integral plasma membrane protein found on nearly all cells of the body; MHC class II: integral plasma membrane protein that is normally restricted to cells of the immune system.

Monoclonal antibody (MAb): antibody of a single sequence and, hence, specific for a single epitope made by a clonal population of artificially immortalized B effector cells (a hybridoma).

Naïve B or T cell: a B or T cell that has not yet encountered its specific antigen.

Non-self: any antigen that is not normally present in that individual.

Paratope: region of a B or T cell receptor that recognizes and binds a specific epitope.

Self (antigen): any antigen normally present in that individual.

T cell receptor (TCR): integral plasma membrane protein of T cells that recognizes a T cell epitope; the epitope is a peptide complexed with an MHC class I or MHC class II protein on the surface of another cell of the same person; also known as antigen or epitope presentation.

T effector cell: short-lived cells that directly act on other cells of the body.

the intestinal tract, the urinogenital tract, and the conjunctiva of the eye. The mucosal system is important as most micro-organisms, including viruses, gain entry through these sites. Mucosal surfaces are particularly vulnerable because they are extensive in area, and their physiological activities require them to be composed of naked epithelium, unlike the skin which is impermeable to viruses unless broken. The systemic and mucosal immune systems are distinct in that cells and antibodies providing systemic immunity do not circulate to the mucosae. However, the mucosae have their own reservoir of lymphocytes, and these are stimulated by antigens to provide a localized immune response. This has implications for vaccine-induced immunity (see Chapter 26).

When considering the adaptive immune system it must be borne in mind that this essential protective system must be able to recognize and respond to the challenge of potential infection by a vast array of different pathogens that we encounter many times throughout our lives. Despite this immense challenge, our existence is evidence that it is successful in defending us against these agents and it is only in the minority of situations where it fails catastrophically.

14.2 Cell-mediated immunity

A large number of different T cells are involved in cell-mediated immunity (CMI) which can be broadly separated into two types: the T-helper (T_H) cells and cytotoxic T (T_C; also called cytotoxic T lymphocytes, CTL) cells. T_H cells are identified by the presence of a cellular protein called CD4 embedded in the plasma membrane and are called *CD4$^+$ T_H cells*. T_C cells are identified by the presence of a protein called CD8 on the cell surface and are referred to as *CD8$^+$ T_H cells*. CD4$^+$ T_H cells can be divided into a

number of different types that are defined on the basis of their location in the body (circulating or resident), the cytokine and chemokine chemical signal molecules that they use and/or respond to, and the function(s) that they perform. However, some of these defining features are present in more than one subset of cells and so the population of T_H cells is a very heterogeneous mixture with cells providing specific activities at certain points in the response. This includes a small subset of T_H cells that can also attack and kill antigen presenting cells and are therefore formally also cytotoxic T cells. Similarly, T_C cells can be divided into groups of cells with somewhat different functions. A full consideration of all of these different cell types is beyond the scope of this text and we will consider only the major players in this discussion.

Some tissues, such as the respiratory tract, benefit from the presence of resident cells involved in CMI. This provides a more immediate response than would be possible by reliance on a circulating cell encountering a site of infection. The trigger for initiating the CMI is detection of a foreign antigen by a T cell. When this occurs, the T cell sends out signals in the form of chemokines and cytokines that attract additional immune cells to the site of infection and scale-up the response. This then sets in motion a cascade of events with activation of a range of different processes that attack the virus-infected cells, limiting the scope of the infection and ultimately eliminating it. This is referred to as the *inflammatory response* that occurs following infection. The inflammatory response often brings with it some of the symptoms of disease that are experienced by the infected individual.

All T cells carry on their surface a receptor, the *T cell receptor* or *TCR* that recognizes a specific antigen and binds to a specific epitope site within it. Each cell displays a slightly different TCR, this diversity being created by imprecise recombination in individual cells that brings together alternative coding segments of the TCR genes and creates variable sequences at the junctions. This somatic TCR diversity is the basis for the differing antigen specificity of individual T cells. The TCR recognizes only peptides. The peptides are always found complexed with cellular plasma membrane proteins, called major histocompatibility complex (MHC) proteins, and a TCR only recognizes its specific, or *cognate*, peptide when it is complexed with an MHC protein. All cells, except red blood cells, express MHC class I proteins, but only a few cell types, notably dendritic cells and B lymphocytes, express MHC class II proteins. There are three major loci encoding MHC class I proteins (HLA-A, HLA-B, HLA-C in humans, often called HLA antigens); each encodes a single polypeptide that forms a dimer with the same β_2-microglobulin protein. A further three loci encode MHC class II proteins (HLA-DP, HLA-DQ, HLA-DR in humans); these proteins are dimers of α and β polypeptides encoded by the three loci. The genes that code for MHC proteins are highly polymorphic, and are expressed co-dominantly which provides the immense variety that is required to recognize the many different antigens we may be exposed to during infections. The MHC proteins combine specifically with their cognate peptide by recognizing two terminally situated amino acid residues that anchor the peptide in a groove on the MHC protein. The epitope presented to T cells by this complex is the peptide sequence situated between the anchor residues. Thus an MHC protein can present an infinite number of epitopes providing they all have the appropriate anchor residues. MHC class I proteins complex with a peptide of 8–10 amino acids and MHC class II proteins with peptides of 17–22 amino acids. MHC proteins bind cognate peptide within the cell; unless complexed with a peptide, MHC proteins cannot mature and be displayed on the cell surface.

After first meeting their cognate antigen, T cells migrate to one of the lymphoid centres. There are many of these loosely organized aggregates of cells, and the best known are the *lymph nodes*, located at strategic sites around the body (e.g. at the junction of limbs and torso, torso and head), and the *Peyer's patches* in the gut wall. Here they undergo clonal expansion and differentiate into *effector* cells, often also referred to as *activated* cells, that all target the same antigen. Activated T cells patrol the body and interact with their antigen when it is encountered. In this way, the response is not limited to the physical location where it was initiated. The activated T cells are short lived and do not divide further and this imposes a limit on the timeframe of their action unless antigen stimulation continues. However, a subset of these cells becomes long-lived *memory* cells (see below) that form the basis for future responses against the same antigen. Thus, the immune cells adapt to the infection and give the name to this aspect of the immune response. The receptors also evolve by somatic mutation to better fit their cognate epitope but this takes a few days, during which time the virus can multiply and may cause clinical disease. After the virus is defeated, the population of T effector cells falls but not to the original level, as an expanded population of memory cells remains.

CD4$^+$ T helper cells

The major function of activated CD4$^+$ T cells is *help* and only these T_H cells can provide it. Help is a positive regulatory function which is essential for the activation of all T and B lymphocytes when they bind their cognate foreign epitope. Thus the entire immune response depends on the T_H cells. This is also the main type of cell infected by human immunodeficiency virus (HIV) and explains why their destruction is so devastating (see

Chapter 21). The first step in the process is that a dendritic cell takes up the invading virus, or virus proteins secreted from infected cells, by phagocytosis. It is important to note that the proteins are acquired from outside of the dendritic cell and are not synthesized within the dendritic cell by an infecting virus. Following phagocytosis the proteins are degraded in lysosomes and the short peptides that are produced are loaded onto MHC class II molecules in the cytoplasm and this complex, together with β2 microglobulin, is transported to the plasma membrane where the peptide is made available (presented) for binding to a specific TCR. Each T cell has many copies of a single TCR which recognizes a specific peptide antigen. A CD4$^+$ T$_H$ cell carrying an antigen-specific TCR on its surface can only recognize its peptide epitope if the peptide fragment containing it is complexed with an MHC class II protein on the surface of the presenting cell (Fig. 14.1). When the antigen presenting cell

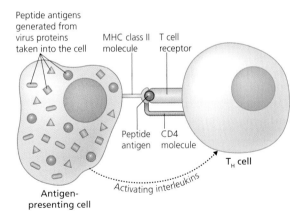

Fig. 14.1 Diagram of the activation of a naïve CD4$^+$ T$_H$ cell. An antigen presenting cell, typically a dendritic cell, presents a short virus peptide antigen on its surface in a complex with an MHC class II molecule. This complex is recognized by a specific T cell receptor on the T$_H$ cell surface in association with the CD4 molecule. The interaction causes the antigen presenting cell to release interleukins that activate the T$_H$ cell.

(APC) and the T$_H$ cell interact in this way, the APC secretes cytokines, particularly interleukin (IL)-1 that activates the previously naïve T$_H$ cell. The activated T$_H$ cell then migrates to the lymphoid centre where it is clonally expanded and memory cells are established. The activated T$_H$ cell in turn secretes a range of cytokines that stimulate other cells of the immune system. At this point, the T$_H$ cell can mature into any of several subclasses, adopting one of a number of functions, of which the T$_H$1 and T$_H$2 classes are particularly important.

When a naïve T$_H$ cell encounters an APC for the first time the determination of whether it becomes a T$_H$1 or T$_H$2 cell is achieved through a complex series of processes that occur in response to cytokines. The details of all of these processes are not fully determined and an overview is summarized in Fig. 14.2. When the naïve T$_H$ cell interacts with the APC, the next event is determined by the presence of a signalling molecule. The signalling molecules will be produced by the innate immune response of cells in the local vicinity and the concentration of each will be determined by those cells. If sufficient IFNγ is present at the key moment the T$_H$ cell is activated, this initiates a pathway that leads to the generation of a T$_H$1 cell. This also requires the presence of IL-12. The T$_H$1 cell secretes a spectrum of cytokines including IFNγ and TNFβ. The IFNγ provides positive stimulation to generate more T$_H$1 cells when new T$_H$ cells interact with more APCs. An alternative pathway to generate T$_H$2 cells is stimulated by the presence of an as yet unknown factor. In the presence of IL-4 the T$_H$2 cell is generated and this produces a spectrum of cytokines that differ from those produced by T$_H$1 cell. One of these is further IL-4 that provides a positive feedback to stimulate production of more T$_H$2 cells. The generation of these two types of T$_H$ cells is also affected by negative feedback in that IFNγ and IL-12 which stimulate T$_H$1 cell production also inhibit T$_H$2 cell production. Similarly, Il-4 and IL-10 inhibit

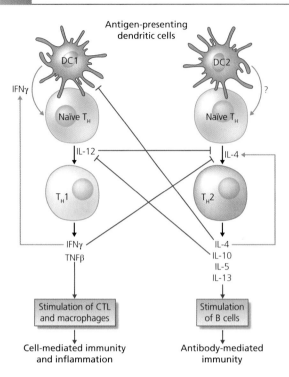

The T_H1 and T_H2 cells have different functions and so the ratio of each present will ultimately determine the nature of the adaptive immune response that is seen, which in turn will determine the outcome of the infection. T_H1 cells stimulate the activity of cytotoxic T cells and macrophages which form part of the inflammatory response to infection in conjunction with any inflammatory elements stimulated by the innate immune system. T_H2 cells stimulate the action of B cells as part of antibody-mediated immunity. The balance between these two arms is critical not only for the initial immune reaction but also for future reactions as the memory cells generated by the initial response will determine the balance of response if the same virus is encountered at a later time: if the initial encounter generates, for example, a strong T_H1 response any subsequent encounter with the same virus will lead to the same response (Box 14.2).

Fig. 14.2 Diagram of the key events in the production of T_H1 and T_H2 cells following interaction with antigen presenting dendritic cells. The production of cells in the two lineages is enhanced (green arrows) or inhibited (red lines) by the presence of cytokine signals including IFNγ, and IL-4, 10 and 12. The T_H1 and T_H2 cells in turn produce a spectrum of different cytokines and hence stimulate either T_C cells and macrophages (T_H1 cells) or B cells (T_H2 cells).

the generation of T_H1 cells. The result of these control processes is that both cell types are produced but the ratio of each is determined by the strength of the initial inducing signal and the ability of each lineage to reduce the production of the other. The balance is therefore determined by the environment in which the initial interaction between an APC and a naïve T_H cell takes place and the local concentrations of several cytokines, some of which are produced by the infected cells and others by activated CD8+ T cells attracted to the site of the infection.

CD8+ cytotoxic T cells

Just as for the CD4+ T cells, the CD8+ cytotoxic T cells also contain multiple copies of a single TCR on their surface and these are specific for specific short peptide sequences. In this case, however, the short 8–10 amino acid peptide antigens are presented to the TCR in association with MHC class I molecules rather than MHC class II and, unlike the situation with MHC class II presentation to T_H cells, the peptide antigens associated with MHC class I molecules are produced inside the cell as a result of virus infection. The virus proteins are degraded by cellular proteosomal complexes and any peptide fragments with the correct anchor residues are taken into the endoplasmic reticulum by a TAP transporter complex comprising four molecules made up of two proteins, TAP-1 and TAP-2. The TAP transporter complex, in association with additional cell proteins, then loads the peptide

Box 14.2

The importance of a balanced T$_H$1 and T$_H$2 response to infection

Respiratory syncytial virus (a pneumoviruses that is a member of the Paramyxovirus family) is a major cause of hospitalization of very young children worldwide. The children most severely affected are those with cardiopulmonary disorders, or immunosuppression or other complications. In the mid 1960s, an inactivated vaccine preparation was given to a group of 3-week-old children who were monitored to investigate protection against a natural infection. The results of this trial were disastrous with many of the children experiencing very severe disease when they subsequently became infected, considerably greater than would have been expected normally and, tragically, two children died.

It was not until decades later that the explanation for this was discovered. The particular batch of vaccine used (lot 100) was used to inoculate mice and their immune response was studied. It became clear that the vaccine induced a very strong T$_H$2 response which was quite different to the response to a natural infection that is predominantly of T$_H$1 cells. When the immunized mice were later inoculated with infectious virus the T cell response was almost exclusively biased towards the production of T$_H$2 cells. The result is that the cytotoxic T cells that are required to remove infected cells were not produced in sufficient quantity to control the infection.

The conclusion is that in the original trial the children responded to the vaccine with a T$_H$2-biased response. This will have stimulated a B cell response to the virus antigens with little if any T$_C$ cell response and created an immune memory of this. When the children were infected naturally with RSV at a later date the immune memory revived a T$_H$2-biased response that suppressed the natural induction of a T$_H$1 response, and the infection was able to run out of control with devastating consequences. This emphasizes not only the different types of response but the importance of a balanced immune response to allow a resilient defence against infection.

onto the MHC class I molecule. The MHC molecule plus the peptide and β2 microglobulin then move to the plasma membrane to present the peptide antigen to T cells. The T cell binds to the antigen held by the MHC class I molecule in association with the CD8 molecule (Fig. 14.3). The binding to the TCR activates the T cell and stimulates clonal expansion and the new cells travel through the circulatory system to seek out infected cells that present this antigen on their surface. The clonal expansion is enhanced by the action of several cytokines including IL-2 produced by T$_H$1 cells. Some of the new T$_C$ cells will become memory cells to ensure a rapid and robust response if the same antigen is encountered again at any point in the future.

Once bound, the TCR holds the T cell firmly to the infected cell which it targets for destruction. The T$_C$ cells attack the infected cells by releasing secretory granules containing the cytotoxic molecules perforin, granulysin and several granzymes. After secretion, the granules disassemble in a degranulation process that releases their toxic contents. The perforin creates small localized pores, and the granulysin creates small holes, in the plasma membrane of the infected cell that allow entry of the granzymes. The granzymes are serine protease

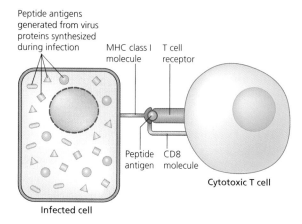

Peptide antigens generated from virus proteins synthesized during infection

MHC class I molecule

T cell receptor

Peptide antigen

CD8 molecule

Cytotoxic T cell

Infected cell

Fig. 14.3 Diagram of the activation of a naïve CD8$^+$ T$_C$ cell. An infected cell presents a short virus peptide antigen on its surface in a complex with an MHC class I molecule. This complex is recognized by a specific T cell receptor on the cytotoxic T cell surface in association with the CD8 molecule. Recognition leads to activation of the CTL and killing of the infected cell.

enzymes whose activity within the cell activates the apoptotic pathway leading to cell death (see Chapter 13). T$_C$ cells can also induce apoptosis in the infected cells they target by another, less effective, mechanism. For this, the activated T$_C$ cells express the protein FAS ligand (FasL) on their surface. This binds to the Fas molecules found on the surface of the adjacent infected cells. The Fas/FasL complex activates the extrinsic apoptotic pathway leading to the death of the cell.

14.3 Antibody-mediated humoral immunity

Circulating non-dividing B cells express B cell receptors (BCR) on their surface. These are epitope-specific receptors and each B cell contains only one type of receptor with specificity for just one epitope. Unlike the TCR which can bind only to peptides, the BCR can

be specific for almost any type of molecule including peptides, nucleic acids, lipids, carbohydrates and environmental and synthetic chemicals. BCR are therefore able, in principle, to react with any type of foreign molecule an organism is likely to meet. However, during virus infections these are most likely to be peptides. As with the TCR, each individual B cell displays a different BCR, produced by somatic recombination, and allelic exclusion prevents a cell from expressing protein from its second BCR gene once functional BCR expression has been achieved from the first allele.

When the BCR on the surface of a B cell recognizes and binds its epitope, the B lymphocyte migrates to the nearest lymph node where it undergoes clonal expansion and differentiation. These processes are promoted by the action of cytokines secreted by T$_H$2 cells (see Section 14.2). One of the key events that is triggered is reorganization of DNA sequences by recombination to generate genes that synthesize antibodies that contain the variable regions of the BCR that together comprise the *paratope*, the region which interacts with the epitope. A second critical event is mutation of DNA encoding the variable regions of the BCR paratope; this creates further diversity within the BCRs of the activated cell pool. The mutation of the paratope encoding sequences proceeds at a stunningly high frequency (and is referred to as an example of hypermutation). This creates a huge pool of cells each with a variant paratope, and Darwinian evolution and natural selection follows. B cells with the best fitting paratope are preferentially stimulated by binding the cognate epitope; in turn, their BCR mutates and affinity and antibody activity is further increased.

The nature of human antibodies

Antibodies, also known as immunoglobulins (Ig), can take two forms: membrane-bound or

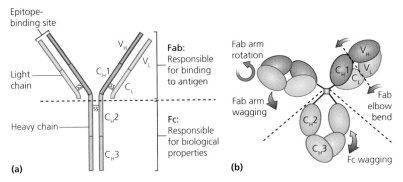

Fig. 14.4 Generalized immunoglobulin molecule, showing (a) the outline structure, consisting of two identical dimers formed of heavy (H) and light (L) polypeptides. Note the sequence-variable (V) region of the H and the L chains. Each contains 3 hypervariable sequences of 10–20 residues (not shown) that are folded together to form the unique epitope-binding site (paratope). The remaining H and L sequences are relatively constant. The constant region is subdivided into domains with sequence homology (C_H1 to C_H3). IgM and IgE both have an additional domain (C_H4) compared to IgA, IgD and IgG. The molecule is also divided into an N-terminal half that binds epitope (FAb), and a C-terminal half (Fc) that is reactive with various cell mediators. IgM and IgA form 5-mers and 2-mers respectively linked by their C' ends with a joining (J) polypeptide. (b) The globular domains, with arrows indicating the flexibility which allows the molecule to bind to two (identical) epitopes, which can be different distances apart in three dimensions. SS, disulphide bond.

secreted. This is determined by the presence or absence of hydrophobic domains to anchor the antibodies in the plasma membrane of the B cell. The basic unit of an antibody comprises two heavy and two light chains. These can be considered as Y-shaped molecules, with paratopes at the two tips and a constant Fc domain forming the stalk. The structure of an antibody is shown in Fig. 14.4. Since each antibody contains two identical paratopes they can therefore bind two copies of the target epitope, i.e. they are functionally bivalent. In some antibody types, copies of the basic unit of two H and two L chains join together to form multimers. Each additional antibody in the complex increases the number of epitopes that can be bound.

Human antibodies exist in five different isotypes. The isotypes are differentiated by the nature of their heavy chain that are called alpha (α), delta (δ), epsilon (ε), gamma (γ) and mu (μ), and these define the isotypes IgA, IgD, IgE, IgG and IgM, respectively. Their light chains are

either kappa (κ) or lambda (λ). Antibodies can also be distinguished by the time at which they are seen during and after an infection, their location in the body and whether they act as monomeric antibodies or as multimeric antibodies. Some isotypes can be further subdivided into types. Importantly, BCRs are composed of surface-bound IgD or IgM antibodies; thus the BCR is encoded by the Ig gene loci. In this way, the receptor specificity of a B-cell is tightly linked to the specificity of the antibody it can produce. Some of the key features of antibody isotypes are given in Table 14.1. In terms of antibody-mediated immunity to virus infections the most important isotypes are IgG, IgA and IgM and these are described in more detail in Box 14.3.

Following interaction of the BCR on a naïve B cell with its cognate antigen and the processes of clonal expansion and differentiation in lymph glands the first antibody isotype produced is IgM. Subsequently, the B cells

Table 14.1

Features of antibody isotypes.

Isotype	No. of types	Function	Active form
IgA	2	Found in mucosa of the respiratory tract, gut, and urogenital tract. Also found in saliva and tears	Dimer
IgD	1	Membrane-bound and found on the surface of naïve B cells	Monomer
IgE	1	Primarily binds to allergens to trigger allergic responses. Involved in protection against parasitic worms	Monomer
IgG	4	Provides the majority of circulating antibody providing humoral immunity. Can cross the placenta to provide temporary protection to the foetus and into early life	Monomer
IgM	1	Found as both membrane-bound form on B cells as the BCR, and secreted form. The earliest isotype to appear following infection	Pentamer

Box 14.3

Properties of the major immunoglobulins

- **IgM** is composed of five IgM monomers covalently linked at their C-termini. It binds its epitope with low affinity but, being pentameric, the overall IgM avidity (binding strength) is high. It is found only in blood. IgM and IgG are the only antibodies capable of activating the complement system (see below). IgM is the first class of antibody synthesized when lymphocytes are activated, so detection of virus-specific IgM in serum is diagnostic of a recent infection. Contributes to neutralization of virus infectivity.
- **IgG** is the collective name for IgG1, IgG2, IgG3 and IgG4, which each have distinct γ chains synthesized from separate genes and differ in specific ways that allow them to combat a variety of infections. IgG is a monomer; its small size gives it mobility and allows it to leave the blood stream and enter tissues. It can also enter the body of the foetus by crossing the placenta and protects it from infection. The immune system of the newborn infant develops slowly and maternal IgG remains in circulation for several months until immunity is functional. IgG activates complement and binds to Fc receptor proteins on the surface of phagocytic cells. Antigen stimulation of mucosal surfaces gives rise to the synthesis of 'local IgG'. IgG is important for virus neutralization.
- **IgA** is a key defender of the mucosal surfaces, and exceeds IgG in both total amount produced and in local concentration. It has the unique property of being secreted across the mucosal epithelium, so that it occurs outside the body in the lumen of the gut, respiratory tract and urinogenital tract. Maternal IgA is secreted into milk and protects the gut of the newborn from infection. IgA is particularly resistant to degradation by digestive enzymes. The most abundant type of IgA is a dimer, but monomers also exist. Both neutralize virus infectivity.

undergo *isotype switching* (also referred to as immunoglobulin class switching) in which the isotype class of antibody produced changes from IgM to IgG, IgA or IgE. Class switching is achieved through a process of genetic recombination in which the regions of cellular DNA encoding the constant region of the original immunoglobulin class heavy chain are deleted and the region encoding the new class of heavy chain is brought together with the DNA regions encoding the original variable regions to generate a functional immunoglobulin gene. Since the variable regions stay the same the antigenic specificity of the different antibody classes is identical. This allows the daughter cells derived from the same activated B cell to produce antibodies of different isotypes or subtypes. The activated, or *effector*, B cells will now produce antibodies that act against the current infection.

Activated B cells undergo hypermutation to refine the paratope affinity for the antigen, in a mechanism similar to that described for activated T cells (see Section 14.2). Some of these high affinity mutated B cells do not mature into plasma cells but become memory cells instead. These have an affinity-enhanced BCR, and on reinfection by the *same* virus there is now a sufficiently large population of cognate memory cells to rapidly provide effector cells that can repulse the infection before it can cause disease. Such cells constitute *immunological memory*. Memory cells closely resemble lymphocytes in morphology and lack of effector functions. They require stimulation by the cognate epitope just like lymphocytes, but are activated more readily than B lymphocytes as they respond to antigen at a lower receptor: antigen ratio.

The antiviral activities of antibodies

Antibodies produced by activated B cells are usually secreted from the cell and circulate freely in serum, though some are synthesized as membrane-bound antibodies that are inserted into the plasma membrane of the B cell. Both types can bind their specific antigen and induce the responses that will lead to the elimination of the virus, either by directly binding to the virus or by binding to infected cells expressing virus proteins on their surface.

Immunoglobulins may combat virus infectivity in more than one way. Some immunoglobulins are directly *neutralizing*, i.e. they can bind to viruses and cause them to lose infectivity. Other immunoglobulins are *non-neutralizing* and they attach to *non-neutralizing epitopes*. These act indirectly, either by inducing phagocytosis of bound virus via their Fc domains or by recruiting complement. The activation of *complement* by IgG or IgM bound to viruses can enhance the activity of neutralizing antibody, or enable non-neutralizing antibody to become neutralizing. Complement is a nine-component system of soluble proteins found in blood. Activation of the first complement component activates the next component by proteolytic cleavage and so on, with progressive amplification in a classic biochemical cascade system. Usually, each component is cleaved into two parts, one of which adheres to the antigen–antibody complex or close to where the antibody has bound, and a second diffusible part, which forms a chemical gradient and attracts cells of the innate immune system into the vicinity of antigen. The final stage is the insertion of pore structures (the membrane attack complex) into the cell membrane, which, in sufficient number, create an ionic imbalance and kill that cell. Thus complement-enhanced neutralization of non-enveloped viruses is brought about by a build-up of complement proteins on the surface of the virus that sterically prevent attachment of the virus to its cell receptors, while enveloped viruses in addition can be permeabilized by having complement pores inserted into their lipid bilayers that permit the entry of nucleases, etc.

However, while complement is important in combating certain bacterial infections, it is not certain that it is essential to anti-viral defence in vivo. For example, people who have a congenital complement deficiency do not have an increased number or severity of virus infections. For these individuals it may be that other aspects of their immune response are enhanced and compensate for the lack of complement activity.

An antibody molecule is approximately the same size as a single membrane protein and, not surprisingly, does not harm an infected cell even when many molecules are bound to virus antigen on the cell surface. Also, antibodies cannot cross membranes and enter cells unless they are bound to a virus and carried in by the virus entry process. However, IgG or IgM bound to infected cells through interaction with their antigens can activate complement, and in sufficient amounts this can lyse the cell as described above. Because cells can repair minor damage, it needs about 10^5 complement pore structures inserted into the plasma membrane to kill a cell. Alternatively, bound IgG can act as a ligand for phagocytes that have Fc receptor proteins on their surface, and this leads to the phagocytosis and destruction of the infected cell. In this way, the adaptive immune system endows the innate immune system with antigen specificity.

Virus neutralization by antibody

Neutralization is the loss of infectivity which ensues when antibody binds to a cognate epitope on the virus particle. Viruses are unusual as neutralization is usually mediated by antibody alone, whereas larger organisms, like bacterial cells, also require the action of secondary effectors such as complement as described above. However, not all antibodies which bind to a virus particle are capable of neutralizing its infectivity. Neutralization is an epitope-specific phenomenon.

Early work assumed that neutralizing antibody acted solely by preventing virus from attaching to receptors on the cell surface. However, while this was the mechanism operating with most rhinovirus-specific MAbs, the majority of influenza A virus- and poliovirus-specific neutralizing MAbs did not block cell attachment. Thus there is more than one way of killing a virus with antibody. The mechanism of neutralization is antibody-specific, so a virus can be neutralized in a variety of different ways, and neutralization is determined largely by the epitope to which the antibody binds. Surprisingly, no antibody is made to the attachment sites of influenza A virus, poliovirus or rhinovirus. These sites are contained in depressions on the surface of the virus, where they are hidden from the immune system, presumably as a result of evolution trying to evade the host's immune response. Rhinovirus-specific neutralizing antibody attaches to and bridges amino acids on either side of the rhinovirus attachment site and blocks it indirectly (Fig. 14.5). Antibody attached to virions can be visualized by EM as a fuzzy outer layer. However, antibody can be diluted so that it is no longer detected by EM but can still thoroughly neutralize infectivity. So few molecules of antibody bound to a virus

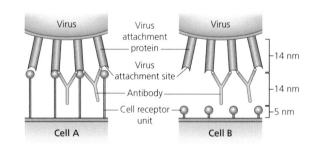

Fig. 14.5 Hypothetical scheme based on the length of the cell receptor to explain inhibition by antibody of attachment of virus to cell B but not cell A.

particle are unlikely to interfere with attachment. Interference with attachment of influenza virus would be particularly inefficient, as there is a high density (500–1000) of trimeric attachment proteins (the haemagglutinin) per virion, and an IgG molecule is slightly smaller than one haemagglutinin protein. In addition, the existence of non-neutralizing antibodies demonstrates just how specific the neutralization reaction is, and means that antibodies bound to the surface of a virus do not necessarily interfere with the attachment process. Additionally, a non-neutralizing antibody bound close to a neutralizing epitope can sterically prevent the binding of neutralizing antibody, and allow the virus to evade the immune response.

So how do viruses lose infectivity? Antibodies (MAbs) to certain epitopes do indeed neutralize by blocking attachment of a virion to its cell receptors, but equally it has been demonstrated that neutralizing MAbs to other specific epitopes does not inhibit virus attachment. In the case of poliovirus, the neutralized virion attaches to cells, is taken up in an endocytic vesicle, but is unable to uncoat. In a second example, an influenza virus attaches to cells, is endocytosed, but fusion of the viral and cell membranes does not take place. Current thinking is that there are as many mechanisms of neutralization as there are processes that a virus has to undergo before its genome can enter a cell and be expressed. This range of neutralizing mechanisms means that any definition of the mechanism of neutralization must include the virus and the MAb, and this may change depending on other factors including the antibody isotype, number of antibody molecules per virion, cell receptor, and the cell type.

For non-enveloped particles, non-neutralizing antibodies can also lead to inactivation of the virus through a different route. In this case, the virus particle becomes coated with antibodies and the virus:antibody complex enters the cell. The cytoplasmic TRIM21 protein, which forms part of the innate immune system, has a very high affinity for the Fc portion of the antibody. When TRIM21 binds the antibody:virus complex is targeted to the cellular proteasome where it is degraded (see Section 13.4). In effect, this mechanism renders the virus non-infectious following its recognition by a non-neutralizing antibody.

There are other conditions under which virions can bind neutralizing antibody and yet not be neutralized. This occurs when the concentration of antibody is too low, or the affinity of antibody is so low that it dissociates rapidly, or because virions are aggregated and protected from contact with antibody.

Sources of antibody

Antibodies are usually obtained from venous blood as an antiserum several weeks after an infection has resolved (convalescent serum) or after at least two immunizations of a person or an experimental animal – usually a mouse, rat, rabbit or guinea pig. Antibodies may be formed to multiple viral epitopes present on each of several different virus proteins, and they join the pool of antibodies to other foreign antigens that the animal has experienced earlier. A further complication is that the isotype distribution of antibody in serum differs from that of antibodies present in mucosal secretions – in serum IgG > IgA, and in secretions IgA > IgG, so care must be taken in extrapolating from one situation to the other (Box 14.4).

Understandably, analysis of such a complex mixture of antibodies is difficult. For unambiguous analysis of viral epitopes, monoclonal antibodies (MAbs) are used. An antibody synthesizing B cell makes an antibody of just one sequence (i.e. it is monoclonal), but such cells do not divide, and single cells do not make enough antibody to be of practical

---

Box 14.4

Sources of antibody for virus neutralization

- Serum is the fluid portion of clotted blood, i.e. blood less cells and clotting components.
- Plasma is the fluid portion of blood in which clotting has been inhibited and cells removed.
- Serum containing a defined antibody reactivity is called an antiserum.
- Serum and plasma contain a population of polyclonal antibodies, i.e. from many different B cells.
- Mucosal antibody can be relatively easily obtained from nasal secretion (collected by chemically irritating the nasal cavity), milk, and from extracts of faeces.
- A monoclonal antibody (MAb) is the product of a single immortalized B cell that has formed a clonal population.
- All MAbs from a given cell clone have exactly the same properties – including the same sequence, epitope, affinity, and isotype.
- In effect, an antiserum is a mixed population of monoclonal antibodies, and may well contain several MAbs that react with different epitopes on the same antigen. In this case, antibody reactivity will be an average of all the reacting antibodies.
- In contrast, reaction of an antigen with a single MAb is unambiguous, so MAbs are valuable reagents for research and bioassays.

---

use. However, in 1975 Köhler and Milstein devised a means to immortalize an antibody-synthesizing cell by fusing it with a cell from a B cell tumour which no longer makes its own antibody. The resulting hybrid cell line (*hybridoma*) can then be grown to large numbers in the laboratory. Each cloned cell line synthesizes antibody of a single sequence, a *monoclonal antibody* or MAb. In practice, a crude mixture of B cells is used in the fusion reaction and many clones are generated. The desired hybridoma is then identified in this clone library by the reaction of its antibody with the desired antigen.

14.4 Virus evasion of adaptive immunity

Since both the T cell and B cell elements of adaptive immunity rely on recognition of specific epitopes, any mutation that changes a crucial amino acid within the epitope will mean that the pre-established immune memory against that epitope will be rendered ineffective. The mutated virus is then referred to as an *escape mutant*. As discussed in Chapter 4, viruses undergo mutation to produce a quasispecies population from which a successful mutant can be selected by environmental pressure – in this case, the pressure is provided by the immune system and an escape mutant will have a growth advantage over the original infecting wild-type virus as it will not be targeted by the pre-existing adaptive immune system. The adaptive immune system will recognize the escape mutant as an invading virus, but must initiate a completely new response beginning with naïve cells. In the situation where an antibody epitope has been mutated to generate a neutralizing antibody escape mutant, the wild-type virus will be neutralized and the escape mutant replaces it in the population. None of the progeny of an antibody-escape mutant is neutralized either. The location of the mutation can be determined by comparing sequences of the wild-type and mutant genes encoding the neutralization protein. Analysis of several escape mutants can identify the epitopes and antigenic sites within a protein. In some rarer situations, a neutralizing antibody escape mutant which has a mutation that does not alter the epitope but still abrogates neutralization can arise. This type of mutation probably acts by preventing a downstream event in the neutralization pathway.

Many large DNA viruses encode genes whose functions are to interfere with the adaptive immune response, particularly cellular immunity. Human adenovirus type 2 E3/19K protein specifically prevents the transport of newly-synthesized MHC class I molecules to the cell surface, which results in inhibition of peptide presentation to $CD8^+$ T_C cells. This therefore blunts the cellular immune response to infection. However, this tactic carries with it some risks. Natural killer cells of the innate immune system patrol and seek out cells with reduced MHC class I expression on the cell surface and target them for destruction. Thus, in principle the adenovirus infected cell that becomes less of a target for T_C cells will become a target instead for natural killer cells. Adenovirus type 2 anticipates this as the multifunctional E3/19K protein also suppresses recognition of infected cells by natural killer cells. Several herpesviruses also prevent MHC class I presentation of antigens but achieve this by inhibition of the antigen transporter protein (TAP) that loads the peptides on to the MHC molecule. A separate protein prevents natural killer cell destruction of these cells.

Viruses may also employ some mechanisms to distract the adaptive immune system. An example of this is where a virus produces large amounts of a highly antigenic protein which is secreted from an infected cell. This protein will bind circulating antibody, reducing the levels of antibody in the immediate vicinity of the infected cell and so potentially reducing the impact of the adaptive immune response on the infection.

14.5 Age and adaptive immunity

The very young and the very old suffer a greater number of episodes of infectious disease than people of other ages, and these infections are sometimes of greater severity. This arises because the young and the old have less developed or less active immune responses respectively. The development of the immune system lags behind other body systems, although an infant gets some protection from maternal transplacental IgG and from immunoglobulins (mainly secretory IgA) from maternal milk. However, by the age of 9 months maternal IgG has disappeared. The developing immune response increases quantitatively, and by 12 months an infant produces 60% of the adult level IgG but only 20% of its adult level IgA; it takes until around 2 years of age for the system to be effectively mature. In the elderly, the immune system, like other body systems, declines with age. With immunological memory the elderly have less of a problem in combating repeat infections, but their immune system is less able to deal with new infections. These age-related aspects pose a special problem when trying to enlist the immune response to protect against infection, e.g. when delivering antigen for vaccinations.

14.6 Interaction between the innate and adaptive immune systems

All innate and adaptive immune responses act in an interlocking and orchestrated fashion. The primary aim of the innate immune system is to prevent the establishment of an infection. This is achieved in a number of ways but part of the innate immune response involves the use of cytokine signals (see Chapter 13). These cytokines set in motion an inflammatory response, attracting cells of the adaptive immune system that initiate the various elements described above. Additionally, several cytokines influence the nature of the adaptive immune response by determining whether a T_H1 or T_H2 response is established and by promoting isotype switching in B cells. Many of these interacting factors exert a concentration-dependent effect that results in a very fine level of control of the details of the final response that is achieved. A simplified outline of the interacting components of the innate and adaptive immune systems that play a role in combatting a virus infection is shown in Fig. 14.6.

Recent studies have suggested that the innate immune system may retain some memory of stimulation within natural killer cells; however, it is not easy to determine which viruses may have been successfully defeated by the innate immune response. In contrast, the adaptive immune system establishes a lifelong memory of the pathogens that it has encountered. This memory ensures that a second encounter is much more likely to be effectively and successfully defeated. It also allows us to investigate our immunological past history. This history may have consequences for the outcome of an infection such as was seen with the reduced frequency of disease in the elderly when infected with the 2009 pandemic influenza virus (see Chapter 20).

The concerted actions of the innate and adaptive immune systems provide a range of mechanisms by which an infection can be countered. Whilst all are important, it is likely that one aspect of the response will be key for an individual in overcoming a particular infection, even though many other responses may be stimulated. Only the key response is effective, the other responses being less effective or completely ineffective. However, it is extremely difficult to determine which element of the immune system is the key response needed to overcome a particular virus infection and a different answer may be found for different individual circumstances.

Key points

- The immune system is a complex interacting mixture of cells and soluble components that has evolved to protect us from infection.
- Viruses that evade the innate immune system are dealt with by the adaptive immune system, comprising T cells and B cells (antibodies). Each cell of the adaptive immune system is specific for one epitope only.
- Cells of the adaptive immune system are inactive and each specificity is present in low number until stimulated by specific epitopes and, as a result, the adaptive immune system takes some days to reach its peak. Thus, a virus causing a primary infection can establish infection and clinical disease.
- Adaptive immunity has subsystems that operate semi-independently in the body (systemic immunity) and at mucosal surfaces (mucosal immunity) – the main route of infection for all micro-organisms. It is necessary to stimulate the mucosae directly to get mucosal immunity.
- Adaptive immunity functions through epitope-specific receptors: BCRs and TCRs.
- Antibodies are soluble BCRs that can recognize any type of molecule; they can directly neutralize virus infectivity, but can also signal to other mediators to kill cells.
- TCRs are cell-surface proteins that recognize only peptides complexed with MHC proteins.
- Stimulation of naïve T cells and B cells produces immunological memory – a

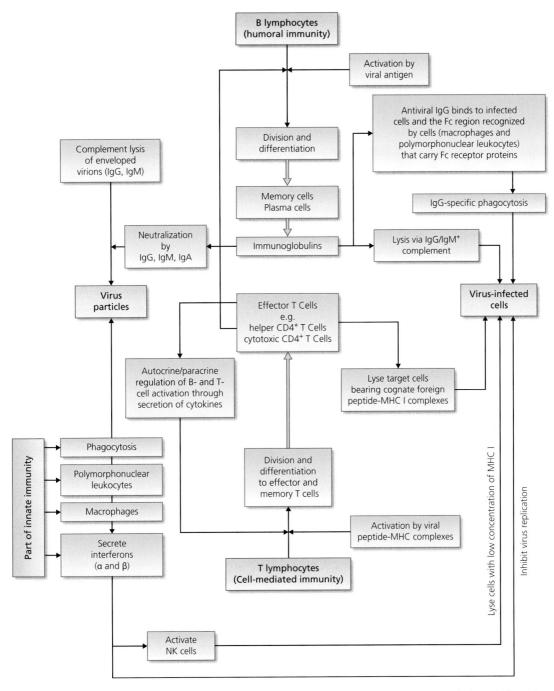

Fig. 14.6   Summary of some of the responses of the immune system to viruses and virus-infected cells.

heightened cellular presence that can respond with a shorter lag period than naïve cells, which is effective in preventing a second infection by the same agent.

- Viruses have evolved a variety of mechanisms for evading immune responses; without them there would probably be no viral disease and, depending on their efficacy, a virus remains in the body for a few days or a lifetime.
- The immune system has evolved layers of protection that operate at different times and different locations, and by different means. Thus it can be envisioned as several separate but communicating systems.

Questions

- Discuss the ways in which the humoral (antibody) arm of an adaptive immune response can counteract virus infection.
- What roles do T-cells play in the adaptive immune response?

Further reading

Bauer, D., Tampé, R. 2002. Herpes viral proteins blocking the transporter associated with antigen processing TAP – from genes to function and structure. *Current Topics in Microbiology and Immunology* 269, 87–99.

Braciale, T. J., Hahn, Y. S. 2013. Immunity to viruses. *Immunological Reviews* 255, 5–12.

Burton, D. R., Williamson, R. A., Parren, P. W. H. I. 2000. Antibody and virus: binding and neutralization. *Virology* 270, 1–3.

Farber, D. L., Yudanin, N. A., Restifo, N. P. 2014. Human memory T cells: generation, compartmentalization and homeostasis. *Nature Reviews Immunology* 14, 24–35.

Globerson, A., Effros, R. B. 2000. Ageing of lymphocytes and lymphocytes in the aged. *Immunology Today* 21, 515–521.

Jankovic, D., Liu, Z., Gause, W. C. 2001. Th1- and Th2-cell commitment during infectious disease: asymmetry in divergent pathways. *Trends in Immunology* 22, 450–457.

Male, D., Brostoff, J., Roth, D., Roitt, I. 2012. *Immunology*, 8th edn. Chichester, John Wiley & Sons.

Owen, J., Punt, J., Stranford, S. 2012. *Kuby Immunology*, 7th edn. W. H. Freeman, New York.

Schmolke, M., García-Sastre, A. 2010. Evasion of innate and adaptive immune responses by influenza A virus. *Cellular Microbiology* 12, 873–880.

Su, L. F., Davis, M. M. 2013. Antiviral memory phenotype T cells in unexposed adults. *Immunological Reviews* 255, 95–109.

Chapter 15

Interactions between animal viruses and cells

The properties of a virus and the cell it infects together determine if the cell lives, dies, or remains infected for ever.

Chapter 15 Outline

15.1 Acutely cytopathogenic infections
15.2 Persistent infections
15.3 Latent infections
15.4 Transforming infections
15.5 Abortive infections
15.6 Null infections
15.7 How do animal viruses kill cells?

The common perception of virus infections is that the only outcome is death of the infected cell through lysis. However, there are several possible outcomes of the interaction between a virus and a host cell, ranging from no infection to a long-lived infection. Virus–cell interactions can be classified into acutely cytopathogenic, persistent, latent, transforming, abortive, and null infections. Initial studies rely on analysis of infections of cells in the laboratory and the data obtained are used to pave the way for the eventual understanding of infection of the whole organism (Chapter 16). Two points should be borne in mind:

1. that a prerequisite of any of these types of infection is the initial interaction between a virus and its receptor on the surface of the host cell and hence any cell lacking the receptor is automatically resistant to infection;
2. that both virus and cell play a vital role in determining the outcome of the interaction. A virus may exhibit, for example, an acutely cytopathogenic infection in one cell type and latency in another.

This last point illustrates the intimate nature of the relationship between a virus and the cell it infects.

Introduction to Modern Virology, Seventh Edition. N. J. Dimmock, A. J. Easton and K. N. Leppard.
© 2016 John Wiley & Sons, Ltd. Published 2016 by John Wiley & Sons, Ltd.

15.1 Acutely cytopathogenic infections

Acutely cytopathogenic infections are those that result in cell death. These have also been called 'lytic' infections, but this term is not entirely accurate, as in some infections cells die without being lysed, i.e. by apoptosis or programmed cell death (see Chapter 13). The viruses which cause acutely cytopathogenic infections are the ones most commonly studied in the laboratory, since cell killing is an easy effect to observe. In these infections, production of infectious progeny can usually be monitored without difficulty, and the time scale is usually measured in hours. The one-step growth curve (Section 1.3) describes the essential features of any eukaryotic or prokaryotic virus–host interaction which results in cell death, and Fig. 15.1 shows the successive appearance of

cell-associated virus infectivity, infectivity that has been released from the cell into the tissue culture fluid, and the cytopathic effect (CPE). The pathological effects are always last to appear, just as in human infections (Chapter 16). CPE is observed under the microscope and is often manifested as a change in cell shape from a flattened cell morphology, adherent to a culture dish, to a rounded-up structure. Quite early on in infection, viruses often inhibit the synthesis of cellular proteins, DNA or RNA, but frequently cell death occurs sooner than can be explained by these inhibitory events. Exactly how an animal cell is killed in most cases is still not certain (see Section 15.7), but it is not associated with lysis by any lytic enzyme equivalent to lysozyme, an activity which is restricted to a minority of bacteriophages and bacteria. Lysis is one of the ways that a non-enveloped virus may be liberated from the infected cell (e.g. poliovirus, Fig. 15.1). In other infections, virus release is achieved without cell lysis; instead, viruses enter a cytoplasmic vesicle which then releases its contents into the tissue culture fluid by fusion with the plasma membrane. All membrane bound viruses bud from cellular membranes (see Section 12.6).

15.2 Persistent infections

Persistent infections result in the continuous production of infectious virus and this is achieved either by the survival of the infected cell or by a situation in which a minority of cells are initially infected and the spread of virus is limited, so that cell death is counterbalanced by new cells produced by cell division, i.e. no net loss within the culture. Persistent infections result from a balance struck between the virus and its host as a result of either: (i) the interaction of virus and cells alone; (ii) the interaction of virus and cells with antibody or interferon acting to limit virus production; (iii) the interaction of virus and cells with the

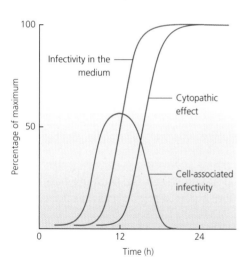

Fig. 15.1 An acutely cytopathogenic infection of a HeLa cell line by poliovirus (picornavirus; class 4). Cells were inoculated with an m.o.i. (multiplicity of infection) of 10 infectious units per cell, so that nearly all cells were infected and a one-step growth curve resulted. Note that cell-associated infectivity declines at later times as cells die.

production of defective-interfering (DI) virus (see Section 8.2); or (iv) a combination of these events.

The suggested explanation for the ability of certain virus–cell combinations to establish a persistent infection is that viruses evolve to a state of co-existence with their host. In other words, the virus gains no advantage in killing its host, rather the reverse as the virus depends absolutely on the host for its very survival.

Persistent infections resulting from the virus–cell interaction

Simian virus (SV) 5 (also called parainfluenza virus type 5) causes an acutely cytopathogenic infection with cell death in the baby hamster kidney (BHK) cell line (Fig. 15.2a), but when it infects a monolayer of monkey kidney (MK) cells, it establishes a persistent infection. The virus multiplies at the same rate in both cell types, and multiplies with a classical one-step growth curve in MK cells (Fig. 15.2b), but the MK cells show no CPE, remain healthy, and produce progeny virus continuously (Fig. 15.2c). Infection by SV5 does not damage

the MK cell, in the sense that it does not perturb the synthesis of cellular protein, RNA or DNA, and cell division continues as normal. Calculations show that the SV5 infection of MK cells makes little demand on the host's resources, e.g. total viral RNA synthesis is <1% of cellular RNA synthesis (even though each cell is producing about 150,000 particles/day). Thus, in monkey cells SV5 causes a harmless persistent infection, while in BHK cells it causes an acutely cytopathogenic infection from which all cells die. The origin of the cell is thus all important in determining the outcome of this relationship.

Persistent infections can also arise when viruses are able to inhibit the apoptotic response that normally gives rise to an acutely cytopathogenic event. Many viruses express gene products which can inhibit apoptosis (Box 15.1). For example, human cytomegalovirus (a herpesvirus) encodes a protein called UL37x1 which inhibits apoptosis of infected cells, permitting the virus to establish a long-lasting infection. Since the persistent infection allows the virus to multiply for a long period of time, it is to the advantage of the virus to have evolved the means to suppress apoptosis.

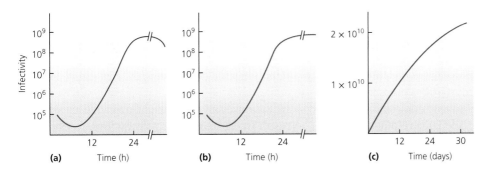

Fig. 15.2   Different types of infection caused by simian virus 5 in BHK (baby hamster kidney) cells and MK (monkey kidney) cells. (a) The acutely cytopathogenic infection in BHK cells. Note that virus yield drops after 24 h. (b) The initial one-step growth curve in MK cells which kills no cells and becomes persistent. (c) The cumulative yield from cells infected in (b) over 30 days. Cells grow normally during this infection and have to be subcultured at intervals of approximately 4 days.

> **Box 15.1**
>
> **Evidence that viruses can inhibit apoptosis to establish a persistent infection**
>
> Infection of normal cell lines with Sindbis virus (a togavirus) results in lysis. However, infection of primary cultures of neurons results in the establishment of a persistent infection. Analysis of gene expression from the neuronal cell genome showed that the Sindbis virus infection activated the expression of the host cell gene, *bcl*-2. This gene is responsible for preventing apoptosis in normal cells. By stimulating *bcl*-2 expression Sindbis virus prevents the normal apoptotic response of the cell to infection and the virus is able to persist.

Persistent infections resulting from interactions between viruses, cells and interferon, or viruses, cells and antibody

A persistent infection may be established as a result of an equilibrium between the generation of new infectious particles and the antiviral effect of interferon (see Section 13.1). Interferon supresses virus growth, and when only a few cells in a culture are infected, naturally-produced interferon will induce an antiviral state in the uninfected cells. As a result, virus production is significantly reduced, with an associated reduction in interferon induction. The reduction in interferon levels allows the virus to multiply, infecting a proportion of the cells in culture, until the level of interferon response rises to a point where it inhibits virus production again. In this way, a cycle of virus production and repression is set up and a steady-state can be achieved.

A persistently infected culture may also be established in the laboratory when only a few cells are infected initially, and a small amount of specific neutralizing antibody (low-avidity antibody is most effective) is added to the medium. The antibody decreases the amount of progeny virus available to reinfect cells. As with the interferon situation, the result is that the rate of infection and, hence, cell death is matched or exceeded by the division of non-infected cells, so that on balance the cell population survives. The antibody must be constantly added to ensure sufficient is present as new virus is being produced. This is a question of establishing a dynamic equilibrium tilted in favour of the cell. Of course, the overall net production of cells is less than in an uninfected culture but, in an animal, cell division would be upregulated by the normal homeostatic mechanisms that control cell number. Indeed, this situation is thought to mimic certain sorts of persistent infections in the whole animal (see Chapter 16).

Persistent infections resulting from interactions between viruses, cells and defective-interfering (DI) virus

DI genomes are produced by all viruses as a result of errors in replication which delete a large part of the viral genome (see Section 8.2), making them *defective* for replication. The DI genome retains the sequences that are needed for recognition by viral polymerases and for the packaging of the genome into a virus particle, but little or nothing else. Thus, the DI genome is replicated only in a cell that is infected with infectious virus of the type from which the DI genome was generated, as this is needed to supply replicative enzymes and structural proteins. In this sense, the DI genome is parasitic on infectious virus. The DI virus particles that result from this collaboration are usually indistinguishable in appearance from infectious particles. *Interference* between the DI and infectious virus comes about, in part, because more copies of the shorter DI genome can be made in the same time it takes to synthesize the full-length genome. For example, if the DI genome represents one-tenth of the infectious genome, then for every full length genome synthesized, 10 copies of the DI genome will be made. Since viral polymerases are made in modest amounts, the large number of DI genomes produced will eventually sequester all the polymerase and synthesis of infectious genomes and virus particles will cease. The DI genomes will also compete with the helper virus for proteins required to package the genome into particles and, again, their greater number will provide an advantage for the DI genomes. Thus, as the infection proceeds, the yield of fully infectious helper virus declines. However, since the DI genomes require the presence of helper virus the production of DI genomes will also decline due to the reduced availability of essential proteins. In practice, the situation is more complex, as the generation of DI genomes is very much dependent on the type of cell infected, so that both cell and virus contribute to a balanced situation.

Persistent infections result when there is an equilibrium between the multiplication of infectious virus, the multiplication of DI virus, and cell division. In the beginning, a DI genome is generated *de novo*, and the increase in amount of that DI genome follows that of infectious virus, upon which it is dependent (Fig. 15.3). Initially, there is sufficient polymerase to allow replication of both infectious and DI genomes, so there is no interference. Interference is

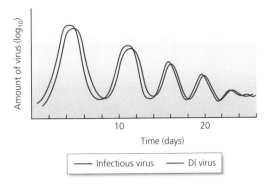

Fig. 15.3 A persistent infection established as a result of interaction in susceptible cells between a virus that normally causes an acutely cytopathogenic infection, and defective-interfering virus derived from it. The dynamic cycles of production of infectious and DI virus eventually give way to a low-level steady state persistent infection.

apparent only when the number of DI genomes is so great that they sequester the majority of the polymerase proteins. At this point, there is interference with the multiplication of infectious virus, which in turn leads to a concomitant decrease in the dependent DI virus. When the amount of infectious virus reaches a low point, cell numbers recover. As they do so, infectivity increases in the relative absence of DI virus, but then the cycle of events is repeated. In this very dynamic way, infectious virus persists under conditions where, without DI virus, it would produce a short-lived acutely cytopathogenic infection. In some systems, the oscillation within the cycles gets progressively smaller, until there is only a low-level production of infectious virus and DI virus and a steady-state persistent infection results (Fig. 15.3).

15.3 Latent infections

The term 'latent' is defined as existing, but not exhibited. Thus, in the context of a

Table 15.1				
Examples of latent infections.				
Virus		**Synthesis of:**		
	State of virus genome	**At least one transcribed RNA***	**Viral protein(s)**	**Infectious progeny**
Herpes simplex virus	Episomal	+	−	−
Epstein-Barr virus	Episomal	+	+	−
Adeno-associated virus	Integrated	+	+	−

*The amount and nature of gene expression is limited but varies between virus systems, and is strictly controlled.

virus-infected cell, this means that the viral genome is present but no infectious progeny are produced. However, latency is an active infection and some virus-coded products are always expressed (Table 15.1). Lysogeny by temperate bacteriophages is clearly a latent infection, and in animal cells, for example, herpesviruses and adeno-associated virus (AAV) exhibit latency. Our understanding of the molecular control processes involved in latency is incomplete but is increasing (Chapter 17). Bacterial viruses which achieve latency and AAV (Box 15.2) do so by integrating a DNA copy of their genome into the host's genome. This ensures that the viral genome will be replicated together with host chromosomal DNA and transmitted to daughter cells, and will be protected from degradation by nucleases. On the other hand, the DNA of herpesviruses is not integrated, but remains episomal, although the normally linear molecule is circularized.

It is in fact inaccurate to call a virus latent, as latent infections always commence as acutely cytopathogenic infections. This initial infection is converted into a latent infection, and then delicate molecular controls operate to maintain the latent state. Eventually, latency is broken by certain external stimuli, and there is full expression of the virus genome with production

---

Box 15.2

Evidence that adeno-associated virus establishes a latent infection by integration into the host genome

- The virus genome can be detected as a tandem, head-to-tail, pair integrated in the host chromosome.
- Analysis of the integrated DNA shows that it is present at the 19q13.3-qtr region of human chromosome 19.
- Site-specific integration is dependent on the presence of the virus Rep protein which binds to the virus genome terminal repeat region. In the absence of the Rep protein integration is seen to be random.

of infectious virus, giving the virus the opportunity of infecting new hosts. Typically, the latent state is only broken in a subset of the cells in a latently-infected culture and the remainder retain the virus in the latent state. Thus latency can be seen as an evolutionary strategy to remain in a host for a long period of time. In humans, latent infections with herpesviruses last for a lifetime, interspersed with periodic reverses into an acutely cytopathogenic infection. The presence of a latent virus can sometimes be shown by using labelled antibody specific for the relevant viral proteins or, in all cases, by polymerase chain reaction (PCR) amplification of virus genome sequences.

Epstein-Barr virus (EBV), a member of the herpes virus family, becomes latent when it infects B lymphocytes (non-dividing, non-antibody-producing cells that become antibody-synthesizing cells when activated by antigen) in vitro. Its very large, 172 kilobase-pairs (kbp) of linear double-stranded DNA is circularized and not integrated. As a result of infection the B cell is stimulated into continuous cell division. The viral DNA replicates semi-conservatively, once each cell cycle, and each molecule segregates to a daughter cell like a host chromosome. During latency, up to nine of the 100 viral genes are expressed. However, about 1 in 10^3–10^6 cells convert spontaneously to the acute phase of infection, where the full set of genes is expressed, and infectious virus is synthesized and released. EBV is also associated with various malignancies (see Section 25.6).

AAV has a small, single-stranded linear DNA genome of 4680 nucleotides. This is an unusual virus in that it is usually dependent on coinfection of the same cells with an adenovirus or a herpes virus for its replication (see Section 7.6). In the absence of a helper virus, AAV becomes latent with the help of the virus-encoded Rep protein, integrating into chromosome 19 of the host genome, and is replicated as part of the host genome. There is minimal gene expression during latency. On superinfection with helper virus, and in the presence of Rep, latency is broken. Viral DNA is excised from the host genome and the production of infectious AAV commences.

15.4 Transforming infections

As a result of infection with a variety of DNA viruses or some retroviruses, a cell may undergo more rapid multiplication than the other cells in the same culture, concomitant with a change in a wide variety of its properties, i.e. it is transformed. This is often preceded by integration of at least part of the viral genome with that of the host. Chapter 25 deals in detail with transformation and other aspects of viruses associated with cancer. However, transformation is not the same as carcinogenesis which refers to the production of a tumour. Transformation is a rare event generated by a virus that usually causes acutely cytopathogenic or persistent infections and in cell culture there is no production of a tumour which can only occur in an animal. Nevertheless, transformation is a significant event which may reflect some of the changes in cells that can predispose them to becoming cancer cells if introduced into an animal. In cell culture, the result of transformation is that one immortalized cell can take over the whole population. For DNA viruses, transformation has no evident evolutionary significance, as the transformed cell contains only a fragment of the genome and cannot give rise to any progeny (Table 15.2).

15.5 Abortive infections

Any cell that possesses the appropriate receptors will be infected by a virus, but different cells do not replicate that virus with

Table 15.2
Types of virus that (very rarely) cause transformation of the infected cell.

Family	Genome	Proportion of genome integrated	Progeny
Retrovirus	RNA	Whole	+
Polyomavirus	DNA	Part	−
Papillomavirus	DNA	Part	−
Adenovirus	DNA	Part	−
Herpesvirus	DNA	None; episomal	−
Hepadnavirus	DNA	Part	−

equal efficiency. There may be a reduction in the total yield of virus particles (sometimes to zero), or the quality of the progeny may be deficient, with the production of poorly infectious particles or a large proportion of non-infectious particles as shown by their particle: infectivity ratio. Both the reduction in yield and the reduction in quality of the particles reflect a defect in the production or processing of some components necessary for multiplication, be it DNA, RNA or protein. One example is avian influenza virus growing in the mouse L cell line, in which both the amount of progeny and its specific infectivity are reduced, probably due to the synthesis of insufficient virion RNA. Another is infection of other non-permissive cells with human influenza viruses which gives rise to normal yields of virions, which are non-infectious. This is because these cells lack the type of protease required to cleave the haemagglutinin precursor protein into HA1 and HA2 (see Section 12.8). This can be reversed by adding small amounts of trypsin to the culture or to released virions. Abortive infections present difficulties when trying to propagate viruses, but have been used to advantage when research into the nature of the defect has furthered understanding of productive infections. In natural infections, abortiveness contributes to the character of the infection by effectively restricting virus to only those cells in which a productive infection takes place.

15.6 Null infections

This category represents cells which do not have the appropriate receptors for a particular virus, and thus cannot interact with a virus particle. Often this is the sole block to infection, as shown when infectious viral nucleic acids are artificially introduced into such cells in the laboratory and produce progeny virions.

15.7 How do animal viruses kill cells?

The recognition that animal viruses often kill the cells in which they multiply gives the simplest criterion of infectivity, that of viral CPEs (see Section 5.1). However, such killing is not a property of the virus alone, but of a specific virus–cell interaction, and as exemplified by SV5 above, a virus does not necessarily kill the cell in which it multiplies. Surprisingly, it is still by no means clear exactly how a cell is killed, except that every virus needs to undergo at least part of its multiplication cycle for killing to occur. Thus, it seems that a toxic effect is produced by the virus and that viruses with different replication strategies are likely to invoke different mechanisms of toxicity. One problem in these studies is in distinguishing between an effect on a cell function which operates early enough to be responsible for toxicity and those which appear late on and are a consequence of those toxic effects.

The most commonly-studied form of cell death is that referred to as *necrotic*. Necrosis is associated with changes in cell structure and appearance, particularly rounding and swelling. This is accompanied by one or more of loss of membrane integrity, lysosomal leakage and random degradation of the cell DNA. Ultimately, the cells rupture. The alternative is that cells may die as a result of stimulation of the apoptotic pathway which is an altogether more controlled process that is governed by host cell mechanisms that respond to the infection (Section 13.5).

The process of cell death, either necrotic or apoptotic, is usually associated with virus-directed or virus-induced inhibition of synthesis of host proteins, RNA and DNA or perturbation of normal homeostatic processes (Table 15.3). Cells in which macromolecular synthesis is switched off do not die immediately. Thus, the mechanisms discussed below, with the exception of apoptosis and the upset in Na^+/K^+ balance, resemble death by slow starvation rather than acute poisoning. Presumably this is to the advantage of the virus as it provides time to produce more virus particles.

Prevention of host cell transcription during a virus infection inevitably leads to the death of the cell. Inhibition of host transcription is seen with several viruses, including examples of the rhabdovirus family such as vesicular stomatitis virus (VSV) and enteroviruses such as poliovirus. Both VSV and poliovirus target the TATA binding protein (TBP), a component of the cellular TFIID protein complex which is an essential transcription initiation factor used in the expression of genes transcribed by RNA polymerase II. The poliovirus 3C protease, in addition to its essential role in cleaving the poliovirus polyprotein (Section 11.3), specifically targets TBP and cleaves it, rendering it inactive and thereby preventing transcription of cellular genes. The VSV M protein which forms part of the virus particle is also found in the nucleus of infected cells where it specifically inactivates TFIID activity by an as yet unknown mechanism.

Inhibition of host cell protein synthesis can occur by a number of mechanisms. By far the simplest is when the virus generates an overwhelming amount of mRNA which can then successfully compete with the cellular

Table 15.3

Some suggested mechanisms of viral cytopathology.

Mechanism	Virus
Inhibition of host transcription	rhabdoviruses, poliovirus
Competing out of cellular mRNA by excess viral mRNA	Semliki Forest virus
Loss of ability to initiate translation of cellular mRNA	poliovirus, reovirus, influenza virus, adenovirus
Degradation of cellular mRNA	influenza virus, herpes virus
Failure to transport mRNA out of the nucleus	adenovirus, influenza virus
Apoptosis	sindbis virus, Semliki Forest virus, influenza A, B, C viruses, HIV-1*, adenovirus, measles virus
Imbalance in intracellular ion concentrations	Semliki Forest virus, rotavirus

*HIV-1, human immunodeficiency virus type 1.

mRNA for ribosomes by weight of numbers. This is seen with Semliki Forest virus which makes very large quantities of a subgenomic 26S mRNA to encode the virus structural proteins (Section 11.4). In contrast, the inhibition of cellular mRNA translation by poliovirus takes a more subtle form. The poliovirus inhibition results from inactivation of the translation initiation factors which are responsible for recognizing capped messenger RNAs (mRNAs). Poliovirus mRNA itself is not capped and relies on a special mechanism of translation initiation involving an internal ribosome entry site (IRES) to allow direct translation of the RNA without the need for a cap structure. The poliovirus translation is therefore unaffected by its inactivation of cap recognition (Section 11.3). A less direct effect is seen with the ability of herpesviruses and influenza virus to promote the degradation of cellular mRNA. For herpesviruses this is a general degradation of mRNA, including that of the virus but as this is in great abundance the effect on the virus is considerably smaller than on the host cell.

Cellular mRNAs are synthesized in the nucleus and are then transported into the cytoplasm for translation in a process orchestrated by a number of cellular proteins. Several viruses such as adenoviruses and influenza virus have been shown to inhibit the transport of some cellular mRNAs during infection. The process by which this differential inhibition of transport of some cellular mRNAs is not completely understood but the effect is to reduce the expression of several host cell genes.

Inhibition of cellular homeostatic processes or critical events in the cell cycle frequently leads to the induction of apoptosis. During many virus infections, normal synthesis of cellular DNA is inhibited. This is rarely a direct effect but is more frequently a consequence of inhibition of cell gene expression that prevents the synthesis of the necessary DNA polymerase components. It is therefore often most obvious in infections where the viruses inhibit transcription and/or translation of mRNA. Perturbation of DNA synthesis is a key stimulator of apoptosis through the p53 pathway (Section 13.5). A striking example of disruption of homeostasis is seen with Semliki Forest virus infection. In Semliki Forest virus-infected cells, there is evidence that the virus inhibits host protein synthesis by affecting the plasma membrane Na^+/K^+ pump, which controls ion balance. As a result, intracellular Na^+ concentration increases to a level where viral but not cellular mRNA is translated. The rotavirus non-structural protein, nsP4, has been shown in a number of studies to affect the ion balance in infected cells by altering the permeability of the plasma membrane and other cell membranes, and is toxic even when expressed in a cell by itself. It has thus been claimed as the first known viral endotoxin. While the ionic imbalance will itself lead to death of the cell it will also stimulate an apoptotic response.

It is not clear what advantage, if any, accrues to the virus in killing its host cell, as most progeny viruses leave the cell by exocytosis or by budding from the cell membrane. However, there is a distinction to be made between cell death that leads to the death of the animal host, and death of cells (e.g. in the gut) that can readily be replaced by the host – often without the host realising that they are infected. The latter does no harm, but the former suggests that the infection represents a very new virus–host interaction which is poorly evolved in terms of survival of virus and host (Section 4.5).

Key points

- All animal virus–cell interactions can be classified under just six headings.
- Classification depends on the type of cell infected, and as a result a virus can be classified in more than one category.

- Any definition of the virus infection should specify the circumstances of the infection.
- Virus infection can lead to the death of a cell by several mechanisms.

Questions

- Describe three long-lasting interactions between viruses and cells in culture and consider the mechanisms by which these occur.
- Discuss the potential outcomes of an acute cytopathogenic infection of a cell and describe the processes which are responsible for death of the host cell, when this occurs.

Further reading

Carrasco, L. 1995. Modification of membrane permeability by animal viruses. *Advances in Virus Research* 45, 61–112.

Levine, B., Huang, Q., Isaacs, J. T., Reed, J. C., Griffin, D. E., Hardwick, J. M. 1993. *Conversion of lytic to persistent alphavirus infection by the bcl-2 cellular oncogene. Nature (London)* 361, 739–742.

Lyles, D. S. 2000. Cytopathogenesis and inhibition of host gene expression by RNA viruses. *Microbiology and Molecular Biology Reviews* 64, 709–724.

O'Brien, V. 1998. Viruses and apoptosis. *Journal of General Virology* 79, 1833–1845.

Roulston, A., Marcellus, R. C., Branton, P. E. 1999. Viruses and apopotosis. *Annual Review of Microbiology* 53, 577–628.

Chapter 16
Animal virus–host interactions

When an animal is infected with a virus, several outcomes are possible, ranging from no disease through short periods of disease to long-lasting disease states. The various types of interaction can be placed in one of seven possible categories.

Chapter 16 Outline

16.1 Cause and effect: Koch's postulates
16.2 A classification of virus–host interactions
16.3 Acute infections
16.4 Subclinical infections
16.5 Persistent and chronic infections
16.6 Latent infections
16.7 Slowly progressive diseases
16.8 Virus-induced tumours

The study of animal virus–host interactions is essential for our understanding of the capacity of viruses to cause disease. The aim is to identify important stages in the infectious process that may offer possibilities for intervention. It must always be borne in mind that viruses are parasites and that the biological success of a virus depends absolutely upon the success of the host species. Hence, the evolutionary strategy of a virus in nature must take into account that it is counterproductive for the virus to eliminate the host species (although killing a small proportion will not affect the host species' survival) or to impair their reproductive ability that would reduce the number of new

potential hosts. Viruses demonstrate a number of different, and often complex, interactions with whole animals. A number of different factors are involved in determining the ultimate outcome of the virus–animal interactions. Many of these interactions involve components of the immune system which are described in Chapters 13 and 14.

16.1 Cause and effect: Koch's postulates

In 1877, the bacteriologist Robert Koch first enunciated a set of criteria for deciding if an infectious agent was responsible for causing a particular disease and was not merely a passenger. These have stood the test of time and, with the appearance of previously unknown diseases, are equally relevant (with some modifications) today. In essence, Koch postulated that: (i) the suspected agent must be present in particular tissues in every case of the disease; (ii) the agent must be isolated and grown in pure culture; (iii) pure preparations of the agent must cause the same disease when they are introduced into

Introduction to Modern Virology, Seventh Edition. N. J. Dimmock, A. J. Easton and K. N. Leppard.
© 2016 John Wiley & Sons, Ltd. Published 2016 by John Wiley & Sons, Ltd.

healthy subjects. As understanding of pathogenesis grew the following modifications were made:

Postulate 1. Koch originally thought that the agent should not be present in the body in the absence of the disease; this was abandoned when he realized that there were asymptomatic carriers of cholera and typhoid bacteria. As will be seen below, many viruses also cause such subclinical infections.

Postulate 2. Viruses were not known in Koch's time and some still cannot be grown in culture today, so this postulate was modified for viruses to say that it is sufficient to demonstrate that bacteria-free filtrates induce disease and/or stimulate the synthesis of agent-specific antibodies.

Postulate 3. Clearly, it is impossible to fulfil the third postulate when dealing with serious disease in humans, although accidental infection can sometimes provide the necessary evidence. For example, unfortunate laboratory accidents have directly demonstrated that human immunodeficiency virus (HIV) is responsible for the acquired immune deficiency syndrome (AIDS).

16.2 A classification of virus–host interactions

A classification of the various types of virus–host interactions with some named examples is given in Table 16.1. However, the table is only intended as a guide, as there is a continuous spectrum of virus–host interactions, and divisions are imposed solely for the convenience of classification. The categories are distinguished on four criteria: the production of infectious progeny; whether or not the virus kills its host cell; if there are observable clinical signs or symptoms; and the duration of

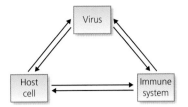

Fig. 16.1 Diagram to show the three-way interactions that decide the outcome of infection.

infection. It can be seen that cell death does not necessarily correlate with disease, as many types of cell can be replaced without harming the individual. It also appears that the duration of infection correlates inversely with the need for efficient transmission. Finally, it is also apparent that a single virus can appear in several of the categories, depending on the nature of its interaction with its host.

It is important to appreciate that a three-way interaction between virus, host cell and immune system determines the outcome of all infections (Fig. 16.1). Thus, in acute and subclinical infections, the balance favours the host (i.e. the virus is cleared from the body), whereas in persistent and chronic infections it is tilted towards the virus (which is not cleared for a long time and can continuously produce new virus that can infect new hosts throughout that time). Persistent, chronic and latent infections have in common the feature that the immune system cannot clear the viruses responsible from the body. However, all virus infections interact with the immune system and modulate one or more of its functions to some degree, so it never operates at its full unimpeded potential.

Immune-mediated disease

In attempting to classify virus *infections* of animals, it must be borne in mind that this is different to, but inextricably entwined with, virus *disease*, e.g. the only difference between an

Table 16.1

A classification of virus infections at the level of the whole organism.

Type of infection	Production of infectious progeny	Cell death	Clinical signs of disease	Duration of infection*	Transmission	Examples**
Acute	+	+	+	Short	Efficient	Measles virus, poliovirus (1 % of infections)
Sub-clinical	+	+	–	Short	Efficient	Poliovirus (99% of infections)
Persistent	+	– or +	–	Long (+ immune defect?)	Many opportunities	Rubella virus
Chronic	+	+	+	Long (+ immune defect?)	Many opportunities	Hepatitis B virus
Latent	–	–	–	Long	Many opportunities	Herpes viruses
Slowly progressive disease						
(a)	+	+	Eventually	Long	Many opportunities	HIV-1, TSE agents
(b)	–	+	Eventually	Long	None	Measles virus
Tumourigenic	+	–	+	Long	Many opportunities	Retroviruses, Epstein-Barr virus
	–	–	+	Long	None	Hepatitis B virus, human papillomaviruses

*Short, approximately 3 weeks or less; long, up to a lifetime.
**Examples are given of viruses that have the given type of infection at some point of their life history, and are not intended to convey that they cannot also be classified elsewhere, e.g. herpesviruses switch between latency and acute infection (see text).
HIV-1 human immunodeficiency virus type 1; SSPE, subacute sclerosing panencephalitis; TSE, transmissible spongiform encephalopathy (includes bovine spongiform encephalopathy and Creutzfeld-Jacob disease) – these are not virus infection but prion diseases (see Chapter 28).

Box 16.1

Evidence of involvement of the immune system in virus-associated disease

Early indications that the immune system may contribute to disease rather than universally reduce the impact of an infection were most dramatically seen in studies with mice infected with lymphocytic choriomeningitis virus (an arenavirus) which causes a fatal neurological infection in adult animals. When infected mice were immunosuppressed up to three days after the infection, 90% of the infected animals survived and the infection was converted from an acute infection into a persistent one. This demonstrates both the role of the immune system in the development of a fatal disease in these animals and also emphasizes that the establishment of persistent infections has an immunological component. In slightly less dramatic situations, reduction of the immune response can lead to an increase in the time from infection to death in fatal infections; in immunosuppressed mice the time to death following a lethal infection with Venezuelan equine encephalitis virus (a flavivirus) was extended by two days compared to normal animals, and with tick-borne encephalitis virus or dengue virus type 2 (both alphaviruses) the time to death in immunosuppressed compared to normal mice was extended by up to six days.

---

acute infection and a subclinical infection, and the main difference between chronic and persistent infections is that a disease may or may not ensue. The disease associated with an infection may be caused directly by the virus destroying target cells (as seen with poliovirus and motor neurons), but an important component of disease results from the immune response to that infection. At first sight this seems strange when the immune system is meant to be fighting the infection, but it is one of the consequences of that response, and is brought about by the many and conflicting demands put upon it; other less common potential consequences are allergic reactions and autoimmunity. The immune system is immensely powerful, and if this is not properly directed or controlled it can cause catastrophic, and potentially fatal, damage to the body (Box 16.1). Indeed, there are extreme responses that occur in certain virus infections of animals that result in death. In between are a variety of clinical signs and symptoms, such as the measles rash (due to deposition of antigen-

antibody complexes in the skin), that is the result of the immune response rather than the virus. It is not just the adaptive immune response that plays a role in this way, as elements of the innate immune response may also contribute to the signs and symptoms of disease (Chapter 13).

Infection of organisms can be described in terms of clinical signs and symptoms (Box 16.2) and by a variety of laboratory tests. Without the latter, no identification of the causative agent is complete.

Box 16.2

Definitions of signs and symptoms

Clinical signs: attributes of infection that are objectively assessed, such as elevated body temperature.

Clinical symptoms: attributes of infection that are subjectively assessed, such as pain.

16.3 Acute infections

Acute infections are short-lived infections where the infecting virus is rapidly and completely cleared from the body. Acute infections result in the presentation of a disease. These infections are analogous to acutely cytopathogenic infections in vitro, except that the infecting dose of virus is always small in comparison with an experimental infection of a cell culture, and that the virus therefore goes through many rounds of replication, spreading from the first cells infected to new susceptible cells. Thus, the time-scale is measured in days rather than hours. During many infections, viruses may enter the bloodstream and circulate around the body in a process called *viraemia*. During a viraemic infection the virus comes in contact with many different organs that it may infect. However, most viruses do not cause generalized infections but attack particular organs or tissues, known as *target organs* or *target tissues*. Thus, hepatitis viruses infect the liver, and influenza viruses infect the respiratory tract, and the reverse never occurs. This specificity is achieved largely through the presence of specific cell receptors on only certain cells in the body, as discussed for in vitro infections in Chapter 15, but there may also be intracellular restrictions on infection.

An acute infection begins with infection of one or a few cells, infectious progeny are produced and the infected cells die either directly from the action of the virus or indirectly due to the host immune response (see Section 15.7). Further cycles of multiplication follow with increasing numbers of infected cells and eventually the first signs and symptoms of

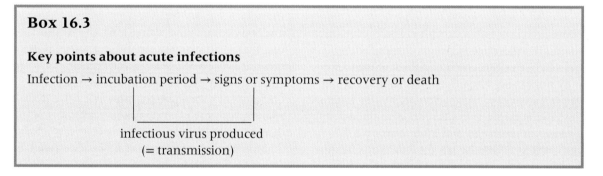

Box 16.3

Key points about acute infections

Infection → incubation period → signs or symptoms → recovery or death

infectious virus produced
(= transmission)

Box 16.4

Evidence of an acute common cold infection caused by a rhinovirus

Virus infections are measured by clinical signs and laboratory tests for the presence of virus. Following infection with a rhinovirus there is no fever or rise in body temperature (see Fig. 16.2). Respiratory signs are detected only three days after the infection and can be assessed by the simple process of measuring the mass of paper tissues used per day before and after infection – each used only once. The respiratory signs persist for two days before declining. Laboratory signs include virus isolation, and the generation of specific neutralizing antibody after 20 days. Virus can be detected at least one day before the onset of symptoms and antibody can be seen several days after the clinical signs.

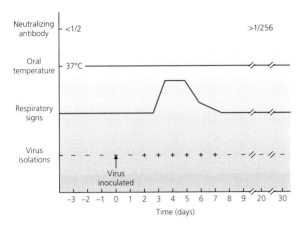

Fig. 16.2 A representation of an experimental infection in humans caused by a rhinovirus following deliberate intranasal inoculation.

disease appear. Consequently, the infection has been progressing for several days before we are even aware of it. Often, virus is being shed too before we are aware that an infection is present, which will enhance the spread of the virus to the susceptible people we meet. Fortunately, most people recover from acute virus infections within a few days or weeks. Acute infections are cleared by the actions of the innate and adaptive immune systems. There are many different acute infections and the essentials are summarized in Box 16.3.

An excellent example of an acute infection is seen with measles virus because almost all the infections that it causes are acute. By comparison, only 1% of poliovirus infections are acute, and 99% are subclinical (see Section

16.4). The example shown in Fig. 16.2 is of that familiar infection, the common cold, caused by one of the 100 or so different serotypes of rhinovirus (a picornavirus; see Box 16.4).

16.4 Subclinical infections

Subclinical infections are also known as inapparent or silent infections. These are the commonest infections and, as their name implies, there are no signs or symptoms of disease. The absence of a disease separates subclinical infections from acute infections to which they are similar in all other regards. Evidence of infection comes only from laboratory isolation of the virus or by a post-infection rise in virus-specific antibody. Any virus which causes a subclinical infection has evolved a favourable equilibrium with its host. The enteroviruses (picornavirus family), which multiply in the gut, are one such group and a classic example is poliovirus, which causes no symptoms in over 99% of infections. A subclinical infection is arguably the expression of a highly-evolved relationship between a virus and its *natural* host, since there are many examples of such a virus causing lethal disease when it infects a different host, e.g. yellow fever virus (a flavivirus) causes a subclinical infection in Old World monkeys but results in a severe infection in humans and is fatal in some New World monkeys. Subclinical infections have the same duration and are cleared by the same means as acute infections (Box 16.5).

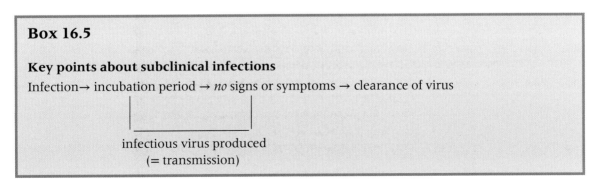

16.5 Persistent and chronic infections

Both persistent and chronic infections are acute or subclinical infections that are not terminated by the immune response (see Table 16.1). However, quite frequently persistent infections infect few cells and produce low levels of progeny virus, whereas chronic infections are more active (Box 16.6). The key question is why is the immune response ineffective? In general, it appears that viruses that cause persistent and chronic infections have major inhibitory effects on some aspects of the immune response itself, or on the expression of major histocompatibility complex (MHC) proteins. Alternatively, persistent or chronic infections result when an individual with an abnormality of the immune system contracts a virus that would otherwise cause an acute or

Box 16.6

Key points about persistent and chronic infections

- Most result from the virus modulating the immune system in some way (e.g. immunosuppression, down-regulation of interferons, down-regulation of MHC proteins, infection early in life); a few result from congenital immunodeficiency.
- Because chronic infections yield large numbers of infected cells and large amounts of viral antigen for years, immune responses can have pathological rather than beneficial consequences.

Box 16.7

Evidence that an incomplete immune response is associated with the establishment of persistent and chronic infections

Persistent and chronic infections are frequently associated with incomplete immune responses including:

- Immunosuppression by infection and inactivation of functions of macrophages, B cells, T cells, etc.
- Immune deficiency of the host, e.g. agammaglobulinaemia leading to persistent infection with live poliovirus vaccine viruses.
- Infection of foetus or neonate before it has a fully competent immune system, resulting in immunological tolerance, e.g. infection of the foetus *in utero* by rubella virus.
- Viruses that are poorly immunogenic, and/or fail to synthesize or display enough antigen on the cell surface.
- Viruses that synthesize excess soluble antigen that binds all available neutralizing antibody.
- Viruses that stimulate non-neutralizing antibody that binds to virions and blocks neutralizing antibody.
- Viruses that stimulate insufficient interferon (and cells that synthesize low amounts of interferon).
- Viruses that generate mutants through inaccurate transcription (includes synthesis of antigenic variants and defective interfering (DI) genomes).

subclinical infection. Whatever the cause, the end result is that virus particles and infected cells are not cleared. There are many scenarios which have this effect and some are listed in Box 16.7. It is important to appreciate the three-way interaction between virus, host cell and immune system that determines the outcome of all infections (Fig. 16.1). In persistent and chronic infections, this balance is tipped more to the virus, whereas in acute and subclinical infections it favours the host. Many viruses have specific functions that downregulate the expression of specific immune responses, including type 1 interferon. Downregulation does not result in complete inhibition but refers to a measurable reduction and this makes it harder to analyze. For example, interferon must be present in a sufficiently high concentration to be effective and if it falls below this threshold its effect is seriously impaired. In addition, viruses vary in their inherent sensitivity to the antiviral activity of interferon, so a level that inhibits one virus may not be effective against another.

A classic example of a chronic infection occurs when lymphocytic choriomeningitis virus (an arenavirus) is inoculated into newborn mice. The animals become immunologically tolerant to the viral antigens, and virus can be found in large quantities in the circulation and every tissue including the brain. One unexplained feature is that tolerance is rarely complete, but presumably the reduced immune response is unable to cope with the infection. Similarly, little interferon is made, although the viruses are sensitive to its action. Neonatally-infected animals remain healthy. In contrast, when adult animals are infected they experience an acute infection during which they mount an extremely strong T cell response to the virus. This T cell response results in significant damage to the cells of the host animal and is lethal; effectively the animals experience an immunologically-driven death.

Another example of a chronic infection occurs when humans are infected with hepatitis

Table 16.2 Infection of *adults* by hepatitis B virus: chances of progression of liver infection and disease.	
Approximate percentage of infected* adults who develop the type of infection listed	**Type of infection**
100% ↓	Acute or Subclinical ↓
10% ↓	Chronic ↓
1% ↓	Liver cirrhosis ↓
0.1%	Liver cancer (primary hepatocellular carcinoma)

*Figures are expressed as a proportion of all people shown to be infected. Current estimates are that approximately 30% of people exposed to hepatitis B virus become infected.

B virus (HBV, a hepadnavirus), one of the causative agents of viral hepatitis which is considered in Chapter 22. After an initial acute infection of an adult there is disease with liver damage, and most infections are completely cleared. However, in a small proportion of people (10%), virus persists in the liver for a lifetime, although most of these infections are subclinical (Table 16.2). This contrasts with the situation in the Far East where HBV is endemic (i.e. most people are infected), and is transmitted from a carrier mother to her children early in life, possibly via infected saliva. The younger the age at which infection takes place, the higher is the chance of a persistent infection resulting, and in very young infants, it rises to over 90% (Fig. 16.3).

HBV is able to persist by down-regulating MHC class I proteins on the surface of infected liver cells (hepatocytes), with the result that CD8$^+$ cytotoxic T cells are unable to act. Increased expression of MHC class I proteins

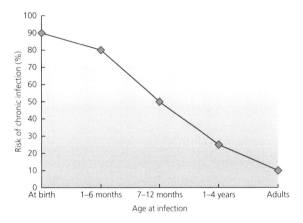

Fig. 16.3 The risk of becoming a carrier who is chronically infected with hepatitis B virus (HBV) depends on the age at which a person is infected.

stimulated by the action of type 1 interferon leads to a reduction of the number of infected cells through the action of CD8$^+$ cytotoxic T cells (see Chapter 13), but unless all infected cells are destroyed the relief is only transient. A small proportion of the persistent infections evolve to become chronic, with further liver pathology which can develop into total liver failure. This probably results from the balance tilting in favour of the CD8$^+$ cytotoxic T cells which then destroy a large number of infected liver cells. In addition, during chronic infection, co-circulation of large amounts of virus antigens and virus-specific antibody causes the formation of antigen-antibody complexes. When deposited in the kidney, these circulating complexes can activate complement and result in the destruction of kidney tissue leading to immune complex disease. Hence, the appearance of disease in this instance depends not on the cytopathic effects of virus but upon the relative proportions of viral antigens and virus-specific antibody. Finally a very small proportion of the chronic cases (0.1%) develop liver cancer (see Section 25.9) with part of the viral genome integrated into the host DNA. The example of HBV shows how one virus can cause four different types of infection depending on host cell-virus-immune response balance.

16.6 Latent infections

Latent infections are dealt with in detail in Chapter 17. By definition, all latent infections start as acute or subclinical infections. This is followed by establishment of a latent state during which no infectious virus is present, and such infections are always subclinical. Subsequently, there are usually periodic reactivations of an acute or subclinical infection followed by return to the latent infection (Box 16.8). The classic examples of viruses which establish latent infections are the

Box 16.8

Key points about latent infections

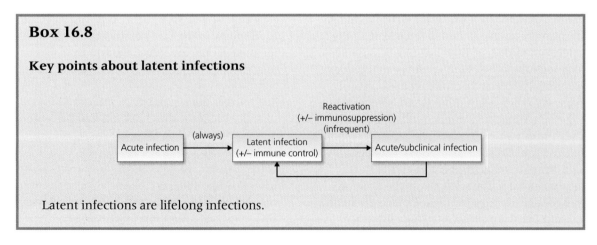

Latent infections are lifelong infections.

herpesviruses, a diverse and ubiquitous family which causes latent infections in many different animal species. Several herpesviruses cause common childhood infections. All become latent and such infections endure for life, making these viruses probably the best adapted of all to coexistence with their host. The action of the immune system is important for some viruses in maintaining their latent state. All of us are latently infected with several different human herpesviruses. One of the commonest and most successful is the human herpes simplex virus type 1 (HSV-1), because by 2 years of age nearly everyone has been infected. This is caused by contact with infectious saliva (e.g. when the baby is kissed by an infected adult). HIV-1 is also able to adopt a latent state in some cell types.

16.7 Slowly progressive diseases

As their name implies, these are diseases that take many years to manifest, while virus multiplication proceeds at the normal rate. There are two categories: slowly progressive diseases caused by viruses and those classed as spongiform encephalopathies. The viruses are subdivided into those that make infectious progeny throughout (e.g. HIV-1) and those whose genomes become defective during the long incubation period and hence are non-infectious (e.g. measles virus causing subacute sclerosing panencephalitis or SSPE). The transmissible spongiform encephalopathies are believed to be caused by a novel type of infectious agent, and are dealt with in Chapter 28.

Slowly progressive virus diseases that are infectious

The classic example of a slowly progressing virus disease during which infectious virus is produced is that of HIV-1 which predominantly infects CD4$^+$ T cells. HIV-1 infection is considered in Chapter 21. From the detailed statistics that are only available for infections in the developed world, a typical adult HIV-1 infection starts with acute influenza-like symptoms. Untreated, the disease is then quiescent for an average of 10 years, although the virus continues to multiply to high titre (10^{10} virions produced per day), with high turnover. During this time there is a progressive loss of CD4$^+$ T cells, and these eventually reach such a low level that there is a virtual collapse of the immune response. The patient is then overwhelmed by various endogenous viral, bacterial and fungal infections that are normally kept in check by the immune system. In the developed world these are commonly *Pneumocystis carinii* (a yeast-like organism) or cytomegalovirus.

Slowly progressive virus diseases that are non-infectious

Infection with measles virus (a paramyxovirus) demonstrates a different slowly progressive disease. Before a vaccine was available, most children contracted an acute measles virus infection, which was then cleared from the body by the immune system in around three weeks. However, in a very small proportion, the virus established itself in the brain and, after a long incubation period, caused a pattern of degenerative changes in brain function, including loss of higher brain activity, and inevitable progression to death (Box 16.9). Two different, but related diseases result from the neural infection. These are measles inclusion body encephalitis (MIBE) and subacute sclerosing panencephalitis (SSPE). MIBE occurs in approximately 1 in 2000 cases of measles infection. It is a chronic progressive encephalitis occurring in children and young adults and is associated with a long-lasting measles virus infection. MIBE can be fatal with the average

Box 16.9

Comparison of acute and slowly progressive measles virus infections

Acute measles virus infection

- A common acute childhood infection worldwide before mass immunization was established
- Respiratory transmission
- Systemic infection
- 100% disease – a diagnostic smooth skin rash

Sub-acute sclerosing panencephalitis (SSPE)

- Very rare, affecting about 6–22 per 10^6 cases of acute measles
- A sequel to acute measles
- Brain infection and disease
- Incubation period of 2–6 years
- Associated with measles infection early in life (<2 y old)
- Invariably fatal

survival time after diagnosis being three years. It is often seen in immunocompromised patients and is associated with early (<2 years of age) primary infection. SSPE occurs in approximately 1 in 1 million cases. It is a slow progressing neurological degenerative disease associated with a long-term persistent measles virus infection leading to death many years after the primary infection.

In SSPE at post-mortem, areas of 'hardening' or 'sclerosing' of brain tissue can be seen which give the disease its name. Why only certain individuals contract the disease is not understood. Infectious viruses cannot be isolated from SSPE patient tissue, although

virus can be transferred to susceptible cells by co-cultivation with SSPE brain extracts. The virus generated by co-culture grows extremely poorly and the virus produced spreads only very inefficiently between cells. Genomes of MIBE and SSPE-measles viruses obtained from the brains of infected individuals have now been sequenced, and these have accumulated many mutations and produce a defective internal virion protein (matrix). The inability to synthesize a functional matrix protein is the reason for poor infectivity. The nature of the presumed defect in the immune response which fails to clear the initial acute infection is not known.

16.8 Virus-induced tumours

The involvement of viruses in cancer is considered in Chapter 25. Most viruses that cause tumours have DNA as their primary genetic material in the cell or, in the case of the retrovirus family, have a DNA component in their replication cycle. A number of viruses whose principal interactions with their hosts can be classified into one or more of the categories described in Sections 16.3–16.7 can also, as a rare consequence of infection, trigger a sequence of events leading to the formation of a tumour. The in vitro parallel of this is the morphological transformation of cells in culture following virus infection. The viruses associated with tumours have in common the feature that they all cause an acute infection initially that is not cleared, and remain in the body in a variety of ways. Then, depending on the circumstances of the infection and the co-factors experienced in life, a cancer may eventually be manifested. All cancers are multifactorial.

DNA 'tumour' viruses normally cause an acute infection and it is rare that a tumour results, e.g. nearly everyone is infected as a child or young adult by Epstein–Barr virus

(EBV; a herpesvirus), which is subclinical or causes infectious mononucleosis (glandular fever) respectively, and yet occurrence of the tumours (Burkitt's lymphoma, nasopharyngeal carcinoma) with which the virus is associated is very rare. Demonstration of tumourigenicity in the laboratory is usually made under 'unnatural' circumstances which are known to be favourable to the development of the tumours. Important factors are genetic attributes and age at which infection takes place; certain inbred lines of animals develop tumours more readily than others, and young animals are more susceptible because their immune system is not fully mature. If no tumours result, it may be necessary to transfer cells transformed by the virus in vitro into the animal to demonstrate viral tumorigenicity.

Virus-associated tumours fall into two groups: those that produce infectious virus and those that do not. The latter are the more common and are caused by viruses from several different families. Such tumours either contain viruses which are unable to multiply or whose multiplication is in some way repressed, or they retain only a fragment of the viral genome. However, some tumours may be caused by a 'hit-and-run' infection which leaves no trace of the virus behind.

As with other virus–host interactions that have been discussed above, the induction of tumours depends upon the balance of a complex situation. This is represented in Fig. 16.4 where, initially, infection may be acute, persistent or latent. The transformation/tumorigenesis process initiated by infection requires additional factors, such as chemical carcinogens. How often transformation of cells takes place in the animal is not known, but it is likely that this is frequent and that on most occasions the immune system recognizes and destroys the transformed cell. The apoptosis mechanisms present in all cells also act as a powerful protection against tumour development. Tumour formation probably

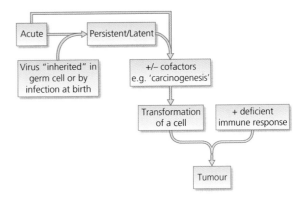

Fig. 16.4 The relationship between virus infection and tumourigenesis.

requires that the immune system is in some way deficient, a state perhaps induced by infection or by ageing. The mouse polyomavirus is a good example of this. It produces tumours only in inbred mouse strains where a protein expressed from an endogenous mouse retrovirus in the germline has caused deletion of a large subset of T cells, resulting in a deficient response to polyoma antigens.

Key points

- There are 7 categories of infection at the whole organism level: acute, subclinical, persistent, chronic, latent, slowly progressive, and tumorigenic.
- Any one virus may be found in more than one category, thus the categorization of a virus and its infection must include the circumstances of the infection.
- The majority of infections result in a satisfactory end for both the host species (if not the individual) and the virus. Infections are dealt with by the host's immune system so that the host recovers and can reproduce and provide new non-immune hosts for the virus. Either before adaptive immunity has matured or by evading immune responses, the virus multiplies and perpetuates itself by

being transmitted to other susceptible individuals.

- Infections that kill (or reproductively incapacitate) large sections of the host species are rare, and these result from an evolutionarily new virus–host interaction (such as HIV-1).

Questions

- Discuss the proposal that the outcome of virus infection is regulated both by the virus and by the host.
- Discuss the features of persistent and chronic infections and the factors that predispose viruses towards these outcomes.

Further reading

Buchanan, R., Bonthius, D. J. 2012. Measles virus and associated central nervous system sequelae. *Seminars in Pediatric Neurology* 19, 107–114.

Casadevall, A., Pirofski, L.-A. 1999. Host-pathogen interactions: redefining the basic concepts of virulence and pathogenicity. *Infection and Immunity* 67, 3703–3713.

Craighead, J. E. 2000. *Pathology and Pathogenesis of Human Viral Disease*. Academic Press, New York.

Griffin, D. E., Lin, W. H., Pan, C. H. 2012. Measles virus, immune control, and persistence. *FEMS Microbiology Reviews* 36, 649–662.

Mims, C. A., Nash, A., Stephen, J. 2000. *Mims' Pathogenesis of Infectious Disease*, 5th edn. Academic Press, London.

Ondondo, B. O. 2014. Fallen angels or risen apes? A tale of the intricate complexities of imbalanced immune responses in the pathogenesis and progression of immune-mediated and viral cancers. *Frontiers in Immunology* 5, 90.

Roizman, B., Whitley, R. J. 2013. An inquiry into the molecular basis of HSV latency and reactivation. *Annual Reviews of Microbiology* 67, 355–374.

Chapter 17

Mechanisms in virus latency

Latent infections are an important facet of the interaction of certain viruses with their hosts. The ability to establish a latent infection means that the virus can maintain its genome in the host for the entire life of the host, being reactivated periodically to produce new viruses which can infect new hosts. This can be seen as a highly evolved virus–host interaction. As James Lovelock, originator of the Gaia hypothesis, put it: "An inefficient virus kills its host. A clever virus stays with it."

Chapter 17 Outline

17.1. The latent interaction of virus and host
17.2. Gene expression and the lytic and lysogenic life of bacteriophage λ
17.3. Herpes simplex virus latency
17.4. Epstein-Barr virus latency
17.5. Latency in other herpesviruses
17.6. HIV-1 latency

A common perception of virus infection is that the only possible outcome is an immediate and rapid production of progeny virus particles. While this is the most common result of infection it is not the only one, as described in Chapter 16. Of the possibilities described there, one of the most interesting is that where certain viruses enter a different type of replicative cycle in which the host cell survives, either for a fixed period or indefinitely. During the infection the virus is quiescent, with no progeny being produced and no signs or symptoms of disease in the host. This is termed a latent infection. The appreciation that some viruses have the capacity to establish a latent infection has significant implications for our understanding of their associated diseases and potential therapies. It also has relevance to the potential use of animal viruses, such as the defective parvovirus adeno-associated virus, as vectors for gene therapy. The establishment of latency represents the ultimate parasitic interaction in virology, with the virus able to co-exist with a host which, in most cases, gains no obvious advantage.

17.1 The latent interaction of virus and host

All latent infections begin with an acute infection, which may or may not be associated with a disease. However, the virus is not eliminated from the body but instead reappears periodically with episodes of production of infectious virus particles, which again may or may not be associated with disease (Section 16.6). The new particles produced when latency 'breaks down' are then available to infect other hosts. The diseases experienced at the beginning of the process and in the reappearances may differ; a classic example is

Introduction to Modern Virology, Seventh Edition. N. J. Dimmock, A. J. Easton and K. N. Leppard.
© 2016 John Wiley & Sons, Ltd. Published 2016 by John Wiley & Sons, Ltd.

varicella zoster virus, a herpesvirus, which causes chicken pox following the initial infection but shingles in the same individual when it reappears later from the latent state. Following the initial infectious event the virus establishes an intimate association with the host cell and this stage may last for a very extended period of time, even for the lifetime of the host. This offers the virus an opportunity to lie dormant until a new susceptible host presents itself. The new host may be separated by a generation or more from the original infected individual, so latency has important implications for virus survival and transmission in the host population.

The group of viruses for which latent infections are best known are the herpesviruses. Herpesviruses infect a wide range of hosts, with eight viruses able to infect humans: herpes simplex virus (HSV) types 1 and 2, Epstein-Barr virus (EBV), varicella zoster virus (VZV), cytomegalovirus (CMV) and human herpesviruses 6, 7 and 8 (also known as Kaposi's sarcoma herpesvirus). The herpesviruses which infect mammals all establish latent infections as part of their normal processes, and once the latent infection is established the virus remains associated with

the host for the remainder of its life. It is likely that the herpesviruses infecting other animals can also establish latent infections but there is little information about these. While in the latent state, the herpesviruses do not produce any infectious progeny but at various times the virus may be reactivated, go through a complete replication cycle, and produce progeny viruses. Reactivation typically occurs in only a small fraction of the total latently infected cell population at any given time. The stimuli which induce reactivation are very varied and while some, such as the presence of fever, are associated with reactivation in many people, others can be very individual for each infected person.

The first indication that two alternative outcomes of infection by the same virus are possible was seen with bacteriophage λ (Box 17.1). This was followed by the discovery of other bacteriophage which can also adopt two different types of infectious process. Infections leading to the immediate production of new infectious phage with the death of the host cell are referred to as *lytic* infections, and those that permit survival of the host (at least until a lytic infection is induced) are referred to as *lysogenic infections*. Bacteria carrying quiescent bacteriophage genomes are called lysogens.

Box 17.1

Evidence that infection with bacteriophage lambda can have two outcomes

Following infection of a culture of *E. coli* with phage λ, most of the bacteria in the culture lyse, releasing more infectious phage. This is the result of a lytic infection. However, a small number of bacteria survive the infection and can be propagated. The surviving bacteria are resistant to infection by phage λ and also by related phage. When grown in culture, the resistant individual bacteria can spontaneously produce infectious phage. Treatment of the resistant bacteria with ultraviolet irradiation at levels sufficient to stop the culture growing induces this event in all cells, causing lysis, with release of large amounts of phage λ. These observations indicate that the phage λ introduced initially is present in the resistant bacteria and can be propagated from one generation to the next. However, the phage must be present in a non-infectious form which can spontaneously revert to produce virus.

Bacteriophage that are able to adopt either a lytic or a lysogenic replication cycle are called temperate phage. Several studies have shown that the phage DNA is inserted in the bacterial host genome during lysogeny. In this state, the virus DNA is replicated with the host chromosome and is called a prophage. The bacteria carrying the phage DNA are called lysogens. Phage λ remains a key example of latency and our understanding of the molecular events which occur during both lytic and lysogenic replication have formed the basis for our understanding of gene expression and latency.

17.2 Gene expression and the lytic and lysogenic life of bacteriophage λ

Bacteriophage λ has two alternative states of interaction with its E. coli *host, lytic and lysogenic, each of which is meta-stable and maintained through a series of transcription induction/repression control loops that exert positive feedback. The lysogenic state corresponds to latency. The establishment of latency is determined by the physiological state of the host cell, as is the induction of a latently infected cell back into the lytic pathway.*

Phage λ is a paradigm for the processes involved in latency and will be considered in detail first here. When phage λ enters a bacterium it must initiate either a lytic or a lysogenic cycle. For the latter to occur, the phage DNA must insert itself into the host chromosome and be maintained. In order to understand fully the mechanism by which lysogeny is established and maintained, and how the virus can regain the lytic replication cycle from the lysogenic state, it is necessary first to consider the genes involved in a lytic infection.

Following attachment, the phage λ linear genomic DNA is introduced by an injection mechanism similar to that used by phage T2 (Section 6.5) into the *E. coli* and is immediately converted into a covalently closed dsDNA circle by host enzymes. Circularization is possible because the phage lambda dsDNA genome contains 12 bases of single-stranded DNA at either end of the linear molecule which are complementary. The single-stranded regions are called *cos* sequences and on infection they come together by base-pairing to form a circular DNA molecule which is covalently closed by the host DNA ligase enzyme (Fig. 17.1). Having circularized, the phage λ genes are then switched on and off in a tightly regulated, coordinated manner which determines the outcome of the infection. During lytic replication, the circular phage DNA replicates independently of the host DNA and does not integrate into the genome of the host.

The phage λ genome contains two principal promoters which are recognized by the host cell DNA-dependent RNA polymerase which immediately begins transcription of mRNA. One of the promoters (P_L, Fig. 17.2) directs transcription in a leftwards direction and generates an mRNA that terminates at the end of the gene encoding the N protein. The other promoter (P_R) directs transcription in a rightwards direction to encode the protein Cro. However, termination of transcription at the end of the *cro* gene is not absolute and some mRNAs extend through the *cII*, *O* and *P* genes. The proteins are then translated from the polycistronic mRNA. The N protein causes the RNA polymerase to transcribe through the regions of DNA at the ends of the *N*, *cro* and *P* genes, where it had previously stopped, to generate polycistronic mRNAs encoding several additional proteins. The mRNA from P_L extends through the *N*, *cIII*, *xis* and *int* genes, and mRNA from P_R extends through the *cro*, *cII*, *O*, *P* and *Q* genes. Thus, N protein acts as a transcriptional anti-terminator and allows expression of

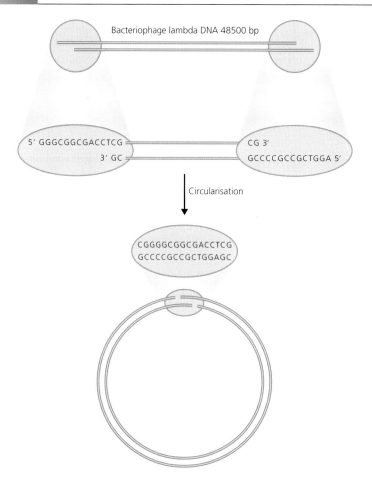

Fig. 17.1 Formation of circles of phage λ DNA after infection. The base composition of the complementary cos regions are shown. In λ these are extensions to the 5′ ends of the genome.

additional genes. These events occur before phage λ DNA synthesis and are referred to as immediate early (*N* and *cro*) and early (*cIII, xis* and *int* from P_L, and *Q* gene from P_R) gene expression.

The Cro and Q proteins are important for the next phase of gene expression. This occurs after phage λ DNA synthesis and is therefore, by definition, a late event in the replication cycle. Immediately on infection, and at the same time as P_L and P_R are utilized, host cell RNA polymerase recognizes a third promoter region in phage λ DNA, $P_R′$, located immediately after

the *Q* gene. However, transcription is terminated just downstream to synthesize a very short mRNA, even in the presence of the N protein. This short mRNA does not encode a protein. The Q protein is an anti-terminator which causes the RNA polymerase molecules initiating transcription at $P_R′$ to ignore the termination signal and to continue mRNA synthesis. The resulting mRNA is extremely long and extends through the genes encoding the structural proteins which make up the phage head and tail (Fig. 17.2). At the same time as the Q protein is exerting its activity, the

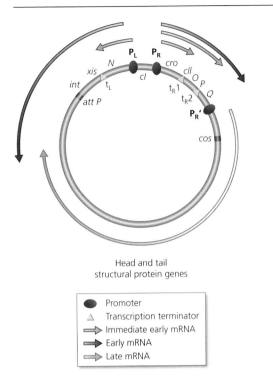

Head and tail
structural protein genes

⬤	Promoter
△	Transcription terminator
⇒	Immediate early mRNA
⇒	Early mRNA
⇒	Late mRNA

Fig. 17.2 The circular map of the bacteriophage λ genome. The genes involved in the lytic cycle are indicated. The positions of the phage promoters and transcriptional terminators involved in the lytic cycle are shown together with the mRNAs produced at immediate early, early and late times. The sequence attP is relevant to lysogenic replication.

Cro protein is also at work. The Cro protein binds to the phage λ DNA at the operator elements (O_L and O_R, respectively) of the promoters P_L and P_R. By doing this, Cro inhibits transcription from these promoters, stopping production of the early mRNAs. Sufficient Q protein is present to ensure that P_R', which is not affected by Cro protein, remains active. The result is that the only mRNA found at late times encodes the structural proteins which can package the newly-synthesized phage λ DNA into progeny virions utilising the *cos* site (Section 12.5).

Establishment of lysogeny

The initial events in the process of lysogeny are identical to those seen in a lytic infection. The circularized phage DNA is transcribed from the two major promoters, P_L and P_R, and also from P_R'. Since transcription from P_R' makes only a small mRNA which does not encode a protein and it is not involved in the lysogeny process it will not be considered further. Transcription from P_L and P_R makes the mRNAs encoding the N and Cro proteins, with small amounts of the cII, O and P proteins. Subsequently, the anti-terminator action of the N protein results in the synthesis of cIII, Xis, Int and Q proteins and larger quantities of cII, O and P, as described above.

Diverging from the pathway of lytic gene expression, the cII protein acts as a gene activator, directing the host RNA polymerase to begin transcription at two promoters which would otherwise be inactive. These are P_{RE} (promoter for repressor expression) and P_{int} (Fig. 17.3a). The mRNA initiated at P_{int} directs the synthesis of the Int protein which is responsible for integrating the phage λDNA into the host chromosome (see below). The use of the P_{int} promoter ensures expression of larger quantities of Int protein than can be generated from the polycistronic mRNA transcribed from P_L.

P_{RE} directs transcription of a single gene, *cI*. The cI protein, usually referred to as the cI (or lambda) repressor, is critical in lysogeny and phage with mutations in the *cI* gene can only replicate lytically. The cI protein binds to O_L and O_R, inhibiting transcription from P_L and P_R and preventing production of the early proteins. By inhibiting synthesis of the early proteins, the cI protein prevents the subsequent appearance of the late, structural proteins and, consequently, of infectious particles. In order to ensure that its own synthesis is not prevented by an absence of cII protein, the cI protein also directs RNA

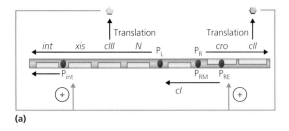

(a)

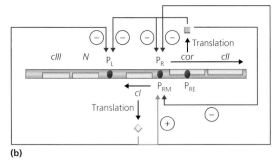

(b)

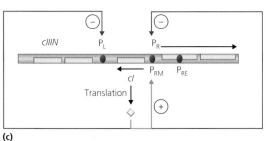

(c)

Fig. 17.3   The region of the bacteriophage λ genome encoding the genes responsible for lysogeny. (a) The cII protein activates the P_{RE} and P_{int} promoters to produce mRNAs for the cI and Int proteins, respectively. Note the mRNA from P_{RE} contains sequences antisense to the *cro* gene. (b) The cI protein binds O_L and O_R, inactivating P_L and P_R, respectively while activating P_{RM}, ensuring its own continued synthesis throughout lysogeny. The Cro protein also inhibits P_L and P_R activity. (c) The steady state of *cI* gene expression during lysogeny.

polymerase to an additional promoter, P_{RM} (promoter for repressor maintenance), which initiates transcription of the *cI* gene alone (Fig. 17.3b).

The action of the cII protein in activating expression from P_{RE} has an additional

consequence. The mRNA transcribed from P_{RE} contains some sequences that are antisense to those of the mRNA from P_R encoding the Cro protein. Hybridization of these will prevent translation of the *cro* gene mRNA and reduce the level of Cro protein in the bacterium.

The final result of the process described above is that the phage λ DNA is integrated in the host chromosome and the only gene which is active is that encoding the cI protein. The cI protein, by inhibiting P_L and P_R, prevents the expression of any other phage genes. By activating P_{RM}, the cI protein ensures transcription of its own gene and therefore its own continued synthesis so stabilizing the lysogenic state (Fig. 17.3c).

The choice between the lytic and lysogenic pathways

As described above, the initial steps in both the lytic and lysogenic pathways are the same. While the lytic process occurs much more frequently, it is clear that some circumstances must favour lysogeny. Several aspects of what determines the choice between lysis and lysogeny remain unclear but a key factor is the balance between the various repressors and activators produced early in the phage's interaction with the bacterium and, amongst these, the role of the cII protein is critical. If cII is very active, cI protein will be produced in large amounts. This will efficiently inhibit synthesis of all genes except its own, and lysogeny will result. On the other hand, if the cII protein is poorly expressed or has low activity, very low levels of cI protein will be present. The Cro protein will inhibit the activity of P_L and P_R and the synthesis of the cII protein, amongst others, will be significantly reduced. Without sufficient cII protein P_{RE} will not be strongly activated and synthesis of the cI protein will, in turn, be further reduced. In this case the lytic pathway will follow.

The activity of the cII protein is determined, at least in part, by environmental factors. The cII protein is susceptible to proteolytic degradation by host enzymes. The levels of host proteases are affected by many factors, especially growth conditions. Healthy bacteria grown in rich medium contain high levels of proteases and, when infected, are more likely to support lytic replication. In contrast, poorly growing bacteria have lower levels of proteases and so will encourage establishment of lysogeny. It should be clear that it is to the advantage of the phage to undertake a lytic infection in healthy bacteria which will contain high levels of ATP and are equipped to synthesize the virus proteins. When bacteria are nutritionally deficient it is to the advantage of the phage to establish itself as a lysogen and wait until conditions improve.

The cIII protein also plays a role in the choice between lysis and lysogeny. The function of cIII is to protect cII from proteolytic degradation. This protection is not complete since cII can be destroyed even in the presence of cIII. However, if cIII is absent or not functional cII is almost always degraded and the phage can only undergo a lytic lifestyle.

Integration of bacteriophage λ DNA during lysogeny

During lysogeny the phage λ DNA integrates into the genome of the host. This contrasts with the situation during a lytic infection when the phage DNA replicates independently. The integration of phage λ DNA into the host chromosome is carried out by the Int protein. The integrated DNA is referred to as a prophage and genetic mapping studies showed that there is only one site of integration, adjacent to the galactose (*gal*) operon in the *E. coli* genome at a position referred to as the lambda attachment (*att* λ) locus. In 1962, Campbell suggested that

crossing-over between the circularized phage DNA and the host genome results in insertion of the entire phage genome as a linear structure into that of the bacterium by reciprocal recombination (Fig. 17.4a). The position of the recombination site in the phage DNA lies downstream of the *int* gene and the region is referred to as the *att P* site (Fig. 17.2). The bacterial and phage attachment sites, *att* λ and *att P*, contain identical tracts of 15 base pairs indicating that homologous recombination occurs during the integration event. The actual site of the crossover for both integration and excision must take place within, or at the boundaries of, this common sequence (Fig. 17.4b).

Induction and excision of integrated DNA

Early observations showed that ultraviolet irradiation of a lysogenic culture of *E. coli* stimulates the phage to relinquish the lysogenic state and to enter a lytic mode of replication (Box 17.1). This is termed induction of the lysogen. Several other stimuli, such as treatment with potent mutagens, also induce the phage and a common feature of these treatments is that they cause significant changes in the host cell, and in particular expression from an array of genes not previously active. This is referred to as the *E. coli* SOS response and the role of the new gene products is to protect the bacterium against the adverse effects of the stimulus. Just as lysogeny may allow the phage to wait for the host to provide an optimum environment for replication, induction may be seen as a means by which such a phage can escape from a host whose survival is threatened.

A critical component in the SOS response is the Rec A protein. The function of the Rec A protein is to mediate recombination between DNA molecules. However, when the bacterium

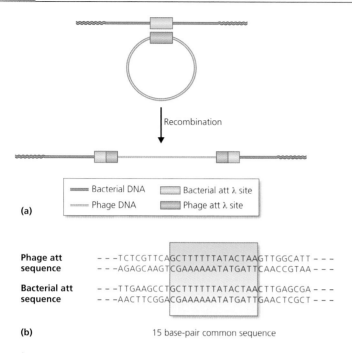

Phage att
sequence
```
– – –TCTCGTTCAGCTTTTTTATACTAAGTTGGCATT – – –
– – –AGAGCAAGTCGAAAAAATATGATTCAACCGTAA – – –
```

Bacterial att
sequence
```
– – –TTGAAGCCTGCTTTTTTATACTAACTTGAGCGA – – –
– – –AACTTCGGACGAAAAAATATGATTGAACTCGCT – – –
```

(b)   15 base-pair common sequence

Fig. 17.4 (a) Proposed model for the insertion of phage λ DNA into the bacterial genome by reciprocal recombination between phage and host DNA. (b) The common sequence in the bacterial and λ attachment sites.

is subjected to stress, such as irradiation with ultraviolet light, the Rec A protein alters its activity to become a specific protease. The primary target of the proteolytic Rec A protein is a repressor called Lex A. The Lex A protein represses expression from a range of genes and its cleavage removes this repression giving expression of the genes. Rec A protein also cleaves the phage λ cI repressor protein and when this occurs the promoters P_L and P_R become functional, transcribing the *N, cro* and other genes described above. Because the activity of the altered Rec A protein will continually prevent accumulation of functional cI protein, only lytic replication can occur even if the cII protein is active.

As well as re-establishing the lytic pattern of phage λ gene expression, escape from lysogeny also requires that the λ genome is excised from the host genome and becomes independent again. Reversal of the recombination event between the phage and host genomes which generated the prophage results in the regeneration of the circular phage genome. However, while the integration event requires only the Int protein, for excision to occur the *xis* and *int* genes must both be transcribed as both proteins are required to act together. The mRNA encoding the Xis and Int proteins is transcribed from the reactivated P_L promoter. The excision process is very precise, generating an exact copy of the lambda genome. However, on occasion, the excision process can go wrong resulting in the virus DNA carrying with it a portion of the adjacent cellular DNA. This cell DNA can then be transmitted to other *E. coli* by infection (see Box 17.2).

Box 17.2

Specialized transduction

The process of excision of the lysogen genome is usually very precise but occasional errors can occur, such that the region excised includes a small amount of bacterial DNA, with a corresponding piece of phage DNA deleted. If the portion of bacterial DNA is located to the left of the prophage, it may include some or all of the genes of the *gal* operon, with a compensatory loss of some of the genes at the right end of the prophage. This incomplete phage chromosome then serves as the template for replication, such that essentially all of the phage progeny issuing from that bacterium carry the genes for galactose utilization and have lost some of the phage genes. Infection of a Gal⁻ bacterium with such a phage, under conditions allowing lysogenization, can confer the ability to ferment galactose, since the necessary genes become inserted into the bacterial chromosome. The process of transferring genes from one host to another in a virus is called transduction and since phage λ is able to transfer only the genes immediately adjacent to the *att* λ region of the bacterial chromosome it is referred to as a specialized transducing phage. Many transducing phage are defective in replication, since they may lack some essential phage genes, and these can only be propagated if a normal ('helper') phage is present to supply the missing gene functions.

Immunity to superinfection

An unusual feature of lysogeny is that a bacterial lysogen is immune to superinfection by a second phage of the same type that it carries, but it is usually not immune to other, independently isolated temperate phages. Phages that induce immunity to one another are termed 'homoimmune' while those that do not are called 'heteroimmune'. The immunity that the lysogens display is very different to the immunity seen in animals. Analysis of the genetic factors involved in the specificity of immunity elicited by phage λ showed that the *cI* gene is responsible.

Lysogens continuously express the *cI* gene and contain significant levels of functional cI repressor until induced. When the DNA of a superinfecting, homoimmune phage enters the lysogenic bacterium the cI repressor in the cell binds to the O_L and O_R sequences on the incoming phage genome and prevents phage gene expression, aborting the infection. If the superinfecting DNA is from a heteroimmune phage, the cI repressor of the lysogen cannot bind the incoming DNA and the second phage is able to initiate an infection.

The benefits of lysogeny

The fact that temperate phages carry so many genes involved in lysogeny indicates that lysogeny is likely to have some advantages for them. A significant advantage would be to provide a method for a phage which has infected a host low in energy and synthetic capacity to persist in the bacterium until conditions improve. The phage can then 'reappear' when conditions have improved.

It is important that in the lysogenic state the phage does not represent a significant disadvantage for the survival of the host. In fact, in contrast to the situation with latent animal viruses where no advantage for the host is known, lysogeny often carries a strong selective value for the bacterium, since temperate phages frequently confer new characteristics on the host. This phenomenon manifests itself in many ways and is referred to as lysogenic conversion.

An interesting example of lysogenic conversion is that observed in *Corynebacterium diphtheriae*. Strains of this bacterium, which cause the serious childhood disease diphtheria, carry the diphtheria toxin gene (and are called toxigenic). Infection of non-toxigenic *Corynebacterium* strains with phage β isolated from virulent, toxigenic bacilli of the same species produces lysogens which acquire the ability to synthesize toxin. Loss of the prophage results in loss of toxin production as the structural gene for toxin is carried by the phage itself.

17.3 Herpes simplex virus latency

HSV infects a very high proportion of individuals and establishes lifelong latency in all those infected. Latency is maintained through the combined action of viral microRNAs which downregulate gene expression that would commit to productive infection and the host immune response that recognizes and responds to signs of reactivation before a clinically apparent event can occur.

HSV-1 establishes latency in the cell bodies of sensory neurons that innervate the area of the epithelium where the primary infection is occurring (Box 17.3). An important question is whether latent HSV-1 DNA is integrated into the host genome. This question was technically difficult to answer because only a small proportion of neurons in the dorsal root ganglion are infected. Estimates indicate that between 1 in 6 to 1 in 60 neurons in a ganglion contain an HSV-1 genome. The DNA in the latently infected neurons differs from the linear DNA in the virus particle. It is not integrated but is circularized and exists as a free, circular molecule in the cell nucleus.

Gene expression during HSV-1 latency and reactivation

In contrast to the extensive and well-characterized pattern of HSV-1 lytic gene

---

Box 17.3

Evidence for HSV-1 latency in sensory nerve ganglia

- In mice experimentally infected with HSV-1 via the eye, explanted trigeminal ganglia contain no detectable infectious virus but, if placed in culture together with cells susceptible to HSV-1 productive infection, virus re-emerges.
- In trigeminal ganglia from corpses of people who have died of causes unrelated to HSV-1 infection, HSV-1 DNA is detectable in samples from those who are seropositive for HSV-1 infection but not those who are seronegative.
- When explanted ganglion DNA is probed for sequences at the termini of the linear genome, these are found in a single 'junction' fragment with the individual terminal fragments being undetectable, indicating that the genome has circularized.

expression described in Section 10.5, the activity of the genome in a latently infected neuron is very limited. The primary reason why the lytic program does not occur in these cells is a failure to express the α (immediate early) proteins which would initiate the cascade of expression of the genes seen in an acute infection. In neural cells latently infected with HSV-1, the only viral RNAs which can be detected are the latency-associated transcripts (LATs) and small regulatory non-coding microRNAs (miRNAs). The LATs are transcribed from the opposite strand of the genome to the α gene, ICP0 and are actually stable introns, produced by splicing of longer RNAs; LATs do not encode proteins. The precise role of the LATs in latency has been identified from a range of studies (Box 17.4).

Many details of the process of HSV-1 latency remain to be discovered but the accumulated information has been formulated into a model. The first step is that, following infection of the mucosal cells at the site of the primary infection, the virus enters the sensory neurons that innervate the area. Here the virus continues to replicate. However, in some of the neuronal cells the circumstances favour the establishment of a latent infection. Two key factors play an important role. Firstly, when the HSV-1 DNA is introduced into the cell it rapidly forms a complex with host cell histones and associated enzymes and, importantly, also with the cell promyelocytic leukaemia (PML) protein. The PML protein is part of the cell's innate defence mechanism and the result of the formation of a chromatin-like structure associated with PML protein is to suppress virus gene expression. The second factor is that in neurons the HSV-1 protein VP16 and its cellular co-factor HCF (host cell factor) are found in the cytoplasm rather than in the nucleus where they are normally located in cells undergoing an acute infection. HCF is a nuclear protein in most cell types, but is cytoplasmic in sensory neurons. It has also been suggested that VP16 is inefficiently packaged into virus particles in infected neurons, possibly due to inefficient transport to the site of virus particle assembly in these cells. As successive rounds of replication

---

Box 17.4

Evidence suggesting the involvement of LAT RNA in HSV latency

- LAT$^-$ mutant viruses induce a greater level of apoptosis in rabbit ganglia than LAT$^+$ virus, suggesting that LATs help to prevent neuronal cell death following infection.
- Infection of mice with a LAT$^-$ virus results in higher levels of neuronal death during establishment of latency than is seen with LAT$^+$ virus, indicating a role for LATs in maintaining neuronal cell survival.
- The number of mouse sensory gangliar neurons expressing virus antigens is higher in mice infected with a LAT$^-$ mutant virus than when infected with a LAT$^+$ virus, suggesting that LATs inhibit expression of virus genes involved in lytic infections.
- Production of immediate early and early virus mRNA is increased in mouse neurons following infection with a LAT$^-$ mutant virus when compared to infection with a LAT$^+$ virus, suggesting a role for LATs in suppressing virus gene expression.
- Cells expressing LATs suppress the production of immediate early mRNAs.
- LATs suppress the host cell apoptosis, preventing self-programmed cell death.

in neurons occur, the level of VP16 in particles is progressively reduced. VP16 and HCF are required for the expression of the HSV-1 α genes, particularly ICP0 which activates the gene expression cascade necessary for a complete replication cycle. In the absence of VP16 or with no VP16/HCF complex in the nucleus to stimulate virus gene expression the balance is tipped to favour the establishment of latency.

Having established a state favouring latency, the LATs and miRNAs are expressed. During the latent state there is some transcription from the HSV-1 genome. The LATs and miRNAs are able to maintain the latent state by directing degradation of any residual virus mRNAs, particularly the mRNA encoding ICP0, using the normal cellular systems that remove unwanted RNA. In effect, the LATs and miRNAs act as a repressor of virus gene expression. Mutants with a defective ICP0 gene reactivate poorly from latency, which confirms the importance of miRNA/LAT control of ICP0 levels in this context. In cell culture models maintenance of HSV-1 latency also appears to involve internal cellular signals involving phosphoinositol 3-kinase in response to nerve growth factors.

Reactivation of HSV-1 from latency occurs in response to a number of possible stimuli. The molecular events that occur at this time are not well understood but are likely to be due to one of two possible scenarios: either there is a huge induction of virus gene expression which overwhelms the LAT and miRNA control, or the LATs and miRNAs fail to degrade the low levels of virus mRNA that are constantly produced. A combination of both of these is possible. The result is that the ICP0 protein is, in a few cells, produced in sufficient quantity to activate the remaining α genes which initiates a full replication cycle with production of new virus particles. A model for the gene expression patterns in HSV-1 latency is shown in Fig. 17.5.

HSV-1 latency and the immune system

An acute HSV-1 infection is resolved by the immune system, raising the question of how the virus is able to establish a latent infection despite this response. The main answer is one of timing. By the time the immune system has succeeded in clearing the acute infection the virus has already travelled up the sensory nerve supplying the infected area and become latent in the cell body of one or more of the sensory neurons that form the dorsal root ganglia. Since each dorsal root ganglion supplies innervation to a specific area, the latent infection is established only in the ganglion supplying the original area of infection. For HSV-1 infecting the oral region this is the trigeminal cranial nerve. All acute HSV-1 infections become latent.

Reactivation of HSV-1 occurs regularly several times each year in about 10% of infected people. It usually results in the familiar pathology of a cold sore in the epithelium around the mouth, the blisters of which contain infectious virus that can be transmitted to any non-immune person, usually an infant. However, up to 40% of reactivations are subclinical and thus virus can be transmitted inadvertently. When latency is broken, the virus multiplies in one of the neurons in which it is latent, and infectious virus particles descend the sensory nerve to infect the epithelium that the nerve supplies. Thus, because the virus is confined by the sensory nerve, the reactivated acute infection reoccurs each time in one of the areas of the skin supplied by this nerve. In addition, reactivated infection is usually monolateral as the nerves are paired with each supplying only the right or the left side of the body, and virus does not move from one side to the other.

An important question is why people with a normal, functional immune system suffer

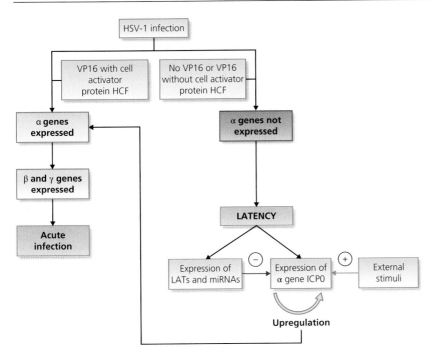

Fig. 17.5   Molecular control of latency in herpes simplex virus type 1 infection. Of 72 viral genes, only the latency-associated transcripts (LATs) are synthesized during latency. External factors that can break the latent state upregulate the synthesis of the viral non-structural protein, ICP0, which then enhances its own expression. This activates expression of the α genes and unlocks the pathway to acute infection (see text).

repeated acute infections after virus reactivation. Although the immune system does not prevent the cold sores, they are short-lived, indicating that the immune system does respond and clear them. The importance of T cell involvement in control of HSV-1 infection is clearly seen when studying individuals who have a T cell deficiency, for whom the virus is life-threatening. During reactivation there are two stages of immune involvement (Fig. 17.6). Firstly, there is a short-term, virus-mediated evasion of the immune response that allows the reactivation to get started and, secondly, there is immune activation which allows the infection to be resolved. Initially in the early stages of reactivation, expression of MHC I proteins in infected cells is inhibited. The main way in which this is achieved is that the immediate

early viral protein (ICP47) prevents the translocation of processed peptides into the endoplasmic reticulum, and hence the formation of MHC-peptide complexes. As the virus begins to express its genes, another mechanism comes into play and the virion protein that shuts off host protein synthesis prevents the synthesis of MHC polypeptides. Thus, virus peptides are not presented at the cell surface and virus-specific CD8$^+$ T memory cells are not activated (see Section 14.2). In due course, CD4$^+$ cells and NK cells move into the cold sore lesion where lysis of infected cells is taking place. Viral antigen is processed and presented as MHC class II protein-virus peptide complexes on the surface of antigen presenting cells (local dendritic cells and macrophages). The CD4$^+$ T cells are activated and, in turn,

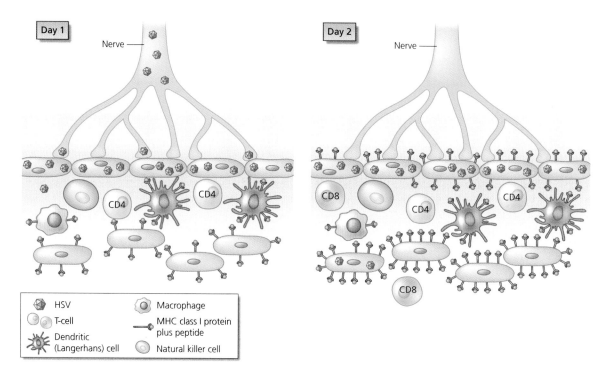

Fig. 17.6 Development and resolution of a cold sore from reactivated HSV-1. On day 1, reactivated virus is released from cells of the dorsal root ganglion and transported down the sensory neuron. It then infects adjacent epithelial cells. Expression of major histocompatibility complex (MHC) I protein complexed with viral peptide on the cell surface is inhibited (see text) and virus-specific memory $CD8^+$ T cells are not activated. NK cells and virus-specific $CD4^+$ T cells move gradually into the lesion and release cytokines. By day 2, they are releasing enough γ-interferon to up-regulate the expression of MHC class I protein. The $CD8^+$ memory T cells are then activated and virus-infected cells are cleared. (From Posavad et al., 1998 *Nature Medicine* 4, 381.)

through the cytokines that T cells secrete, NK cells are also activated. By day 2, both $CD4^+$ T cells and NK cells are producing γ-interferon. This increases the expression of MHC I proteins in infected epithelial cells, opposing the inhibitory effects imposed by the virus. $CD8^+$ T memory cells are then activated, virus-infected cells are killed, and the infection resolves. This very delicate relationship between HSV-1 and the host allows both to prosper.

17.4 Epstein-Barr virus latency

Epstein-Barr virus (EBV) normally causes a subclinical infection in infants. However, if infection is delayed until adulthood, infectious mononucleosis (IM or glandular fever) results. The symptoms of glandular fever take several weeks to appear after infection and normally last for only two or three weeks. After the symptoms disappear the virus remains in the body, establishing a latent infection in B lymphocytes. During the acute infection EBV expresses its genes in a controlled, coordinated pattern similar to the situation with HSV-1, though the nature of the genes is different. During latency, the genome is circularized and it is maintained in this state, which is known as an episome. Between one and nine proteins are expressed (see Section 25.7), depending on the growth state of the cells and whether or not latency is fully established, as well as two non-coding RNAs and several miRNAs. One of these protein-coding genes is formed through the ligation of the two ends of the linear genome to form a circle (see Section 7.3), which links the promoter with the coding regions of the gene. Transcription covers more than half of the 172 kbp genome, with extensive differential splicing to produce alternative mature mRNAs from the primary transcripts. This fact emphasizes the issues of control that arise in maintaining the EBV latent state. The primary transcripts produced during latency include the sequences of a great many of the genes expressed only in the acute programme, but these are discarded by splicing and so no protein production occurs; instead, expression is focused on the small number of latency-specific genes.

Unlike HSV-1 latency, where DNA replication does not occur, EBV has a specific replication mechanism which is used during latency establishment and maintenance; this mechanism is completely distinct from that operating during a productive EBV or HSV-1 infection (Section 7.3). Only one of the latency proteins, an origin-recognition DNA-binding protein EBNA-1, is required directly for DNA replication; this protein also links EBV episomes to host chromosomes to ensure they are maintained during cell division. The remainder of the latency proteins appear to be involved in modulating cell signalling and cell cycle control. The role of latent replication is to maintain the viral genome in an expanding pool of B lymphocytes which are stimulated to divide upon infection. Subsequently, latency is maintained in a pool of resting, memory B cells.

EBV can be reactivated from the latent state in individual cells by a variety of stimuli. The reactivation is associated with a switch into a productive pattern of gene expression and DNA replication. Expression of two virus-encoded transcriptional activators, BZLF-1 and BRLF-1, is thought to be the event that commits a latently infected cell to make this switch. In latently infected B lymphocyte cell lines, a few cells are actively producing virus at any given time. Similar sporadic reactivation is also thought to occur in vivo.

17.5 Latency in other herpesviruses

Other important examples of human herpes viruses are shown in Table 17.1. Herpes simplex virus type 2 can be sexually transmitted, and its cycle of acute → latent → acute infection is very similar to that of HSV-1, but centres on the genital area. In fact, the locations of infection by HSV-1 and HSV-2 are interchangeable and which virus is resident depends on the site of the initial acute infection. This in turn determines which dorsal root ganglion becomes the site of the latent virus. With a genital primary infection, the virus becomes latent in the sacral dorsal root ganglion that innervates that area. There is antigenic cross-reactivity between the two viruses such that an initial infection with HSV-2 after establishment of a latent HSV-1 infection is significantly less pathogenic than in someone who has not previously acquired HSV-1. This, however, is not sufficient to prevent the second infection from establishing latency.

Varicella-zoster virus (VZV) causes acute chicken pox of children and shingles (zoster) on

Table 17.1

The establishment of latent infections in man with herpes viruses and the breakdown of latency (reactivation).

Virus	Primary acute infection	Site of latency	Stimulus for reactivation	Reactivated acute infection
HSV-1	Stomatitis: infection of the mouth and tongue	Dorsal root ganglion of the trigeminal (cranial) nerve	e.g. strong sunlight, menstruation, stress	Cold sore*
HSV-2	Genital lesions	Dorsal root ganglion of the sacral region of the spinal cord	Not known, though probably similar to HSV-1	Genital lesions (and infection of neonates)
VZV	Chicken-pox: generalized infection with fluid-filled vesicles over the body surface	Any dorsal root ganglion of the central nervous system	Release of immune control, e.g. in the elderly	Zoster (shingles)
EBV	Child: subclinical Adult: glandular fever (IM: acute)***	B cells and possibly throat epithelium	Not known; frequent	Subclinical**
CMV	Prenatal:**** Child: subclinical Adult: subclinical	Salivary glands and probably other sites	Frequent	In all body fluids, especially during the immunosuppression that accompanies pregnancy. A major cause of death in AIDS and transplant surgery

*10–40% of reactivations are subclinical; varies between individuals.
**EBV also causes cancer: nasopharyngeal carcinoma and Burkitt's lymphoma (see Chapter 25).
***IM: an example of a more severe clinical disease that occurs when primary infection takes place after childhood – a common microbiological problem of the (over-) sanitized world. 'Mononucleosis' refers to the uncommonly large numbers of mononuclear cells (mainly lymphocytes) that are found in the blood.
****The foetus is only at risk when its mother gets a primary infection.
Abbreviations: AIDS, acquired immune deficiency syndrome; CMV, cytomegalovirus ('cytomegalo' refers to the characteristic swollen cell cytopathology caused by CMV to cells of the kidney, lungs and liver); EBV, Epstein-Barr virus; HSV, herpes simplex virus; IM, infectious mononucleosis; VZV, varicella-zoster virus.

reactivation. Apart from chicken pox being generalized over the entire surface of the body, the transition from acute to latent infection occurs exactly as with HSV-1. Because of the generalized nature of the initial infection, almost any of the brain and spinal dorsal root ganglia can be latently infected. Symptomatic reactivation is a rare event which is normally restricted to a single ganglion, and the lesions formed by the reactivated virus are restricted to the precise segment of the body that is innervated by the nerve in which virus was latent. Unlike HSV latency, there are specific virus proteins made in latently infected neurons. Latent VZV is maintained in the presence of antibody and a T cell response, both of which may contribute to maintaining virus in its latent state. Immunosuppression (including the natural waning of immunity that accompanies ageing) results in virus reactivation.

Another ubiquitous infection is cytomegalovirus (CMV). This is normally a subclinical childhood infection that becomes latent in the salivary glands. It reactivates frequently but reactivations are controlled by the immune system and are completely subclinical. However, the infection can become generalized and life-threatening when there is some immunosuppression, such as that induced to avoid organ rejection during transplantation surgery, and as occurs during HIV-1 infection. In addition, although adult infection is subclinical, the virus can pass through the placenta and infect the foetus. As a result, some foetuses suffer severe and permanent brain damage (including microcephaly). This infection is thought to be the result of the immunosuppression that occurs naturally during pregnancy and helps to prevent the foetus (a foreign graft) from being rejected by the maternal immune system. The immunosuppression may also result in an enhanced severity of disease.

17.6 HIV-1 latency

Following the initial infection with HIV-1, the affected individual carries the virus for the remainder of their life. Unlike the situation with the herpesviruses, during the latent state most of the HIV-1 genomes continue to express the full range of genes and to produce infectious virus, with approximately 10^{10} virions produced per day. The virus is also under constant attack by the immune response. However, there is an underlying minority of HIV-1 genomes that become truly latent in resting CD4$^+$ T lymphocytes, producing no progeny virus until activated. In the latent state,

no virus genes are expressed and the virus genome, which is of course integrated within the genome of the host cell, is replicated along with the cellular DNA. Estimates suggest that approximately 10^6 to 10^7 CD4$^+$ T lymphocytes per person may be infected in this way. The latent virus can reactivate at any time though the stimuli for this are not well understood. This is important as most active virus multiplication, but not the latent virus infection, can be eliminated by chemotherapy (see Section 21.6). Thus, the latent infection acts as a reservoir from which virus can reappear if antiretroviral therapy is stopped. When this occurs, the levels of infectious virus in the body often exceed those prior to the treatment and there is a greatly enhanced likelihood of drug-resistant viruses appearing due to selection pressures. This means that the current therapy regime has to be maintained for life, with the attendant difficulties of expense, toxicity, non-compliance, or the eventual breakthrough of resistant mutants.

Key points

- Latency is an outcome of infection for certain viruses and is a state defined by the presence of viral genome without the production of infectious virus, but in which it is possible to reactivate a lytic infection.
- Among animal viruses, all herpesviruses have the ability to establish latent infections.
- The pattern of gene expression during latency is different to that seen during an acute infection: fewer virus genes, and sometimes no genes, are expressed during latency.
- In phage λ the establishment of latency is determined by a balance between the action of gene activators and gene repressors.
- In HSV-1 only RNA from the LATs and micro RNAs are seen during latency. In the other

herpesviruses, several virus proteins may be produced during latency.

- Latency requires suppression of the early steps in the cascade of lytic gene expression.
- Following reactivation, infectious virus is produced from latently infected cells.
- In animals, the immune system is key in dealing with reactivation of infection.

Questions

- Consider the control of lysogeny in bacteriophage lambda and compare this with the gene expression of herpesviruses during the establishment and maintenance of latency.
- Discuss the potential outcomes of infection of humans with herpesviruses and describe the processes which control and determine the various stages of the infection.

Further reading

Johnson, A. D., Poteete, A. R., Lauer, G., Sauer, R. T., Ackers, G. K., Ptashne, M. 1981. λ repressor and cro-components of an efficient molecular switch. *Nature (London)* 294, 217–223.

Lieberman, P. M. 2013. Keeping it quiet: chromatin control of gammaherpesvirus latency. *Nature Reviews Microbiology* 11, 863–875.

Mitchell, B. M., Bloom, D. C., Cohrs, R. J., Gilden, D. H., Kennedy, P. G. 2003. Herpes simplex virus-1 and varicella-zoster virus latency in ganglia. *Journal of Neurovirology* 9, 194–204.

Ptashne, M. 1986. *A Genetic Switch*. Blackwell Scientific Publications & Cell Press, Palo Alto.

Ptashne, M., Jeffrey, A., Johnson, A. D., Maurer, R., Meyer, B. J., Pabo, C. O., Roberts, T. M., Sauer, R. T. 1980. How the λ repressor and *cro* work. *Cell* 19, 1–11.

Roizman, B., Whitley, R. J. 2013. An inquiry into the molecular basis of HSV latency and reactivation. *Annual Reviews of Microbiology* 67, 355–374.

Tsurumi, T., Fujita, M., Kudoh, A. 2005. Latent and lytic Epstein-Barr virus replication strategies. *Reviews in Medical Virology* 17, 3 Palo Alto: 17.

Chapter 18
Transmission of viruses

Viruses are intracellular parasites and have to find a new host before the original host mounts an effective immune response or dies. However, virus infectivity is intrinsically unstable and transmission has to be achieved usually within a few hours after release from the infected cell. All successful viruses have solved the transmission problem.

Chapter 18 Outline

As discussed in Chapter 16, virus infections of an individual host can result in widely differing outcomes. These range from complete resolution of the infection within a few days or weeks (i.e. clearance of virus from the host), through long-term persistence of the virus in the host – with or without impacts on host physiology – that may last for the life of the host, to the rapid death of the host as a result of the infection. Common amongst all these scenarios is that, at some point, a virus must achieve transfer from an infected host to an uninfected susceptible host in order to survive since, even if a virus achieves lifetime persistence in its host, it will die with that host. This process of virus transfer between hosts is referred to as virus transmission.

Virus transmission between individual animals occurs via a limited number of routes that are considered in detail in this chapter. However, while these routes can be investigated in the laboratory, the elements that are important in natural transmission are often difficult to investigate and details are poorly understood. In populations, virus transmission is an element within epidemiology, the science of defining and quantifying the factors that determine disease incidence rates. Understanding routes of virus transmission, and their relative significance, is a crucial part of developing strategies to control the spread of infection.

18.1 Virus transmission cycles

Virus transmission results in the sequential infection of a series of host individuals. From the perspective of a virus, this process can be thought of as the virus passing through a transient cell-associated replicative phase to achieve an increase in the amount of free virus present. In other words, virus outside a host

Introduction to Modern Virology, Seventh Edition. N. J. Dimmock, A. J. Easton and K. N. Leppard.
© 2016 John Wiley & Sons, Ltd. Published 2016 by John Wiley & Sons, Ltd.

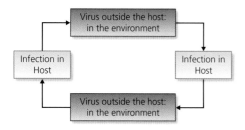

Fig. 18.1 Virus transmission cycles. A virus spends time both within and outside its host. Some viruses spend most of their existence in the environment whilst others are predominantly host-associated.

achieves entry, passes through its replicative cycle and produces progeny that are then released from the host. However, outside the host, the infectivity of many viruses is inherently unstable. So, after being shed from a host organism it is imperative that such viruses quickly and efficiently encounter a new susceptible host to initiate a fresh infection. It

may be more accurate in these cases to think of the virus as primarily host-associated, with a linear pathway from one host to another via a transient non-host phase. In reality, for all viruses these events constitute a cycle (Fig. 18.1) in which some viruses spend most of their time outside of their hosts, e.g. phages that infect marine cyanobacteria, whilst others spend most of their time within hosts, e.g. herpesviruses that achieve lifelong persistence in animals.

The characteristics of a virus' transmission cycle may vary seasonally. Many prominent viral infections of humans show pronounced seasonality as measured by the incidence of disease, e.g. influenza virus and respiratory syncytial virus (Fig. 18.2). Typically, there is a winter peak of disease in temperate latitudes with more continuous disease burdens in the tropics, but the precise timings of these peaks are reproducible and distinct for each virus. Among the parainfluenza viruses, types 1 and 4

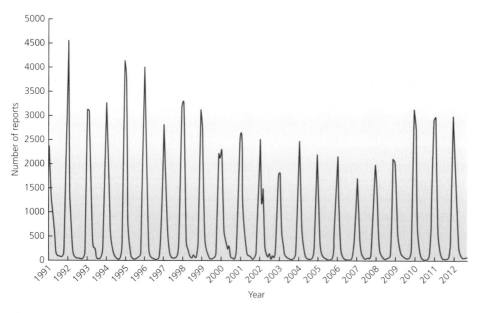

Fig. 18.2 Reported cases of respiratory syncytial virus in the UK from 1991 to 2013. Data taken from Public Health England.

show an autumn peak whilst type 3 peaks in the late spring. The factors affecting seasonality are very poorly understood but are thought to include climate effects on virus survival outside the host and climate-related variation in host behaviour (increased exposure to virus from indoor living), both of which will affect probability of transmission, and variation in the potency of protective immune responses, perhaps related to light levels (see Section 19.2). None of these factors readily explains why there are such clear differences in the timing of the annual peak incidence of different viruses. There is also no explanation of where such viruses are during the seasons when they are not causing disease. Either a low level of transmission continues that is sufficient to sustain the virus, or else transmission continues at a higher rate but with much reduced pathogenic consequences such that most cases go unrecognized.

18.2 Barriers to transmission

To be said to have infected an individual organism, a virus must do more than be taken up into it; it must also enter a cell and begin the process of virus replication. Essentially, all cell types in all organisms are susceptible to infection by one or more specific viruses. However, they are not all equally accessible to virus that is invading a host from the exterior, and they are not all susceptible to all viruses that encounter them. Thinking about humans and other mammals, the bulk of their external body surface is covered in a protective layer of dead skin cells. This layer forms a physical barrier to infection; to gain access to living cells that might support an infection, a virus must exploit breaches in this barrier – either natural ones such as the nose and mouth, or else

injuries that puncture the skin. These represent the available portals for virus to access the host. For plants, the surface of exposed leaf and stem tissue is similarly protected against virus infection by the plant cell walls; the vast majority of plant virus infections occur via insect damage or wind abrasion of this surface that exposes the susceptible cells underneath. Finally, bacterial cells are typically also surrounded by a cell wall and viruses that infect these organisms either have injection mechanisms to get their genetic material into the cell through the cell wall or else have evolved to target more vulnerable structures that emerge from the cell surface, such as pili.

Once a virus has breached the physical barriers to its entry, there are further biological barriers to be overcome. Firstly, it must encounter and interact with a cell that is susceptible to infection. Unless the virus is directly introduced into the cell from the environment (applicable to plants and bacteria), such a cell must have the necessary surface receptors for entry (see Chapter 6). It must also provide the necessary host factors on which that virus depends; most viruses use specific host proteins in their replication cycles in addition to general functions such as protein synthesis and energy production. Only some cell types will have both the receptors and intracellular factors required for infection so, for the virus to succeed, these cells must be accessible via the route of entry. Thus organisms will continually encounter viruses that are not adapted for that host or via routes that are not compatible with an infection occurring. Secondly, the virus will have to overcome innate and intrinsic immunity (see Chapter 13). These systems will be deployed against a virus from the very earliest stages of its encounter with a host. Their action means that only a fraction of the viruses that get past the physical barriers actually establish an infection which can produce virus.

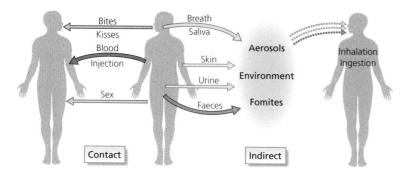

Fig. 18.3 Routes of horizontal transmission in humans. Virus may be shed into the environment and then acquired by another individual or else acquired by direct contact.

18.3 Routes of horizontal transmission in animals

Horizontal transmission occurs between individuals of a species by a variety of routes. Each virus has evolved to utilize particular route(s) of transmission, but this does not prevent a virus from transmitting via an atypical route under special circumstances.

With much of the body surface normally refractory to virus infection, it has been possible to define a relatively restricted number of specific routes by which virus can gain access to the body and initiate infection (Fig. 18.3; Box 18.1). In each case, the virus gains access to specific tissues that are defined by the route of entry. Having established infection there, a virus may achieve spread to other tissues in the body, which may then provide the major source of virus for onward transmission, or it may remain confined near the site of infection and reproduce there. All routes of transmission between two individuals of the same species, other than specific routes of transmission from mother to baby (see Section 18.4), are known collectively as *horizontal transmission*.

Box 18.1

Summary of horizontal transmission routes

- Respiratory route: entry via nose and mouth, e.g. rhinovirus, influenza virus
- Faecal-oral route: entry via mouth into the gastrointestinal tract, e.g. poliovirus, rotavirus
- Conjunctival route: entry into eye, e.g. some adenoviruses, possibly other respiratory viruses
- Via saliva or urine: e.g. Epstein-Barr virus, cytomegalovirus, Lassa fever virus
- Via fomites: e.g. hepatitis A virus
- Sexual route: entry into body via sexual activity, e.g. HIV-1, HBV, papillomavirus
- Mechanical route: entry into body via skin puncture, e.g. papillomavirus, arboviruses, HIV-1, HBV

Infection via the respiratory tract

Respiratory (or airborne) transmission occurs most commonly when virus that is exhaled by an infected individual is inhaled by another individual who is susceptible to infection by that virus. Classic examples of this are the spread of rhinoviruses (common cold; see Chapter 19) and influenza (see Chapter 20). Entry via this route can also occur for certain viruses through inhalation of urine-contaminated dust particles (see below). While many of the viruses spread via the respiratory route cause respiratory infections, others cause generalized infections, such as measles and chicken pox; smallpox, now eradicated, was also contracted by this route.

Progeny virus from replication in the respiratory tract escapes from the host in liquid droplets that result from our normal activities, such as talking, coughing and sneezing. These aerosols are inhaled by others and give rise to a 'droplet infection' of a susceptible individual. The size of droplets is important, as very large (>10 μm) diameter droplets rapidly fall to the ground, while the very small droplets (<0.3 μm) dry very quickly; dessication frequently causes virus inactivation although some virus types withstand it and so their infectivity can persist on surfaces where droplets fall. Thus, the middle-sized droplets are those that transmit infection most efficiently. Such droplets can be inhaled and their precise size will determine where they become trapped in the recipient respiratory tract; there are baffles lining the nasal cavities that remove the larger airborne particles, medium-sized droplets get as far as the trachea (throat), while the smaller droplets may penetrate deep into the lung (Fig. 19.3). However, even arrival at the target tissue does not guarantee infection as particles may be trapped on the overlying mucous and be driven upwards again by the cilia lining the tubes of the respiratory tract in a process known as mucociliary flow.

Infection via the gastrointestinal tract

When virus infects via the gastrointestinal tract, this is often termed 'faecal-oral' transmission. This is because transmission must be cyclical: having gained access to susceptible cells and initiated productive infection, the progeny particles must exit the host in a way that permits their subsequent uptake by new hosts to reach the equivalent susceptible tissues. Because there is constant movement of material through the gut, the way in which virus produced there can exit the body most readily is in the faeces. This then contaminates the water supply and indirectly also contaminates food that is washed in that water, so that it can be ingested by a new host to complete the cycle. For a virus to succeed in transmission by the faecal-oral route, its particles must retain infectivity during passage through the acid environment of the stomach. Enteroviruses such as poliovirus, as well as hepatitis A virus and rotaviruses, all transmit via this route.

Since viruses transmitted via the faecal-oral route are excreted in faeces, their spread is favoured by poor sanitation and poor personal hygiene. Many of these viruses grow to very high titres (10^9 infectious units (IU)/ml faeces), meaning that even microscopic particles/droplets of faecal material contain significant amounts of infectivity. Some virus infections of the gut (rotavirus) are directly associated with diarrhoea which, though not proven, could be an adaptation to improving virus transmission. Many infections via the faecal-oral route occur in early childhood – not surprising when it is seen how small children investigate strange objects by putting them in their mouths and that their personal hygiene is not well developed.

With improvements in sanitation it would be expected that there would be a reduction of disease caused by enteric viruses, and this is the experience. However, there have been some unpleasant surprises. In conditions of poor sanitation, poliovirus infects very young children but results usually in a subclinical gut infection. With improved sanitation, poliovirus infection is delayed, with older children and young adults becoming infected for the first time rather than the very young. These infections of slightly later life are then associated with an increase in the incidence of paralytic poliomyelitis. A very similar age-related difference is also seen in the pathogenesis of hepatitis A virus (see Chapter 22).

Infection via the eye

The eye, especially the conjunctiva (the layer of cells covering the eye), is vulnerable to infection by several viruses, including herpes simplex virus type 1 and several of the human adenoviruses. Since material is not routinely introduced into the eyes (unlike the respiratory or gastrointestinal tracts), transmission depends on the accidental introduction of virus, most likely by rubbing with hands that have virus on the skin. Hands in turn can pick up virus from surfaces onto which virus has fallen (fomites, see below) after droplet exhalation or other forms of excretion. Also, it is now believed that some respiratory viruses are spread directly by contact of infectious droplets with the conjunctiva. Natural drainage to the throat would then carry virus to the respiratory tract.

Transmission via saliva or urine

Saliva is a vehicle for transmission of several viruses. At one level this may be regarded as a continuum with respiratory transmission, whereby infectious aerosols are created and

inhaled. But certain viruses also transmit by close contact with saliva, e.g. during play or by kissing. For viruses such as cytomegalovirus (CMV) and Epstein-Barr virus (both herpesviruses), transfer of saliva is thought to be the major route of transmission.

Urine has traditionally been viewed as being normally sterile. However, there are several examples of viruses that spread in humans via urine, following replication in the kidney. CMV and polyomaviruses such as BK and JC virus are excreted in the urine of small children and may be transmitted by this route, though for CMV, saliva is probably the more important fluid for transmission. Urine is also significant in some zoonotic transmissions (see Section 18.5). Lassa fever (an African arenavirus) and sin nombre (a North American bunyavirus), are both excreted in the urine, and possibly faeces, of their rodent hosts and infect humans via fomites (see below) or by inhalation.

Transmission via fomites

Fomites are solid objects whose surfaces can be contaminated with virus and which provide a source of infection. Secretions/excretions of an infected individual can come into contact with the surface of such objects and the virus they contain will be left on the surface as the material dries. Whilst dessication will kill many viruses, others survive for extended periods. For fomites to be a source of onward transmission of such a stable virus, some host action is then required to bring the virus from the surface into a portal of access to the body that is compatible with the virus encountering susceptible cells (Section 18.2). For faecal-oral transmission, this may be via a child placing the object into its mouth, or a person handling the object before eating, etc. Hepatitis A virus is particularly known for fomite transmission from contaminated bedding.

Infection via the genital tract

Sexual transmission is the term applied to a variety of transmission scenarios where virus moves between hosts as a result of sexual activity. One way in which viruses can be transmitted through sex is by contact. A good example of this is herpes simplex virus type 2, which causes lesions on the external genitalia in men and women which provide a source of infectious virus for transfer to a sexual partner. A subset of the papillomaviruses also infects either the external genitalia or the female genital tract and can be transferred during sex. A second aspect of sexual transmission is the transfer of viruses within body fluids that are exchanged. Both human immunodeficiency virus (HIV; see Chapter 21) and hepatitis B virus (HBV; Chapter 22) are examples of viruses that are transmitted in this way. Both papillomavirus infections in the genital tract, and infections by another sexually transmitted virus, Kaposi's sarcoma herpes virus (KSHV), are associated with the development of cancers (Chapter 25).

Viruses can be transmitted both by heterosexual intercourse, either from male to female or from female to male, and by homosexual intercourse. Sexual transmission of viruses is affected by sociosexual behaviour and spread of these viruses is increased by having multiple partners. For HIV-1 the risk of infection is marginally less (by two- to threefold) if the female is the carrier. This may be because ejaculate contains not only virus but also around 10^6 lymphocytes, some of which may be infected and producing virus. Virus is also spread by homosexual behaviour and the risk for receptive anal intercourse (which is 1/300 to 1/1000) is about the same as that in male–female intercourse when the male is infected. For KSHV, transmission of which is less well understood, male homosexuals appears to be at particular risk of infection.

Infection through the skin

Whilst one group of papillomaviruses infects mucosal surfaces in the genital tract, another group infects the skin. To get through the keratinized protective layer, the virus exploits pre-existing damage to reach the dividing cells in the basal epithelial cell layer. Viruses of this type remain restricted to the skin and cause skin warts. More severe skin damage that penetrates to the blood opens the way to systemic infections by a number of other viruses. Both HIV and HBV can infect very readily via this route, as can hepatitis C virus. Once in the blood, these viruses can move around the body until they encounter susceptible cells. As discussed in Section 18.5, there is also a large class of viruses (from various phylogenetic groups) known as arboviruses that are introduced into an animal host by the bite of an insect that penetrates the bloodstream.

18.4 Vertical transmission

Vertical transmission occurs between mother and foetus/baby in utero, during delivery or breast feeding. Infection in a foetus can have severe consequences because of the absence of fully functional immune responses.

The particular relationship between mother and infant offers some novel routes of virus transfer, which are described as *vertical transmission*. Firstly, intrauterine transmission can occur, most commonly transplacental infection. Secondly, there is the risk of virus transfer to the child on contact with virus lesions or infectious body fluids during natural delivery. Finally, virus transfer can occur in breast milk during feeding. Viruses of humans which infect the mother and can be passed

Box 18.2

Examples of vertical transmission of viruses in humans

Virus	Possible modes of infection			Possible adverse outcomes		
	Trans-placental	During birth*	After birth†	Death of foetus	Clinical disease after birth	Persistent infection
Rubella	+	−	−	+	Congenital abnormality	+
CMV (Primary maternal)	+	−	−	?	Congenital abnormality	+
CMV (Reactivated maternal)	−	+ (2% of all babies)	+	na	−	+
HIV-1	+	+ Up to 15% of babies born to infected mothers without preventive treatment	+	−	AIDS	+
HBV	+	+	+	−	−	+
HSV (genital)	(+)	+	−	+	Herpes lesions	+
HPV (various types)	−	+	−	−	−	+

*A Caesarean delivery can help avoid infection; †from breast milk; (+) small minority of cases; na, not applicable; ?, not known.

Abbreviations: CMV, cytomegalovirus; HBV, hepatitis B virus; HIV-1, human immunodeficiency virus type 1; HSV, herpes simplex viruses; HPV, human papillomavirus.

vertically are listed in Box 18.2. Transmission of virus from mother to child via all those other routes that can occur between any two individuals is known as horizontal transmission (Section 18.3).

Rubella virus (German measles) infection is manifest in children and adults at worst as a mild skin rash; often infection is subclinical. However, in pregnant women, the virus can cross the placenta and multiply in the foetus. As a result, the foetus can die or can be born with serious congenital malformations that affect the cardiovascular and central nervous systems, the eyes and hearing. The risk of these malformations arising from maternal rubella virus infection is high in the early stages of pregnancy (up to 80%) and decreases as foetal development proceeds to become almost no risk by the fifth month of pregnancy or later. Fortunately, there is an excellent live attenuated vaccine against rubella virus (Chapter 26) that is given to young children and protects them through adulthood so that such transmission is very rare in vaccinated populations. HIV vertical transmission to neonates was a major feature of the early years of the pandemic, but the risk of this can be dramatically reduced by antiretroviral therapy for the mother pre-birth and by avoiding breast feeding (see Chapter 21).

18.5 Vector-borne viruses and zoonotic transmission

Arboviruses replicate in both an insect vector and an animal host, and must be carried between individual animals by insect bites. Both arboviruses and other viruses sometimes transmit to unusual host species in which the virus does not replicate very well but may still cause disease; when this unusual host is human, this is known as a zoonosis.

Many viruses are carried between individual vertebrate hosts by biting insect vectors, including mosquitoes, midges, ticks, etc; these are known collectively as arboviruses (Chapter 23). These viruses are introduced into a new host by a bite from an insect that has fed previously on an infected individual and thus picked up the virus. In most cases, the insect is more than a passive carrier; the virus replicates within it and this amplification is necessary for onward transmission. Examples of arboviruses affecting humans include sandfly fever (Fig. 18.4a), tick-borne encephalitis and dengue viruses. There are also viruses of domesticated animals that are spread by this route, including bluetongue virus (an orbivirus) and Schmallenberg virus, which was first recognized in European livestock in 2011. Transmission by an insect also opens up the possibility of zoonotic transmission (see below) if an insect can feed on individuals of more than one host species and hence move virus between

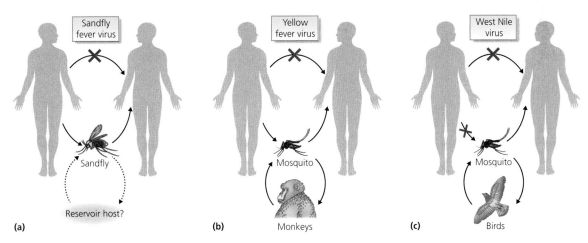

Fig. 18.4 Variations on arbovirus transmission cycles. (a) Sandfly fever virus replicates in sandflies and can transmit transovarially in them. Sandflies can also infect humans, and be infected by biting an already-infected human; there may or may not be another animal reservoir. (b) Yellow fever virus replicates within monkeys and mosquitoes and is spread between monkeys by mosquito bites in a sylvatic cycle; mosquito bites can also infect humans, in whom replication is sufficient to reinfect mosquitoes and so sustain urban transmission cycles also. (c) West Nile virus replication is sustained in birds and transmitted between them by mosquitoes. Infected mosquitoes can infect humans and horses as dead-end hosts.

Box 18.3

Evidence that yellow fever virus is transmitted by a mosquito

While it was known at the end of the 19th century that an insect was responsible for the transmission of yellow fever virus it was not until the late 1920s that definitive evidence that yellow fever virus is transmitted by mosquitoes was presented:

- The infection was transmitted between monkeys or from infected patients to monkeys by injection of blood.
- The mosquito *Aedes aegypti* was able to transmit the infection between monkeys.
- Infected mosquitos were able to transmit the virus throughout their lives, confirming that the virus was able to replicate in the insect.
- A single bite from an infected mosquito was sufficient to transmit the infection to monkeys.

---

them. In a few cases, a virus may be able to replicate and transmit within more than one animal species, and move between them via suitable vectors; Yellow fever virus can infect both monkeys and humans via appropriate mosquito vector species (Fig. 18.4b; Box 18.3).

Viruses may also be transmitted from their natural host into an unusual host without achieving sustained onward transmission between individuals of the unusual host species. When virus transmits to humans in this way, it is termed 'zoonotic transmission'. Although the resulting human infection is not transmissible or only poorly transmissible to further humans, often the disease resulting from such infections is severe (see Chapter 24). Irrespective of the severity of pathogenesis that is observed, this situation is termed a *dead-end infection*, meaning that the virus has not succeeded in reproducing in that host sufficiently to achieve onward transmission. The recipient host in such cases is known as a *dead-end host* whilst the natural host is termed the *reservoir host*.

Examples of zoonotic infections include rabies, which is transmitted in saliva by a bite from an infected animal (bat, dog, raccoon, fox, etc.), and various bunya- and arenaviruses that are transmitted via rodent urine. A further example is West Nile virus, an arbovirus that is transmitted to humans and horses from its reservoir host which is wild birds (Fig. 18.4c). Describing an infection as 'dead-end' is not an absolute for any virus. Individuals infected with West Nile or rabies can be a source of onwards transmission in unusual circumstances, most notably if they become blood or organ donors, and human intervention, as with Ebola virus, can stop a zoonosis developing into sustained transmission.

All viruses have a definable host range and, for most viruses, this range is quite restricted. Typically, a virus that infects species A will not infect species B. The probability that a virus will be able to spread from one species to another is greatest for host species that are closely related. If a zoonotic virus does establish itself in the human population with ongoing transmission between individuals, a process that will require virus evolution through mutation and selection (Chapter 4), then this will no longer be described as a zoonotic infection but rather as a new human infection, as with influenza virus following introduction into the human population from its aquatic bird reservoir. Several human viruses are known to have originated as zoonoses that then became established in humans. Both HIV and measles virus are examples of this phenomenon. HIV transmitted to humans from chimpanzees and other simian species (Chapter 21), whilst measles derived from rinderpest, a virus of cattle (Chapter 4). There are few good examples of viruses that are known to circulate freely in multiple host species, but one would be foot and mouth disease virus that infects cattle, sheep, goats, etc. Some human viruses may also be able to infect other great ape species.

18.6 Epidemiology of virus infections

Epidemiology is the study of the parameters that underlie and explain the incidence of disease in a population. This is not exclusive to infection, but does have major applications in understanding the population dynamics of infectious disease, and the success of intervention strategies such as vaccination that are designed to control such disease. As such, epidemiology is closely linked with studies of virus transmission, since it is the transmission characteristics of a virus that will determine the speed and extent of its spread in a population.

A very important parameter in considering how a virus will spread is how contagious it is. In other words, how easy is it to catch? This is partly determined by its route of transmission, but also by intrinsic features of virus biology that dictate how likely it is that an individual infectious virus particle – having gained access to the host – will actually succeed in initiating an infection. Contagiousness can be quantified experimentally if a suitable host system is available in terms of the minimum infectious dose 50 (MID_{50}), the size of infectious dose that will infect 50% of host individuals which receive such a dose. In epidemiological terms, thinking at a population level, contagiousness is

reflected in the average infectiousness of an infected individual, which is quantified as the basic reproductive number, R_0, which is the average number of new infections established from an already-infected individual. However, such an average conceals wide variations in the actual number of infectious transmissions from one infected person as compared with another. If R_0 is greater than 1, then the epidemic will grow in size (Fig. 18.5).

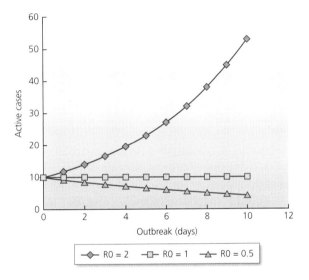

Fig. 18.5 Progress of a hypothetical virus outbreak with different values of R_0. If R_0 is greater than 1, the number of active cases will increase exponentially as time progresses. If R_0 is 1, then the number of transmission events each day exactly equals the number of cases who recover to be non-infectious (or die) so there is no net change in the number of cases. If R_0 is less than 1 then the number of cases will decline to zero over time. Arbitrary numbers of cases (10) and duration of infectiousness (6 days) were chosen for day zero of the outbreak to illustrate these trends. The continued exponential growth of an epidemic where $R_0 > 1$ presumes an infinite supply of individuals who are susceptible to infection, which will not be the case in practice. Thus, such an epidemic will eventually burn out unless migration or new births renew the local population of susceptibles at a sufficient rate.

Several factors will contribute to the observed transmission dynamics of a particular virus. The first is the extent and duration of virus production in the host. The more virus that is shed from a typical host, and the longer the time period of that shedding, the more likely it is that the infection will be acquired by other individuals. A second factor is route of transmission; self evidently, far more people per day will be exposed to potential respiratory transmission than will be exposed to sexual transmission from an infected individual. A third factor is the frequency of susceptible individuals in the population – some may have immunity to the virus already. Finally, there will be variation at the individual host level. Genetic variation will alter the intrinsic susceptibility to and productivity of an infection, leading to the concept of super-spreaders – individuals who are unusually good at transmitting a virus, while behaviour decisions such as frequency of hand-washing also contribute to the overall probability of infection being spread in a population.

The goal of intervention during an outbreak of infection is to manipulate these characteristics to reduce the value of R_0 to below 1. Ways of achieving this include vaccination to reduce the fraction of susceptible individuals, quarantine of infected individuals and restrictions on population movement and assembly, all of which reduce the frequency of potentially infectious contacts. For non-human animal and plant infections, a very important control strategy is culling – the destruction of infected hosts to reduce the reservoir of infectivity. For arboviruses, control of insect populations can achieve the same protective effect for humans.

Virus epidemiology is both informed by and limited by the means used to detect infection in individuals. The most straightforward of these is to recognize infection by a specific set of clinical signs and symptoms; a formal diagnosis may then be made using appropriate tests (see Chapter 5). However, it needs to be understood that, for many viruses, only a subset of those infected actually show clinical signs. Sometimes, the proportion suffering inapparent or asymptomatic infection represents the majority of infected individuals. From a virus transmission perspective, such individuals may represent a very important source of virus for onwards transmission so control measures proposed to counter a virus epidemic have to reflect this. The ongoing global attempt to eradicate poliovirus by vaccination is hampered by the fact that only a small fraction of infectious individuals show clinical signs.

18.7 Sustaining infection in populations

Typically, once an animal host has cleared a virus infection, meaning that there is no longer virus in the system, then that host will be immune to future encounters with that virus. It is this fundamental principle that is exploited to achieve long-lived protection from infection by vaccination (see Chapter 26). So a virus that is antigenically stable will progressively exhaust its supply of new hosts in the existing population and will rely on new births to supply susceptible individuals to which it can spread and so maintain itself. For such viruses there is therefore a minimum host population size below which the supply of new susceptible is not sufficient and the virus will die out unless reintroduced from outside. Measles virus is an example of this (Chapter 4).

In the face of this problem, one evolutionary solution is for a virus to establish a stable persistent association with its hosts so it is not cleared by immune responses. This strategy dramatically reduces the minimum rate of host population renewal that is needed for the virus to be sustained in that population. It is probably no coincidence therefore that many of the antigenically stable viruses that infect animals and humans have developed such a strategy; for example all of the herpesviruses establish latent or persistent infections with opportunities for transmission throughout the life of the host.

The alternative strategy for a virus to sustain itself is to achieve ongoing change in its antigenic character. In this way, the immunity established in an individual during resolution of a previous infection will not protect it against further encounters with that virus. After a suitable period of time has elapsed, during which the virus is infecting other individuals and acquiring altered antigenic sites by random mutation, the virus will now appear 'new' to the immune system and so will have another chance to infect the same host. This strategy is typically found in viruses with RNA genomes, since their replication cycles involve the action of RNA-dependent RNA polymerases that are unable to proofread and so have a much higher error rate than DNA polymerases; influenza A virus is a classic example of a virus that is maintained through this mechanism (see Chapter 20). It is, however, important to note that not all RNA viruses show antigenic variation despite having error-prone replication; both polio and measles viruses are antigenically stable. Virus evolution is considered in detail in Chapter 4.

Key points

- Not all individual encounters with a virus result in transmission.
- Most infections are spread horizontally (person–person) but vertical spread from mother-to-foetus or baby is an important route for some viruses.
- Most viruses are spread by the respiratory or faecal-oral routes of infection.
- Viruses are also spread via urine, by sexual behaviour, or mechanically; a few are contracted by infection of the eye.

- A virus may be usually transmitted by one particular route, but it may spread by other routes under certain circumstances.
- Arboviruses are spread from person to person, or from other species into humans, by biting insects.
- Some infections are zoonoses in which a virus is spread from a non-human animal to humans without sustained onward transmission between humans.
- R_0 measures the average number of infections initiated by an infected individual; an epidemic will spread if $R_0 > 1$.

Questions

- Explain why respiratory and faecal-oral routes are significant transmission routes for many animal viruses.
- Discuss the role of insect vectors in virus transmission.
- Discuss the proposition that horizontal transmission is more important in maintaining viruses in a population than vertical transmission.

Further reading

Chan, J. F. W., To, K. K. W., Tse, H., Jin, D. Y., Yuen, K. Y. 2013. Interspecies transmission and emergence of novel viruses: lessons from bats and birds. *Trends in Microbiology* 21, 544–555.

Fisman, D. 2013. Seasonality of viral infections: mechanisms and unknowns. *Clinical Microbiology and Infection* 18, 946–954.

Killingley, B., Nguyen-Van-Tam, J. 2013. Routes of influenza transmission. *Influenza and Other Respiratory Viruses* 7, 42–51.

Krauss, H., Weber, A., Appel, M., Enders, B., Isenberg, H. D., Schiefer, H. G., Slenczka, W., von Graevenitz, A., Zahner, H. 2003. *Zoonoses: Infectious Diseases Tranmissible from Animals to Humans*, 3rd edn. Herndon, VA, ASM Press.

Kruse, H., et al. 2004. Wildlife as a source of zoonotic infection. *Emerging Infectious Diseases* 10, 2067–2072.

Mims, C. A., Nash, A., Stephen, J. 2000. *Mims' Pathogenesis of Infectious Disease*, 5th edn. London, Academic Press.

Rodriguez-Lazaro, D., Cook, N., Ruggeri, F. M. et al. 2012. Virus hazards from food, water and other contaminated environments. *FEMS Microbiology Reviews* 36, 786–814.

Part IV

Viruses and human disease

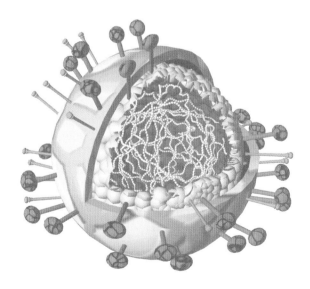

Chapter 19

Human viral disease: an overview

Between them, the viruses that infect humans can target, and cause disease in, essentially every tissue in the body. Even closely-related viruses can cause markedly different diseases, both in severity and type. Thus, an understanding of viral disease processes must be developed individually for each virus.

Since the discovery of viruses and their role in human disease, there have been many significant advances in our understanding of these agents and progress in both the prevention and treatment of infection. However, despite this progress, viral diseases continue to be a major component of the morbidity and premature mortality burden in the human population worldwide. There is also the constant threat of the emergence of new viruses to afflict us (see Chapter 24). Human immunodeficiency virus (HIV), for example, represents one of the greatest viral threats to our species that we have seen; it is considered in Chapter 21. This chapter gives an overview of human viral disease, focusing principally on acute viral infections (those from which we suffer transiently and then recover), which are our most common experience of viral disease. It explores the factors that govern the nature, severity and geographic distribution of viral disease and some aspects of viral pathogenesis in specific examples.

19.1 A survey of human viral pathogens

We now know that viruses are capable of affecting all the major tissues and systems in the human body. The diseases they cause range from trivial to life-threatening. The outcome of infection by a virus is not absolute but can vary considerably in nature and severity between individuals. Table 19.1 contains a brief survey of

Introduction to Modern Virology, Seventh Edition. N. J. Dimmock, A. J. Easton and K. N. Leppard.
© 2016 John Wiley & Sons, Ltd. Published 2016 by John Wiley & Sons, Ltd.

Table 19.1

The principal disease-causing viruses of humans.

Major target organ/system*	Virus (Family; Genus; Species)
Blood and lymphoid system	*Herpesviridae*; Lymphocryptovirus; Epstein-Barr virus (HHV4)
	Parvoviridae; Erythrovirus; human parvovirus B19
	Retroviridae; Deltaretrovirus; primate T-lymphotropic virus 1
	Retroviridae; Lentivirus; human immunodeficiency viruses 1 & 2
Eye	*Adenoviridae*; Mastadenovirus; human adenovirus B
	Herpesviridae; Simplexvirus; herpes simplex virus 1 (HHV1)
Foetus (infection *in utero*)	*Herpesviridae*; Cytomegalovirus; human cytomegalovirus (HHV5)
	Togaviridae; Rubivirus; rubella virus (German measles)
Gastrointestinal tract	*Adenoviridae*; Mastadenovirus; human adenovirus A, F
	Astroviridae; Mammastrovirus; human astrovirus
	Caliciviridae; Norovirus; Norwalk virus
	Reoviridae; Rotavirus; rotavirus A
Genital tract	*Herpesviridae*; Simplexvirus; herpes simplex virus 1 & 2 (HHV1, 2)
	Papillomaviridae; Alphapapillomavirus; human papillomaviruses (especially 6, 11, 16, 18, 31)
Heart	*Picornaviridae*; Enterovirus; Coxsackie B viruses
Liver	*Picornaviridae*; Hepatovirus; hepatitis A virus
	Hepadnaviridae; Orthohepadnavirus; hepatitis B virus
	Flaviviridae; Hepacivirus; hepatitis C virus
	Hepeviridae; Hepevirus; hepatitis E virus
Multisystem/haemorrhagic fevers	*Arenaviridae*; Arenavirus; Lassa fever virus
	Bunyaviridae; Nairovirus; Crimean-Congo haemorrhagic fever virus
	Bunyaviridae; Hantavirus; Hantaan virus
	Bunyaviridae; Hantavirus; sin nombre virus
	Filoviridae; Ebolavirus; several Ebola viruses
	Filoviridae; Marburgvirus; Marburg virus
	Flaviviridae; Flavivirus; dengue virus
	Flaviviridae; Flavivirus; yellow fever virus
Muscle and joints	*Togaviridae*; Alphavirus; Ross River virus
Nervous system	*Bunyaviridae*; Orthobunyavirus; California encephalitis virus
	Bunyaviridae; Phlebovirus; Sandfly fever virus
	Flaviviridae; Flavivirus; Japanese encephalitis virus
	Flaviviridae; Flavivirus; Tick-borne encephalitis virus
	Flaviviridae; Flavivirus; West Nile virus
	Herpesviridae; Simplexvirus; herpes simplex virus 1, 2 (HHV1, 2)
	Herpesviridae; Varicellovirus; varicella-zoster virus (HHV3)
	Paramyxoviridae; Morbillivirus; measles virus
	Picornaviridae; Enterovirus; polio virus
	Rhabdoviridae; Lyssavirus; rabies virus
Respiratory tract	*Adenoviridae*; Mastadenovirus; human adenovirus B, C, D, E
	Coronaviridae; Coronavirus; human coronavirus, SARS coronavirus
	Orthomyxoviridae; Influenzavirus A, B; influenza A, B viruses
	Paramyxoviridae; Morbillivirus; measles virus
	Paramyxoviridae; Pneumovirus; respiratory syncytial virus

Table 19.1
(*Continued*)

Major target organ/system*	Virus (Family; Genus; Species)
Skin	*Paramyxoviridae*; Respirovirus; parainfluenza 1,2,3 *Picornaviridae*; Rhinovirus; human rhinoviruses *Herpesviridae*; Varicellovirus; varicella-zoster virus (HHV3) *Papillomaviridae*; Beta-, Gamma- and Mupapillomavirus; human papillomaviruses (especially 1, 2, 4)
Testes	*Paramyxoviridae*; Rubulavirus; mumps virus

*Viruses can cause disease in more than one system and therefore appear more than once in the table. In some cases, viruses listed as causing disease in a system do so in only a small proportion of infected people.

this spectrum of pathogens, classified by their principal clinical manifestation. Some of these are considered in more detail in the following sections and chapters of this text.

19.2 Factors affecting the relative incidence of viral disease

The likelihood of an individual suffering from a particular viral disease depends on a number of interrelated factors. These include: geographic distribution of the virus and any necessary vector species; climate; human and vector migration patterns; socio-economic conditions, particularly the quality of nutrition and sanitation; individual genotype, especially the MHC haplotype since it determines the immune response; and the age at infection.

Viruses cannot be avoided; virus infections are a fact of life. All individuals inevitably encounter viruses that infect them and make them unwell. However, the extent to which the threat of viral disease should be a cause for concern to an individual will vary greatly depending on where in the world they live, the nature of their community, their socioeconomic circumstances, and their genetics.

The effect of location on viral disease incidence

Many viruses are found throughout the world, often with high rates of seroprevalence (i.e. most people in the community have been infected). However, some are much more restricted in their range, and in many cases this reflects a restricted range for an obligatory vector species. For example, infection by sin nombre virus, a hantavirus, is seen only in the southwestern USA, reflecting the distribution of its vector, the deermouse. Other hantaviruses occupy specific geographic niches based on the location of their specific rodent hosts. Viruses can also be constrained in range by geographic isolation. Historically, it is documented that the exploration of the globe by Europeans spread measles and variola (smallpox) viruses into communities that had apparently never experienced them before and, as a result, epidemics with high mortality were initiated in indigenous populations. In retrospect, the same form of geographic constraint can now be seen to apply to the mosquito-borne West Nile virus (WNV; see Section 23.5), which has emerged in North America since 1999 from its previous range in southern Europe and Africa. The mosquitoes of North America are clearly quite capable of transmitting WNV, they just had not been exposed to it prior to this introduction event.

The prevailing climate also affects the probability of acquiring particular viral diseases. Arbovirus infections (viruses transmitted by biting arthropods, see Section 18.5) are essentially limited to the less seasonal areas of the globe because the colder winters experienced elsewhere restrict the ability of the vector species to survive. Climate also affects the survival of virus while outside a host, since particle stability can depend on hydration/humidity (see Section 18.1). Viral genomes may also be vulnerable to inactivation by ultraviolet radiation in sunlight. Finally, climate influences human behaviour and physiology. As discussed in Chapter 18, several respiratory virus infections, such as influenza A and respiratory syncytial viruses, show annual winter peaks of incidence (Fig. 18.2). Maybe this is because we tend to spend more time in confined spaces with poor air circulation at this time of the year, although there are also respiratory virus infections with peaks of disease incidence at other times. Perhaps also, the reduced sunlight intensity affects our ability to counter infections. The light-sensitive pineal gland in the centre of the brain secretes melatonin and controls our circadian rhythms; low light is recognized to cause seasonal affective disorder (SAD) in some people. This is a complex syndrome of depression and malaise, and mental state is now thought to affect the functioning of the immune system. So, it may be that the seasonal peaks are as much a reflection of variation in host physiology as they are due to other factors.

The effect of environment on viral disease incidence

Any virus that is cleared from a host, with the host becoming immune to subsequent infection, requires access to a sufficient number of new hosts in order to perpetuate itself in a community, as already discussed for measles virus (see Section 4.5). Below this critical host community size, the virus will die out in that community until reintroduced from outside. Thus, the size of a community and its population density – whether a person lives in an urban or rural area – can determine whether or not they will be exposed to a virus. This of course assumes little movement of people in and out of the community, which is no longer true for much of the world, but in isolated areas this factor is still relevant. However, even with extensive population mixing bringing viruses into communities regularly, the greater population density in towns and cities still increases an individual's exposure to some viruses. Conversely, people in rural areas may be more vulnerable to zoonotic infections, because of an increased likelihood of encountering other species and the viruses that they carry.

Beyond the size of the community, another important factor determining the risk of virus exposure is the socio-economic status of the community members. Many important viral pathogens, such as rotavirus, poliovirus and hepatitis A virus, are spread by the faecal-oral route. Thus, rates of infection are determined by the quality of sanitation and the availability of clean drinking water. Nutritional status also affects the severity of viral disease. For example, the possibility of a fatal outcome of measles virus infection is increased by poor nutrition. Indirectly, socio-economic factors also influence another important parameter: the age at which you become infected. Living in a poorer household results in earlier exposure of individuals to a range of different viruses, such as Epstein-Barr virus, poliovirus, hepatitis A virus and hepatitis B virus. Intriguingly, in each of these cases the acute diseases that arise from infection early in life are less severe and/or a greater proportion of infections are subclinical than with later exposure to the virus. Finally, prostitution, which is often associated with regions of deprivation and low income, exacerbates the spread of sexually transmitted infections such as HIV (see Chapter 21).

Individual factors in viral disease susceptibility: genetics

In addition to all the above external factors, which can create differences between individuals in their probability of suffering viral disease, there are also intrinsic individual characteristics that are important. Most obviously, the ability of the immune system to respond to a virus is not the same for everyone. Apart from the case of people who have recognizable problems with their immune systems, each of us has a different ability to respond to specific antigens, determined by the precise nature of the MHC antigens that our genomes encode (see Chapter 13). MHC antigens are produced from highly polymorphic gene loci, so the combination of MHC antigens varies greatly between individuals. There are also differences between communities, with particular MHC alleles being common in one part of the world and rare in another. Natural selection produces MHC alleles in a population that are suited to respond to the prevailing antigens. When a new antigen (e.g. a new virus) enters a population, perhaps only a small proportion of people will be able to mount a strong immune response to it. If the new virus kills those less able to mount a response, then the remaining individuals, and the MHC alleles they carry, will expand to form an increasing proportion of the population. This

phenomenon may partly explain the severity of disease that typically arises from newly-introduced viruses in a population.

Specific alleles at other loci can also affect viral infection. HIV requires a co-receptor known as CCR5 for successful infection (see Section 21.3). Some people are relatively resistant to HIV infection because they are homozygous for an allele of the CCR5 gene that cannot express a functional protein. The frequency of this allele is highest in North European and West Asian populations. As the HIV pandemic continues, natural selection can be expected to increase the frequency of this CCR5 allele, and of alleles at any other loci that favour the survival of HIV-infected individuals. A particular gene, IFITM3, has also been shown to affect the severity of influenza A virus infection (Section 20.5).

19.3 Factors determining the nature and severity of viral disease

Why a virus causes the disease that it does is one of the most important questions in virology, and one of the most difficult to answer. Individual virus-specific features are likely to be crucial in each case. Factors that are thought generally to be relevant include the cell-type tropism of the virus, the amount of virus in the infecting dose, its portal of entry into the body and the relationship that the virus establishes with the host immune system.

One does not have to look far into the diversity of human viral pathogens to see the difficulty in making general or predictive statements about why viruses cause the specific diseases that they do. Within a single virus family, such as the adenoviruses, there are viruses causing respiratory tract infections ranging from mild to severe, and other viruses that cause infections of the eye, gut or urinary tract. Equally, viruses from at least three distinct families with very different molecular biology and particle structures can all cause the acute symptoms of hepatitis (liver infection). Thus there is no feature of a novel virus that can be observed outside of a human host and then related to knowledge about other viruses to predict the nature and/or severity of the disease that the new virus causes; each virus must be studied independently.

Starting from first principles, one plausible determinant of the disease-causing capacity of a virus is its interaction with cell-surface receptors. As discussed in detail in Chapter 6, a virus generally cannot infect a cell successfully in the absence of its specific receptor so the distribution of the receptor around the tissues of the body will act as a restriction on the range of tissues that can be infected and, hence, on the number of systems in the body where signs and symptoms of infection might be experienced. In fact, many of the receptors used by viruses (see Table 6.1) are ubiquitous molecules present on many different cell types. However, there are some examples of receptor tropism correlating with disease, such as hepatitis B virus, which appears only able to enter and to infect primary hepatocytes (Fig. 19.1a). In a variation of this concept, influenza A virus can only bind its receptor and enter cells following proteolytic activation (see Chapter 20); the limited tissue distribution of suitable activating proteases is important in restricting most strains of influenza virus to the respiratory tract.

Another potential determinant of the disease profile of a virus is restriction by portal of entry. Viruses enter our bodies by one of a few routes (see Section 18.3). When they do so, they will encounter certain tissues before others. If these are receptive to infection then this is where the infection will begin and where signs and symptoms might be most likely to occur (Fig. 19.1b). Thus, viruses such as rhinovirus, which are transmitted via aerosols, are inhaled and cause symptoms of upper respiratory tract infection despite having a receptor that is displayed on various cell types (see Section 19.6). Equally, viruses that are restricted by portal of entry initially, such as measles virus, may have the ability to spread systemically from this site of infection and cause symptoms elsewhere (Fig. 19.1c; see Section 19.7). However, it is also possible for a virus to cause no effects following infection at its portal of entry, but to cause disease at a secondary site. For example, poliovirus can cause paralytic disease following neuroinvasion after an inapparent infection of the gut.

Having gained access to a particular tissue in the body, the severity of the resulting symptoms is still unpredictable; this depends crucially on the interaction of the virus with the host immune system. At one level, there is clear evidence that the immune response serves to limit the spread of virus in the body and, hence, the severity of disease. Herpes simplex virus, for example, whilst generally giving mild symptoms, causes a systemic and life-threatening infection when acquired pre- or neonatally when the immune system is not mature, and common adenoviruses that cause minimal disease in healthy individuals can cause severe pathology in immunosuppressed transplant patients. However, it is also the case that the immune response causes many of the general signs and symptoms of a virus infection (see Section 19.4) and can also be responsible for more specific aspects of pathogenesis, e.g. hepatitis (see Chapter 22). Many viruses encode proteins whose function is to reduce the impact of specific aspects of the host immune response (see Chapters 13 and 14). For example, human adenovirus 5 makes a protein that blocks MHC class I antigen maturation, and other proteins that block apoptosis which is part of the innate response to infection. When adenovirus pathogenesis is studied in an animal model, removal of these functions from the virus exacerbates disease. In other words, having a stronger immune response to the infection makes the disease worse in this case.

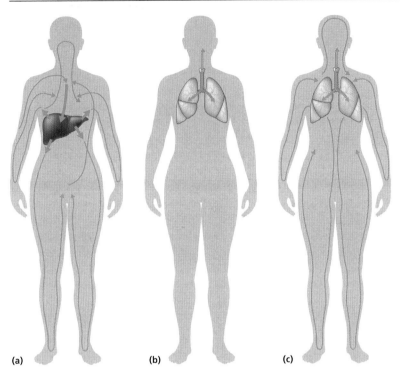

Fig. 19.1 Three scenarios for the relationship between the route of entry of a virus, its circulation in the body, site of replication and sites of disease manifestation. (a) Virus enters and circulates widely but can only infect cells in one location, where it replicates and causes disease, e.g. hepatitis B virus, with entry directly into the blood or by sexual contact and replication in the liver. (b) Virus infects cells at the portal of entry, where it replicates causing signs and symptoms, but physical barriers prevent spread elsewhere despite its use of a cell receptor that is widely distributed, e.g. rhinovirus, respiratory syncytial virus with entry and exit via the respiratory tract. (c) Virus infects cells at the portal of entry, where it replicates and then breaches local barriers to spread systemically, giving a variety of signs and symptoms, e.g. measles virus, which again enters and exits via the respiratory tract. Blue, movement of infecting virus; orange, movement of progeny virus.

19.4 Common signs and symptoms of viral infection

The classic signs and symptoms of a virus infection, fever and malaise, are caused by the host response to infection, not by the virus damaging cells of the host.

Fever (elevated body temperature), aches and pains, and general malaise are features of a variety of different viral infections. Often these occur in advance of more virus-specific signs and symptoms, and they are all associated with activation of the innate immune response. A key element in the innate response is the production of interferon α/β and the subsequent generation of an antiviral state (see Chapter 13). When the antiviral properties of interferon were recognized, attempts were made to use it as a general treatment for viral infections but with limited success. In fact, it was found that administering interferon generated many of the same signs and symptoms as are caused by viruses themselves. It is our body responding to a virus infection that usually generates the first signs and symptoms we are aware of, not virus-mediated destruction of cells.

There are close links between the innate and adaptive arms of the immune response (see Chapters 13 and 14). One of these is the production of the cytokine interleukin 1β (IL-1β) by macrophages and other cells in response to interferon α/β which then stimulates T-cell activation by antigen, potentiating the response. Studies with the poxvirus, vaccinia, in an animal model have demonstrated that the release of this cytokine is also responsible for the fever (pyrexia) associated with infection. Mice infected with wild-type vaccinia actually show a reduction in body temperature during the infection but a mutant virus that is unable to make a protein called B15R instead causes an increase in temperature. In other words, the B15R protein is used by vaccinia virus to block the fever response of the host. B15R is now known to bind and inactivate IL-1β, implicating this cytokine as a key mediator of temperature elevation. The downstream effects of IL-1β also include heightened sensitivity to inflammatory pain, which may be linked to the generalized aching associated with many viral infections.

19.5 Acute viral infection 1: gastrointestinal infections

The classic specific signs of viral infection of the gastrointestinal (GI) tract are vomiting and/or diarrhoea. These signs are both manifestations of dysfunction in the GI tract that results from the virus infection.

Rotaviruses pass through the stomach following ingestion and infect the mature enterocytes that form the epithelium at the tips of the villi in the upper part of the small intestine. This leads to loss of these cells and so-called blunting of the villi (Fig. 19.2). Later, infection spreads progressively to the mid and lower parts of the small intestine. Ultimately, new enterocytes produced by division of progenitor cells in the adjacent crypts move up to replace the lost cells and restore the villi.

Several factors contribute to rotavirus disease. The intestinal epithelium is responsible for absorption of nutrients from the gut contents. When this process is perturbed, the osmotic balance across the epithelium is lost and there is an outflow of water and electrolytes from the body into the lumen of the gut, resulting in diarrhoea. Loss of mature enterocytes will clearly impair the absorptive capacity of the epithelium and could explain the clinical signs of rotavirus infection. However, it is clear from animal studies that diarrhoea is seen significantly before overt villus blunting and that there must be other factors causing disease. Indeed, infection is characterized by active secretion of fluid into the gut lumen rather than simple failure of absorption. This is caused by a viral protein, NSP4, which acts as a toxin in a manner similar to certain bacterial endotoxins (Box 19.1). There is also evidence that infection induces hyperactivity in the enteric nervous system that actively contributes to disease.

Rotavirus disease is typically most severe in babies and young children. Signs of rotavirus diarrhoea begin within 2–3 days of infection and continue for 4–7 days. Treatment is straightforwardly provided by oral rehydration with an osmotically balanced saline/glucose solution. However, in the absence of such treatment, the loss of fluids can prove life-threatening, especially for young children who are not well nourished. A study in 2003 estimated that globally there were about

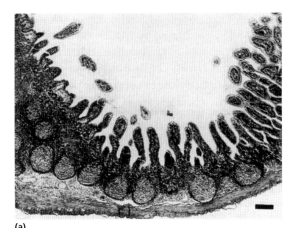

(a)

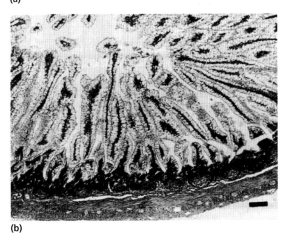

(b)

Fig. 19.2 (a) Section from the ileum of a piglet 72 hr post-infection with rotavirus. (b) Section from the ileum of a control, mock-infected piglet. Note the shortening of the villi and the absence of vacuoles (intracellular white areas) in the villous cells in (a) as compared with (b). These are indicative of normal active absorptive function. The bar represents 50 μm. (Adapted from Ward, L.A., Rosen, B.I., Yuan, L., Saif, L.J. (1996) *Journal of General Virology* 77, 1431–1441, with permission.)

140 million rotavirus infections per year in children under 5 years old, approximately 440,000 of which resulted in death; more than 80% of these deaths occurred in developing countries.

Box 19.1

Evidence for a secretory component in rotaviral pathogenesis

- Acute diarrhoea is seen in both babies and experimentally infected animals at a time when damage to the intestinal villi is minimal.
- Conversely, in experimental animals, recovery from diarrhoea is seen at a time when overt damage to the small intestine is still apparent.
- The rotavirus NSP4 protein (produced in infected cells but not included in virus particles) can cause diarrhoea when introduced into the gut of young experimental animals.

Whilst rotavirus infection can cause vomiting as well as diarrhoea, usually at the onset of clinical disease, the noroviruses cause vomiting specifically. Noroviruses are commonly known for causing 'winter vomiting disease'. Outbreaks of infection occur particularly in relatively confined communities, such as in hospital wards and on cruise ships. Again, the virus passes through the stomach and infects the epithelium of the small intestine, with damage to the villi and features of malabsorption from the gut. The vomiting is thought to be due to delayed emptying of the stomach, which results from impaired gastric motor function caused by the infection. Signs and symptoms begin within 24 hours after infection and continue for 1–2 days. Severe dehydration is less likely than for rotavirus infection because of the shorter duration of disease, but treatment is similar if required.

19.6 Acute viral infection 2: respiratory infections

The classic specific signs of respiratory tract infections are sore throat, coughing and sneezing. These reflect the effects of infection on the local production of inflammatory mediators, also the production and movement of secretions in the airway. Coughs and sneezes create aerosols that may enhance the likelihood of transmission though normal exhalation is often sufficient.

A large number of different viruses cause respiratory tract infections and disease in humans (see Section 19.1). The severity of respiratory signs and symptoms in these infections broadly relates to how deep into the respiratory tract the infection penetrates. Thus rhinoviruses normally infect only the nasal epithelium and result in mild upper respiratory tract (URT) disease whilst respiratory syncytial (RS) virus more usually reaches the lung where it can cause bronchiolitis and/or pneumonia. The severity of non-specific symptoms also varies considerably. Fever is not normally associated with rhinovirus infections, except in infants, but is a significant feature in influenza A virus infections. Influenza A virus infection is considered in detail in Chapter 20. Rhinoviruses are the major cause of what is described as 'the common cold', a mild URT infection, with coronaviruses also making a significant contribution to the common cold burden.

Rhinovirus infection does not cause extensive cytopathology in the affected tissues. Instead, it is thought that the effects of infection on respiratory tract function are mediated by host responses. Sneezing is an early sign of infection and represents a reflex response to irritation in the nasal epithelium and nasopharynx. The sneeze reflex is triggered by the trigeminal nerve, which innervates these areas, and is stimulated by histamine which is released as part of an early innate response to infection. Sneezing in turn stimulates production of the watery secretions that are characteristic of the early stages of a cold.

Coughing is another common sign of URT infection, usually occurring later in the course of infection. Spontaneous cough is triggered by stimulation of the vagus nerve, the sensitivity of which is increased by bradykinin, an inflammatory mediator released in response to infection. Bradykinin is a vasodilator and it functions in inflammation to increase the access of immune cells to the inflammatory source. Enlargement of the blood vessels in the nose by bradykinin constricts the nasal passages and leads to the sensation of nasal blockage. Interestingly, this blockage affects only one side of the nose at a time and represents an exacerbation of the normal cycling of airflow between one side of the nose and the other whereby, over a period of around 4 hrs, the two passages alternate in carrying the bulk of the airflow to the lungs. Nasal blockage as a response to infection may serve to raise the temperature of the nose which may limit replication of some viruses. Bradykinin also heightens pain responses and is considered to be responsible for the sensation of sore throat, which is another early feature of many URT infections.

The vagus nerve innervates the respiratory tract below the larynx so spontaneous coughing as a sign of infection indicates spread of infection downwards into the respiratory tree. Coughing can also be a voluntary response to the build-up of mucus in the airway. The respiratory epithelium cells are ciliated and the cilia beat in a regular pattern similar to a 'Mexican Wave' that forms a conveyor, known as the respiratory escalator, which is constantly moving the layer of mucus that coats and protects the cells upwards towards the nose (see Fig. 19.3). If the ciliated cells become damaged

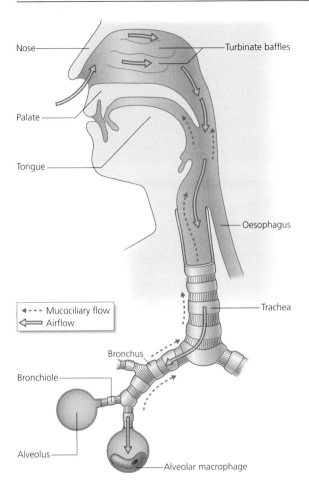

Fig. 19.3 Route taken by aerosol droplets containing virus particles that are inhaled into the respiratory tract. Note the defenses designed to trap and remove small particles: turbinate baffles, the mucociliary flow that runs counter to inspired air, and macrophages. The airways become progressively narrower (here much foreshortened) and end in alveoli, each of which is formed by a single cell. An alveolar macrophage can be seen in one of the alveoli. (Adapted from Mims and White (1984) Viral Pathogenesis and Immunology. Oxford: Blackwell Scientific Publications.)

by infection or the quality of the mucus changes making it too thick to move, then the mucus accumulates and has to be removed by coughing.

RS virus initially infects the epithelium of the nasopharynx but, within 1 to 3 days, can move down to the lower respiratory tract where it causes coughing and wheezing. Respiratory distress is seen due to underlying bronchiolitis and/or pneumonia when the virus infects the lungs. The infection causes loss of epithelial cells in the bronchioles, hyperproliferation in compensation for this loss, disruption of cilial beating accompanied with excess mucus secretion, and a strong inflammatory response with cellular infiltration into the infected tissues. Together, these cause consolidation of portions of the lung with what can be a dramatic loss of air exchange function in those areas (Fig. 19.4). Patients suffering these severe symptoms of RS virus infection require hospitalization for supportive care. Treatment with an aerosol of the antiviral compound, ribavirin, is sometime used in these cases (Chapter 27).

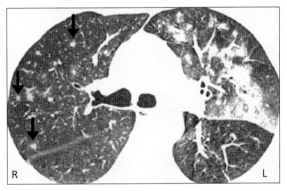

Fig. 19.4 An example of lung consolidation caused by respiratory syncytial virus infection in an immunosuppressed patient. The lungs have been imaged by high resolution CT scan. There are nodular lesions in the right lung (R, arrows), while the left lung (L) shows areas of consolidation (white) where no air is present. (Reproduced with permission from D. L. Escuissato et al. (2005) *American Journal of Roentgenology* 185, 608–615.)

19.7 Acute viral infection 3: systemic spread

Some viruses cause little or no disease associated with replication at the primary site of infection, but spread around the body and cause disease elsewhere.

Some viruses are best known for the signs and symptoms that arise from their systemic spread around the body rather than for disease caused by infection at the point of entry. Two good examples are measles virus, and varicella-zoster virus (VZV) which causes chicken pox on primary infection. Both these viruses cause skin rashes that can reach all extremities of the body, indicative of systemic spread, but both actually initiate infection in the respiratory tract (Fig. 19.1c). Measles virus can, in fact, cause significant symptoms of respiratory tract infection prior to the rash appearing, especially in immunocompromised children, but VZV does not. Other viruses can also spread to specific secondary sites where disease consequences are manifest, e.g. polio virus neuroinvasion from the gut to cause paralytic disease.

VZV and measles virus both replicate initially in cells of the respiratory epithelium. From there, they gain access to cells of the immune system, particularly monocytes. Since these circulate around the body, they provide the means by which virus can spread away from the respiratory tract. Both viruses then replicate further in lymphoid tissues before spreading to the skin and, for measles virus, many other organs. In the skin, VZV replicates in epithelial cells, causing the fluid-filled vesicular lesions which contain infectious virus that are characteristic of chicken pox. VZV then spreads on into sensory neurons where it becomes latent. Measles virus replicates in endothelial cells before spreading to the overlying

epidermis. The measles rash, which is a smooth discolouration of the skin, coincides with the onset of an antibody response. It reflects immune complex formation in the skin, and fades as virus is cleared. In the same way, Koplik's spots, a diagnostic feature of measles virus infection, form on mucosal surfaces of the mouth. Thus, while measles virus and VZV both spread systemically from a respiratory infection and cause skin rashes, these rashes are completely different; the VZV lesions contain infectious virus while the measles rash does not.

What determines whether or not a virus can spread away from its site of initial infection? As just discussed, being able to infect cells that move around the body is one factor. Another is exactly how the virus exits the cells it originally infects. Many viruses initially infect epithelial cells because these are the cells that are first encountered at each of the principal routes of entry of virus into the body. Internal epithelia are polarized, which means that the cell sheet has directionality with the two surfaces being functionally distinct. The apical surface faces the lumen of e.g. the gut or airway while the basolateral surface faces into the body, interfacing with connective tissue, the blood and lymphatic systems. Many viruses exit epithelial cells preferentially to one or other surface (Fig. 19.5 and Fig. 12.8). Clearly, a virus that exits apically, as is normally the case with influenza A virus for example, has less opportunity for systemic spread than one exiting basolaterally. Viruses that are released basolaterally into the circulation generate what is known as a *viremia*.

19.8 Acute viral disease: conclusions

Viruses establish infection initially at sites in the body that are determined by characteristics of the virus and its route of entry. Disease signs

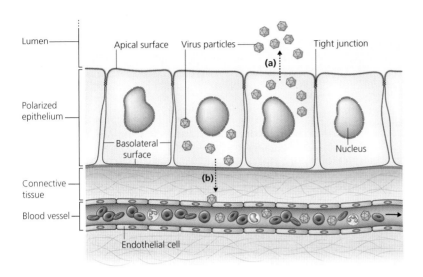

Fig. 19.5   A simplified diagram of a polarized epithelium showing virus exiting an infected epithelial cell either (a) via its apical surface to reach the lumen of the organ, or (b) via the basolateral surface to reach the vascular system. The apical and basolateral surfaces of the epithelial cells are separated by tight junctions that connect the cytoplasm of adjacent cells and prevent movement of membrane proteins between the two surfaces. Not to scale. See Section 19.7 for further details.

and symptoms may arise from this infection and/or from the results of subsequent virus spread to other parts of the body. Wherever in the body specific indicators of virus infection occur, it is reasonable to infer that the virus is present in that tissue. However, it is not always the case that these signs and symptoms are caused by the virus damaging cells of the tissue. Often, the infection is not itself grossly cytolytic but instead, the host immune response to infection leads to tissue damage and consequent dysfunction and overt symptoms.

Key points

- Many different viruses are significant human pathogens.
- All parts of the body can be affected by viral pathogens.

- Similar viruses can cause different diseases and similar diseases can be caused by very different viruses.
- The nature and severity of signs and symptoms caused by a virus can vary greatly between infected individuals.
- The chance of an individual being infected by specific viruses depends on their genetics, which part of the world they live in and the nature of their community.
- The nature of the disease caused by a virus is determined by a series of factors, including the distribution of its cell receptor, its portal of entry into the body, and whether or not it has the ability to spread systemically.
- Viruses can achieve systemic spread either by infecting circulating cells, or by achieving release of particles into the blood from a site of infection.
- Symptoms of acute viral infection are frequently due to the host immune response

to infection rather than to a cytopathic effect of the virus.

- Specific symptoms of acute viral infection result from transient dysfunction in the affected organs or systems of the body.

Questions

- Discuss the factors that affect the incidence of viral disease in developed versus developing countries.
- Discuss the factors that influence the nature and severity of disease during an acute virus infection.

Further reading

Collier, L., Oxford, J., Kellam, P. 2011. *Human Virology*, 4th edn. Oxford, Oxford University Press, chapters 4, 5, 7.

Eccles, R. 2005. Understanding the symptoms of the common cold and influenza. *Lancet Infectious Diseases* 5, 718–725.

Mims, C. A., Nash, A., Stephen, J. 2000. *Mims' Pathogenesis of Infectious Disease*, 5th edn. London and New York, Academic Press, chapters 2, 5, 8, 11.

Nathanson, N. 2007. *Viral Pathogenesis and Immunity*, 2nd edn. London and New York, Academic Press, chapters 7, 13.

Parashar, U. D., Hummelman, E. G., Bresee, J. S., Miller, M. A., Glass, R. I. 2003. Global illness and deaths caused by rotavirus disease in children. *Emerging Infectious Diseases* 9, 565–572.

Ramig, R. 2004. Pathogenesis of intestinal and systemic rotavirus infection. *Journal of Virology* 78, 10213–20.

Chapter 20

Influenza virus infection

Influenza A viruses

- *Members of the orthomyxovirus family.*
- *Comprise 144 viruses (subtypes) that naturally infect aquatic birds in which they cause subclinical gut infections and are spread by the faecal-oral route. Additional subtypes have been found in bats.*
- *Can evolve to infect man, via domesticated poultry; cause respiratory disease that is spread by droplets; cause serious morbidity and mortality, although can be subclinical.*
- *Initial introduction of a new influenza virus into humans is a called an* antigenic shift; *virus then undergoes gradual mutational change or* antigenic drift – *Darwinian evolution in response to positive pressure from virus-specific antibody. Antigenic shift and drift refer to changes in the two major surface proteins of the virus particle.*

Influenza virus infections have become such a common and integral part of our lives that any respiratory infection that causes discomfort is typically referred to as an episode of 'flu. In reality, most of these infections are probably not caused by influenza viruses but by more benign agents. True influenza virus infections can pose a serious threat. Influenza virus infection in humans occurs in seasonal epidemics and is estimated to result in up to 5 million cases of serious disease with 250,000–500,000 deaths each year. Understanding the origins of human influenza virus and the elements of the complex process of influenza virus infection and disease are important steps in establishing preventative and therapeutic treatments.

20.1 The origins of human influenza viruses

All members of the orthomyxovirus family to which influenza viruses belong have segmented genomes comprising six, seven or eight discrete single-stranded negative sense RNA molecules. There are three groups of influenza virus, types A, B and C: the influenza A and B viruses have genomes comprising eight segments and the influenza C viruses have seven segments. These three influenza types are distinguished by the antigenicity of their internal virion nucleoproteins and all three cause respiratory disease in humans. Type A viruses cause

Introduction to Modern Virology, Seventh Edition. N. J. Dimmock, A. J. Easton and K. N. Leppard.
© 2016 John Wiley & Sons, Ltd. Published 2016 by John Wiley & Sons, Ltd.

occasional worldwide epidemics (pandemics) of influenza, and both type A and B viruses cause annual seasonal epidemics. Type C viruses cause only minor upper respiratory illness and will not be discussed further. In terms of natural history, the primary reservoir hosts of influenza A viruses that have been shown to infect humans are wild aquatic birds (such as ducks, terns and shore birds). Recently, two additional subgroups of influenza A virus have been identified in South American bats; however, these have several differences from the avian-derived influenza A viruses and there is no evidence that they can infect humans so they will not be considered further. Influenza B viruses infect only humans.

As discussed in Chapter 4, different subtypes of influenza A virus are defined by their haemagglutinin (H) and neuraminidase (N) genes, and 144 different combinations of H and N have been identified in avian hosts. However, only a very limited range of influenza A subtypes have entered the human population since the late 19th century and most avian influenza viruses do not cause infections in humans. To cause disease in humans, avian influenza viruses require evolutionary changes to efficiently infect and spread in the human population, as considered below. Despite this barrier there have been some cases of zoonotic infection with avian viruses. In almost all cases, the infected individuals have been exposed through direct contact with infected domestic poultry. While many of these zoonotic viruses, such as influenza H7N2, H7N3 and H7N7, have resulted in no, mild, or only very rare serious disease, some have consistently caused serious disease. An avian influenza A H5N1 virus was first associated with serious disease in humans in 1997 and continues to cause sporadic infections with a mortality rate of approximately 60%. In March 2013, a previously unknown avian influenza A H7N9 virus was found to have infected humans. In these cases, approximately 20–25% of infected

people died. As with most other zoonotic infections these viruses have not spread efficiently in the human population but their appearance emphasizes the threat that new influenza viruses may pose.

In humans, influenza A and B viruses cause disease only in the winter, usually January and February in the Northern hemisphere, and June and July in the Southern hemisphere. At the equator, influenza virus is present at a low level throughout the year. The cause of this periodicity, which is also a feature of many other viruses (see Section 18.1), is not known. In addition to aquatic birds and humans, a limited number of other species are naturally infected with type A influenza viruses. These and the possible directions of transmission are shown in Fig. 20.1.

Two mechanisms of influenza A virus evolution

The principles of virus evolution are discussed in Chapter 4. As with all viruses, influenza undergoes evolutionary changes from which successful mutants are selected in response to a variety of environmental pressures. The nature and rapidity of these changes are what have made influenza such a successful organism. Influenza A viruses undergo two types of evolutionary change that alter their major surface glycoproteins. These are *antigenic shift*, arising from reassortment of the genome segments following a dual infection of a single cell (Section 4.4) and mutation that is referred to in influenza as *antigenic drift* (Section 4.2). Since the start of modern virology there have been five shift events (Fig. 4.3). Antigenic shift, by definition, introduces a virus into a population that has little or no pre-existing immunity, so it is usually associated with an explosive pandemic with high levels of disease and potentially deaths, although the number of these varies greatly between different shift

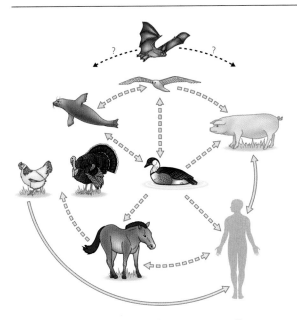

Fig. 20.1 Animal species that are naturally infected with influenza A virus subtypes. Wild birds of the sea and shore form the natural reservoir for the viruses that ultimately cause serious disease in humans. Known routes of transmission are indicated by continuous arrows and probable routes of transmission by broken arrows. The routes of interspecies transmission, if any, of the influenza A viruses found in bats are not yet known.

events. Approximate mortality figures are included in Fig. 4.3.

There are 16 subtypes of influenza A virus haemagglutinin and 9 subtypes of neuraminidase found in nature in wild aquatic birds, such as ducks, terns, and shore birds that form the natural reservoir of influenza A viruses (Section 4.4). These can, in principle, combine to form 144 different subtypes that could find their way into the human population. In fact, only very few have succeeded (Fig. 4.3). These avian viruses are tropic for the gut (not the respiratory tract), cause a subclinical infection, and are evolutionarily stable. Many of these birds migrate enormous distances (e.g. from Siberia

to Australia) and spread virus as they go via infected faeces. However, like most viruses, avian influenza virus is species-specific and does not readily infect other bird species let alone mammals. Thus, the virus has to acquire at least some of the necessary characteristics to allow infection of humans by progressive mutation before the jump to humans can occur.

It is thought that the first link in the chain is the infection of free-range domestic poultry (mainly chickens, ducks, turkeys, and quail) by migrating wild birds. This is usually an avirulent infection restricted to the respiratory tract, although it is not understood how the switch from a gut to respiratory infection comes about. The infection passes from bird to bird until mutations take place in the protease cleavage site of the haemagglutinin precursor polypeptide. Cleavage is necessary for virus infectivity. Normally, the HA can only be cleaved by a protease that is present solely in the respiratory tract, but after an increase in the number of basic amino acid residues in the HA cleavage site it can be cleaved by another protease that is widely distributed throughout tissues, and this allows the virus to grow throughout the body and cause serious disease. The infection of domestic poultry brings the virus into close contact with humans and this may increase the likelihood of transfer between species. At this stage, the virus is able to multiply in humans but not spread from human to human (Fig. 20.2). In humans, the polybasic HA cleavage site does not confer a growth advantage for the virus.

Another variation in the evolutionary progression of virus from wild birds to humans may involve domestic pigs. In rural areas, poultry and pigs are often kept together, giving ample opportunity for the crucial bird-to-mammal adaptation. At this point, the virus has two evolutionary options: it can continue to accumulate mutations and become better adapted to humans, or it can recombine with a human influenza virus strain and so acquire

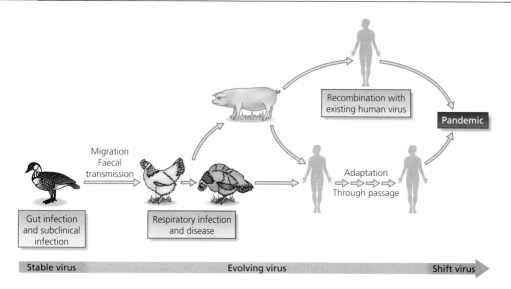

Fig. 20.2 Summary of the events leading up to an antigenic shift of human influenza virus. The evolutionary time scale is not known but probably takes several years.

genes that are already fully adapted in a process of antigenic shift. In fact, it may do both. Antigenic shift that results in a pandemic is most commonly preceded by an abrupt change in haemagglutinin subtype and, apart from the 1968 virus, also by a change in neuraminidase subtype. In 2001, a new H1N2 reassortant virus was isolated in many countries in mainly young people. However, this virus has not spread widely, presumably because it is sensitive to existing immunity to its parent human H1N1 and human H3N2 viruses.

In April 2009, a new strain of H1N1 virus was reported which was antigenically distinct from the normal seasonal H1N1 virus that had been circulating since 1977 (Fig. 4.3). Subsequent analysis identified the first known human case to have occurred in Mexico on 11 March 2009. This new strain of virus spread rapidly, and on 11 June 2009 the World Health Organisation declared a pandemic level 6 – a full pandemic – to be occurring. Within a year, infections with this new virus were reported in 214 countries. The pandemic status was declared over on 10 August 2010 by which time

the majority of the world's population had been infected. As with all new influenza virus strains, the origin of this new strain is not completely understood but it is proposed to have arisen by a complex series of reassortment events between a number of different influenza viruses infecting different animals (Fig. 20.3 and Box 20.1). Fortunately, this new strain did not cause the devastating disease that was associated with the 1918 pandemic (Box 20.3) with only approximately 18,000 confirmed deaths worldwide in the first year of the pandemic. This is a lower level of mortality than seen with previous pandemics. The pattern of infection and disease was also different to previous pandemics with almost twice the level of infections in children less than 15 years old than the levels of infections in adults. This may have been due to residual immunity arising from infection with earlier H1N1 strains in the older population (Section 20.6).

Transmission studies in animals using genetically modified viruses have shown that both the HA and NA proteins contribute to efficient airborne transmission of influenza virus. This is consistent with our knowledge of the receptor preference for viruses and the role of the NA protein in virus exit for the infected cell (Section 20.3). Surprisingly, the pandemic strain of virus that appeared in 2009 did not demonstrate a major change in HA or NA subtype and was of the H1N1 subtype that was already present. However, the 2009 virus contained sufficient differences to the H1N1 virus circulating previously to allow it to establish an infection in people already exposed to the earlier virus. Antigenic drift generates the viruses that cause annual epidemics (or seasonal flu), and is the result of gradual evolution under the positive selection pressure of neutralizing antibody. A new shift virus immediately starts to undergo continuous antigenic drift. Influenza B viruses only undergo drift and this is believed to be because they have no animal reservoir.

Antigenic drift is at least as important in causing the observed patterns of human influenza as antigenic shift. The ability of influenza virus to form a quasispecies (Section 4.2) provides a population of viruses within an infected person from which a mutant that is resistant to the neutralizing antibodies found in the next host can be selected by the pressure of the immune system. The 'new' virus is referred to as an escape mutant and may contain only one or two nucleotide differences from the original infecting isolate. This virus will have an advantage in that it will be able to infect individuals who had previously been infected by, and had become immune to, the original 'parental' strain and so will survive until

---

Box 20.1

The origin of the 2009 pandemic influenza virus

The precise sequence of reassortment events that resulted in the generation of the 2009 pandemic H1N1 strain is not completely understood, but the key players have been identified by careful deduction, made possible by the surveillance of influenza virus strains in humans and a variety of animal species over many decades. The timings of the various events are also not entirely clear. The intricate pattern of genetic exchange that the generation of this virus demonstrates shows the close interaction between different influenza viruses and the capacity of this virus for dramatic evolutionary change which poses a threat to those it infects.

The 2009 H1N1 pandemic virus had its origins in the 1918 H1N1 virus which infected both humans and pigs. Just as with the human virus, the pig virus was maintained in the swine population undergoing antigenic drift and it was referred to as 'classical swine' virus. In 1979, a new H1N1 virus, referred to as Eurasian 'avian-like' H1N1 virus, appeared in pigs and began co-circulating with the classical swine virus. In the 1990s, the classical swine virus underwent reassortment with the circulating human H3N2 virus and an avian strain to produce a number of viruses including H1N2 and H3N2 swine viruses, which each contained genome segments from all three 'parental' viruses. These new viruses have continued to circulate in the North American pig population from that time to now. A further genetic reassortment between the North American H1N2 virus and the Eurasian 'avian-like' H1N1 virus appears to have produced the 2009 pandemic H1N1 strain which entered the human population in late 2008 or early 2009. This new virus therefore contains a mosaic of genome segments from at least four different viruses that have come together in a unique combination (Fig. 20.3).

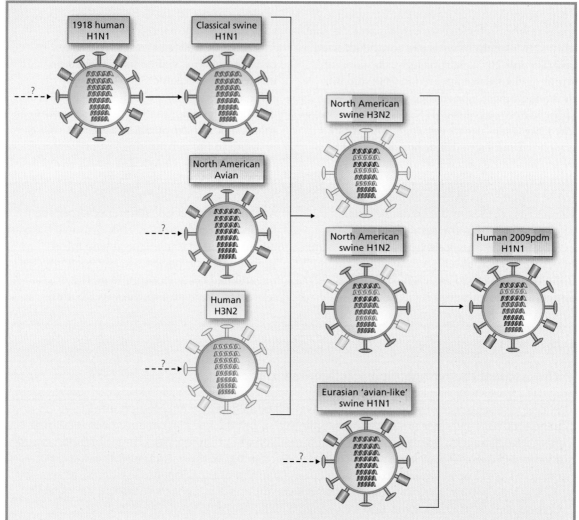

Fig. 20.3 Proposed origin of the 2009 pandemic influenza virus H1N1 strain. The colours indicate the origins of the various genome segments and the HA and NA proteins. See Box 20.1 for explanation.

immunity in the population reaches a level that limits the availability of new hosts. This selection process is repeated as immunity to each successive escape mutant arises.

The name 'antigenic drift' is very apt as the process is a gradual one of accumulating mutations (Fig. 20.4). It has been determined that a virus must acquire on average ≥4 amino acid substitutions in ≥2 antigenic sites to be able

to infect a person who was previously infected with the 'parental' virus from which the drift variant arose. In practice, a drift variant that can cause significant disease, infecting a large proportion of the population anew, becomes predominant approximately every four years. At the same time, the 'old' strain is less likely to find a susceptible person to infect so the proportion of people infected with this virus

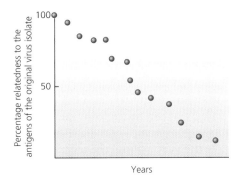

Fig. 20.4 Diagram showing antigenic drift of type A influenza virus in humans. This could represent mutations within either the HA or NA genes. Each point is a virus strain isolated in a different year.

declines. The result is that the 'old' strain is rapidly and completely replaced by the 'new' strain. Influenza B viruses also undergo antigenic drift, but this is slower process. Currently, two distinct genetic lineages of influenza B virus are circulating in the human population. These belong to the B/Victoria and B/Yamagata lineages that are antigenically distinct and diverged from each other in the 1980s. The reason for the differences in the

pattern of antigenic drift of influenza A and B viruses is not known.

As soon as a new shift virus appears and infects people, it begins to drift (Fig. 20.5). Drift happens on a global scale. Antigenic drift of influenza A viruses is linear due to the dominating effects of favourable mutations (Fig. 20.6). It can only be theorized how drift takes place as it is assumed that drift variants arise from virus circulating in the previous year. It is generally believed that variants are selected by antibody (Box 20.2).

The phenomena of antigenic shift and drift in influenza have significant implications for human infection and they are monitored by a worldwide network of laboratories, coordinated by the WHO, that isolate and classify currently circulating influenza viruses. In this way, new strains can be quickly spotted and the vaccine changed appropriately.

20.2 Influenza virus replication

The processes of influenza virus entry (Section 6.2), replication (Section 8.5), gene

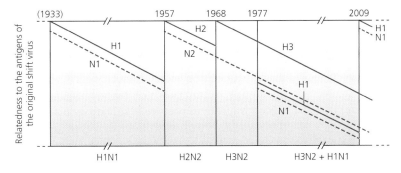

Fig. 20.5 Diagram showing the course of antigenic shift and drift of influenza A viruses in humans. The first virus, isolated in 1933, was H1N1. This arose by antigenic drift from the 1918 virus. Other shift viruses appeared in 1957 (H2N2) and 1968 (H3N2). A 1950 H1N1 virus reappeared in 1977. A new H1N1 appeared in 2009. Drift is shown schematically. The 1957 N2 was acquired by the H3N2 shift virus, and has drifted from 1957 to the present day.

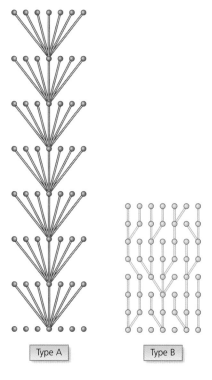

Type A   Type B

Fig. 20.6 Model of antigenic drift of influenza type A and type B viruses. Points on the same level represent drift variants that arise in the same year. The branch length indicates the relative change in antigenicity from virus in the preceding year. Drift is shown for an arbitrary 7-year period. See text for further discussion.

expression (Section 11.5) and virion assembly (Sections 12.6 and 12.7) have been discussed previously. Influenza A viruses contain eight genome segments with the potential to encode up to 15 proteins, though some strains do not synthesize the full range of proteins. The products of translation of the primary mRNA transcript and the additional proteins with the mechanisms by which they arise are shown in Table 20.1. While the functions of many of the influenza proteins are known, this is unlikely to be the complete list of proteins and additional proteins are likely to be discovered.

20.3 Influenza virus infection and disease

Typical influenza in humans is a lower respiratory infection with fever and muscular aches and pains, but can range from the subclinical to pneumonia (where the lungs fill with fluid). In the very young, the elderly and people with underlying clinical problems of the heart, lungs, and kidneys, as well as in diabetics and the immunosuppressed, influenza can be life-threatening. Influenza thus poses a serious disease threat for the human population where changes in the mortality rate associated with emergence of a new strain can have devastating consequences: a brief history of the terrible 1918 pandemic is shown in Box 20.3.

Influenza virus infection can also result in neurological disease with estimates of the incidence in young children ranging from approximately 5% to 20% of those hospitalized. The incidence of this complication in adults is considerably less. The neurological symptoms are encephalopathy, encephalitis and meningoencephalitis. Little is known about the process of infection that leads to these neurological symptoms.***

Influenza virus infection of the respiratory tract is associated with an inflammatory response in which the host immune system promotes the movement of immune cells into the lung. The first immune cells which encounter the virus in the lung are likely to be macrophages that reside there. These cells send out signals to other cells of the immune system which migrate to the site of infection to attack the invading virus, with neutrophils being the most abundant in this response. The signals are then amplified by the incoming cells and also by the innate immune response of the infected lung cells including the production of a proinflammatory response following activation of Toll-like receptor 3 (TLR3) and the RIG-I/mda5 pathways by the double-stranded virus

Box 20.2

Demonstrating influenza virus antigenic drift in the laboratory

- Antigenic drift is modelled in the laboratory using a neutralizing monoclonal antibody (MAb) specific for the HA. Virus and MAb are mixed together and inoculated into an embryonated chicken's egg. This is a surprisingly efficient process and, after just one passage, the progeny virus is no longer neutralized by the selecting MAb. This 'drift' virus is also known as an *escape mutant,* and represents the growth to dominance in the quasispecies of an antigenic variant virus that already existed in the inoculum. Sequencing shows there is a single amino acid substitution in the expected antigenic site. If the drift virus is subjected to a MAb to a different epitope, the process repeats (Fig. A). However, if two or more MAbs are mixed together, no progeny virus – mutant or wild-type – is produced at all. Thus, drift can only take place if an antiserum effectively contains antibody to only a single epitope.

Fig. A   Demonstration of antigenic drift in the laboratory: a neutralizing monoclonal antibody selects a population of influenza virus escape mutants that is no longer neutralized by that MAb. Repeating with a second MAb produces virus that carries two amino acid substitutions.

- Influenza virus HA molecules have 5 antigenic sites (sometimes one is hidden by a carbohydrate group). In H3 viruses these are labelled A–E. Each site comprises around 10 epitopes, thus there are 50 epitopes in all, and in theory the immune system should make antibodies to all of these. However, when tested by immunizing rabbits with virus and measuring the amount of antibody made to individual epitopes, one epitope in site B was dominant, and there were only traces of antibody to two other epitopes (Fig. B). The amounts varied between animals. In mice the results were similar but the dominant epitope was in site A. Thus the immune system responds selectively to foreign epitopes.
- While some antisera were completely neutralizing, others were shown to select escape mutants like a MAb. This suggests that drift operates because certain individuals have an antibody response which is highly biased to one epitope. The derivation of the ≥4 amino acid substitutions in ≥2 sites, referred to in the text, could therefore be achieved as shown in Fig. C. The biased antibody response may be genetically controlled, and drift variants may arise in a sub-population with the relevant antibody response. Further, since adults suffer repeated influenza infections, they are likely to make a complex antibody response that cannot give rise to drift variants. The simplest antibody response is likely to occur in children after their first infection, and this may be responsible for the selection of drift variants.

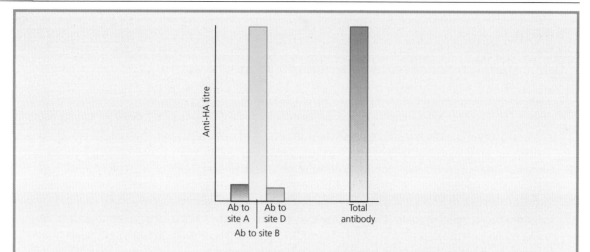

Fig. B An epitope-biased serum antibody response in a rabbit injected with influenza A virus. Nearly all the HA-specific antibody is accounted for by the response to a single epitope in antigenic site B. (Adapted from Lambkin and Dimmock, 1995, *Journal of General Virology* 76, 889–897.)

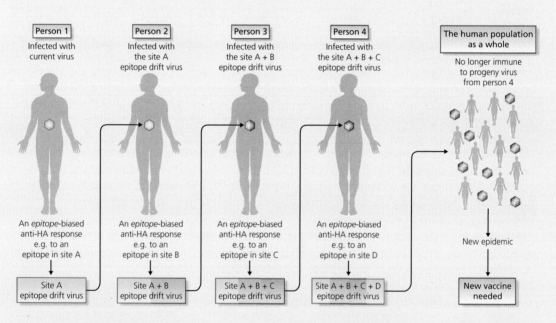

Fig. C How antigenic drift may occur in nature, bearing in mind that clinically significant drift viruses (that cause epidemics) have ≥4 amino acid residues changed in ≥2 antigenic sites present on the haemagglutinin protein. Note that people with biased antibody responses may be uncommon and that there may be any number of non-selective infections of other individuals occurring between the four persons shown here. A similar process could occur for drift of the NA protein.

Table 20.1
Proteins potentially encoded by the eight genome segments of influenza A viruses.

Segment	Protein	Function	Mechanism by which additional protein is generated
1	PB2	Component of virus polymerase	N/A
2	PB1	Component of virus polymerase	N/A
	PB1-F2	Virulence factor	Leaky scanning
	PB1-N40	Unknown	Leaky scanning
3	PA	Component of virus polymerase	N/A
	PA-X	Virulence factor	Ribosomal frame shift
	PA-N155	Unknown function in replication	Leaky scanning
	PA-N182	Unknown function in replication	Leaky scanning
4	HA	Haemagglutinin	N/A
5	NP	Formation of helical ribonucleoprotein complex	N/A
6	NA	Neuraminidase	N/A
7	M1	Matrix protein	N/A
	M2	Ion channel protein	mRNA splicing
8	NS1	Multifunctional	N/A
	NS2	Nuclear export of viral ribonucleoprotein complex	mRNA splicing

---

Box 20.3

1918 influenza

The origin of the 1918 H1N1 shift virus remains unknown despite its common names of 'swine flu' and 'Spanish flu' which infer a fixed source. The current view is that the virus did not arise from an antigenic shift involving exchange of genome segments between viruses around at that time but instead was introduced into the human population directly from a bird, though the virus did not resemble current avian flu viruses and the species of bird from which it originated is unknown. Following the introduction into humans the virus is believed to have evolved over a number of years prior to the appearance of the pandemic virus in 1918. The geographic origin of the virus also remains unknown.

The new virus was able to infect both humans and pigs. The subsequent pandemic was unusual in several ways. It is estimated that in just one year it killed approximately 40 million people worldwide. (The enormity of this can be appreciated when compared to the 36 million that have died in 30 years of the AIDS pandemic.) A marked feature of the pandemic was the exceptionally high mortality in young adults who comprised almost half of the fatalities. The overall mortality rate was 25 times greater than in other pandemics (2.5% compared with 0.1%). The infection occurred in three successive waves, with the first in the spring of 1918 followed by a second, more deadly wave in the late months of 1918 and a final wave of slightly reduced mortality in the early months of 1919.

After its appearance in 1918 the H1N1 virus remained in the human population undergoing antigenic drift until 1957 when a new pandemic strain appeared (Fig. 4.2).

RNA that is generated by virus replication (Chapter 13). As the adaptive immune response begins, virus-specific T cells will be produced and these will also migrate to the site of infection where the cytotoxic T cells will play an essential role in limiting the spread of infection prior to eliminating the infected cells and ultimately clearing the infection (Chapter 14). These immune responses come with a cost to the host due to destruction of cells and the generation of cell debris and mucus in the respiratory tract that impair lung function. This means that the inflammatory response contributes to the symptoms experienced by the infected individual. If the host response is rapid and efficient the extent of virus replication may be restricted, resulting in a relatively short-lived and minor disease. If the virus is not restrained, either because the host response is slow or impaired in some way, the infection can generate a life-threatening disease. However, an over-enthusiastic immune response can also lead to dire consequences due to catastrophic and potentially irrecoverable immune-mediated damage (Box 20.4).

Virus receptors and cell tropism

The first event in initiating an influenza virus infection is the attachment of the virus to a host cell. This is determined by the presence of appropriate receptor molecules on the surface of the cell. The influenza A virus HA protein binds to a sialic acid, N-acetylneuraminic acid (NANA), that is terminally linked to a carbohydrate moiety of a glycoprotein or glycolipid. HA proteins of avian influenza viruses bind preferentially to the terminal sialic acid residue that is attached via an $\alpha2,3$ linkage while human influenza viruses prefer to bind to $\alpha2,6$ linked NANA residues. This preference

---

Box 20.4

The consequences of immune-mediated damage

As indicated in Box 20.3, a striking feature of the 1918 H1N1 pandemic was that a very large proportion of the deaths occurred in young adults. A second feature was that these deaths occurred over a very short time frame; contemporary records of the time refer to young people being apparently healthy one day and dead the next! Even allowing for some degree of exaggeration it is clear that the time from onset of symptoms to death was often extremely short. This is very unusual. Our current understanding from animal studies is that the 1918 virus was particularly potent at inducing a rapid and profound cytokine response in infected people. This is now referred to as a 'cytokine storm' to indicate the rapidity and strength of response. The result is that the immune response of the affected individuals was such that the respiratory tract was flooded with inflammatory cells. These brought with them the associated host cell damage that is always seen, but in these cases the damage was of such a severity and occurred over such a short time that the unfortunate individuals could not cope. The ability of the influenza virus PB1-F2 protein to enhance the severity of disease associated with concurrent respiratory bacterial infections, described in Section 20.4, may also have played a role in the disease associated with the 1918 influenza pandemic.

determines the cell tropism of the virus and is key to the ability of flu viruses to transfer between species and to the outcome of the infection.

The gastrointestinal and respiratory tracts of birds contain high levels of α2,3 linked NANA with only low levels of α2,6 linkages. The situation is reversed in the human respiratory tract with a predominance of α2,6 linked NANA. In humans, the α2,6 linked NANA is found predominantly in the upper respiratory tract and the α2,3 linked NA is predominantly expressed on bronchial epithelial cells and the type II pneumocytes of the alveoli deep in the lung. Thus, while seasonal human viruses readily find suitable host cells in the upper respiratory tract when they first enter the body, avian viruses find only a few cells in the nasopharynx which they can infect and must penetrate deep into the lower respiratory tract to encounter a high number of cells containing suitable receptors. This reduces the efficiency of transmission of avian viruses to humans.

At the end of the influenza replication cycle the virus must be released efficiently from the infected cell. The virus exits the cell by budding at sites where the HA and NA proteins have been inserted into the plasma membrane. The NA protein is essential for the efficient release of the virus. The NA cleaves the NANA sialic acid in the local vicinity. This helps to ensure that when the virus particles are produced by budding they do not immediately reattach to the surface of the cell from which they have come but instead spread to the surrounding cells or are expelled from the respiratory tract. For the virus release to be most efficient it is important that the NA has a specificity for the same NANA molecules found on the surface of the infected cell, i.e. it must have a specificity similar to that of the HA which initiated the infection. This is likely to be the reason that mutation of both the HA and NA genes of avian viruses must occur to allow cross-species

infection. This may also impose a further selection barrier on influenza virus evolution.

20.4 Virus determinants of disease

Following infection, influenza virus undergoes replication in the cells of the respiratory tract. The outcome of infection is the result of a combination of the effects of the virus itself and the response of the host to infection. The precise balance of these effects determines the outcome at an individual level.

Once it has gained entry to the cells of the respiratory tract the replication of influenza virus leads to cell death. Several viral protein activities contribute to this. The PA-X protein serves to shut off host functions whilst leaving virus gene expression unscathed, an effect which ultimately leads to the death of the cell. The virus NS1 protein prevents the export of cell mRNA from the nucleus, leading to a reduction in host cell protein synthesis which also compromises cell viability. The loss of the cells contributes to the signs and symptoms of disease. In addition, several virus proteins interfere with the innate defence mechanisms of the host. Prime targets for these virus countermeasures are the host antiviral interferon system and the cell-driven process of apoptosis that leads to programmed cell death (Chapter 13). These are inhibited by the PB1-F2 and NS1 proteins, respectively. While the virus functions that inhibit the innate immune response contribute to the success of the infection, paradoxically they may also reduce some of the symptoms of infection that are contributed by the host response considered below.

The PB1-F2 protein also exhibits a fascinating effect on the severity of bacterial infections that occur at the same time as the virus infection. The PB1-F2 protein isolated from an H3N2 strain that appeared early in the pandemic of 1968 induces a proinflammatory response with induction of many cytokines that activate inflammatory cells, resulting in enhanced damage in the respiratory tract. The introduction of a bacterial respiratory pathogen such as *Streptococcus pneumoniae* at the same time as the virus infection leads to a very severe bacterial disease. In contrast, the PB1-F2 protein from a virus of the same lineage isolated almost 30 years later has no proinflammatory effect and generates an antibacterial effect. This suggests that the PB1-F2 protein plays an important role in the severity of disease seen early in a pandemic. As the virus acquires mutations with time, the severity of disease, caused directly by the virus and also by secondary infections, is reduced due to changes in the PB1-F2 protein. This attribute of enhancing the severity of associated respiratory bacterial infections may also have contributed to the severity of the response to infection with the 1918 pandemic virus. A summary of the virus proteins that interfere with the innate immune system is shown in Table 20.2.

Table 20.2

Influenza A virus proteins that interfere with the host innate immune response.

Protein	Function
HA	Cell tropism
NA	Cell tropism
PB1-F2	Promotion of apoptosis
	Regulation of host interferon response
	Modulating host susceptibility to bacterial infection
PA-X	Shut-off of host functions
NS1	Inhibition of host interferon response
	Inhibition of host mRNA processing

20.5 Host factors in influenza virus disease

As indicated above, the host innate immune response plays a role in the development of disease associated with influenza virus infection, as it does with all virus infections to some degree. In addition to this, there is clear strong evidence that the genetic make-up of the host also contributes to the outcome of infection. An analysis of family members of people who died following an influenza infection showed that blood relatives were also highly likely to succumb to influenza infection. The nature of these genetic factors in humans remains largely unknown but there are some clues to the existence and identity of some of them.

Early studies in mice identified the *Mx1* gene that is expressed in response to interferon and encodes a guanosine triphosphate-metabolizing protein which generates resistance to influenza and other virus infection. Different *Mx1* alleles confer different levels of virus resistance. The human genome contains a homologue of *Mx1* call *MxA*. There are suggestions that allelic variation in the *MxA* genes may affect the outcome of infection.

In recent years, it has become clear that the genetics of the immune system plays a role in the outcome of infection by many pathogens, and influenza virus is no exception. Genetic variation in the interferon-induced transmembrane protein-3 (IFITM3) gene has also been associated with the severity of influenza virus disease. Initial studies showed that mice lacking the IFITM3 gene developed very severe disease following infection. Subsequently, a specific variant in the human genome, called rs12252-C, was shown to be associated with enhanced influenza disease. The rs12252 variant generates an IFITM3 protein which is shorter due to loss of a segment of the carboxy terminus. People homozygous for the rs12252-C allele have on average a six-fold greater risk for severe disease than those lacking the allele. The frequency of the rs12252-C allele is very low in Caucasian populations but is very high in people of Han Chinese descent, resulting in a ten-fold higher risk of severe influenza disease in the Chinese population than in Northern Europeans. As indicated above, TLR3 recognizes dsRNA during influenza infection and activates the proinflammatory response that is intended to reduce and ultimately contribute to the elimination of infection. It has been shown that allelic variation in the TLR3 gene is associated with enhanced influenza virus-associated encephalopathy, a neurological disease that results periodically from infection.

20.6 The immune response and influenza virus

Following infection and activation of the innate immune response the adaptive immune response is stimulated. This is an essential response to eliminate the virus and to establish immunological memory that will provide protection against subsequent infection by the same virus. The adaptive immune response is the environmental force that drives seasonal influenza virus evolution through antigenic drift. Chapter 14 considers the general features of the adaptive immune response and here we will focus on the specific responses seen following influenza virus infection.

The cellular immune response to influenza virus

The recovery from influenza virus infection requires the activity of both $CD4^+$ helper T cells (T_H) and $CD8^+$ cytotoxic T cells (T_c). Following

infection, respiratory tract dendritic cells carry antigen to the draining lymph nodes where they encounter precursor T cells of both sorts. The dendritic cells present small peptide fragments derived from influenza virus proteins on their surface in association with MHC class I and MHC class II molecules to stimulate the differentiation and activation of naïve $CD8^+$ T cells and naïve $CD4^+$ T cells, respectively. These activated T cells migrate to the site of infection to exert their activities targeted against infected cells.

Depending on the environment in which they are activated, the activated T_H cells differentiate into either T_H1 or T_H2 cells. The T_H1 cells promote the activity of the activated T_c cells while the T_H2 cells promote the activation and differentiation of B cells. Some $CD4^+$ T cells can also lyse target cells similar to the effect of T_c cells. The T_c cells are the major agents of attack to eliminate the virus. When the activated T_c cells encounter an infected cell that is presenting the appropriate virus antigen on its surface in association with MHC class I proteins it binds and attacks the cell as described in Chapter 14. The result is that the infected cell dies, either by lysis or through an apoptotic response, reducing the virus load in the lung.

A proportion of the activated cells in the lymph glands become memory $CD4^+$ and $CD8^+$ T cells that form the basis for future immunity by ensuring that stimulation with the same antigens produces a rapid and robust amplification of activated T cells. Some of these memory cells become resident in the lung which also helps to ensure a rapid response if the virus is encountered again.

The antigens that stimulate T cells come predominantly from the more conserved internal proteins of the virus such as the NP, PA, PB2 and M proteins. The high conservation of the amino acid sequences of these proteins in many different influenza virus strains should provide a degree of cross-protection against infection with different influenza viruses. However, long-lived cross-reactive heterotypic protection is rarely seen. An exception was found with the 2009 pandemic H1N1 virus where older people who had been infected decades previously with an H1N1 virus derived from the 1918 pandemic strain retained some level of cross protection.

The humoral immune response to influenza virus

The primary antigens that stimulate the humoral response leading to virus-specific antibodies are the HA and NA glycoproteins found on the external surface of the virus particle. These proteins are also expressed on the surface of infected cells and present a readily-accessible target for antibodies. Whilst antibody binding to cell surface antigens can stimulate destruction of the cell, influenza virus-specific antibodies are not thought to play a major role in elimination of an already-established infection. Rather, they represent a defence against future attack by the same or a sufficiently closely-related virus.

The HA protein contains five antigenic sites that are recognized by antibodies. These are located in the globular head adjacent to the receptor binding site of the HA protein (Fig. 6.3). Antibody binding to the HA protein on the virus particle sterically inhibits the binding of the virus to the sialic acid target and so prevents infection being initiated. The NA protein is involved in the last stage of the virus replication cycle where it enhances release of virus particles from the infected cell. The anti-NA antibody, therefore, can only inhibit this rather late step and these are generally rather weakly neutralizing. This may explain the primary importance of the anti-HA antibodies which provide the strongest correlation with protection from infection.

Whilst it is accepted that the anti-influenza antibodies are strain-specific and must match the infecting strain to protect against infection, cross-protection against different strains has been seen, albeit rarely. This was observed most recently with the 2009 pandemic. Just as was seen with some cross-reacting CTL epitopes, antibodies raised against H1N1 viruses that circulated in the 1950s were able to cross-react with the 2009 H1N1 pdm strain and provide some protection from infection. This, together with the T_c cell cross-reactivity, provided some protection for the older population that had been exposed to the 1950s virus. This protection was not complete but was sufficient to be detected and accounted for the unusually lower levels of infection in the elderly during the 2009–10 pandemic.

20.7 Anti-influenza treatment

Preventative treatment for influenza is available in the form of two different vaccines. One of these is an inactivated or 'killed' vaccine that is prepared by destroying the infectivity of the viruses in the preparation. This is delivered by subcutaneous injection and stimulates the production of circulating antibodies directed against the virus strains used in the preparation. While this vaccine works well it does not induce the full range of immune responses seen with a virus infection; specifically, it does not induce any local antibody and T cell-mediated immunity in the respiratory tract. The killed vaccine is particularly useful for those people for whom a fully infectious virus vaccine may pose a risk, such as the elderly. The second vaccine that is currently approved for use is a 'live' vaccine containing fully infectious viruses which are attenuated in their capacity to cause influenza disease. The live vaccine has the advantage of inducing the full range of immune responses seen with a natural infection, including antibody and T cell-mediated immunity in the respiratory tract. Vaccination is considered in more detail in Chapter 26.

For both vaccines, the key antigens are the influenza virus HA and NA glycoproteins. As indicated above, the presence of HA-specific antibodies provides significant selective pressure that selects virus variants mutated in the antigenic sites as the virus evolves to escape antibody neutralization. Thus, vaccinated individuals become susceptible to the newly-adapted virus strains. For this reason, it is necessary to monitor the currently circulating strains and reformulate the vaccines to ensure that they match antigenically the most prevalent circulating strains. As of 2014, the inactivated vaccine contains either three (a trivalent vaccine) or four viruses (a quadrivalent vaccine), one H1N1 virus, one H3N2 virus and either one or two influenza B viruses to represent the viruses posing a threat. Typically, the inactivated vaccine is reformulated twice a year, once for the northern hemisphere and once for the southern hemisphere because they experience their influenza virus seasons out of phase with each other and significant virus evolution can occur in the interim. The live vaccine also contains four viruses comprising one H1N1 virus, one H3N2 virus and two influenza B viruses.

In addition to prevention of influenza by vaccines, a number of small molecule compounds have been produced that can inhibit influenza virus. These are shown in Table 20.3.

Two influenza virus proteins have been targeted by these molecules: the M2 ion channel protein that is essential for the uncoating step early in the infectious cycle and the NA protein which is involved in virus particle release from infected cells. While these compounds have been valuable in the battle against influenza their usefulness is compromised by the ability of the virus to

Table 20.3	
Small molecule anti-influenza virus inhibitors. The target proteins for the molecules are indicated.	
Inhibitor	**Target protein**
Amantadine	M2 protein (flu A only)
Rimantadine	M2 protein (flu A only)
Oseltamivir	NA protein (flu A and flu B)
Zanamivir	NA protein (flu A and flu B)

evolve rapidly to escape any inhibitory pressures. This is seen with the increasing prevalence of antiviral resistant strains of influenza that creates an imperative to find alternative drugs for treatment of this disease. Antiviral therapy is discussed further in Chapter 27.

Key points

- Influenza A and B viruses are responsible for annual epidemics (seasonal flu) and influenza A viruses are responsible for periodic pandemics.
- Influenza disease is the result of direct actions of the virus and of the host immune response to infection.
- Influenza virus has evolved mechanisms to counteract the host immune response.
- Host genetics play a role in determining the outcome of infection.
- Humoral and cellular immune responses are involved in prevention and clearance of infection, respectively.
- Antibodies directed against the HA protein play a major role in prevention of infection.
- Vaccines are available to prevent influenza virus infection. These take two forms: inactivated and live vaccines. The vaccines have to be reformulated each year to provide protection against the currently circulating virus strains.

- Small molecule chemical inhibitors that act against influenza virus can be used to treat disease. However, the virus mutates readily to generate resistant strains.

Questions

- Describe the origin of human influenza virus and discuss how the virus evolves to generate new viruses that cause worldwide pandemics and seasonal infections.
- Discuss the role of the immune response in the generation of influenza disease.

Further reading

de Graaf, M., Fouchier, R. A. 2014. Role of receptor binding specificity in influenza A virus transmission and pathogenesis. *EMBO* 33, 823–841.

Kuiken, T., Riteau, B., Fouchier, R. A., Rimmelzwaan, G. F. 2012. Pathogenesis of influenza virus infections: the good, the bad and the ugly. *Current Opinions in Virology* 2, 276–286.

Lin T. Y., Brass, A. L. 2013. Host genetic determinants of influenza pathogenicity. *Current Opinions in Virology* 3, 531–536.

Sun, J. and Braciale, T. J. 2013. Role of T cell immunity in recovery from influenza virus infection. *Current Opinions in Virology* 3, 425–429.

Taubenberger, J. K., Kash, J. C. 2010. Influenza virus evolution, host adaptation, and pandemic formation. *Cell Host and Microbe* 7, 440–451.

van de Sandt, C. E., Kreijtz, J. H., Rimmelzwaan, G. F. 2012. Evasion of influenza A viruses from innate and adaptive immune responses. *Viruses* 4, 1438–1476.

Chapter 21

HIV and AIDS

Acquired immune deficiency syndrome (AIDS) and the viruses now known to cause this syndrome (HIV-1 and HIV-2) were recognized in the 1980s. In the absence of treatment, and normally after many years of infection, HIV infection damages the immune system to the point where the body cannot combat common infections and AIDS results. One type of HIV-1 (group M) has caused a global pandemic.

Chapter 21 Outline

21.1 Origins and spread of the HIV pandemic
21.2 Molecular biology of HIV
21.3 HIV transmission and tropism
21.4 Course of HIV infection: pathogenesis and disease
21.5 Immunological abnormalities during HIV infection
21.6 Prevention and control of HIV infection

During the past 35 years, human immunodeficiency virus (HIV) type 1 has become a common infection of mankind. There is also a second, less common, less virulent, but closely-related virus, HIV-2. Both viruses infect cells of the immune system, mainly helper T lymphocytes and macrophages (Chapter 14). If there is no treatment, a typical infection shows a long, essentially symptomless incubation period of around 10 years. By that time, the numbers of helper T lymphocytes have declined to the point where the immune system can no longer function efficiently, resulting in the immunodeficiency which gives these viruses their name. The consequence of this immunodeficiency is that the affected person can no longer resist certain normally harmless micro-organisms (viruses, bacteria and fungi) which then cause overt clinical disease. A collection of diseases such as this, unrelated except for a common underlying cause, is called a syndrome – hence the term 'acquired immune deficiency syndrome' or AIDS. Untreated, HIV infection is almost always lethal.

Immense progress has been made in understanding HIV and its associated disease through intensive research although, at the time of writing, no curative treatment or effective vaccine has yet been achieved. However, HIV research is a rapidly-moving field, so contemporary literature should be consulted to update the knowledge detailed here.

21.1 Origins and spread of the HIV pandemic

HIV was discovered as the cause of AIDS in the early 1980s. It is now recognized that HIV-1 entered the human population from chimpanzees in the early part of the 20th century, as a zoonosis. Since then, particular

Introduction to Modern Virology, Seventh Edition. N. J. Dimmock, A. J. Easton and K. N. Leppard.
© 2016 John Wiley & Sons, Ltd. Published 2016 by John Wiley & Sons, Ltd.

HIV-1 subtypes have spread from central Africa to cause a pandemic that is affecting humans in all parts of the globe. HIV is a retrovirus and is classified within the lentivirus genus, together with a number of related viruses that infect other animal species.

Discovery of HIV

The first signs of the HIV pandemic began in 1981 with an official report of a cluster of cases of pneumonia due to *Pneumocystis carinii* infection (PCP) in previously healthy young male homosexuals in Los Angeles. Publicity surrounding this report led rapidly to the recognition of similar cases involving this and other pathogens, and to the definition of much larger clusters of cases of Kaposi's sarcoma (KS), a disease now known to be caused by infection by a herpesvirus (see Section 25.6), again in young homosexual men. It was soon shown that the underlying cause of these opportunistic infections, which were very rare in the general population, was a severe generalized immune deficiency in the patients, many of whom died rapidly. PCP and KS became two of the defining conditions of AIDS.

Definition of the causative agent of AIDS followed quickly from the first definition of the syndrome. The virus now known as HIV-1 was isolated by Françoise Barré-Sinoussi and Luc Montagnier in 1983 from patient samples taken either from male homosexuals or from haemophiliacs, another demographic group that was significantly affected in the early part of the pandemic. With the identification of the virus, reagents were soon developed for virus detection. Using these reagents, the earliest proven case of infection by HIV-1 is currently a person in central Africa in 1959, from whom blood samples were stored at the time. A second related virus, HIV-2, was identified in 1986. Whilst it too causes AIDS, its spread so far has been more limited, largely confined to West Africa; the vast majority of infections globally are caused by HIV-1.

Origins of HIV-1 and HIV-2

It is certain that HIV-1 and HIV-2 arrived in the human population as zoonoses. These viruses are related to a group of more than 40 simian immunodeficiency viruses (SIV), each of which (with one exception) has a relatively benign non-pathogenic interaction with its natural primate host. That exception is SIV_{mac}, a virus passaged under laboratory conditions in cynomolgus macaques in which it causes AIDS-like pathology. SIV_{mac} is now understood to be the result of an interspecies transmission of a virus (SIV_{sm}) whose natural host is sooty mangabeys and so, like HIV, is a virus that has recently crossed to a new host. As discussed later, enhanced pathogenicity is typical of viruses when they infect a new host species (see also Section 4.5).

Four distinct lineages of HIV-1 sequence have been defined as groups M, N, O and P. Group M is the predominant lineage worldwide, whilst groups N–P have not spread beyond small populations in west-central Africa. Each of these four lineages is the result of a separate zoonotic transfer into humans. Detailed study of the sequences of many HIV isolates in comparison with the SIV family has shown that HIV-1 groups M and N were derived from SIV_{cpz}, a virus of the chimpanzee *Pan troglodytes troglodytes*; SIV_{cpz} in turn is a recombinant of SIVs from two other species. Similarly, HIV-1 groups O and P were derived from an SIV found in western lowland gorillas whilst HIV-2, like SIV_{mac}, is derived from SIV_{sm}. Looking at the available sequence information, it is estimated that the transfer into humans of the predominant HIV-1 group M lineage occurred in or near Kinshasa in the Democratic Republic of Congo around 1920.

The description of molecular diversity in HIV-1 has been further refined by allocating genome sequences in group M to one of several subtypes (also known as clades) A–D, F–H, J–K based upon closer sequence similarity. A number of recombinants between these subtypes also exist and are circulating in humans. The frequency of occurrence of the various groups and subtypes of HIV-1 in different populations is not uniform across the globe (Fig. 21.1). Notably, subtype B predominates in North America and western Europe while subtype C is dominant in southern and eastern Africa, and south Asia. These differences in frequency distribution reflect founder effects – the chance initiation of new foci of infection in different geographic areas through movement of infected individuals away from the epicentre of the pandemic in central Africa where the original zoonotic transfers occurred. As expected therefore, there is much greater diversity of HIV-1 subtypes found in the area where HIV-1 originated than in any other part of the globe.

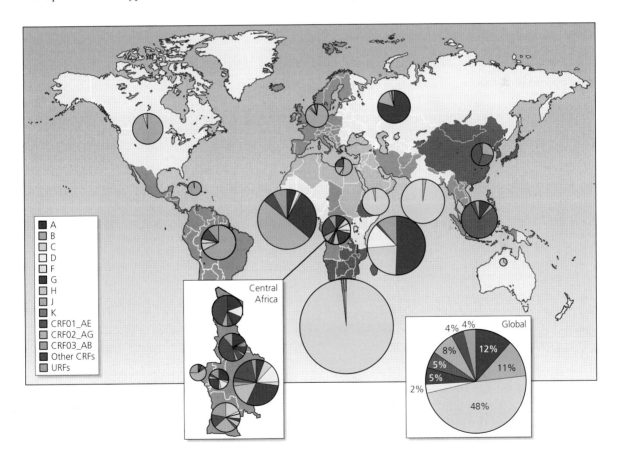

Fig. 21.1   World map showing the relative contribution of different HIV-1 subtypes to the total burden of infection globally (inset) and in different regions (main map) from 2004–7. The overall size of each pie chart reflects the numbers of infected people in each region, with the coloured segments of the charts representing the particular subtypes (A–K) and recombinant forms (CRF) present. The situation in Central Africa is represented in greater detail in an inset which shows the greater heterogeneity in the virus present in this region as compared with other parts of the world. [Reproduced with permission from *Trends in Molecular Medicine*, Hemelaar, J. vol 18, 182–192 ©2012, with permission from Elsevier.]

Table 21.1
Some typical members of the lentivirus genus, their hosts, and diseases.

Virus	Host	Main target cell	Clinical outcome
Visna–maedi	Sheep*	Macrophage	Pneumonia (= maedi) or chronic demyelinating paralysis (= visna); little immunosuppression
Visna–maedi	Goats	Mammary macrophage	Arthritis; rarely encephalitis; little immunosuppression
Equine infectious anaemia	Horses	Macrophage	Recurrent fever; anaemia; weight loss, little immunosuppression
Bovine immunodeficiency	Cattle	Not known	Weakness; poor health
Feline immunodeficiency	Cats	CD4+ T cell	AIDS
Simian immunodeficiency†	Monkeys	CD4+ T cell	Subclinical or AIDS
Human immunodeficiency type 1	Humans	CD4+ T cell	AIDS
Human immunodeficiency type 2	Humans	CD4+ T cell	AIDS

*The same virus can cause two different diseases in sheep.

†Different SIV strains are named after the species of monkey from which they were first isolated, e.g. SIV_{man} from the mandrill, SIV_{agm} from the African green monkey, SIV_{sm} from the sooty mangabey. These are all old-world African monkeys, and SIV strains cause no disease in their natural host. However, although it does not naturally infect Asian macaque monkeys, SIV_{sm} does do so under experimental conditions and causes AIDS. SIV_{mac} is thought to be SIV_{sm} which accidentally infected a laboratory macaque. (This is a further example of a virus interaction with its natural host being relatively benign compared to that with a new host.).

The lentivirus genus of retroviruses

HIV-1 and HIV-2 are typical members of the lentivirus genus of the retrovirus family (Baltimore class 6). The name is derived from the Latin *lente*, which refers to the slow onset of the disease. However, there is nothing slow about the rate of virus multiplication. There are several well-characterized lentiviruses (Table 21.1) that infect a number of different vertebrate species. Their infections have a number of common features (Box 21.1).

21.2 Molecular biology of HIV

HIV-1 and HIV-2 are members of the retrovirus family. These viruses show all the general molecular features of retrovirus life cycles (reverse transcription, integration, gene expression) described in Chapters 9 and 10 but with additional important features described here.

> **Box 21.1**
>
> **Common features of lentivirus infections**
>
> - Infection of bone marrow-derived cells.
> - Integration of DNA copy of the viral genome into host DNA.
> - Persistent viraemia.
> - Lifelong infection.
> - Prolonged subclinical infection.
> - Weak neutralizing antibody responses.
> - Continuous virus mutation and antigenic drift.
> - Neuropathology.

Gene expression

As with all retroviruses, HIV-1 and HIV-2 have RNA genomes but replicate via a dsDNA intermediate (Chapter 9); the reverse transcriptase enzyme that carries out this conversion was the first target of antiviral drugs used to treat HIV infection (Chapter 27). They also express the Gag, Pol and Env proteins that are the hallmark of all retroviruses (Section 10.8) but, in addition, these viruses express a set of accessory gene products from clusters of open reading frames between Pol and Env, and 3′ to Env. In HIV-1 these are Tat, Rev, Vif, Vpr, Vpu and Nef (Fig. 21.2); in HIV-2, Vpu is missing and another gene, Vpx, is present.

The presence of the Tat and Rev genes allows HIV-1 to express its genome with temporal phases. Initially after infection, cell transcription factors result in the synthesis of a small amount of full-length viral RNA from integrated HIV-1 DNA. This RNA is then multiply spliced, removing introns that cover the bulk of the Gag, Pol and Env genes to form small mRNAs encoding Tat and Rev, which are both RNA-binding proteins. Tat binds to the trans-activation response (TAR) element, a sequence in the nascent RNA that is encoded just downstream from the transcription start site, and up-regulates transcription by increasing processivity of RNA polymerase II (Box 21.2). This increases Rev expression so that it accumulates to a critical level that then down-regulates the cytoplasmic accumulation of multiply-spliced RNAs (i.e. mRNA encoding Tat and Rev) and favours the accumulation of non-spliced and several singly-spliced mRNAs, which encode the structural proteins Gag, Gag-Pol, Env, Vif, Vpu and Vpr (Fig. 21.3); this molecular switch creates two distinct phases of gene expression. Rev achieves this by binding to an RNA sequence called the Rev response element (RRE) that is present only in non-spliced and singly-spliced mRNAs and facilitating the transport of these mRNAs from the nucleus (Box 21.3).

Virus structure and structural proteins

The mature HIV-1 particle has a characteristic cone-shaped core (Fig. 12.9; Fig. 21.4a), which is formed of the p24 protein (CA) that is encoded within Gag. This particle morphology develops after cleavage of the Gag and Gag-Pol

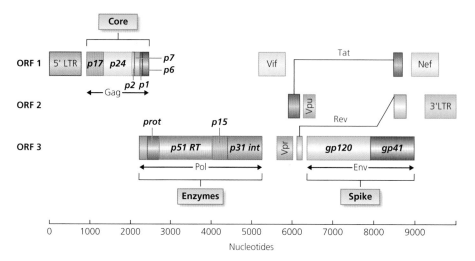

Fig. 21.2 HIV-1 genome organization. The scale indicates the genome size in nucleotides. Note that all three open reading frames (ORFs) are used, but only simultaneously in the region of nucleotide 8500. *Gag, pol,* and *env* are transcribed and translated to give polyproteins which are then cleaved by proteases to form the virion core proteins, enzymes, and spike proteins as shown. The nomenclature 'p17' indicates a polypeptide having an M_r of 17 000, etc. Genes not aligned require a frame shift for expression. Tat and Rev proteins are expressed from spliced RNAs. HIV-2 has a similar genome organization, but gene vpr is in the position occupied by vpu in HIV-1, and a distinct gene, vpx, is present in the location where HIV-1 vpr is found.

Box 21.2

Evidence for the mechanism of action of Tat protein

- Unlike other DNA sequences that upregulate transcription by binding transcription factors, the TAR element has to be positioned downstream of the transcription start site and oriented as found in HIV in order to act. This observation suggests that the TAR element might be working as RNA.
- Purified Tat protein binds the RNA form of TAR, but not to the DNA form of the sequence (as found in the DNA being transcribed), confirming that Tat works through RNA.
- Tat does not alter the rate of synthesis of HIV RNA close to the promoter but does greatly increase the rate of transcription measured at other locations further along the genome; i.e. it has no effect on RNA polymerase II initiation rate but it increases the processivity of the enzyme (how efficiently it holds onto its template and synthesizes a long RNA).
- The Tat-TAR complex binds a host cell protein kinase (a complex of Cyclin T and cdk9) that phosphorylates RNA polymerase II, so increasing its processivity.

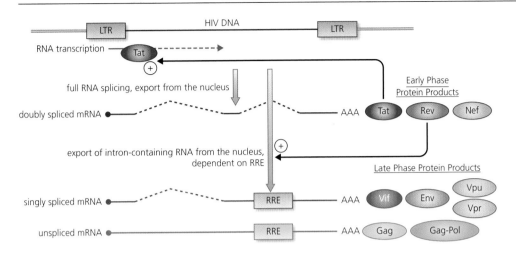

Fig. 21.3 Gene expression from HIV-1 provirus DNA integrated into the cellular genome occurs in two phases: early expression of regulatory genes and late expression of structural genes. Tat protein from the early phase causes activation of transcription. This increases the level of Rev protein to the point where it can switch cytoplasmic mRNA production from the doubly-spliced class to the singly- and unspliced classes, so permitting production of the proteins needed for progeny virions.

Box 21.3

Evidence for the mechanism of action of the Rev protein

- Comparing cytoplasmic mRNA from wild-type and Rev-defective mutant HIV infected cells, the full-length and singly-spliced mRNA that encode Gag, Gag-Pol, and Env are selectively depressed by the absence of Rev while multiply-spliced mRNAs are not affected (Fig. 21.3). Since all the mRNAs come from the same promoter, this Rev effect must be achieved post-transcriptionally.
- Rev protein binds with high affinity to a specific RNA sequence (the RRE) that is found in the Env reading frame of HIV mRNA (and hence is present only in full-length and singly-spliced mRNAs).
- The RRE can confer Rev-dependence for cytoplasmic accumulation on a heterologous mRNA produced from a suitable reporter gene construct. Thus Rev binding to RNA is required for it to enhance cytoplasmic mRNA accumulation.
- Rev protein is located in the nucleus as assessed by immunofluorescence. If two cells, one expressing Rev and the other not, are fused together to form a heterokaryon (one cell with two nuclei), and further protein synthesis is blocked with a chemical inhibitor, then Rev can be seen to accumulate in the second nucleus. Thus Rev can move from one nucleus to the other via the cytoplasm, reflecting an ability to move reversibly between nucleus and cytoplasm in a normal cell, a process known as shuttling.
- The sequence in Rev protein that is required for shuttling, the nuclear export signal (NES), is essential for Rev-dependent mRNA accumulation. Via the NES, Rev binds to components of a host cell nuclear export pathway and so brings out of the nucleus any RNAs bound to it.

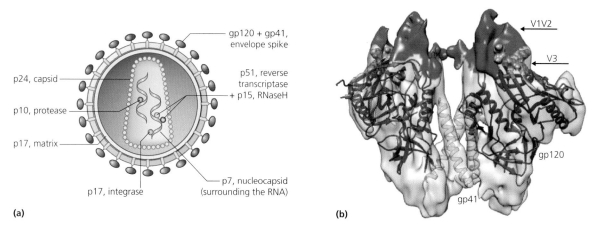

Fig. 21.4 (a) Schematic representation of the HIV-1 virion, with major proteins identified. (b) A cryo-EM structure of a soluble form of gp120:gp41 trimeric complex in which gp41 was truncated before its transmembrane anchor region and the complex stabilized by other modifications. The red ribbon secondary structure elements derive from gp120 while the turquoise elements derive from gp41. The space filling densities of the V1V2 and V3 variable loops are shown in purple and orange respectively, with the remainder of the complex in light blue. Not present in the molecule analyzed here, the gp41 sequence extends downwards from the turquoise helices via its principal transmembrane segment to a cytoplasmic/intravirion tail of around 100 residues. Some data suggest that this tail may in fact loop back through the viral envelope to expose an epitope of the tail outside the virion. (Reprinted by permission from Macmillan Publishers Ltd: Bartesaghi, A., Merk, A., Borgnia, M. J., Milne, J.L.S. and Subramaniam, S., vol 20, 1352–1357 ©2013.)

precursor polyproteins by the virion protease within assembled immature particles (Section 12.8). The particle also possesses a lipid envelope that contains approximately 72 glycoprotein spikes, each comprising a trimer of the gp120:gp41 Env protein complex. Within these spikes, gp41 provides the transmembrane core of the protein while the gp120 components are entirely outside the envelope. Gp120 interacts with the viral receptors whilst gp41 provides the membrane fusion function for virus entry (Section 6.2). The Env trimers are acquired with the lipid envelope when virus buds from the plasma membrane, so infected cells also display Env trimers on their surface. Because the association between gp120 and the gp41 core of the protein is non-covalent, gp120 is readily shed from infected cells into the circulating fluid with potential effects on immune function and

pathogenesis (Section 21.5). The lipid envelope also contains a number of host proteins that are abundant in the plasma membrane from which HIV-1 particles bud.

The Env gp120 sequence is highly variable between HIV isolates. This variation is driven by the pressure of immune selection, working in the context of the error-prone replication mechanism of virus replication that creates a quasispecies on which selection can operate (see Section 4.2). HIV is thus constantly changing, even within a single infected individual, and so continually escapes the antibody responses that are mounted against it. Comparisons show that sequence variation is concentrated particularly in a series of hypervariable regions (V1–V5) within gp120, interspersed with somewhat more conserved regions. Based on their variability, these V regions were predicted to form surface loops on the fully folded and assembled protein. The surface of the folded protein is also heavily glycosylated, which offers the virus another means of shielding potentially antigenic sites from detection. The structure of the gp120:gp41 complex has been extensively studied, both to understand how it performs its critical functions in virus entry (which are potential targets for antiviral therapy) and to try to define its antigenic characteristics for the purpose of vaccine development. There are now high-resolution structures available of gp120:gp41 complexes that have come from X-ray crystallography and cryo-electron microscopic reconstruction techniques applied to stabilized derivatives of the mature protein (Fig. 21.4b).

The viral enzymes encoded by the Pol gene, reverse transcriptase, integrase and protease, are present within the mature HIV-1 capsid. Multiple copies of the accessory proteins Vif, Vpr and Nef, are also found within the particle as well as certain cellular proteins, most notably cyclophilin A (CypA) which is a peptidyl prolyl cis-trans isomerase that has specific binding affinity for Gag CA protein. The Vpr protein is multifunctional, having several effects on the biology of the infected cell, but the most significant role of encapsidated Vpr is to facilitate the transport of the viral pre-integration complex (the residual capsid containing a DNA copy of the genome RNA formed by reverse transcription; Section 9.5) to the nucleus. It is this function that allows HIV to infect non-dividing cells; retroviruses that lack a Vpr function have to rely on the breakdown of the nuclear envelope during cell division to access the nucleus.

The HIV-1 Nef protein was originally named for a supposed negative effect on viral gene expression but is now known to have a positive effect on infection. As well as being present in the particle, it is one of the first viral proteins to be expressed in a newly-infected cell. The protein has a myristoylated N-terminus which allows it to associate with membranes. Its key functions involve the down-regulation of selected host proteins from the plasma membrane, including CD4 and MHC Class I antigens. The effect of this is to render the infected cell less susceptible to cytotoxic cell killing and, hence, to favour virus production. Nef also modulates cell signalling pathways to mimic T-cell activation, which upregulates HIV gene expression and hence virus production.

Intrinsic immunity to HIV

As explained in Section 13.5, viruses have to contend with an array of intrinsic antiviral defence molecules in the early stages of infection, even before an innate immune response is mounted by the host. At least three such factors act against HIV-1 and/or other primate lentiviruses: TRIM5α, APOBEC3G and tetherin. In turn, the viruses have evolved countermeasures to mitigate the action of these factors.

TRIM5α has specific affinity for a binding site that is formed by assembled Gag CA monomers within the capsids of primate immunodeficiency viruses and other retroviruses. Binding of TRIM5α promotes premature or inappropriate capsid disassembly and the ubiquitin ligase activity of TRIM5α may target virion components for degradation. TRIM5α is also thought to act as a pattern recognition receptor so that its binding to an HIV capsid can trigger innate immune signalling (Section 13.5). The interaction between TRIM5α and a viral capsid can be both host species-specific and virus-specific, and evidence from studying the evolution of SIV from sooty mangabeys into the pathogenic SIV that causes AIDS-like disease in laboratory macaques suggests that the potency of this interaction is a significant determinant of host range for these viruses (Box 21.4). In humans, TRIM5α has relatively low binding affinity for HIV-1 capsids, so has limited effects on infection. However,

Box 21.4

Evidence that TRIM5α is a host-range determinant for primate immunodeficiency viruses

Laboratory populations of macaques are polymorphic in their TRIM5α loci, with three prevalent alleles, designated 'Q', 'TFP' and 'CypA'. The last of these alleles is a fusion between TRIM5α and CypA genes that effectively replaces the C-terminal part of TRIM5α, where the Q/TFP polymorphism is found, with a CypA coding sequence.

Infection of macaques with a strain of SIV from sooty mangabeys (SIV$_{sm}$) gave rise to 'set point' viral titres that were strongly correlated with the TRIM5α genotype of individual animals: the most susceptible animals were Q/Q homozygotes and the least susceptible were TFP/CypA heterozygotes (no CypA homozygotes were present in the tested population).

During the course of infection in highly-resistant animals, a virus variant emerged with a mutation in the Gag CA protein at the site of interaction with TRIM5α (the CypA binding loop in HIV-1 Gag CA), that was resistant to TRIM5α inhibition, and this correlated with an increase in virus load.

SIV$_{mac}$ is a pathogenic isolate derived from SIV$_{sm}$ by repeated passage in macaques in the laboratory. Reverting the sequence of SIV$_{mac}$ Gag CA in the CypA/TRIM5α binding loop to match the sequence found in the ancestral SIV$_{sm}$ rendered the virus sensitive again to the TFP and CypA alleles of TRIM5α that are also restrictive for SIV$_{sm}$.

Data from Kirmaier, A. et al. *PLoS Biology* 8: e1000462 (2010).

removal of TRIM5α from cells in culture does modestly increase virus production and there are polymorphisms in human TRIM5α that have been linked with increased susceptibility to HIV-1 infection and/or accelerated rates of disease progression, both of which suggest that there is some restrictive effect of TRIM5α on HIV-1 in at least some people. One reason why human TRIM5α has only limited effects on HIV-1 may be related to the ability of HIV-1 capsids to bind host CypA protein via the same loop of the Gag CA protein to which TRIM5α binds.

Vif (virion infectivity factor) protects the HIV genome from the action of APOBEC3G, a host protein that is acquired from the cell which produced the virion. APOBEC3G is a member of a family of cytidine deaminases, enzymes that convert C residues in RNA into U residues; their effect on an RNA virus genome is to cause

hypermutation that destroys its capacity to encode functional proteins. Vif blocks this activity by targeting APOBEC3G for ubiquitin-mediated proteasomal degradation. There is also evidence that APOBEC proteins exert a repressive effect on HIV-1 replication independent of their enzymatic activities. The action of Vif was first revealed in experiments that demonstrated that Vif-defective genomes formed infectious HIV virions normally but that those virions showed dramatic differences in their ability to infect different cell lines, all of which were equally infectable by wild-type HIV. From this it was inferred that Vif was needed to counteract a host factor, now known to be APOBEC3G, which was present in some cell lines and not others.

Finally, tetherin has been recognized as a potent repressive factor for HIV-1 replication. It

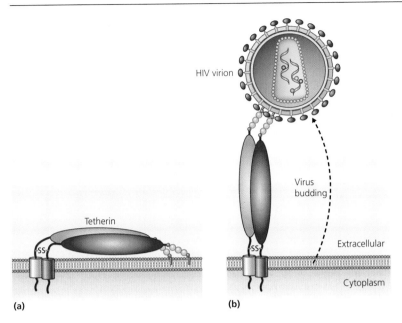

Fig. 21.5 Schematic representation of the topology of tetherin protein dimers in (a) an uninfected cell and (b) an HIV-infected cell, in which it prevents virus release by anchoring exiting virions to the producer cell. In HIV-1 infection, this effect is opposed by the action of the viral Vpu protein to allow more efficient spread of virus between cells.

is a cell-surface protein that, in addition to being anchored in the plasma membrane by a hydrophobic membrane-spanning region, also has a second membrane anchor in the form of a GPI (glycosyl phosphatidyl inositol) modification at the C terminus of its extracellular domain; it also dimerizes via its extracellular domains which become linked by disulphide bonds (Fig. 21.5a). Possession of a GPI anchor causes the protein to locate in lipid rafts within the plasma membrane, which are known to be sites for budding of HIV and several other enveloped viruses. Thus, when HIV particles exit the cell, they carry with them the tetherin GPI anchor and so remain attached to the cell by a tetherin bridge (Fig. 21.5b). This antiviral mechanism has no molecular specificity for HIV, so it acts on several types of enveloped viruses; consequently they have evolved trans-acting functions to counteract tetherin action. In HIV-1 infection, the accessory protein Vpu performs this role: it associates with tetherin in the membrane and recruits a cellular ubiquitin ligase, βTrCP (which ordinarily functions in innate immune signalling, Chapter 13), to cause tetherin degradation. HIV-2 does not encode Vpu; instead, tetherin downregulation is achieved by an action of the Env proteins. In a further variation, the various SIVs use their Nef protein to downregulate non-human primate tetherin: Nef cannot act on human tetherin because of sequence difference between it and the tetherin of other primates. Thus, the acquisition of novel tetherin-regulatory functions in HIV can be seen as necessary adaptations that were selected when the viruses transferred to a human host.

21.3 HIV transmission and tropism

> *HIV is transmitted most commonly by sexual activity but can also be introduced into the bloodstream by contaminated needles, etc. or acquired at birth from an infected mother. The virus principally infects a subset of T lymphocytes that carry the CD4 molecule on their surface.*

Routes of HIV transmission

HIV infection can be acquired by three main routes (Chapter 18). The first of these is through the transmission of infected CD4$^+$ T lymphocytes or free virus during sexual activity. In Africa, HIV is spread predominantly by heterosexual contact. In Western countries, male homosexual contact was responsible for the majority of infections early in the pandemic, but now the greatest percentage increase in new infections is due to heterosexual activity. Bisexual activity provides a conduit between hetero- and homosexual people. HIV is spread equally well by heterosexual and by male homosexual activity, but is better spread by infected males than infected females.

Injection using hypodermic needles contaminated with HIV can also cause infection. The virus spreads quickly between injecting drug users who share unsterile equipment, and there is a similar risk with clinical practitioners who do not have effective sterilization available for hypodermic syringes and needles. Before the virus was recognized, some haemophiliacs were accidentally infected by being given clotting factor VIII prepared from blood contaminated with HIV. Heat sterilization and screening of blood donors have now eliminated this risk. Finally, without intervention, 15–25% of babies born to HIV-positive mothers may be infected before or during delivery; this risk is substantially reduced by giving the mother antiviral therapy to reduce her viral load before labour (see Section 27.4). Breastfeeding increases the risk of infection of the baby by a further 15%. However, breastfeeding also confers significant advantages to infants, especially in less economically developed countries. The current advice from UNICEF is for mothers to avoid breastfeeding only if a safe and acceptable alternative is available.

What cell types are infected?

CD4, the primary HIV receptor that is bound by gp120, is a protein present on the surface of helper T cells and cells of the monocyte–macrophage lineage. Its normal function is to recognize and interact with major histocompatibility complex (MHC) class II antigens on target cells. Successful HIV infection also requires the presence of the chemokine receptors, CCR5 or CXCR4, which have been hijacked by the virus as co-receptors.

HIV-1 isolates have been classified as M-tropic or T-tropic based on differences in their cell targets. M-tropic viruses infect both CD4$^+$ macrophages and T cells, and use the CCR5 co-receptor: for this reason they are now more commonly referred to as R5 viruses. T-tropic viruses preferentially infect CD4$^+$ T cells and use the CXCR4 co-receptor and are known as X4 viruses. There are also dual-tropic HIV-1 isolates known as R5X4 viruses. These differences in tropism depend on the gp120 sequence. Most infections take place through the mucosal surface of the genital tract. Virus binds to the surface of a dendritic cell through the surface protein, DC-SIGN, a C-type lectin that specifically binds carbohydrate moieties. HIV-1 may be taken up into an intracellular vesicle, but does not enter the cytoplasm or cause an infection. The dendritic cell then acts as a transporter, carrying the virus to lymph nodes where the virus is transferred to, and infects, a CD4$^+$ T cell. CD4$^+$ macrophages can also be infected but these are found infrequently in lymph nodes and blood.

Although CD4$^+$ T cells are the main target of HIV, other cells that do not express the CD4 protein may also be infected, which suggests that here the virus can use a different receptor molecule. In the central nervous system, it is thought that HIV-1 infects microglial cells, which belong to the same cell lineage as the macrophage. Infection via the mucosal surface of the rectum, resulting from anal intercourse, may also take place through dendritic cells. Unlike the simple retroviruses, HIV-1 can infect resting cells as well as dividing cells, in this case resting CD4$^+$ T cells, as its proviral DNA is transported into the nucleus through associated viral proteins that carry nuclear localization signals. There it is integrated into the genome, but little transcription takes place until the cell is activated by antigen.

Tropism and transmission

The differences in cell tropism of HIV-1 R5 and X4 strains lead to differences in the viruses predominating early and late in the course of HIV infection, when acquired by a mucosal route, where the ability to infect macrophages is an important step in the infectious process. In such situations, R5 strains predominate early in the course of infection. However, the high mutation rate of HIV-1 establishes a quasispecies in which variants emerge that can have X4 tropism. This evolution of the virus has consequences for pathogenesis: X4 strains have higher replication rates and are more pathogenic, their emergence in a patient being associated with an increased rate of decline in CD4$^+$ T cell numbers and more rapid disease progression.

21.4 Course of HIV infection: pathogenesis and disease

HIV-1 infection progressively destroys the helper T cells that are crucial to adaptive immunity. In most people, unless the infection is treated, it progresses largely without symptoms over several years before overt signs of immune deficiency signal the progression to AIDS. However, a subset of people appears able to control the infection naturally, without ongoing antiviral therapy, and progress to AIDS very slowly if at all.

The progression of HIV-1 infection has been most studied in people from the developed world; although infection in developing countries is essentially similar, the disease course may be hastened by other factors such as nutritional status and burden of other infections. The course of HIV-1 infection in a typical untreated adolescent or adult is detailed in Box 21.5. Primary infection is characterized by a brief influenza-like illness and the ensuing persistent infection progresses from category A eventually to full-blown AIDS (category C). These stages are formally defined by the concentration of CD4$^+$ T cells in the circulation; the onset of AIDS is defined by CD4$^+$ T cell concentrations dropping below 200 cells per μl of plasma. An example of the progressive decline in CD4$^+$ T cells in one case study is shown in Fig. 21.6. Although the average incubation period from primary infection to the appearance of symptoms of immunodeficiency is 10 years without treatment, some people will develop disease earlier and some will remain healthy for much longer. The course of infection in babies is much more rapid, death typically ensuing in around two years in the absence of treatment.

Box 21.5

Classification of stages in the progression of HIV infection to AIDS in a typical adult without antiviral chemotherapy

Initially, HIV-1 causes a short, self-limiting flu-like illness that starts a few weeks after infection and lasts only a few days; there is then a period lasting some years of high levels of virus multiplication. Eventually, immunosuppression commences. The stages of infection can be categorized as follows (an individual can be a source of infection for others at all stages):

- Category A: Ranges from primary HIV infection to asymptomatic subclinical infection, to persistent generalized lymphadenopathy (swollen lymph nodes), which indicates that immune suppression has started and is progressing. Without treatment, this stage lasts on average 10 years. Formally defined as having >500 CD4$^+$ T cells/µl of blood (normally 800–1200 cells/µl).
- Category B: Symptomatic people presenting with a selection of conditions not found in category A. These include weight loss, opportunistic infections (e.g. candidiasis, fever, diarrhoea lasting more than one month, and more than one episode of shingles). These conditions suggest a defect in cell-mediated immunity. Most category B people have 200–500 CD4$^+$ T cells/µl blood.
- Category C: Formal diagnosis of AIDS; includes people with more severe infections or cancers. Ranges from mild with constant infections to severe with severe infections, major weight loss, myopathy, peripheral nervous system disease and central nervous system disease (dementia). Most category C people have <200 CD4$^+$ T cells/µl blood.

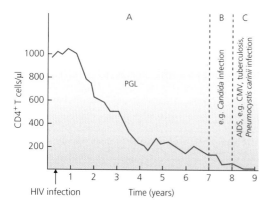

Fig. 21.6 A case-study of AIDS, which ended in death in year 9. The CD4$^+$ T cell loss was exceptionally rapid. This case occurred before anti-HIV therapy was available. PGL, persistent generalized lymphadenopathy; CMV, cytomegalovirus.

HIV pathogenic course in detail

During the initial/primary infection by HIV-1, virus-specific CD4$^+$ T cells are stimulated to proliferate by viral antigens. As virus infects and replicates in activated CD4$^+$ cells, these same cells are preferentially destroyed. At the same time, there is a dramatic expansion of virus-specific CD8$^+$ T cells which coincides with the suppression of viraemia and a recovery in CD4$^+$ T cell numbers. However, CD8$^+$ T cell proliferation is dependent on CD4$^+$ T cell help. Thus there is a fine balance between virus destroying CD4$^+$ T cells and leaving enough CD4$^+$ T cells to help produce virus-specific activated CD8$^+$ cells. It is suggested that this balance determines the plasma virus load (concentration) at the end of the acute phase,

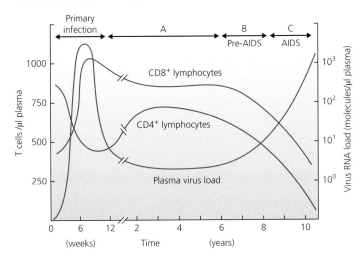

Fig. 21.7 The course of HIV-1 infection, the inverse relationship between infectious virus load and overt disease with T cell concentrations in plasma. A, B, C refer to the categories of infection (see Box 21.5 and text).

also known as the 'set point'. The set point load appears to be a critical determinant of the rate of progression to AIDS; a low set point load means a longer subclinical period, and a high load means rapid progression to AIDS. However, virus loads ultimately rise again as immune function collapses (Fig. 21.7).

The key failure in immune responsiveness underlying AIDS is the loss of the helper function of the CD4+ lymphocytes, which are central to immune responses (Chapter 14). One of the functions of T cells is to control a variety of commensal micro-organisms, a selection of which each of us carries. The immune system is normally unable to evict these micro-organisms but its continuous action ensures they remain subclinical and harmless. However, when T cell function is compromised by HIV-1 infection, these micro-organisms can multiply unchecked and cause chronic disease. People in different parts of the world carry different micro-organisms, so AIDS patients suffer different diseases. For example, *Mycobacterium tuberculosis* is commonly carried in some developing countries. Declining immune function also leaves HIV+ patients vulnerable to more severe

outcomes of infection by external pathogens. If an AIDS patient does not die from such infections, HIV-1 progresses to infect CD4-negative cells and causes disease in the muscles and the peripheral and central nervous systems. 'AIDS dementia' or collapse of brain function is the final stage in this process.

During the intermediate stages of infection, there is relatively little virus and few infected cells in the plasma. Initially, this led to the mistaken view that there was little virus replication during the asymptomatic period. Later, it was discovered that most of the infected cells at this time were to be found in lymphoid tissue, and that although large amounts of virus were being continuously released into the circulation, this was balanced by its efficient removal. Indeed, it is now estimated that an asymptomatic infected person produces approximately 10^{10} HIV-1 virions every day, a colossal total. In addition, there is a small, but vitally important, reservoir of virus in latently-infected resting CD4+ memory T cells (10^3–10^4/person), and possibly other reservoirs, that cannot be shifted by drug therapy because the available drugs all target aspects of the active replication process (see Section 27.4). Latently infected cells that produce no viral proteins on the cell surface are also invisible to the immune system and, because the viral genome is integrated, can divide and produce daughter cells that are also infected.

Long-term non-progressors: natural control of HIV infection

It has been clear for many years that there exists a subset of HIV-1 infected people whose disease course does not match the description above. They maintain low viral loads and good CD4$^+$ cell counts for very extended periods, possibly indefinitely, without any antiviral therapy. These people are known as *long term non-progressors* (LTNP), sometimes also called *viraemic controllers*; they represent around 2–3% of the total infected population. Among LTNPs there is also a small subset whose long-term viral load is so low as to be almost undetectable; they are known as *elite controllers* (EC). These particular patient groups have been carefully studied to try to determine the basis for their benign disease outcome, which resembles the typical course of infection with the intrinsically less pathogenic virus, HIV-2.

There is no single factor that accounts for the unusual course of HIV-1 infection in LTNPs and ECs. Instead, and as might perhaps have been expected, various combinations of viral and host factors contribute to this outcome in different patients. From the virus perspective, infection by a relatively attenuated strain can give rise to LTNP status; a number of different attenuating mutations have been identified in naturally-occurring HIV-1 isolates including a deletion in the Nef gene that prevents the virus from downregulating adaptive immune responses. One significant host factor is an allele of the CCR5 co-receptor that has a deletion of 32 amino acids (CCR5Δ32). This allele is present particularly in Caucasian populations,

in which about 1% are CCR5Δ32 homozygotes who are thus completely resistant to infection by R5 strains because they lack the necessary co-receptor for HIV-1 entry. Of course, a much greater proportion of these populations (9–20%) are heterozygotes and there is considerable evidence that this genotype correlates with LTNP status. Unfortunately, the CCR5Δ32 allele is virtually unknown in native African, East Asians and native American groups. Finally, possession of particular MHC class I alleles, such as B57, is strongly linked with LTNP status. As discussed in Chapter 14, MHC class I antigens are responsible for selectively presenting particular antigenic peptides from invading pathogens to elicit an adaptive response, each allele selecting different peptides for presentation. Thus it can be inferred that B57 and other HLA alleles associated with LTNP status must elicit particularly powerful or favourable immune responses for the control of HIV-1 replication.

21.5 Immunological abnormalities during HIV infection

While loss of CD4$^+$ T cells is the major consequence for the immune system of HIV infection, there are many other alterations to immune cells and functions that can be observed, some of which are listed in Box 21.6. These immunological abnormalities during HIV infection result from one or more of the following causes:

- The virus kills the infected cell, and in so doing removes cells that are important for immune function.
- Envelope protein expressed on the surface of an infected cell attaches to CD4 molecules on non-infected CD4$^+$ T cells and causes the cell membranes to fuse and form a single cell

Box 21.6

Some immunological abnormalities associated with AIDS

Diagnostic abnormalities[*]
CD4$^+$ T cell deficiency
Reduction in levels of all lymphocytes
Lowered cutaneous delayed-type hypersensitivity (a T cell-mediated response)
Non-specific elevation of immunoglobulin concentration in serum

Other abnormalities
Decreased proliferative responses to antigen
Decreased cytotoxic responses to all antigens
Decreased response to new immunogens
Decreased CD8$^+$ T cell cytotoxic response to HIV
Decreased macrophage functions
Production of autoantibody
Decreased NK cell activity
Decreased dendritic cell number and activity
Loss of lymph node structure

[*]Diagnostic abnormalities are always found in AIDS patients; other abnormalities are found irregularly, i.e. they vary between individuals.

(a syncytium). This is a lethal process and fused cells die.

- Infected cells suffer no direct cytopathology but viral proteins expressed on the cell surface are recognized by antibody, and the cells are then attacked by complement or phagocytic cells bearing Fc receptors (see Chapter 14).
- Infected cells proteolytically process viral proteins and present peptide-MHC I or II protein complexes on their surface. These cells are then attacked by CD8$^+$ or CD4$^+$ cytotoxic T cells (see Chapter 14).
- Non-infectious virus or free envelope protein which is released from infected cells can attach to non-infected-CD4$^+$ cells and render those cells liable to immune attack, either by antibody or, if the antigen is processed and presented by MHC proteins, by cytotoxic T cells.

21.6 Prevention and control of HIV infection

Despite huge investment and numerous clinical trials, there is no effective vaccine to protect against HIV infection. However, advances in antiviral therapy now slow progression to AIDS in infected people to the point where it is possible to consider living with HIV indefinitely. These therapies are expensive, which makes effective treatment in some of the worst affected areas of the world very challenging.

Vaccination

No other virus has been subjected to such intense scrutiny as HIV but, as yet, there is no sign of any effective vaccine. At first sight, this is surprising, since the smallpox vaccine has been in use for over 200 years and in the last 30 years some very effective vaccines have been devised and used against other viruses. The problem with HIV is complex. Unlike measles for example, HIV-1 infection does not result in an immune response that can eliminate the virus. So a vaccine for HIV-1 must do much more than reproduce the normal anti-HIV-1 response. Also, the antigenic character of the viral Env proteins, the principal target of a neutralizing antibody response, is constantly changing which makes it very difficult to protect people with a single vaccine. However, some broadly cross-reactive neutralizing monoclonal antibodies have been obtained under laboratory conditions, giving hope that it might be possible to trick the immune system into producing these in response to a vaccine immunogen; there is also interest in T-cell vaccines. A variety of vaccine preparations have been tested in human trials, with some limited evidence of efficacy, but none has yet given a robust indication of long-term protection. We need to understand more about the immune system in general and about the response to HIV in particular before the rational design of an effective HIV vaccine becomes a reality. In this context, understanding the precise character of the immune response in those patients with long-term non-progressor or elite controller status is a very promising avenue.

Antiviral therapy

To be useful as an antiviral drug, a compound must target a viral protein or function that is unique to the virus, i.e. is not present in the uninfected host, so as to minimize the side effects of treatment. The first HIV target against which useful drugs were obtained was the reverse transcriptase, with azidothymidine (AZT, a nucleoside analogue) being the first

anti-HIV therapy to be approved in 1987. Subsequently, the viral protease and integrase enzyme, and the HIV fusion/entry mechanism have also been exploited. The nature of these drugs is considered in more detail in Chapter 27. When AZT was launched, its effect on patients' CD4$^+$ T cell counts was dramatic and there were high hopes of a viable treatment. However, these benefits were short-lived and it quickly became clear that the high rate of virus mutation seen within a patient during HIV infection allowed the virus rapidly to develop resistance to the therapy. Thus, the subsequent development of drugs active against multiple distinct targets in the virus life cycle was crucial. By giving drugs in combination, it proved much more difficult for the virus to evolve to resistance.

The current standard treatment for HIV infections is known as HAART (highly active anti-retrovirus therapy). This involves taking a combination of usually three different compounds with inhibitory activity against the virus, that are chosen to target at least two distinct features of the virus replication cycle simultaneously. This form of therapy commenced in 1994, and under the best conditions can reduce plasma virus load to undetectable levels and restore the CD4$^+$ T cell population and immune function. However, it does not clear virus from resting CD4$^+$ memory cells (and possibly other reservoirs) in which the virus is latent, and even a few days without taking the antiviral drugs sees a rapid rebound in plasma virus levels. What is worse, rebound virus may be drug-resistant and can then be spread to others in the population, further diminishing the effectiveness of the available therapies. Thus it is crucial that HIV therapy is taken continuously in accordance with the required dose regime. Used correctly, these therapies can give a prolonged delay in, or possibly even prevent, the progression of infection to AIDS although they cannot clear the infection completely.

Although the effectiveness of current triple therapies puts an HIV-positive person today in a very much better position than those who became infected early in the pandemic, there are still concerns. A patient will have to be treated for the rest of their life – say 60 years, and it is not known if the drug regimen can be tolerated for this time or if resistant virus will eventually emerge. Also, triple therapy is very expensive, which presents major challenges for treating the majority of infected people who live in less economically developed countries. However, concerted action by manufacturers, governments and the charitable sector has improved the position substantially since 2000 such that perhaps half of those who might benefit from HIV triple drug therapy globally now receive it.

Successful therapy for HIV infection is closely linked with being able first to diagnose the infection and then to monitor the virus load in a patient over time. As noted above, triple therapy has the goal of reducing circulating virus to undetectable levels. Once this has been achieved, any upswing in virus load is an indicator that resistance to therapy is emerging and that the drug combination needs to be changed. For initial diagnosis, possession of serum antibody to the HIV internal antigen, Gag p24, is the main criterion, since this protein provokes the strongest and most reliable antibody response. This can be measured by an ELISA in various formats. A positive result is then followed up with quantitative PCR testing to monitor the load of viral RNA in the circulation. Such techniques are described in Chapter 5.

Prevention of HIV infection

In the absence of any vaccine or curative therapy, avoiding exposure to HIV is paramount. The main route for transmission, already discussed, is through sexual contact. Education programmes have been promulgated worldwide, with advice to have as few sexual partners as possible, and to practise safer sex (whether heterosexual or homosexual) by avoiding the exchange of body fluids and by using condoms as a barrier to infection. The possibility of using topical (largely vaginal) virucides to kill virus is being actively explored. However, there are immense social problems in dealing with any sexually transmitted disease, and these vary around the world according to social and religious conventions. Historical parallels are not encouraging, as it is well documented that many people were undeterred by the risk of contracting syphilis at a time when no treatment was available. The involvement of male and female prostitutes is a particular problem, as, in some areas of the world, a high proportion are infected with HIV.

Key points

- HIV-1 and HIV-2 are typical members of the Lentivirus genus of the Retrovirus family.
- HIV-1 and HIV-2 are newly-emerged human pathogens, which transferred to humans as zoonotic infections from other primate species.
- HIV is primarily a sexually transmitted virus.
- HIV infects cells that carry the CD4 protein on their surface, notably CD4$^+$ T cells which play a central role in many immune responses.
- HIV rarely causes disease directly, but over many years leads to a decline in the immune system (AIDS) that allows a variety of micro-organisms to cause serious infections; it is these that eventually cause death.
- The time to AIDS is proportional to the virus plasma load after the initial acute infection.
- Antiviral drugs are available that inhibit virus multiplication; they act by reducing virus load and can prevent progression to AIDS if used properly.

- No HIV antigen preparation yet tested stimulates effective neutralizing antibody or antiviral T cells; hence there is no HIV vaccine.

Questions

- Describe the origins of HIV-1 and discuss the factors that have led to its emergence as the cause of a global pandemic.
- Explain how HIV-1 infection leads to the development of AIDS.
- Discuss the roles of the accessory proteins of HIV-1 in the virus life cycle.

Further reading

Abraham, L., Fackler, O. T. 2011. HIV-1 Nef: a multifaceted modulator of T cell receptor signalling. *Cell Communication and Signaling* 10, 39. doi: 10.1186/1478-811X-10-39.

Arhel, N. J., Kirchhoff, F. 2009. Implications of Nef: host cell interactions in viral persistence and progression to AIDS. *Current Topics in Microbiology and Immunology* 339, 147–175.

Arias, J. F., Iwabu, Y., Tokunaga, K. 2011. Structural basis for the antiviral activity of BST-2/tetherin and its viral antagonism. *Frontiers in Microbiology* 2, 250. doi: 10.3389/fmicb.2011.00250.

Barre-Sinoussi, F., Chermann, J. C., Rey, F. et al. 1983. Isolation of a T-lymphotropic retrovirus from a patient at risk for acquired immune-deficiency syndrome (AIDS). *Science* 220, 868–871.

Chiodi, F., Weiss, R. A. 2014. Human immunodeficiency virus antibodies and the vaccine problem. *Journal of Internal Medicine* 275, 444–455.

Chowdhury, A., Silvestri, G. 2013. Host-pathogen interaction in HIV infection. *Current Opinion in Immunology* 25, 463–469.

Desimmie, B. A., Delviks-Frankenberry, K. A., Burdick, R. C. et al. 2014. Multiple APOBEC3 restriction factors for HIV-1 and one Vif to rule them all. *Journal of Molecular Biology* 426, 1220–1245.

Este, J. A., Cihlar, T. 2010. Current status and challenges of antiretroviral research and therapy. *Antiviral Research* 85, 25–33.

Hemelaar, J. 2012. The origin and diversity of the HIV-1 pandemic. *Trends in Molecular Medicine* 18, 182–192.

Kirchhoff, F. 2010. Immune evasion and counteraction of restriction factors by HIV-1 and other primate lentiviruses. *Cell Host and Microbe* 8, 55–67.

Lenasi, T., Contreras, X., Peterlin, B. M., 2010. Transcription, splicing and transport of retroviral mRNA. Chapter 6, pp. 161–185, in *Retroviruses: Molecular Biology, Genomics and Pathogenesis*, Kurth, R. and Bannert, N., eds., Caister Academic Press, Norfolk, UK.

Poropatich, K., Sullivan, D. J. Jr. 2011. Human immunodeficiency virus type 1 long-term non-progressors: the viral, genetic and immunological basis for disease non-progression. *Journal of General Virology* 92, 247–268.

Rambaut, A., Posada, D., Crandall, K. A., Holmes, E. C., 2004. The causes and consequences of HIV evolution. *Nature Reviews Genetics* 5, 52–61.

Ruelas, D. S., Green, W. C. 2013. An integrated overview of HIV-1 latency. *Cell* 155, 519–529.

Trono, D., van Lint, C., Rouzioux, C. et al. 2010. HIV persistence and the prospect of long-term drug-free remissions for HIV-infected individuals. *Science* 329, 174–180.

Chapter 22
Viral hepatitis

Viral hepatitis is a global health problem, the current scale of which is second only to that posed by HIV. This problem is associated particularly with chronic infections by hepatitis B and C viruses, which give rise to a huge burden of late onset liver disease.

Chapter 22 Outline

Hepatitis, meaning inflammation of the liver, is a major cause of morbidity in human populations. Although there are non-viral causes of hepatitis, viral infections represent the major cause of this disease. As with infections of other organs, there is a number of unrelated viruses that can infect the liver to cause hepatitis. This is emphasized by the fact that these viruses have been named in series, based on their tissue tropism. Thus there are hepatitis viruses A, B, C, D and E, each of which is classified in a different virus family. Importantly, these can be divided into two groups, based on whether or not they can cause chronic infections (see Section 16.5). Those viruses that can do this routinely (hepatitis B and C viruses) are particularly associated with

chronic or late-stage liver disease and with the emergence of liver cancer (see Section 25.9). This chapter considers each of these viruses in turn, focusing on their disease associations and features of their epidemiology.

22.1 The signs and symptoms of hepatitis

The most widely-recognized signs and symptoms of hepatitis are fatigue and jaundice, which result from impairment of liver function, particularly bile production and blood detoxification. However, acute infections with hepatitis viruses are frequently subclinical or produce only non-specific signs and symptoms.

The liver is the largest organ in the human body, weighing around 1.5 kg in an adult. Its primary role is as part of the digestive system, since nutrient-enriched blood flows from the intestine to the liver via the portal vein where that nutrient is assimilated (Fig. 22.1). It also supports the digestion and

Introduction to Modern Virology, Seventh Edition. N. J. Dimmock, A. J. Easton and K. N. Leppard.
© 2016 John Wiley & Sons, Ltd. Published 2016 by John Wiley & Sons, Ltd.

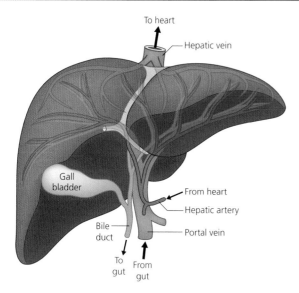

To heart

Hepatic vein

Gall bladder

From heart

Hepatic artery

Bile duct

Portal vein

To gut

From gut

Fig. 22.1 A representation of the flows of blood and bile to and from the human liver. Oxygenated blood arrives via the hepatic artery while the bulk of blood flow into the liver comes from the gut via the portal vein, supplying nutrients for metabolism. Blood returns to the circulation via the hepatic vein. Bile is produced in the liver, stored in the gall bladder and delivered to the gut via the bile duct.

absorption of fats within the gut by synthesizing bile, which is then stored in the gall bladder and released into the gut via the bile duct. The liver is one of the main sites of glycogen storage and so is critical in energy homeostasis, maintaining blood glucose levels within a narrow range by taking up or releasing glucose in response to hormonal signals. It is also a major site of biosynthesis for blood proteins, including albumin and clotting factors. Finally, the liver is a major site of blood detoxification. Aged red blood cells are being continually destroyed in the spleen and this leads to the release of haem breakdown products, bilirubin and biliverdin, into the blood. The liver removes these,

ultimately delivering them into the gut via the bile for excretion. The liver also removes other toxins, such as alcohol, that may be ingested.

The functions of the liver are carried out by hepatocytes, which are arranged in sheets over which the blood flows. The tissue is organized into modules known as portal triads, each of which has branches of the portal vein, hepatic artery and bile duct at its centre. Blood then flows outwards towards the three vertices of the triad, at each of which is a branch of the hepatic vein by which blood returns to the heart. The modular organization allows the liver to regenerate after damage – normal liver mass is restored within a few weeks even after removal of the majority of the tissue. This ability to regenerate is integral to the way viral hepatitis manifests itself.

During hepatitis, caused by virus infection or otherwise, some hepatocytes are lost. For all of the hepatitis viruses, pathology is immune-mediated rather than through direct cytopathogenesis caused by virus replication. As a result, there is a diminution in all liver functions that is broadly speaking proportional to the scale of that loss. For as long as the infection or other inflammatory source persists, hepatocyte loss will continue, counterbalanced by liver regeneration. Once the source is eliminated, then both tissue and function will be restored. Following virus infection, the appearance or not of any overt signs and symptoms of the infection will depend on the extent of the damage. Often, symptoms will extend only to mild fever or malaise, typical non-specific features of many infections (see Section 19.4). However, in more severe cases, the damage affects the production of bile to the point where haem degradation products build up in the blood to toxic levels. These then permeate other body tissues where they colour the skin and whites of the eyes yellow, the characteristic of jaundice.

22.2 Hepatitis A virus infections

Hepatitis A virus is transmitted by the faecal–oral route and is prevalent in areas with inadequate sanitation. Infection rarely results in disease in young children but acute signs and symptoms are more common in infected adults. Infections are normally resolved by the immune response, leading to lifelong immunity to the virus. A very effective vaccine is available.

Hepatitis A virus (HAV), a picornavirus, is transmitted by the faecal–oral route. As such, the risk of infection is greatest in areas with poor sanitation where, consequently, significant numbers will be infected and excreting virus at any given time. HAV causes acute infections of the liver that are normally cleared rather than becoming chronic. Hepatitis, when it occurs, arises a month or more after infection. This is a long period in comparison with viral infections generally, though it is actually a shorter incubation time than for other hepatitis viruses. The virus is a close relative of the enteroviruses which infect the gut, and HAV can sometimes cause gastrointestinal symptoms that generally occur before hepatitis is seen. However, it is not certain whether the virus actually replicates initially in the gut or whether it moves directly to the liver as the primary site of replication. Once it has reached the liver, HAV replicates in hepatocytes and is shed into the bile and hence back into the gut, from which it is excreted for onward transmission.

In common with other viruses causing hepatitis, HAV replication does not appear to kill hepatocytes; instead, the damage to the liver that causes disease is mediated by the host immune response, particularly the cytotoxic T cell response (Box 22.1; Section 14.2). In most adults, HAV infection is self-limiting and virus is eventually cleared with resulting immunity to re-infection (Fig. 22.2). There is no specific therapy; patients are advised to avoid other toxic insults to the liver (e.g. alcohol) to aid recovery. In a small number of people (less than 1%), rapid and extreme tissue destruction occurs, termed *fulminant hepatitis*. This results in liver failure which is fatal unless a transplant is available.

Only a proportion of HAV infections result in clinically apparent hepatitis. Interestingly, this proportion is linked to the age at infection. Thus, in young children (most of whom actually become infected very early in life), fewer than 10% of HAV infections result in hepatitis whilst 70–80% of adults develop

---

Box 22.1

Evidence for immune-mediated pathogenesis in hepatitis A virus infections

- The phase of infection in which virus is shed from the liver largely precedes the onset of symptoms of hepatitis (a manifestation of liver cell death).
- HAV infections in cell culture are largely non-cytopathic.
- Cytotoxic T-cells cloned from liver biopsies obtained from patients with active HAV disease can cause lysis of infected cells.

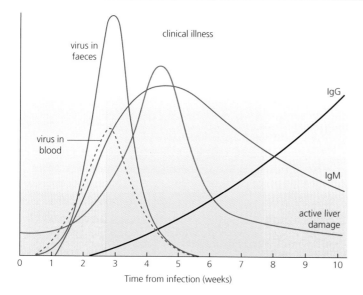

Fig. 22.2 A schematic representation of the course of a typical uncomplicated HAV infection in an adult. Following infection, virus reaches high titres in faeces and is also detectable in the blood (viraemia). The onset of clinical illness coincides with the decline of virus titres and the detection of high titres of HAV-specific antibody, first IgM and later IgG. Active liver damage is measured by the presence in the blood of liver enzymes such as alanine amino-transferase (ALT) and aspartate amino-transferase (AST).

disease. This difference may reflect the beneficial effect of maternal antibody in protecting against disease following HAV infection acquired in infancy. Alternatively, an immature immune system may not respond to HAV infection in the same way as that of an adult.

One consequence of the difference in disease incidence between age groups is that, in countries where HAV is endemic because of poor sanitation, the burden of HAV disease is paradoxically low because the average age at infection is low. HAV-related disease is also rare in highly developed countries because most people do not encounter the virus; in such populations, HAV can cause outbreaks of disease. In between these extremes, as attempts are made to provide clean water and better sewage disposal in poorer countries, so the average age at infection rises and so does the incidence of HAV disease. There is a single serotype of HAV and no animal reservoir; humans are the only species it is known to infect. As a result, it was possible to produce an effective killed vaccine for HAV, which is recommended for people travelling to endemic areas from places where infection is rare.

22.3 Hepatitis E virus infections

Hepatitis E virus (HEV) was only discovered in 1990. First classified as a Calicivirus but now allocated its own family (Hepevirus), it was defined through strenuous efforts to find a viral cause in cases of acute hepatitis that did not produce antibodies to HAV. HEV infections are, like HAV, normally acute and self-limiting with little evidence of long-term persistence except in immunosuppressed patients. Symptomatic infection is most common in adults whilst infection in younger children is more likely to be asymptomatic. Generally, there is the same low risk (about 1%) of developing fulminant hepatitis with HEV as for HAV infections. However, pregnant women seem to be at much greater risk than the rest of the population for such a severe outcome. Estimates obtained from several HEV outbreaks suggest around 20% mortality among women infected in the third trimester of pregnancy. As yet, there is no explanation for this pronounced difference in pathogenicity.

Like HAV, HEV is transmitted by the faecal–oral route and so it might be expected that rates of seropositivity for HEV and the incidence of HEV disease would mirror those for HAV but this is not always the case. There is good concordance for HEV and HAV seropositivity in various age groups in Egypt, with both viruses being acquired early in life; HEV outbreaks are therefore rare. However, more typically, rates of HEV seropositivity are markedly lower than for HAV in children and rise only later in life;

consequently, there are outbreaks of HEV disease in countries such as India. Finally, around 20% of the population of the USA is seropositive for HEV whilst HEV disease there is virtually unknown, in contrast to HAV which caused significant outbreaks prior to vaccination. Although not fully explained, these differences are likely to be related to the existence of multiple antigenically cross-reactive genotypes of HEV that have different properties.

HEV genotypes 1 and 2, which apparently infect only humans and are pathogenic, are present in Africa, the Middle East and Asia while genotypes 3 and 4 infect humans in North and South America, Europe, China, southeast Asia and Australasia (Fig. 22.3). Genotypes 3 and 4 also infect pigs, boar, rabbits and deer; the possibility of other reservoirs cannot be excluded. There is some evidence that genotype 3 at least is intrinsically of low pathogenicity, causing disease in only a small fraction of people it infects unlike genotypes 1 and 2. Aside from rare cases in travellers returning from abroad, HEV seropositivity in the USA and

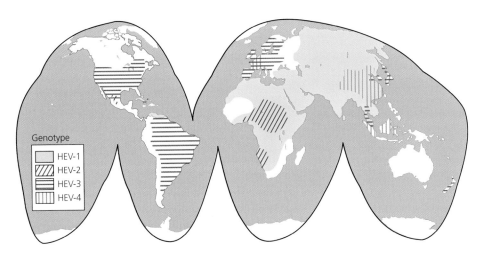

Fig. 22.3 The global distribution of HEV genotypes in humans. Genotypes 1 and 2 are found only in humans while genotypes 3 and 4 are found additionally in other species, especially pigs. Genotype 3 is considered to be globally distributed in its animal reservoirs. (Reproduced with permission from Krain, L. J. et al. 2014. *Clin. Microbiol. Rev.* 27, 139–165.)

other developed countries is attributed to acquisition of genotype 3 infection from pigs. Transmission can be linked to handling pigs or pig meat, but most cases are thought to be due to consuming undercooked pork offal, etc. Because the four genotypes produce cross-protective immune responses, prior infection by genotype 3 could reduce the intensity of human infection by genotypes 1 and 2 in populations where the viruses co-circulate. The probability of prior infection by genotype 3 will depend on the likelihood of exposure to a zoonotic reservoir species or meat from it relative to the risk of exposure to genotype 1 or 2 determined by the quality of sanitation, etc. Thus Egypt exemplifies the simple transmission of a pathogenic virus via the faecal–oral route whilst the situation in India and many other countries is complicated by other factors. A vaccine against HEV that has the potential to reduce the impact of HEV disease was first licensed in China in 2012, but it has not yet been widely adopted.

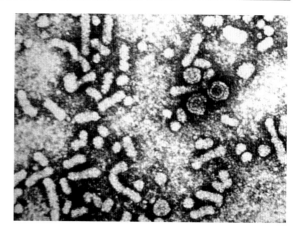

Fig. 22.4 An electron micrograph of particles in serum from an HBV infection. A cluster of three larger particles towards the top right have a pronounced core and are infectious 45 nm HBV virions, also known as 'Dane particles'. The much greater number of smaller round and elongated particles are empty envelopes displaying the same S antigens as are present on the surface of the infectious virions. (Centers for Disease Control, CDC.)

22.4 Hepatitis B virus infections

> *Hepatitis B virus is transmitted at birth, sexually or by blood transfer. Infections, particularly in young children, are normally not resolved by the immune response, leading to chronic infection and the risk of serious liver disease. Around 240 million people are estimated to be chronically infected with this virus. An effective vaccine is available and being deployed in national vaccination programmes in most countries.*

Hepatitis B virus (HBV) is a hepadnavirus that is transmitted variously by close contact, by sexual activity or by transfer of blood, such as from the sharing of needles during i.v. drug use

or iatrogenically in settings with inadequate infection-control practices, or vertically from mother to child. HBV replication and gene expression are considered in Sections 9.8 and 10.9. A significant feature of HBV infection is that, in addition to the presence of infectious virions in the blood, there are also much larger quantities of non-infectious particles that are essentially empty lipid envelopes displaying the three forms of the HBV surface (S) protein (Fig. 22.4). These may function as a decoy to limit the binding of antibody to infectious virions.

Like HAV and HEV, the probability of overt symptoms of hepatitis during acute infection with HBV is very low in young children but rises to become a likely outcome in adults. The time from infection to disease, when it occurs, is typically 2–3 months which is even longer than for HAV, with the individual being infectious through most of that period. Again, pathology is immune-mediated and there is the same low (0.5–1%) probability of a severe fulminant hepatitis developing. Unlike HAV or HEV, there is a significant risk of HBV infections becoming persistent or chronic. The probability of this outcome is >90% among children infected under 1 year of age, falling to <10% in adolescents and adults (Fig. 16.3). In other words, the immune response to HBV resolves infection in the majority of those infected as adults but not in young children. Thus there is an inverse relationship between the probability of clinically apparent acute infection and the probability of immune clearance. All infected people produce antibodies to the viral core (C) protein but antibodies to the S protein are only found in those achieving clearance. Serum markers of the status of an HBV infection are summarized in Table 22.1.

The most significant disease burdens due to HBV result from late-onset consequences of chronic infection, specifically liver cirrhosis and hepatocellular carcinoma. Cirrhosis is the end result of chronic inflammatory insult to the liver that is present over many years. As described in Section 22.1, the loss of hepatocytes to such causes is compensated by regeneration of tissue by cell division. However, the capacity for liver regeneration is finite and, over extended periods, both the speed and scale of regeneration decline. As a result, the liver mass becomes increasingly formed of fibrous scar tissue and liver function declines; ultimately, a liver transplant is the only viable therapy for such end-stage liver disease. Hepatocellular carcinoma is considered in Section 25.9. It is thought that this disease develops in part through the accumulation of mutations in cells as a result of the long-term increased rate of cell division that occurs to compensate for hepatocyte loss, but also through genetic instability resulting from infection and direct effects of HBV proteins on the cells.

A vaccine for HBV, the first viral recombinant subunit vaccine to be produced, has been available since 1982 (see Section 26.4). This vaccine, which is formed of the major viral S protein, is highly effective, its only real drawback being the high cost per dose. Nonetheless, a combination of national government funding and international programs saw population-wide HBV vaccination strategies established in 179 WHO member countries by 2011. In fact, most of the small number of countries not running such programs are developed countries such as the UK where the risk of local HBV transmission is considered to be too low to make such a program cost-effective. However, the vaccine is generally provided to those in high-risk occupations such as healthcare workers. Over time, these programs are expected to break the cycle of high endemicity and high-level chronicity for HBV and lead to substantially reduced end-stage liver disease burdens from this virus.

Once an HBV infection has become chronic, it may be more or less active in terms of

Table 22.1
Serum markers of HBV infection status.

Serum properties	HBV infection status
S antigen negative Anti S antigen negative Anti C antigen negative	Naïve; susceptible to infection
S antigen positive Anti S antigen negative Anti C antigen positive Anti C antigen IgM positive	Acutely infected
S antigen positive Anti S antigen negative Anti C antigen positive	Chronic infection
S antigen negative Anti S antigen positive Anti C antigen positive	Immune due to natural infection
S antigen negative Anti S antigen positive Anti C antigen negative	Immune due to vaccination

production of viral S antigen and infectious virus particles and, consequently, there is variation in the extent of ongoing immune-mediated liver damage in patients. Recall that the distinction between a persistent and a chronic infection is the extent of replication and disease observed in the long term (Section 16.5, which considers HBV as an example). A widely-used marker of a chronic HBV infection that is producing significant amounts of infectious virus is the presence of circulating HBVe antigen; this antigen is produced from the pre-C protein (Section 10.9). An HBVe$^+$ chronic infection may resolve at some point to HBVe$^-$ status with the onset of production of anti-HBVe antibodies. From that point, although production of the viral S antigen will continue, there will be much less infectious virus present and further damage will be reduced. At present, there is no curative therapy available for chronic HBV infection, but in patients with high circulating virus loads, treatment can be appropriate to limit virus replication and so the extent of liver damage; one aim in such cases is to achieve conversion to anti-HBVe antibody positive status. Drugs to target HBV include PEG-stabilized interferons as well as specific reverse transcriptase inhibitors (Section 27.5). The interferon acts to increase the expression of MHC class I molecules which present virus antigens on the surface of infected cells, stimulating cytotoxic T cells to destroy them and so reduce the level of infection (see Section 14.2).

HBV chronic infections are common in many parts of the world. In 2014, WHO estimated that around 240 million people were living with chronic HBV, with about 600,000 deaths per annum attributable to the combined acute and chronic effects of the virus. The fact that chronic infection is the most likely outcome in very young children reinforces a cycle that maintains high levels of persistence in areas of high endemicity, where infection is acquired vertically or by close contact in early childhood.

A female infected at this early age will typically become chronically infected and so is likely to infect her children when they are at a similarly early age. In such areas, 5–10% of the adult population is chronically infected. In contrast, areas of the world where infection is typically acquired in adulthood (sexually, shared needles, etc.) will have both lower absolute levels of HBV seropositivity and far lower proportions of HBV seropositive individuals being chronically infected.

Age at infection is not the only factor that contributes to determining whether an HBV infection will be resolved or become chronic; viral genotype is also important. Based on complete sequence analysis of many isolates, a series of HBV genotypes A–H has been defined. Unsurprisingly, these genotypes are not all equally distributed around the world. Instead, each area has its own distinct dominant genotype(s). Viral variation is clearly a potential source of difference in clinical outcome but identifying viral genotype-related differences in pathogenicity or likelihood of persistence is difficult, since each area of the world also differs in the genomic diversity of its human population and host effects will also contribute to determining these outcomes. However, a 20-year study in chronically infected Alaskan Natives, who have very high rates of chronic infection and who would be expected to be relatively genetically homogeneous, showed that HBV genotype C was particularly associated with long-term chronic infection (Box 22.2). There are also marked HBV genotype differences in the frequency of association with cirrhosis and hepatocellular carcinoma. Just as genotype C is the most likely to persist as chronic infection, so it is also most likely to be associated with these diseases. This genotype is the dominant one in eastern China, though it is also present worldwide at lower levels, which perhaps contributes to the very high rates of hepatocellular carcinoma which are seen in that country.

Box 22.2

Evidence for the association of HBV genotype C with long-term chronic infection

Prospective cohort study of 1536 HBV-infected Alaskan Native people by 6-monthly sampling and analysis over 20 years:

HBV strains from 1158 persons were successfully genotyped; 507 of these people were HBVe antigen-positive at first sampling.
57 of 507 initial HBVe⁺ remained HBVe⁺ at the end of sampling. Those infected with genotype C were significantly more likely still to be HBVe antigen positive at final sampling than those infected by other genotypes:

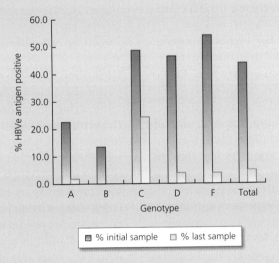

Median ages at first HBVe⁺ sample were: A 12.9; B 12.8; C 16.9; D 8.1; and F 5.8 years. Median follow-up was similar for all genotypes (median 20.8 years, range 18.8–22.3).
The time taken for 50% of HBVe⁺ individuals to clear to HBVe⁻ was significantly longer for genotype C infections, at 17.8 years, than for other genotypes (range 3.3 to 7.5 years).
Data from Livingston, S. E. et al. 2007. *Gastroenterology* 133, 1452–1457

22.5 Hepatitis D virus infections

Hepatitis D virus (HDV) is a highly unusual virus that can be considered parasitic on HBV in that it uses HBV S antigen to provide the protein for its envelope and thus HDV can only replicate and spread within HBV-infected individuals. Consequently, it is not possible to be infected with HDV without also being infected with HBV. Some molecular features of HDV are considered in Section 3.6. Routes of transmission for HDV are the same as for HBV.

The frequency of HDV-positive individuals among those who are HBV-positive varies greatly around the world, with values ranging from very low (0–5%) in countries such as Canada, Argentina, and much of China and S.E. Asia to very high (>60%) in countries such as Columbia, Venezuela, Romania and Kenya; data are not available for all countries. Of course, this is not to say that the *absolute* prevalence of HDV is high in these countries since a high proportion of those who are HBV$^+$ can still be a very low proportion of the total population if the HBV prevalence is low.

Since HDV absolutely depends on HBV for production of infectious progeny, it is not possible to be exposed to HDV without also being exposed to HBV, and infection by HDV alone cannot occur. Thus there are two scenarios for acquiring HDV infection: co-infection with HBV and HDV of an HBV-negative host; and super-infection of an already HBV-positive chronically infected host by further exposure to HBV and HDV together. In the context of co-infection, HDV exacerbates acute disease in adults so that it is far more likely that the infection will be symptomatic; rates of fulminant hepatitis are also increased. The presence of HDV with HBV during infection of an HBV-naïve host also reduces further the chance that such an HBV infection will become persistent or chronic. The super-infection scenario, in contrast, leads to exacerbation of the chronic signs and symptoms of infection without usually provoking any resolution of the chronic HBV infection. Such an infection will display significant ongoing liver damage, more severe than in a typical chronic HBV infection, and the average time to end-stage liver disease will be reduced compared to a chronic HBV alone. Because the membrane proteins of HDV and HBV are identical, the HBV vaccine that comprises this protein will protect equally against HBV and HDV. Thus, with the progress of the global vaccination program against HBV, rates of HDV infection are also falling.

22.6 Hepatitis C virus infections

Hepatitis C virus is transmitted primarily by blood transfer. Infections in all age groups are typically asymptomatic and rarely resolved by the immune response. Around 150 million people are estimated to be chronically infected with this virus and at serious risk of liver disease. The virus is highly variable in sequence and no vaccine exists. Treatment is already possible that can clear HCV infection and a new generation of specific antiviral drugs is emerging.

Hepatitis C virus is a major human pathogen, despite being identified only as recently as 1989. Prior to that date, it had been present as the principal undefined cause of what was termed 'non-A, non-B hepatitis', a disease usually appearing in blood transfusion recipients and which was proven to have an infectious cause by transmitting disease to chimpanzees using patient serum. Infected chimpanzees produce very high titres of virus in the blood and this proved to be the route via which the viral genome could be cloned and sequenced (Box 22.3). Using this sequence information, diagnostic assays for HCV infection were quickly designed, based either on PCR or, by expressing recombinant protein from the cloned sequence, antibody-capture ELISA (Chapter 5). HCV has been classified within the Flavivirus family, similar to yellow fever virus but placed in its own genus, Hepacivirus. An interesting feature of HCV is the basis of its tropism for the liver, which is significantly determined by the virus making use of a host miRNA (miR122), which is specifically expressed in liver, during its replication cycle.

Like HBV, HCV can establish a chronic infection in which ongoing liver damage can lead cumulatively to liver cirrhosis and/or

---

Box 22.3

Cloning and characterization of the HCV genome

Nucleic acid was purified from material pelleted by ultracentrifugation of viraemic plasma from chimpanzees infected with serum from non-A, non-B hepatitis patients.

The nucleic acid was reverse transcribed and then randomly cloned into a phage λ cDNA expression vector to produce a library of clones.

The clone library was screened using antiserum from non-A, non-B patients, to detect those clones that were expressing protein (antigen) to which there was antibody in patient serum but not in healthy controls.

A cDNA fragment from one positive clone was used to screen further libraries, leading eventually to the assembly of a full-length cDNA clone of the HCV RNA genome.

Cloning of HCV was reported by Choo, Q.L. et al. 1989. *Science* 244, 359–362.

---

hepatocellular carcinoma. However, unlike HBV, there is little evidence of any age-related difference in outcome of primary HCV infection and a high proportion of those encountering the virus become chronically infected, usually with limited if any symptoms associated with the primary infection. When primary HCV infection is acutely symptomatic, there is a long incubation time of 6–7 weeks and, as for the other hepatitis viruses, pathogenesis is immune-mediated. The USA Centers for Disease Control (CDC) estimates that, if untreated, about 80% of HCV infections become chronic, most of them leading to ongoing liver disease and that about 5–20% of infections will lead to cirrhosis within 20–30 years. The WHO estimates that around 150 million people worldwide are living with chronic HCV infection and that HCV currently causes 350,000–500,000 deaths per year. Thus, the total health burden of HCV infection is very substantial. A global view of the fractions of different populations that are HCV infected is shown in Fig. 22.5.

Known routes of HCV transmission include i.v. introduction by needle sharing, other blood transfer injury, blood and blood-product transfusion (until 1992 when screening of blood donations became possible), vertical transmission and sexual transmission. These routes are similar to those defined for HBV but, at least in western countries, i.v. transfer through needle-sharing is by far the most significant source of infection identified retrospectively for HCV-positive people in the population. In contrast to HBV, sexual transmission is not efficient and so is a minor route for spread of this virus. The explosion in i.v. drug use in western countries in the middle of the 20th century is thought to have caused the substantial HCV epidemic now affecting those countries. The transmission route responsible for the often significant levels of HCV in other populations is less certain.

Until recently, standard therapy for chronic HCV infection was to give a 48-week course of ribavirin plus PEGylated interferon α and to monitor virus load. This regime can cure infection, as judged by absence of detectable virus that is sustained for 24 weeks after finishing treatment (a sustained virologic response, SVR), though with very pronounced differences between HCV genotypes. Only 40–50% of patients infected with genotype 1,

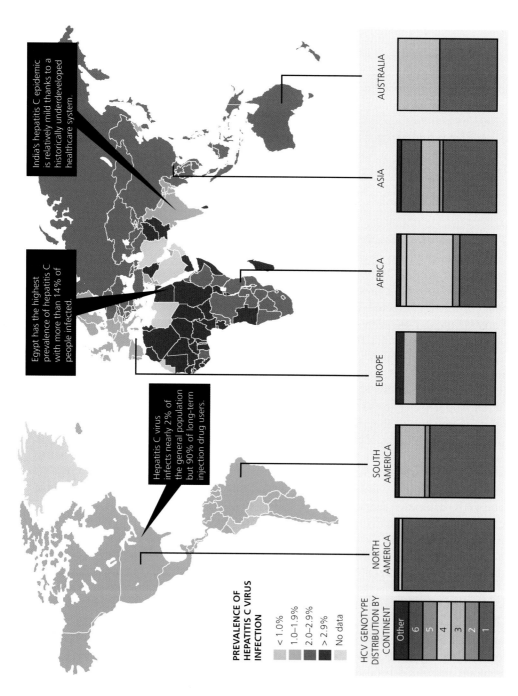

Fig. 22.5 A map showing variation by country in the prevalence of HCV infection. Increasing depth of blue shading indicates increasing fractions of the population are infected. Below the map, histograms show the relative contributions of different HCV genotypes to the burden of infection in different continents. (Reprinted by permission from Macmillan Publishers Ltd: Gravitz, L. *Nature 474*, S2-4, ©2011.)

the most common HCV type affecting western populations (Fig. 22.5), achieve this result whilst clearance of infections with the less common genotypes 2 and 3 is seen in 70–80% of cases. Also, the treatment can produce unpleasant side-effects and so, with the extended duration required, patients do not always manage to complete the course. For these reasons, and in the absence of a vaccine to protect against the virus, there has been substantial effort devoted to the development of antiviral drugs that are active specifically against HCV. These drugs target key enzymes in viral replication, such as the viral NS3/4A protease. The recent introduction of these drugs is achieving substantial improvements in rates of virus clearance for all genotypes. However, HCV is a highly variable virus, with error-prone replication that produces a quasi-species (see Section 4.2), so it may be expected that drug-resistant HCV mutants will be a problem for these drugs as they have been for HIV therapy unless they are deployed in combination. Specific HCV antiviral therapy is considered in detail in Section 27.5.

What factors contribute to determining the outcome of infection by HCV for an individual? There are two broad categories: variation in the virus and variation between hosts. Separating out these two possibilities is very difficult when both host and virus are highly variable as is the case for HCV. However, the importance of individual host genotype to determining HCV infection outcome was demonstrated with a cohort of women who were known to have been exposed to a single source of virus. Despite this homogeneity in virus exposure, they still showed a spectrum of individual outcomes from non-infection (no seroconversion) through spontaneous clearance to long-term chronic infection. Analysis of the MHC Class I haplotypes of these women showed that possessing either of two particular alleles, A-02 and B-27, was strongly associated with spontaneous clearance of HCV whilst some

other alleles were associated with a high risk of HCV chronicity. Such an association should perhaps not be surprising, given the critical importance of MHC antigens to determining the nature and extent of the adaptive immune response (Chapter 14).

Another important host factor determining HCV infection outcome was revealed by genome-wide association studies analyzing the results of PEG-IFN/ribavirin treatment. Possession of one of two alternative bases found in the population at each of at least three distinct sites (single nucleotide polymorphisms; SNPs) located close together on chromosome 19 was found to be strongly linked with a good response to this treatment (Box 22.4). These SNPs are located close to two type III interferon genes, *IL28A* and *IL28B*. Moreover, in untreated individuals, possession of the 'good responder' haplotype is linked with higher basal expression of type III IFN and homozygotes have a greater likelihood of spontaneous clearance of infection. Thus, individual variation in this aspect of the innate immune response is an important determinant of HCV infection outcome and response to treatment. The frequency of the 'good responder' haplotype is not the same in all populations and appears to be particularly low in those of Afro-Caribbean ancestry. This may well explain why people from this ethnic group suffer significantly higher rates of chronic HCV than other ethnic groups.

Key points

- Infection by any of a series of unrelated hepatitis viruses A–E causes inflammation in the liver and immune-mediated destruction of hepatocytes. Except for HEV, mortality from acute infection is less than 1%.
- HAV and HEV typically transmit via the faecal–oral route and cause acute or subclinical self-limiting infections. Such

Box 22.4

Evidence that expression of the IL28A/B genes is linked to clearance of HCV infection

Genome-wide association studies (GWAS) are a powerful tool to reveal associations between genotype and phenotype in complex traits. They scan a large panel of SNPs located across the genome for linkage between individual SNPs and a particular phenotype and can identify multiple genome regions that each makes a contribution to a disease.

Linkage of a phenotype with a SNP does not mean necessarily that the actual nucleotide at that SNP determines the phenotype, but rather that a particular variant of a gene closely linked to that SNP location and therefore moving with it between generations (i.e. nearby in the genome) is important in the phenotype.

A cohort of 1600 chronically HCV-infected people in a trial comparing treatment with two different PEG-IFN preparations plus ribavirin were genotyped at SNPs throughout the genome and each SNP was analyzed for association with the outcome of treatment.

Homozygotes for a C nucleotide at SNP rs12979860 responded well to all treatments in the trial and achieved good rates of clearance whilst those homozygous for T responded poorly (Fig. 22.6a). This SNP is 3 kbp upstream of the *IL28B* gene, which encodes IFNλ3 (Fig. 22.6c). A good response to therapy was also linked with having a T rather than a G at rs8099917, located downstream of *IL28B*.

A second smaller study confirmed the linkage of T at rs8099917 with a good response to HCV therapy and also showed that uninfected T/T homozygotes at this SNP had higher levels of IFNλ expression than G/G homozygotes.

C/C homozygotes at rs12979860 were shown to be more likely than T/T homozygotes to clear an HCV infection spontaneously and so avoid becoming chronically infected (Fig. 22.6b).

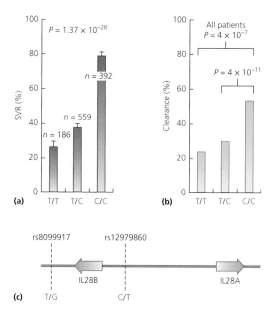

(a) (b) (c)

Fig. 22.6 An allele of the *IL28B* locus determines efficiency of response to HCV infection. (a) The nucleotide identity at SNP rs12979860 determines the effectiveness of PEG-IFN plus ribavirin therapy for chronic HCV infection, measured in terms of the % of patients who showed a sustained virologic response (SVR) to the therapy. Data shown are for all ethnic groups (European-American, African-American and Hispanic) combined; each group showed the same effect when analyzed individually. (Reprinted by permission from Macmillan Publishers Ltd: Ge D. et al., *Nature* 461, 399–401, ©2009.) (b) The nucleotide identity at SNP rs12979860 affects the likelihood of spontaneous clearance of HCV infection without therapy. Data shown are for all ethnic groups (European ancestry, African ancestry) combined; both groups showed the same effect when analyzed individually. (Reprinted by permission from Macmillan Publishers Ltd: Thomas D. et al., Nature 461, 798–801, ©2009.) (c) A schematic representation of an approximately 25 kbp region of human chromosome 19 showing the locations of two relevant SNPs and two genes encoding type III interferon, *IL28A* and *IL28B*. See Box 22.4 for further details.

infections provide lifelong immunity against re-infection.

- HBV and HCV are transmitted in blood and other body fluids; they frequently cause persistent or chronic infections with ongoing liver damage and with the risk of delayed-onset cirrhosis or hepatocellular carcinoma.
- Initial infection by HAV or HBV is very likely to be asymptomatic in young children; initial HCV infection is most commonly asymptomatic in all age groups.
- HDV only infects together with HBV and generally exacerbates the acute or delayed pathology associated with HBV infection.
- HEV causes up to 20% acute mortality specifically in pregnant women. It is also the only one of the viruses to have an animal reservoir.

Questions

- Discuss the factors that influence the incidence of disease associated with infection by the different hepatitis viruses.
- Discuss the proposition that hepatitis C infection is a greater problem for human populations than hepatitis B virus infection.

Further reading

Cuthbert, J. A. 2001. Hepatitis A: old and new. *Clinical Microbiology Reviews* 14, 38–58.

Hughes, S. A., Wedemeyer, H., Harrison, P. M. 2011. Hepatitis delta virus. *Lancet* 378, 73–85.

Lavanchy, D. 2004. Hepatitis B virus epidemiology, disease burden, treatment, and current and emerging prevention and control measures. *Journal of Viral Hepatitis* 11, 97–107.

Kamar, N., Dalton, H. R., Abravanei, F., Izopet, J. 2014. Hepatitis E virus infection. *Clinical Microbiology Reviews* 27, 116–138.

McMahon, B. J. 2009. The influence of hepatitis B genotype and subgenotype on the natural history of chronic hepatitis B. *Hepatology International* 3, 334–342.

Meng, X. J. 2010. Recent advances in hepatitis E virus. *Journal of Viral Hepatitis* 17, 153–161.

Pawlotsky, J.-M. 2004. Pathophysiology of hepatitis C virus infection and related liver disease. *Trends in Microbiology* 12, 96–102.

Perez, V. 2007. Viral hepatitis: historical perspectives from the 20th to the 21st century. *Archives of Medical Research* 38, 593–605.

Rehermann, B. 2013. Pathogenesis of chronic viral hepatitis: differential roles of T cells and NK cells. *Nature Medicine* 19, 859–868.

Chapter 23

Vector-borne infections

Many viruses are transmitted between vertebrate hosts by an arthropod vector. These are referred to as arthropod-borne viruses which is commonly shortened to arboviruses. *Arboviruses are often associated with one or only a few vector species and the geographical distribution of the virus infection is restricted to the habitat of those vector species.*

Many human and animal virus infections are restricted to very defined geographical areas of the world. This geographical restriction and the observation that direct contact between an infected individual and a subsequent victim was not always necessary for transmission of infection led early investigators to the conclusion that an intermediate was responsible for transmission of the virus between people. Following intense study, it was shown that haematophagous (blood sucking) insects such as some species of mosquitos, sand flies, biting midges and ticks can transmit viruses from their natural animal reservoir to a susceptible host of a different species (a zoonotic infection; see Section 18.5). This chapter will consider four examples of arboviruses that have been associated with severe disease in humans.

23.1 Arboviruses and their hosts

Arboviruses are acquired by an insect either directly from its mother whilst still an egg (*in ovo*, vertical transmission, see Section 18.4), or as an adult female when it takes a blood meal from an infected animal or person. The viruses can only be acquired and spread if they establish a viraemic infection in which virus is found in the blood of the infected vertebrate host. As the viruses are present in the bloodstream they are spread widely throughout the body and can therefore potentially infect a wide range of tissues and organs leading to serious disease. In such infections, many features of the resulting diseases are common to all, but specific details differ. For efficient transmission the virus must be present in sufficient quantity to ensure that the insect takes up an infectious dose during the blood meal. Once the virus has been ingested, it typically replicates within the insect without causing any significant disease. When the virus infects members of its animal reservoir it is also

---

Introduction to Modern Virology, Seventh Edition. N. J. Dimmock, A. J. Easton and K. N. Leppard.
© 2016 John Wiley & Sons, Ltd. Published 2016 by John Wiley & Sons, Ltd.

unlikely to cause significant disease as the virus and host will have usually established a mutually acceptable co-existence over a long period of interaction (see Section 4.5). However, this is not always the case and some arboviruses do cause disease in their vertebrate host. When such a virus is introduced into a new species such as a human, the consequences can be extremely serious. Some examples of arboviruses that cause severe diseases in vertebrates are shown in Table 23.1. This list is not comprehensive but it can be seen that members of certain virus families are frequently associated with an insect vector for their transmission.

23.2 Yellow fever virus

Yellow fever virus is a major threat to human health in tropical and sub-tropical regions of South America and Africa. The animal reservoir is monkeys within which the virus does not cause disease. Transmission is by female mosquitos that acquire and introduce the virus during blood meals.

Yellow fever virus has played an important role several times in human history, primarily in times of war in South America and the islands of the Caribbean. Yellow fever was introduced

Table 23.1

Examples of arboviruses associated with severe disease in vertebrate hosts. The nature of the vector and the virus family are also shown.

Vector	Virus	Virus family
Mosquito	Yellow fever virus	Flavivirus
	Dengue virus	Flavivirus
	West Nile virus	Flavivirus
	Chikungunya virus	Flavivirus
	Japanese encephalitis virus	Flavivirus
	Eastern equine encephalomyelitis virus	Togavirus
	Western equine encephalomyelitis virus	Togavirus
	Venezuelan equine encephalomyelitis virus	Togavirus
	Rift Valley fever virus	Bunyavirus
Sand fly	Vesicular stomatitis virus	Rhabdovirus
	Alagoas virus	Rhabdovirus
	Cocal virus	Rhabdovirus
	Chandipura virus	Rhabdovirus
	Sandfly fever Naples virus	Bunyavirus
Midge	African horse sickness virus	Reovirus
	Bluetongue virus	Reovirus
	Schmallenberg virus	Bunyavirus
	Bovine ephemeral fever virus	Rhabdovirus
Tick	Tick-borne encephalitis virus	Flavivirus
	African swine fever virus	Asfarvirus
	Nairobi sheep disease virus	Bunyavirus
	Thogoto virus	Orthomyxovirus
	Louping ill virus	Flavivirus
	Kyasanur Forest disease	Flavivirus

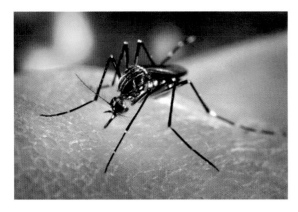

Fig. 23.1 A female *Aedes aegypti* mosquito taking a human blood meal. *Aedes aegypti* is the primary vector for transmission of yellow fever virus in the urban setting.

into the Americas during the years of the slave trade with captives brought from Africa where the virus was present. This trade also brought the primary insect vector, the mosquito *Aedes aegypti* (Fig. 23.1), providing the other necessary component for spread of the virus. The impact of yellow fever on geopolitics was seen dramatically in 1801 when the emperor Napoleon sent 20,000 troops to quell a rebellion in Haiti, which was under the control of France at that time. Within six months, 80% of the men were dead of yellow fever. A further 20,000 troops sent as reinforcements also succumbed and the French were forced to return with only 3000 surviving soldiers. Shortly afterwards, in the Mexican–American war of 1846–47, ten times more soldiers died from yellow fever than were killed in battle. The impact of yellow fever infections on the local populations and on military activities in South America and the Caribbean led to the studies of Walter Reed to identify the route of transmission and the nature of the agent of the disease.

As indicated in Fig. 23.2, Yellow fever virus infections are restricted to regions of tropical South America and sub-Saharan Africa lying

Fig. 23.2 Geographical location of endemic yellow fever virus infections. The regions where naturally-acquired yellow fever virus infections are found are highlighted. The infections only occur in the tropical regions of South America and Africa.

approximately 15° north and 10° south of the equator. This reflects the geographical habitat of the primary mosquito vectors that transmit the virus, and in particular of *Aedes aegypti* as described below. In recent times in excess of 200,000 cases of yellow fever with approximately 30,000 deaths have been recorded each year. Approximately 90% of this disease burden falls in Africa where large outbreaks occur, whilst the remainder are in South America where incidents of infection in urban environments are more sporadic than large scale. An intriguing feature of yellow fever virus is that genetic analysis has shown a clear geographical distinction in strains isolated in different parts of the world. This is particularly surprising given the relatively short time that African and South American viruses have been physically separated, and presumably reflects environmental selection in the different parts of the world.

Transmission vectors of yellow fever virus

The mosquito species involved in transmitting yellow fever virus depends on the environment in which the transmission is taking place. In both Africa and South America, the virus is maintained in an animal reservoir that consists of monkeys which live in jungle treetops. This is referred to as a *sylvatic* (or *jungle*) *cycle* (Fig. 23.3). The infection does not produce a serious disease in monkeys.

In South America, a number of different *Haemagogus* species mosquitos, found only in the jungle, transmit the virus between monkeys. The mosquitos can spread the infection to humans who enter the jungle. When these infected people are taken to urban populations they can transmit the virus to the *Aedes aegypti* mosquito which in turn spreads the virus throughout the community in what is referred to as the 'urban cycle' (Fig. 23.3). In

Africa, the virus is maintained in monkeys almost exclusively by the mosquito *Aedes africanus*. On the fringes of the jungle areas, a number of other *Aedes* species can also acquire the virus and these mosquitos can transmit the infection to human in an intermediate, or savannah, cycle. Infected individuals may then be taken to urban populations where, as in South America, only *Aedes aegypti* transmits the virus between individuals (Fig. 23.3).

Yellow fever virus disease

Approximately one-third of people infected with yellow fever virus become ill, with the remainder experiencing an asymptomatic infection. Of those infected, approximately 80–85% recover and 15–20% die if untreated. During epidemics, case fatality rates of up to 50% have been recorded. Yellow fever has a very short duration and begins with an abrupt onset of fever associated with headache, nausea and muscle pains and, in severely affected individuals, vomiting. This period lasts for only a few days and is followed by a period of remission of symptoms lasting 1–2 days. The final phase, which ensues only in the most severely affected patients and is referred to as the intoxication phase, typically begins 3–6 days after the initial appearance of symptoms. During the intoxication phase fever, nausea and vomiting reappear. There is appearance of petechial haemorrhages (small red spots that arise from broken capillaries – a harbinger of the more severe disease that can follow) particularly in the gums and gastrointestinal tract. The latter is apparent by the appearance of blood during vomiting. Patients present with jaundice from which the disease gets its name. In the latter stages of disease, patients can go on to suffer renal failure, cardiovascular shock and multi-organ failure.

The underlying processes leading to yellow fever are well understood. Following the

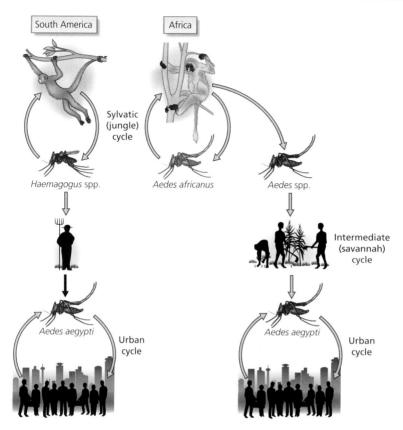

Fig. 23.3 Transmission cycles of yellow fever virus. The virus is maintained in an animal reservoir of tree-living monkeys and is transmitted by a variety of *Haemagogus* species mosquitos in South America and *Aedes africanus* in Africa. In the urban environment, the virus is transmitted by *Aedes aegypti* mosquitos.

initial infection at the site of the mosquito bite, infected cells initiate an innate immune response that leads to the production of pro-inflammatory cytokines (see Chapter 13). The cytokines probably contribute to some of the early symptoms such as fever. Many different cell types are infected at this time, including endothelial cells lining the blood vessels and some local and circulating cells of the immune system that are drawn to the site of infection by the innate immune response. These infected immune cells make their way to local draining lymph nodes, carrying the virus with them. It is at these sites that virus levels amplify hugely and the virus is introduced into the circulatory system, generating a viraemic infection that spreads through the body.

Circulating virus targets the liver as its primary, but not only, site of infection. Initially Kupfer cells, the resident macrophages of the liver, are infected and then the virus spreads to the hepatocytes, especially of the midzone, where the major damage is done. Following infection, huge numbers of hepatocytes die either as a result of the cell apoptotic response (see Section 13.5) or directly due to virus infection (*necrotic death*). In fatal cases, up to 70% of the liver hepatocytes can be killed in these ways. While there is activation of the adaptive immune response, including CD4$^+$ and CD8$^+$ T cells, macrophages and B cells, to counteract the infection there is only a minimal inflammatory response in the liver itself. However, there is a very strong systemic inflammatory response which includes the production of very large amounts of inflammatory cytokines in what is referred to as a '*cytokine storm*'. The cytokines stimulate an inflammatory response that causes damage to many cells throughout the body, including those lining blood vessels, leading to leakage of blood and the characteristic haemorrhaging seen with severe disease. The haemorrhaging is enhanced because the damaged liver fails to produce sufficient quantities of the blood clotting factors that would be needed to stem this leakage.

Prevention and treatment of yellow fever virus infection

The presence of an animal reservoir which is not accessible to treatment means that eradicating yellow fever virus is not feasible. An excellent vaccine to protect against yellow fever virus has been available for over 70 years. This vaccine, discussed in more detail in Section 26.2, has saved countless lives. However, yellow fever continues to threaten many, especially in sub-Saharan Africa. Attempts to reduce infection by the use of bed nets and other barrier methods have had limited success as, unlike most blood feeding insects, *Aedes aegypti* is active during daylight hours. Some success has been seen in reducing sites for breeding of *Aedes aegypti* mosquitos in urban environments but these have had only limited effect in the most badly affected areas of the world. Treatment for infected patients is limited to supportive therapies intended to

address the most severe symptoms. This has some limited effect and prevention remains the main focus of attack for this virus.

23.3 Dengue virus

Dengue virus is considered by the World Health Organisation to be the major virus threat in some Asian and Latin American countries.

Dengue virus infection causes disease ranging from asymptomatic or mild to the severe dengue haemorrhagic fever and dengue shock syndrome. The first clearly-recorded outbreaks occurred in the late 18th century. However, there are records of diseases with similar characteristics in ancient Chinese documents dating from the 3rd and 4th centuries AD in which the association with flying insects was noted. Dengue virus infects monkeys which act as an animal reservoir from which the virus can be transmitted to humans. The virus is maintained in a jungle cycle similar to that described for yellow fever virus. Dengue virus is transmitted between humans primarily by *Aedes aegypti* mosquitos and, to a lesser extent, by *Aedes albopictus*, which occur in all tropical and sub-tropical region of the world. Up to the 1960s, dengue virus infections were largely restricted to only a few countries. Since then the incidence of infection has increased significantly, spreading to over a hundred countries. Dengue virus is endemic in South-East Asia, the Pacific, East and West Africa, the Caribbean and the Americas and is now considered to be a major international disease threat with up to 40% of the world's population potentially at risk. Current estimates suggest that there may be as many as 400 million Dengue virus infections every year; up to 500,000 people, predominantly children, are hospitalized with severe Dengue

Fig. 23.4 Dengue virus surveillance centre in Salvaterra on the Island of Marajo, Brazil. The advice encourages that standing water is removed from plants to reduce breeding sites for mosquitos which transmit the virus.

virus disease of whom approximately 2.5% die. The threat of Dengue is such that awareness and surveillance systems have been established in many parts of the world (Fig. 23.4).

Dengue virus is a member of the Flavivirus family and four major serotypes of the virus have been identified, referred to as DEN-1, DEN-2, DEN-3 and DEN-4. Following infection with one serotype the antibodies generated provide some level of cross-reactivity with the other types but this is limited and short lived. While the presence of reactive antibodies may seem beneficial, in fact their presence can lead to enhanced risk of serious disease as discussed below.

Dengue virus disease

Dengue virus infections lead to a complex disease with a wide range of clinical signs and symptoms. In approximately 95% of cases the disease is either asymptomatic or mild. These infections are very short lived, lasting 2–7 days, and are characterized by a sudden onset of fever. This is accompanied by a variable range

of additional symptoms including headaches localized around the eyes, and muscle and joint pains of potentially significant severity that gave the traditional name of the disease as 'break-bone fever'. There is often a measles-like rash and in the later stages of infection some petechiae are also seen. During this period, virus is found in the blood of patients who can therefore provide an inoculum to feeding mosquitos which can then transmit the virus to others.

In approximately 5% of dengue fever cases the disease progresses to the more severe dengue haemorrhagic fever or dengue shock syndrome, which are life-threatening. These serious consequences are more commonly seen in children, and while 10% of the cases are associated with a primary infection with a single serotype the remaining 90% are associated with a second infection with a different serotype to that which caused the primary infection. Dengue haemorrhagic fever occurs in the 2 to 3 days after the resolution of dengue fever. It is associated with gastrointestinal bleeding and plasma leakage from blood vessels due to loss of integrity of the endothelial cell lining of the vessels. In extreme cases, this may also be accompanied by activation of the blood clotting system that leads to blockage of the smaller blood vessels leading to more damage and leakage. The liver and kidneys are significantly affected, and in extreme cases this can lead to hepatic failure that generates a shock response characteristic of dengue shock syndrome.

The primary cause of the severe dengue disease is dysregulation of the immune system following infection. This is seen with both the cell-mediated immune response and, in the case of a second infection, the humoral response (Chapter 14). The cell-mediated immune response in dengue haemorrhagic fever is accompanied by a rapid and extremely high level of cytokine production. These molecules stimulate many aspects of the immune system including an inflammatory response, bringing immune cells to the sites of vascular damage so causing further damage. In addition, the 'cytokine storm' may enhance innate immune responses such as apoptosis (Section 13.5) in endothelial cells, again enhancing the damage to blood vessels.

A major component of dengue haemorrhagic fever and dengue shock syndrome arises from the presence of antibodies generated to a different dengue virus serotype during a prior infection. Whilst the antibodies that are generated following infection with any of the four dengue virus serotype can cross-react with the other serotypes, these antibodies do not result in neutralization of such a virus. When a subsequent infection occurs by one of these other serotypes, the antigenic similarity of the primary and secondary infecting viruses leads to activation of the immune memory B cells that were generated following the initial infection rather than stimulating a new antibody response. In effect, the immune system is 'fooled' into reacting as though the second infection is simply a reinfection with the same virus. The antibodies produced are thus precisely tailored to the original virus, and while they can bind to the new virus they cannot neutralize its infectivity. Instead, the binding of the non-neutralizing antibody leads to a more extensive infection and *antibody-dependent enhancement* (ADE) of disease. This arises because the Fc component of the antibodies (Section 14.3) bound to the infecting virus is recognized by Fc receptor molecules on the surface of monocytes and macrophages. The function of this interaction is to introduce antibody-bound antigens to phagocytic cells within which they will be destroyed. However, in this case the dengue virus is fully infectious and when it is introduce into the phagocytic cells it undergoes a full replication cycle, killing the cells and producing more virus particles that are released into the circulatory system. The result is greater viraemia and a more severe infection. The combination of these different facets of the infection and immune response leads to devastating consequences for the infected individuals. The vascular damage causes loss of blood and fluid into the peritoneal and chest cavities. In turn, this leads to reduced blood volume and decreased blood supply to vital organs.

Prevention and treatment of dengue virus infection

There is currently no vaccine to prevent dengue virus infection. The presence of an animal reservoir makes elimination of dengue virus unachievable. Consequently, the only preventative measures are aimed at reducing exposure to infected mosquitos. As with yellow fever virus, the use of bed nets does not provide protection in the way that has been successful in reducing exposure to insects transmitting other diseases such as malaria. To date, the only preventative measures available focus on reducing opportunities for the insects to persist in urban areas (Fig. 23.4). Treatment of infected patients is generally supportive to reduce the impact of blood and fluid loss in the circulatory system and to reduce the impact of tissue damage. Early treatment of patients with dengue haemorrhagic fever can reduce mortality from levels of 10% seen with untreated individuals to 1% if treated.

23.4 Chikungunya virus

The threat from Chikungunya virus increased significantly following a change in preference for mosquito species. This extended the potential geographical range of the virus.

Chikungunya virus was first identified in 1952 in Tanzania and was subsequently found to be responsible for several large outbreaks. In 2005–2007, an outbreak on Reunion Island in the Indian Ocean resulted in almost a third of the entire population becoming infected, and this marked a rapid increase in the incidence and geographical range of Chikungunya virus. The virus has now been found in many parts of South-East Asia, with small numbers of cases also reported in Australia and in Italy and France in Europe. The regions of the world considered to be at significant risk of Chikungunya virus infection are shown in Fig. 23.5. The animal reservoir of Chikungunya virus is not clear, but there is evidence suggesting the involvement of rodents, birds and small mammals in Africa. Elsewhere in the world, it is believed that the virus is maintained within the human population by vector-mediated person-to-person spread. Genetic studies have shown that three distinct lineages of Chikungunya virus are circulating in different parts of the world, and these studies have also identified a unique genetic change that has played a role in the increased geographical spread of the virus, as described below.

The maintenance of Chikungunya virus in the population has several similarities with yellow fever virus. In Africa, the virus is maintained in an enzootic sylvatic cycle in

Fig. 23.5 The areas of the world considered most at risk from Chikungunya virus infection are highlighted.

which several animal species including non-human primates, bats, rodents and some bird species, act as a reservoir. The virus does not appear to cause serious disease in these animals. The virus is transmitted between the various hosts by several different *Aedes* mosquito species within which it can replicate. Chikungunya can also be maintained in an urban cycle, which is seen in both Africa and Asia. During the early outbreaks of Chikungunya, the *Aedes aegypti* mosquito was identified as the vector species. However, *Aedes aegypti* mosquitos are relatively rare in the islands of the Indian Ocean and during the Reunion Island outbreak it became clear that the primary vector was *Aedes albopictus* (Fig. 23.6). Genetic analysis of the virus isolates from this outbreak showed that a single amino acid change in the E1 glycoprotein of the virus, changing an alanine at position 266 to valine, was associated with improved fitness of the virus within *Aedes albopictus*. Unfortunately, *Aedes albopictus* has a greater geographical distribution than *Aedes aegypti* and this alteration in insect host preference of the virus has consequently been associated with its increased spread across tropical and subtropical regions of the world. Thus, Chikungunya is an example of a virus where a subtle genetic change can lead to selection of a mutant with characteristics that favour its survival in a new host, leading in turn to an extended geographical range and access to a larger potential host population. Associated with this is the apparent absence of an animal reservoir for Chikungunya virus found in Asia where the virus appears to be maintained by transmission between infected individuals by the insect vector without the need for a sylvatic cycle.

Chikungunya virus disease

The frequency of presentation of disease following infection with Chikungunya virus ranges from approximately 4% to 30%. Infection is generally not life threatening, with mortality estimated to be approximately 1 case in 1000 infected. While the disease, when seen, is self-limiting and typically lasts for 7–10 days, the consequences for affected people can be significant and potentially long lasting. Chikungunya fever is marked by a very rapid onset with fever, nausea, headache and, particularly, joint pains especially in the extremities. In approximately half of patients a rash develops and in children this may be associated with petechial spots. The virus is found in very high titres in the blood where it can reach 10^7 pfu per ml, which contributes to its ability to readily infect the insect vector and so spread between individuals. While the initial disease resolves rapidly, the joint pains that are frequently experienced do not. In many patients, particularly the elderly and those with other associated problems such as arthritis, the joint pains reappear in episodes that can occur many months or years after the initial infection. These recurrences can be very distressing and painful. Despite the high levels of virus found in the blood, neurological involvement is not common and when it occurs is usually seen in

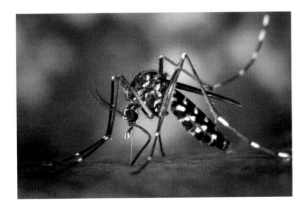

Fig. 23.6 A female *Aedes albopictus* mosquito taking a human blood meal. *Aedes albopictus* has become the primary insect vector for the transmission of Chikungunya virus.

children in the form of seizures and encephalopathies.

The symptoms of Chikungunya fever are a direct consequence of the blood-borne distribution of the virus in the body. Following the initial infection, which probably targets cells in the vicinity of the insect bite that introduces the virus, endothelial cells lining the blood vessels and macrophages in the blood and local tissues become infected. The virus is then carried to secondary lymph glands, probably by migratory macrophages, and is amplified enormously in these glands. A consequence of this is the production of a pro-inflammatory innate immune response with the production of cytokines that activate the adaptive immune system (see Chapters 13 and 14). As the virus yield increases it is spread through the lymphatic system into the circulatory system by which it is disseminated throughout the body. The virus is then carried into the synovial tissue of joints by infected monocytes and macrophages where an inflammatory response is established. It is this response that produces the intense joint pain experienced by patients. Ultimately, the infection is resolved by the adaptive immune response that clears the infection from the body and establishes immunity.

Prevention and treatment of Chikungunya virus infection

No vaccine is available for Chikungunya virus and no alternative form of protective treatment has been discovered. For this reason the major effort for protection is focussed on control of the mosquito vector, as for dengue virus. Similarly, no drugs that attack the virus have been approved for use and treatment is thus restricted to addressing the symptoms of disease. Treatment primarily takes the form of anti-inflammatory drugs to reduce the joint pains that are characteristic of the disease.

23.5 West Nile virus in the USA

West Nile virus was known as an agent of disease in large tropical areas of the Old World. In 1999, it made its first appearance in the USA, after which it spread rapidly across the continent causing significant disease and death in birds, horses and humans. The virus is transmitted between hosts by mosquitoes. The spread of West Nile virus in the USA is a clear example of the risk of introductions of viruses into new territories.

West Nile virus was first isolated in the late 1930s from the blood of a patient exhibiting a mild febrile illness in Omogo in the West Nile District of Uganda. Later, it was found in many parts of Europe, particularly in the Mediterranean region, and also in Africa and India. West Nile virus is a member of the Flavivirus family and is assigned to the Japanese encephalitis antigenic complex that includes many other members such as Japanese encephalitis, Kunjin and St. Louis encephalitis viruses. As the name of the complex suggests, the viruses are frequently associated with acute neurological disease in humans which can be fatal. However, the majority of infections are asymptomatic and only approximately 20% result in overt disease of which a small proportion then develop serious acute neurological complications.

West Nile virus is transmitted from one vertebrate host to another by mosquitoes within which the virus also multiplies. The virus is able to infect a very wide range of invertebrate and vertebrate hosts, including humans and birds. However, in most mammals the virus rarely, if ever, reaches sufficient levels in the blood to be an efficient source of infection for mosquito transfer and it is believed that the natural reservoir are birds of various species, including common species such as crows and magpies. The restriction of the virus to the Eastern hemisphere of the world reflected the habitat distribution of the insect vector and natural physical barriers.

In 1999, West Nile virus was detected during an outbreak of encephalitis in New York, the first time that it had been seen in the Americas. Following the first signs of its presence the virus spread relentlessly across the USA and southern Canada, and after five years was present from the east to the west coast of the continent (Fig. 23.7). A particularly striking feature of West Nile virus infection in the USA was a significant mortality in horses. Infection of these animals almost certainly aided the spread of the virus across the less-densely populated parts of the continent.

To date, a total of over 40,000 human cases of West Nile virus infection have been identified, with over 1800 deaths (Table 23.2). At the same time, the virus has been responsible for countless deaths of birds and other vertebrate species. A consideration of the economic impact of West Nile virus infection in the USA is given in Chapter 29.

The origin of the virus which initiated the introduction of West Nile virus to the USA remains unclear although several possibilities have been considered. From nucleotide

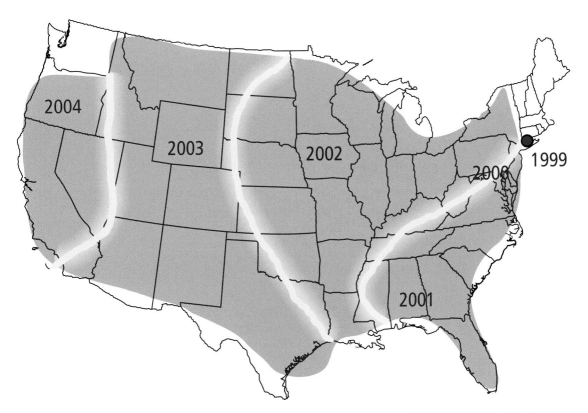

Fig. 23.7   Map showing the spread of West Nile virus from east to west of the USA. Following its identification in New York in 1999, West Nile virus spread south and westwards. The yellow lines show the approximate boundaries of the spreading virus in the years indicated. By 2004, the virus was present from coast to coast.

Table 23.2

Cases of West Nile virus (WNV) infection in humans in the USA. Figures taken from the Centers for Disease Control (Data correct to October 2015).

Year	Total cases	Cases of WNV fever	Cases of WNV neurological disease	WNV deaths
1999	62	3	59	7
2000	21	2	19	2
2001	66	2	64	9
2002	4156	1162	2946	284
2003	9862	6830	2860	264
2004	2539	1269	1142	100
2005	3000	1607	1294	119
2006	4269	2774	1495	177
2007	3630	2403	1227	124
2008	1356	667	689	44
2009	720	334	386	32
2010	1021	392	629	57
2011	712	226	486	43
2012	5674	2801	2873	286
2013	2469	1202	1267	119
2014	2205	858	1347	97
2015	1197	454	743	63
Total	40754	22986	19526	1827

sequence analysis of the virus genomes it is known that it is linked to virus strains circulating in birds in the Eastern Mediterranean region. The most likely explanation is that the virus was introduced from an infected bird imported (possibly illegally) into the USA. From this initial source of infection, native American mosquitoes were exposed to the virus, and from there it was spread to a wide range of vertebrates as the insects fed.

The spread of the West Nile virus infection in the USA has been extremely well documented and is an excellent case study of how a virus can enter a new territory and spread rapidly though a naïve population. The figures presented in Table 23.2 illustrate some important features of its epidemiology. At first sight, the data appear to indicate that the proportions of fever,

neurological disease and deaths have altered during the course of the spread of the virus, but this is not the case and the figures merely reflect the changes in the method of detection. In the early years of the outbreak, cases of neurological disease were those which were most readily identified. As knowledge of the cause of the disease improved, clinicians were more able to identify the fever associated with infection. Ultimately, testing for evidence of infection has improved and become more widespread and so the number of detected cases has risen. The rapid spread of this virus serves to emphasize the risks posed by a virus when it enters a new territory. The case illustrates that we cannot be complacent about viruses even when they appear to have established a pattern and range of infection over a long period of time.

Prevention and treatment of West Nile virus infection

While effective vaccines for prevention of West Nile virus in horses are available there is no human vaccine. No antiviral drugs to treat West Nile virus infections are routinely available and treatment for West Nile disease is limited to supportive therapy to address the symptoms.

Key points

- Arboviruses are transmitted between hosts by insects that take blood meals from vertebrates. These include mosquitos, sand flies, biting midges and ticks.
- The habitat of the insect vector defines the geographical restriction of the virus infections. Changes in preference for insect vector can alter the geographical distribution of the virus.
- Most arboviruses that can infect humans have an animal reservoir within which the virus is maintained. However, some arboviruses, such as Chikungunya virus, may be maintained by vector-mediated human-to-human infection.
- Arboviruses are spread through the body in the circulatory system and therefore can infect a wide range of tissues.
- Yellow fever virus is maintained in the animal reservoir by a sylvatic cycle involving specific mosquito species and in an urban cycle by *Aedes aegypti* mosquitos.
- Yellow fever is primarily the result of destruction of liver cells and damage to vascular tissue.

- Severe dengue virus disease is the result of dysregulation of the immune system.

Questions

- Discuss the role of mosquito vectors in the transmission of yellow fever virus and Chikungunya virus.
- Discuss the underlying processes that play a role in dengue fever and Chikungunya fever.

Further reading

Burt, F. J., Rolph, M. S., Rulli, N. E., Mahalingam, S., Heise, M. T. 2012. Chikungunya: a re-emerging virus. *The Lancet* 379, 662–671.

Carrington, C. V., Auguste, A. J. 2013. Evolutionary and ecological factors underlying the tempo and distribution of yellow fever virus activity. *Infection, Genetics and Evolution* 13, 198–210.

Chawla, P., Yadav, A., Chawla, V. 2014. Clinical implications and treatment of dengue. *Asian Pacific Journal of Tropical Medicine* 7, 169–178.

Monath, T. P. 2008. Treatment of yellow fever. *Antiviral Research* 78, 116–124.

Murray, N. E., Quam, M. B., Wilder-Smith, A. 2013. Epidemiology of dengue: past, present and future prospects. *Clinical Epidemiology* 5, 299–309.

Paessler, S., Walker, D. H. 2013. Pathogenesis of the viral hemorrhagic fevers. *Annual Review of Pathology* 8, 411–440.

Thiberville, S. D., Moyen, N., Dupuis-Maguiraga, L., Nougairede, A., Gould, E. A., Roques, P., de Lamballerie, X. 2013. Chikungunya fever: epidemiology, clinical syndrome, pathogenesis and therapy. *Antiviral Research* 99, 345–370.

Chapter 24

Exotic and emerging viral infections

We are all aware of the importance of virus diseases in our lives. However, the most dramatic impacts are often seen when a previously unknown viral disease is encountered. This can arise either as an introduction of a virus from another species, or the appearance of an entirely new disease. As human activities extend into new areas in the world the risk of exposure to previously unknown viruses increases.

The world's increasing population generates considerable pressure for people to move into regions that were previously unoccupied. Frequently, this population movement is associated with dramatic changes to the environment by deforestation or alteration in irrigation patterns through the construction of waterways and dams. This increased activity in new regions of the world potentially brings humans and their domestic animals into contact with new agents of disease, including previously unknown viruses. These *'exotic'* viruses may infect humans directly or may first infect domestic animals from which humans may acquire the virus. In addition, known viruses may demonstrate changes in their pattern of infection by extending their host range to include humans, or by showing a rapid increase in severity of illness compared to the historical pattern. In all of these cases, the virus is referred to as *'emerging'* or *'re-emerging'*. In an era of rapid transportation that allows people to travel vast distances immediately following infection and before disease symptoms appear, new viruses can potentially be rapidly disseminated across vast areas.

We constantly encounter new viruses and the 20th century saw the appearance of a number of previously unknown viruses affecting humans. Some of these, such as HIV, have had an immense and devastating impact on the human population (see Chapter 21). Several other viruses have appeared which are associated with a high frequency of deaths in infected people. However, with the notable exception of HIV, in most cases the spread of

Introduction to Modern Virology, Seventh Edition. N. J. Dimmock, A. J. Easton and K. N. Leppard.
© 2016 John Wiley & Sons, Ltd. Published 2016 by John Wiley & Sons, Ltd.

these viruses has been very restricted for one reason or another and the total number of infected individuals has remained low. Examples include the filoviruses, Marburg virus and Ebola virus in Africa and the paramyxoviruses, Nipah virus and Hendra virus in South-East Asia and Australia. The dramatic consequences of West Nile virus introduction into the USA at the beginning of the 20th century demonstrate the potential impact of a virus when it invades new territory (see Section 23.5). As we progress in the early years of the 21st century we have already seen the appearance of two previously unknown viruses, Schmallenberg of sheep and the Middle East Respiratory Syndrome (MERS) virus and a concerning increase in the geographical spread of Chikungunya virus into the Americas. It is clear that we should expect more new viruses to appear in the coming years and we can only hope that their effects can be contained.

In this chapter we will consider six examples of emerging viruses that have been chosen to illustrate different facets of the factors which must be considered when new viruses appear.

24.1 Ebola and Marburg viruses: emerging filoviruses

In 1967, reports appeared of a haemorrhagic fever in primate facilities in Marburg, Germany, and Belgrade, Serbia. The disease affected staff handling material from African green monkeys (*Cercopithecus aethiops*) that originated from Uganda. The infection spread to health workers and family members of the affected staff. Of 31 people affected, seven died. The disease was caused by a previously unknown virus, named Marburg virus, which was present in the monkeys captured in Uganda. The monkeys did not have any disease associated with the infection. This outbreak of infection was contained and quickly died out; however, the

virus next appeared in 1975 when three people in South Africa presented with the disease, one of whom died. Two cases of Marburg were then seen in 1980 and one in 1987 in Kenya, with two deaths, before the appearance of an extended outbreak in the Democratic Republic of Congo where, from late 1998 to late 2000, a total of 154 people were infected with 128 deaths (83% mortality). The largest outbreak to date occurred in Northern Angola between October 2004 and July 2005 with a total of 374 positively confirmed cases and 329 deaths (88% of infected individuals). In 2007, four people in Uganda were infected with two deaths. The last recorded cases were in 2008 with a single fatal case in the Netherlands and one case in the USA where the patient survived. In the last two outbreaks all of the infected individuals had recently visited cave systems in Uganda. Based on these outbreaks, overall mortality associated with Marburg virus is estimated at 82%. No effective treatment for Marburg virus infection has been identified.

Analysis of the genome of Marburg virus showed that it comprised a single-stranded negative sense RNA molecule and that the genome organization was different to those of other viruses of this type (Fig. 24.1). It was therefore placed in the new family of *Filoviruses*. Only one other virus has been placed in this family: Ebola virus.

Ebola virus disease first appeared in 1976 in two simultaneous outbreaks: in Nzara, a town in South Sudan, and in Yambuku village in the Democratic Republic of Congo. Yambuku is located near the Ebola River, from which the disease takes its name. The disease was similar to that seen previously with Marburg virus and its serious nature stimulated rapid interest; analysis showed that it was due to a new virus which, though similar in structure and other features to Marburg virus, was quite distinct. Ebola virus is antigenically distinct from Marburg virus and has been placed in a different genus of the Filovirus family to reflect

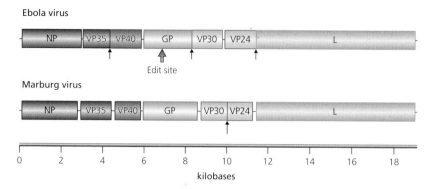

Ebola virus

Marburg virus

Fig. 24.1 The genome organization of Marburg and Ebola virus. The genomes of the filoviruses Marburg virus and Ebola virus comprise a single molecule of negative sense RNA. The genomes transcribe 7 mRNAs, each encoding a single protein. Most mRNAs are separated by an intergenic region of variable length but the mRNAs encoding the VP30 and VP24 proteins of Marburg virus overlap with each other at the position shown by an arrow. Ebola virus mRNAs overlap at three positions (arrowed). The Ebola virus Zaire strain glycoprotein (GP) is translated from an mRNA that contains an additional A residue at the position indicated.

the differences in detail of the molecular biology of both viruses. Since the first report of Ebola infection, a number of outbreaks have occurred in which the human disease has been associated with a very high mortality level (Table 24.1). Until 2014, all of the outbreaks, with mortality ranging from 50–90%, occurred in a relatively restricted region of equatorial Africa but in 2014 an outbreak was identified in Guinea in West Africa and this rapidly spread into the neighbouring countries of Liberia and Sierra Leone. At the time of writing (October 2015), this Ebola outbreak has been the largest to date and only now appears to be ending. A small number of cases originating from the West Africa outbreak were detected in other countries in people who had travelled from the highest areas of infection.

In 1989, a fatal disease in monkeys was reported from a primate centre in Reston, Virginia. This was also shown to be due to Ebola virus but, although serological evidence suggested that some animal handlers may have been infected, this particular isolate did not cause disease in humans. This virus is referred to as the Reston species and is distinct from the other species described to date, Zaire, Sudan,

Bundibugyo and Tai Forest, all of which have caused human pathogenic infection.

For both Marburg and Ebola viruses, transmission between infected and susceptible individuals occurs only via close contact with body fluids, particularly blood, of infected humans or other primates, in which large amounts of virus can be detected. There is no evidence for other transmission routes. The limited route of transmission plays an important role in our ability to restrict the infection, when it appears, by taking appropriate containment measures when dealing with infected patients (and animals). A striking feature is that the Marburg and Ebola outbreaks have occurred in very restricted geographical areas with no spread beyond. This presumably reflects the range of the natural host(s) for the virus from which it is transmitted to humans, which is suspected to be specific species of the *Pteropodidae* family of fruit bats in which they do not appear to cause disease (Fig. 24.2).

Ebola virus and Marburg virus disease

The diseases presented by Marburg and Ebola virus infections in humans are classed as

Table 24.1
Summary of the recorded outbreaks of Ebola virus infection in humans.

Year	Country	Ebola virus species	Cases	Deaths	Mortality (%)
1976	Democratic Republic of Congo	Zaire	318	280	88
	Sudan	Sudan	284	151	53
1977	Democratic Republic of Congo	Zaire	1	1	100
1979	Sudan	Sudan	34	22	65
1994	Gabon	Zaire	52	31	60
	Côte d'Ivoire	Tai Forest	1	0	
1995	Democratic Republic of Congo	Zaire	315	254	81
1996	Gabon	Zaire	91	66	72
	South Africa[1]	Zaire	1	1	100
2000	Uganda	Sudan	425	224	53
2001–2002	Gabon	Zaire	65	53	82
	Congo	Zaire	59	44	75
2003	Congo	Zaire	237	157	66
2004	Sudan	Sudan	17	7	41
2005	Congo	Zaire	12	10	83
2007	Democratic Republic of Congo	Zaire	264	187	71
	Uganda	Bundibugyo	149	37	25
2008	Democratic Republic of Congo	Zaire	32	14	44
2011	Uganda	Sudan	1	1	100
2012	Uganda	Sudan	31	21	68
	Democratic Republic of Congo	Bundibugyo	57	29	51
2014–2015[2]	Guinea	Zaire[3]	3804	2534	67
	Liberia	Zaire[3]	10666	4806	45
	Sierra Leone	Zaire[3]	13945	3955	28
	Secondary affected countries	Zaire[3]	36	15	42
Total			**30899**	**12901**	**42**

[1]The patient was a doctor who treated infected people in Gabon.
[2]Correct at the time of writing. Not all cases have been confirmed as Ebola virus disease.
[3]This virus is closely related to the Zaire strain but is genetically different from the strains that have been identified previously.

haemorrhagic fevers. Both present with very similar signs and symptoms of disease, though some of the precise details of the infections and the underlying causes of symptoms differ slightly. The description here describes the general features of both infections, with some specific information presented for Ebola virus where details of the process of pathogenesis have been determined.

The initial infection is followed by an incubation period of approximately 1 week. The virus is believed to be introduced by direct contact with contaminated material or the body fluids of infected animals or humans. The body fluids of infected humans are known to contain high levels of virus demonstrating a generalized viraemia in which the virus is spread through the body by the circulatory system. The disease begins with an abrupt onset of fever which is often associated with chills and other general symptoms of disease. Analysis of the blood from patients has shown that by the time of onset of

Fig. 24.2 Geographic distribution of members of the *Pteropodidae* family of fruit bats.

disease the virus is already present in high quantities. In patients who progress to the most severe disease, virus levels in blood and other body fluids remain high.

Infected patients may present with a general rash and symptoms of small haemorrhagic events such as petechiae (small spots due to broken capillary blood vessels), ecchymoses (regions of the skin showing discolouration due to ruptured blood vessels immediately under the skin) and persistent bleeding from minor wounds such as needle punctures. These signs are an indication of the severe symptoms that may follow. Within the body, the viruses cause loss of integrity of the endothelial cell layers of the vascular system. With Ebola virus, this is partly due to direct infection of these cells but is also due to indirect effects of infection of other cells. Ebola virus infects many different cell types, including circulating macrophages. The

infected macrophages secrete a range of highly active cytokines which induce responses in adjacent cells, and particularly endothelial cells. The endothelial cells respond to this cytokine 'storm' by inducing apoptosis resulting in cell death (see Section 13.5). The loss of the endothelial cells leads to breakdown of the vascular structure, allowing blood to leak into surrounding tissues. The viruses also infect and kill T and B cells of the immune system and destroy many other lymphoid tissues. The damage to the immune system reduces or prevents clearance of the virus.

As the disease progresses, signs of liver damage can be detected. The loss of vascular integrity leads to reduction of blood pressure and patients develop renal problems that can lead to renal shock which can be fatal. At this time, bleeding from the gastrointestinal tract is often seen. Ultimately, the most severely affected succumb to the effects of the infection, usually in the second week of illness.

For those who survive filovirus infections, the convalescence and recovery is a long process often extending into weeks or months. An unusual feature of Ebola virus infection is that the virus is not cleared from the body quickly, and so is present for long periods even after the symptoms of disease have gone. The practical consequence of this is that recovered patients remain potentially infectious even when they feel well. The virus resides in a variety of sites in the body but the most problematic in terms of ability to be transmitted is the presence of virus in testes which can be transmitted through sexual contact. The extended period of infectivity poses problems for ensuring elimination of the virus and explains some of the sporadic cases at the end of outbreaks.

24.2 Hendra and Nipah viruses: emerging paramyxoviruses

Hendra virus made its first appearance in September 1994 with reports of a severe respiratory disease in racehorses in stables in the suburb of Hendra in Brisbane, Queensland. It was sufficiently severe that 14 of the 21 animals affected died from the infection or were euthanized because of the suffering generated by the disease. During the outbreak, two staff who had been in close contact with the infected horses also became ill with an acute respiratory disease and one of these died. Almost a year later, a third person living in northern Queensland died of the same disease. Subsequent investigation showed that they had worked with sick horses almost 12 months previously at the time of the initial outbreaks and almost certainly contracted the infection at that time. Since the first appearance of Hendra

virus, there have been 47 further outbreaks up to 2013, typically involving small numbers of horses. There have been four further human infections in people with contact with infected horses, leading to two deaths.

Analysis of samples from infected animals identified a previously unknown virus. Determination of the nature of its genome showed that this was a paramyxovirus that resembled members of the Morbillivirus genus such as measles virus, but that it had sufficient differences to justify the creation of a new genus.

In late 1998, an outbreak of acute encephalitis in humans was identified in Malaysia. The outbreak continued into 1999 with 265 cases leading to 105 deaths (a mortality rate of 39%). Most of the patients also presented with signs of respiratory disease. At the same time, 11 similar cases with one death were reported in Singapore. In the Singapore outbreak the affected people were workers in pig abattoirs. The link between pigs and the disease led the Malaysian government to sanction the slaughter of almost 1 million pigs, linked with strict movement controls of the remaining animals. This action appeared to halt the spread of infection in humans, which provided strong circumstantial evidence of a link between pigs and the disease.

A previously unknown virus was isolated from the cerebrospinal fluid of a patient from Sungai Nipah village which provided the name Nipah virus to the new agent. Analysis showed that Nipah virus was extremely similar genetically and structurally to Hendra virus and both were assigned to the new genus named Henipavirus (Fig. 24.3). Subsequently, with the availability of new diagnostic tools, Nipah virus was unambiguously shown to be associated with the encephalitis outbreaks and also with an infection of pigs that has a mortality rate of approximately 5%. It was also shown that a small proportion of people who survived the

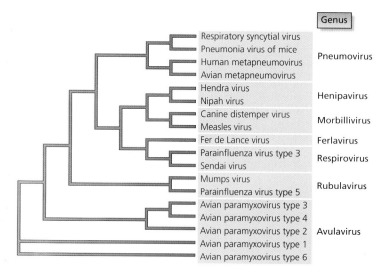

Fig. 24.3 Phylogenetic relationships of selected members of the Paramyxovirus family. The evolutionary relationship of selected members of the Paramyxovirus family based on the conserved polymerase protein. Representative members of the different genera are shown

initial infection succumbed to encephalitis many months later. Since the initial outbreaks, there have been several more Nipah virus outbreaks in South East Asia and India, though in these there has been no further associated infection of pigs. A consideration of the economic impact of Nipah virus infection in Malaysia is given in Section 29.2.

The similarities of Hendra and Nipah viruses are very striking. In the Hendra virus outbreaks and the first Nipah virus outbreak the serious diseases seen in horses and pigs, respectively, strongly suggested that these animals were not the natural host but were intermediate hosts for the virus which then infected humans. However, the lack of an intermediate host in the subsequent Nipah virus infections suggested that humans may also be infected directly from the natural host. Analysis of many animal species in the affected areas quickly identified fruit bats as the most likely potential animal reservoir and natural host of these viruses and this has been borne out by subsequent studies. Both

Hendra and Nipah viruses are harboured by Pteropid fruit bats, just as for the filoviruses. However, given the geographical distribution of the henipavirus infections exclusively in the eastern region of the distribution of members of the *Pteropodidae* family of fruit bats (Fig. 24.2), it is likely that only one or a few species of bats restricted to this region are involved.

Hendra virus and Nipah virus disease

The henipaviruses are transmitted by the respiratory route and the initial infection is in the respiratory tract. Both viruses are able to infect the brain, causing encephalitis in the case of Nipah virus and a range of encephalitis-like symptoms ranging from mild (e.g. headaches) to severe (convulsions and coma) in the case of Hendra virus. The infections of both Nipah and Hendra virus begin in the epithelial cells of the lower respiratory tract. This is accompanied by significant death of alveolar cells in the lung which is also associated with the generation of type 1 interferon and a broad range of pro-inflammatory cytokines that stimulate elements of the immune system. The combination of these various effects is severe respiratory disease with congestion leading to acute respiratory distress. The virus can also infect the endothelial cells of the vascular capillary system of the lungs. The damage caused by infection leads to haemorrhage in the lungs and entry of the virus into the circulatory system where it can circulate throughout the body either as free virus particles or by binding to the surface of leukocytes.

From the circulation the viruses are able to target many organs of the body; those targeted by the henipaviruses include the spleen, kidneys and brain. The damage caused to these tissues is significant and contributes to the observed symptoms. The mechanism by which the viruses enter the central nervous system is not yet fully clear. The blood–brain barrier is usually an effective block to prevent virus access to the brain. However, the inflammatory response induced by infection compromises the integrity of the barrier and this may allow virus infection to occur. Equally, it is possible that some virus enters the brain directly by transport along the olfactory nerve. Once in the brain, the consequences for infected individuals are severe. In patients who survive infection the virus is cleared from the body through the action of cell-mediated immunity (see Section 14.2).

24.3 SARS and MERS: emerging coronaviruses

In late February 2003, a patient presented in Hanoi, Vietnam, with symptoms of a severe,

acute respiratory infection. Early symptoms were similar to those of influenza but it became rapidly clear that this was a novel infection which was considerably more severe than a typical bout of flu. A dramatic situation arose when several health workers in contact with the patient also became ill with similar symptoms. The initial patient was known to have travelled recently from Hong Kong, and in March 2003 almost 50 medical staff there also presented with severe respiratory disease which had been contracted from a patient admitted in mid-February. At the same time, Chinese authorities reported that since the previous November over 300 cases of atypical pneumonia resulting in five deaths had been recorded in the Guangdong Province. Over the following months, as the disease spread it became clear that the Chinese outbreak was the initial stage of an outbreak of a previously unknown virus that had then spread to these other locations. For a period of some weeks the new disease made headlines around the world and led to scenes reminiscent of those seen in the 1918 influenza pandemic. A startling aspect of the disease, termed severe acute respiratory syndrome (SARS), which led to significant alarm, was its rapid spread around the world. The infection appeared in other parts of Asia, Canada and Europe and, by the end of February and within a further month or so, it had been reported in most continents of the world (Table 24.2). This rapid spread was not due to an inherent quality of the virus but to the transport of infected individuals by air travel. Recognizing this, many countries imposed travel restrictions to reduce the risk of introduction of the virus. Between the first cases in China and the last recorded case in July 2003 in Taiwan, it is estimated that 8098 people were infected of whom 774 died, a mortality rate of 9.5%. Since July 2003, there have been no further confirmed cases of SARS.

The transmission of SARS requires close contact between infected and susceptible

Table 24.2		
Transmission of SARS throughout the world.		
First recorded case	**Country (region)**	**Last recorded case**
16 November 2002[*]	China (Guangdong)	7 June 2003
15 February 2003	China (Hong Kong)	22 June 2003
23 February 2003	Vietnam (Hanoi)	27 April 2003
23 February 2003	Canada (Toronto)	2 July 2003
25 February 2003	Singapore	31 May 2003
25 February 2003	Taiwan	**5 July 2003**[*]
2 March 2003	China (Beijing)	18 June 2003
4 March 2003	China (Inner Mongolia)	3 June 2003
8 March 2003	China (Shanxi)	13 June 2003
28 March 2003	Canada (Vancouver)	5 May 2003
1 April 2003	China (Jilin)	29 May 2003
5 April 2003	Mongolia (Ulaanbaatar)	9 May 2003
6 April 2003	Philippines (Manila)	19 May 2003
12 April 2003	China (Shaanxi)	29 May 2003
16 April 2003	China (Tianjin)	28 May 2003
17 April 2003	China (Hubei)	26 May 2003
19 April 2003	China (Jiangsu)	21 May 2003
19 April 2003	China (Hebei)	10 June 2003

[*]The first and last recorded cases are highlighted in bold.

individuals such as is seen in the hospital setting. However, the virus is probably spread by respiratory transmission, in droplets shed by coughing or sneezing and this means that the risk of spread is much higher than seen for Marburg or Ebola viruses. The rapid spread and severity of SARS led to a huge international effort to identify the cause, which was achieved by two independent research groups in May 2003. The virus was shown to be a new member of the coronavirus family, members of which have been known to be associated with respiratory disease in some hosts, including humans. The new virus, called the SARS coronavirus, has now been extensively studied in an attempt to assess its likely risk to humans in the future.

Attempts to identify the origin of SARS rapidly identified the live-animal markets in China as a potential source of infection.

Attention focused particularly on the masked palm civet (often called a civet cat) and racoon dog, within some individuals of which SARS coronavirus was identified. Serological analysis of traders in the animal markets showed that a significant proportion of them carry SARS coronavirus-specific antibodies, suggesting that human infection is a relatively frequent event in these environments, though disease is not common. The conclusion of these studies was that while the animals in the markets were the source of transmission to humans they were not the primary host for the virus. Further studies showed that the SARS coronavirus is closely related to a coronavirus that infects Chinese horseshoe bats. The current hypothesis is that the SARS coronavirus originated in bats from which it infected a number of intermediate mammalian hosts. This transition to new hosts required the virus to undergo mutation;

humans then acquired the virus from these animals. However, the virus appears to have required further mutation in the spike protein genes to permit efficient replication in a human host.

The worldwide spread of the SARS coronavirus due to air travel is the first example of its kind and raises many questions about our ability to control such infections. It also raises many unsolved questions about other viruses such as influenza A and B whose epidemiological pattern of winter infection in both hemispheres does not seem to be affected by being spread around the world by air travel. A consideration of the economic impact of the SARS outbreak is given in Section 29.1.

A report of a fatal case of respiratory disease in the Kingdom of Saudi Arabia in September 2012 marked the appearance of another previously unknown coronavirus capable of causing infections in humans. The incidence of disease has shown periodic peaks and, by October 2015, a total of 1,595 confirmed or suspected cases had been recorded with 571 deaths in the confirmed cases (a confirmed mortality rate of 36%). The total number of cases continues to rise. The countries affected are primarily in the Middle East and Northern Africa with a number of cases recorded in other parts of the world in patients with recent links to the Middle East. Determination of the genome organization of the new virus showed that it was related to the SARS coronavirus and it was named Middle East Respiratory Syndrome (MERS) coronavirus. While the rate of human-to-human spread does not approach that of the SARS coronavirus the severity of the disease has raised concerns about the implications of wider spread.

Early indications have implicated dromedary camels as the host for transmission of the MERS coronavirus to humans. However, as with palm civets and SARS coronavirus, camels are suspected to be intermediate hosts for the MERS coronavirus. Coronaviruses with similarities to the MERS virus have been found in bats in many countries though none contain the MERS virus itself. This suggests that either the MERS virus has undergone mutation to be able to infect new hosts, as was the case with the SARS coronavirus, or that the actual reservoir for MERS coronavirus has not yet been identified. One study identified a coronavirus closely related to the MERS virus in the European hedgehog.

SARS virus and MERS coronavirus disease

Both SARS and MERS coronavirus infections in humans involve severe respiratory disease. Considerably more is known about the SARS disease but both viruses appear to follow similar processes of infection leading to illness. The viruses infect the epithelial cells of the respiratory tract and in SARS infection the virus load in the early stages of the infection is low, rising later to maximal levels around the time of the appearance of symptoms. This feature is unusual and explains why the virus is spread less well in the incubation phase but more efficiently when the patient is symptomatic. This also explains the high risk of transmission of SARS to health workers treating infected patients. MERS virus does not display this feature and transmission appears greatest in the early stages of infection. When the infection with these coronaviruses reaches the lower respiratory tract it results in significant damage, both directly as a result of the virus infection and indirectly due to the induction of a profound inflammatory response. Together these produce haemorrhaging and consolidation (Fig. 19.4) due to the influx of inflammatory cells. The result in the most severe cases is fatal pneumonia frequently associated with additional complications as the viruses also infect other tissues, particularly the kidneys. The viruses can also spread to the central nervous system to produce a range of disease symptoms. There have also been reports of damage to skeletal and cardiac muscle.

24.4 Predicting the future: clues from analysis of the genomes of previously unknown viruses

Predicting the future is an uncertain process. Despite many decades of study of influenza viruses it has not been possible to predict the origin of future pandemic strains and this is likely also to be the case for other emerging viruses. What is certain is that we will continue to encounter previously unknown viruses which may pose a significant threat to human health as our activities expose us to new habitats. Viruses that are relatively benign in their natural hosts are likely to pose a threat to other species if they are able to infect so it is important that we learn more about the spectrum of viruses in the natural world: this is often referred to as the *virome* (Section 5.8). Similarly, we can be confident that the pattern of disease displayed by a number of known viruses will alter and that new areas of the world will experience infections by some of these; the spread of Dengue and Chikungunya viruses is an indication of this. The question arises: what can we do to try to predict where new threats may arise so that we can focus our attention on them?

New developments in technology are beginning to provide us with information that may be useful in identifying places we should look for potential threats. In particular, the ability to sequence all nucleic acids present in animals, including the genomes of the viruses they are infected with, is demonstrating that our current knowledge of viruses is very limited and that there are many more in the world than we have been aware of hitherto. An example of this is the number of viruses identified as infecting bats. To date, sequences of more than 4000 different viruses from greater than 20 virus families have been identified in a range of bat species from different parts of the world. Many of these show only low levels of similarity to the genomes of known viruses. A similar approach with domestic pigs has also identified previously unknown viruses. Our experiences of the viruses considered in this chapter, all of which have an association with bats, demonstrate the importance of further study. As the new generation sequencing approach is extended to other animal species we will undoubtedly discover many more viruses that we must watch carefully for transmission to humans.

Key points

- Emerging viruses arise when humans explore new territories and become exposed to infection.
- Emerging viruses are transmitted to humans from other species in which they typically do not cause serious disease. Transmission often involves an intermediate mammalian host.
- Bats have been implicated as natural reservoirs of several recent emerging viruses.
- New nucleic acid sequencing technologies have demonstrated that the virome is considerably more extensive than previously thought.

Questions

- Discuss the role of bats and intermediate hosts in the emergence of previously unknown viruses.
- Compare the emergence of the henipaviruses with that of the SARS and MERS coronaviruses.

Further reading

Aljofan, M. 2013. Hendra and Nipah infection: emerging paramyxoviruses. *Virus Research* 177, 119–126.

Chen, L., Liu, B., Yang, J., Jin, Q. 2014. DBatVir: the database of bat-associated viruses. *Database 2014*, Article ID bau021, doi: 10.1093/database/bau021.

Coleman, C. M., Frieman, M. B. 2014. Coronaviruses: important emerging human pathogens. *Journal of Virology* 88, 5209–5212.

Croser, E. L., Marsh, G. A. 2013. The changing face of henipaviruses. *Veterinary Microbiology* 167, 151–158.

Hilgenfeld, R., Peiris, M. 2013. From SARS to MERS: 10 years of research on highly pathogenic human coronaviruses. *Antiviral Research* 100, 286–295.

Kortepeter, M. G., Bausch, D. G., Bray, M. 2011. Basic clinical and laboratory features of filoviral hemorrhagic fever. *Journal of Infectious Diseases* 204 Supplement 3: S810–6.

Chapter 25

Carcinogenesis and tumour viruses

It is estimated that viruses are a contributory cause of 20% of all human cancers but, in each case, infection is just one factor that contributes to disease development. As a result, these viruses cause cancer in only a small percentage of the people who become infected. Studying viruses that can cause cancer in people or in laboratory animals has revealed a great deal about the processes that underlie the formation of all cancers, not just those that are caused by viruses.

Chapter 25 Outline

25.1 Immortalization, transformation and tumourigenesis
25.2 Oncogenic viruses
25.3 Polyomaviruses, papillomaviruses and adenoviruses: the small DNA tumour viruses as experimental models
25.4 Papillomaviruses and human cancer
25.5 Polyomaviruses and human cancer
25.6 Herpesvirus involvement in human cancers
25.7 Retroviruses as experimental model tumour viruses
25.8 Retroviruses and naturally-occurring tumours
25.9 Hepatitis viruses and liver cancer
25.10 Prospects for the control of virus-associated cancers

The great majority of viruses of vertebrates are not oncogenic, i.e. they do not have the ability to initiate a cancer. However, for many years it has been recognized that certain viruses can induce tumours in appropriate experimental animals. More recently, good evidence has implicated some virus infections in the development of specific human cancers or naturally-occurring cancers in animals. As a result of the importance of understanding human cancer as a route to finding effective treatments, those viruses that caused such disease experimentally were subjected to intensive study from the 1960s onwards. Our detailed understanding of these viruses today owes much to this driving motivation for research. This chapter describes some of the most important viral associations with human tumours and the underlying mechanisms involved.

Introduction to Modern Virology, Seventh Edition. N. J. Dimmock, A. J. Easton and K. N. Leppard.
© 2016 John Wiley & Sons, Ltd. Published 2016 by John Wiley & Sons, Ltd.

25.1 Immortalization, transformation and tumourigenesis

Normal cells, i.e. cultures established from healthy tissue and then passaged, senesce and die after a fixed number of divisions (about 50). No matter how carefully they are looked after, their lifespan is limited. Immortalized cells, in contrast, can be passaged indefinitely in cell culture, allowing them to form permanent cell lines. A key event in immortalization is the reactivation of telomerase, the enzyme that maintains the specialized DNA sequences at the ends of the chromosomes; the absence of this enzyme in somatic cells leads to a progressive reduction in telomere length at each cell division that is an important trigger for senescence. Transformed cell lines are recognized by further changes in phenotype compared with immortalized cells, such as reduced contact inhibition (meaning the cells now grow over one another) and lower dependence on animal serum for survival in culture. Some members of the same families of viruses that contain tumour-associated viruses are able to immortalize and transform appropriate primary mammalian cells in culture and/or to transform already immortalized cells. Transformation usually results in visible changes in cell appearance, a process known as morphological transformation. The properties of transformed cell lines resemble in many ways those of cell lines isolated from tumour tissue and the process of transformation in cell culture is closely linked with that of tumourigenesis in vivo (Box 25.1).

There is now overwhelming evidence that the mechanisms underlying immortalization, transformation and tumourigenesis involve alteration to a cell's DNA, i.e. mutation. In principle, mutation could lead to these changes through either the loss of functions that are required for normal cell behaviour or the acquisition of functions that disrupt normal behaviour, mediated either by changes to existing genes or the arrival of new genes. Genes whose loss is tumourigenic are known as tumour-suppressor genes, whilst genes that can promote tumour formation if they sustain mutations that either alter the protein product in specific ways or increase its level of expression are known as oncogenes. Oncogenes are involved in controlling cell signalling and

---

Box 25.1

Evidence for a link between transformation and tumourigenesis

- Transformed cells, but not normal or simply immortalized cells, can often form tumours in suitable animal hosts.
- The same genetic changes that have been found to underlie transformation of cells in culture are also found frequently in naturally-occurring tumours.
- Several of the defining properties of transformed cells reflect decreased cell adhesion and increased mobility, which are essential features of invasive malignant cells in vivo.
- Cell lines established from tumours have similar properties to transformed cell lines.
- Where a virus can both transform cells in culture and cause tumours in vivo, the viral genes that are required for the two processes are often the same.

regulating the cell division cycle. Applying the term 'oncogene' to a normal gene that has such activity only when mutated is rather confusing; therefore, the non-mutated forms of these genes are sometimes called proto-oncogenes, to distinguish them from their mutated, oncogenic forms.

The involvement of viruses in oncogenesis can be through altering the expression of proto-oncogenes or tumour suppressor genes and/or through the provision of new virus-encoded functions that are directly oncogenic. Remember, though, that transformation and tumourigenesis are multi-step processes in which the input of viral genes is just one possible event; it is the accumulation of a set of changes that leads to the tumour cell phenotype. These changes may have to occur in a defined order or may accumulate randomly (Fig. 25.1). Thus there may be multiple routes by which a cell can reach a malignant state.

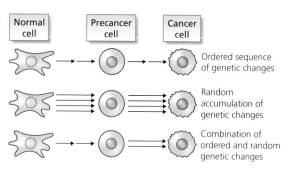

Fig. 25.1 Alternative scenarios for multi-step progression from a normal to a cancer cell occurring by the acquisition of genetic alterations, sometimes including the effect of specific virus genes. Genetic alterations are symbolized by arrows, though the number of arrows shown is not intended to indicate the actual number of events involved in converting a normal cell into a fully malignant cell. The specific nature of early events in the pathways probably predisposes cells to suffer further mutations.

As well as the acquisition by cells of an array of genetic alterations, tumour formation involves complex interactions with the immune system of the host animal. The immune system is important as it can kill cells bearing viral or non-self tumour antigens (even tumour cells arising without virus involvement typically display novel antigens on their surface) and so limit tumour development. Hence, a virus able to cause transformation in culture is not necessarily able to cause malignancy in vivo. However, studying transformation in culture has been very revealing of the events required for the development of malignant potential by cells.

25.2 Oncogenic viruses

A diverse collection of viruses is associated with naturally-occurring or experimental tumours. To have the chance to be a cause of a tumour, a virus must be capable of a non-cytolytic interaction with cells and, if genes from the virus are to be able to contribute to the ongoing maintenance of a tumour cell phenotype during rapid cell division, they must be stably inherited by daughter cells; this means either becoming integrated into the cell genome or being replicated episomally.

All the viruses listed in Table 25.1 are known or suspected to be oncogenic in humans or animals (for the meaning of 'oncogenic' and other cancer-related terms, see Box 25.2). Particularly good evidence links hepatitis B and C viruses (HBV, HCV) with primary hepatocellular carcinoma (HCC), Epstein-Barr virus (EBV) with nasopharyngeal carcinoma and Burkitt's lymphoma, and specific human papillomaviruses (HPV) with cervical cancer. Complete observation of Koch's postulates to

Table 25.1

Viruses known or suspected to have oncogenic properties in humans, plus some selected non-human tumour viruses.

Virus family	Type	Tumour	Animal*	Co-factor/Comment
Adenovirus	Human Ad9	Benign fibroadenoma, sarcoma of breast tissue	Female rat	Newborn only; not natural host
	Human Ad12	Sarcoma	Hamster	Newborn only; not natural host
Flavivirus	Hepatitis C	Primary hepatocellular carcinoma	Humans	? Alcohol, smoking
Hepadnavirus	Hepatitis B	Primary hepatocellular carcinoma	Humans	Age at infection, aflatoxin, alcohol, smoking
Herpesvirus	Epstein–Barr	Burkitt's lymphoma	Humans	Malaria, genome rearrangements
		Nasopharyngeal carcinoma	Humans	Dietary nitrosamines, HLA genotype
		Hodgkin's disease	Humans	?
		Immunoblastic lymphoma	Humans	Immunodeficiency
	HHV-8	Kaposi's sarcoma	Humans	Immunodeficiency, HLA genotype
		Multicentric Castleman's disease	Human	?
		Primary effusion lymphoma	Human	?
	Herpesvirus saimiri	Experimentally induced lymphomas and leukaemias	Owl monkey	Not the natural host
Papillomavirus	Marek's disease virus	T cell lymphoma	Chicken	Species
	Human papilloma type 16, 18, 31, others	Cervical neoplasia	Humans	Smoking
	Bovine papilloma type 4	Warts, fibroepithelioma	Cattle, hamster, rabbit	Age (newborn), consumption of bracken fern
Polyomavirus	SV40	Gliomas, fibrosarcomas (various tumours – now discounted)	Hamsters (Human)	Newborn; not the natural host Laboratory contamination error
	Mouse polyomavirus	Sarcoma, carcinoma, mutiple tissues	Mouse, other rodents	Newborn, mouse strain, HLA genotype
	Merkel cell polyomavirus	Merkel cell tumours	Humans	?
Retrovirus	Human T cell lymphotropic virus type 1	Adult T cell leukaemia, lymphoma	Humans	?
	Rous sarcoma virus	Sarcoma	Chicken	None
	Mouse mammary tumour virus	Adenocarcinoma	Mouse	Strain, gender, hormone status
	Feline leukaemia virus	T cell lymphosarcoma, leukaemia, fibrosarcoma	Cat	?
	Jaagsiekte sheep retrovirus	Pulmonary adenocarcinoma	Sheep	?

*Tumours are restricted to the species mentioned; this is the natural virus host unless indicated otherwise. HLA, human leucocyte antigen (i.e. human form of major histocompatibility complex, or MHC antigen).

Box 25.2

Definitions of some cancer-related terms

Adenocarcinoma A carcinoma developing from cells of a gland.

Benign An adjective used to describe growths which do not infiltrate into surrounding tissues. Opposite of malignant.

Cancer Malignant tumour; a growth that is not encapsulated and which infiltrates into surrounding tissues, the cells of which it replaces by its own. Its cells are spread by the lymphatic vessels to other parts of the body (metastasis). Death is caused by destruction of organs to a degree incompatible with life, extreme debility and anaemia or by haemorrhage.

Carcinogenesis Complex multi-stage process by which a cancer is formed.

Carcinoma A cancer of epithelial tissue.

Fibroadenoma Tumours of mixed cell type originating from fibrous and glandular tissue.

Fibroepithelioma Tumours of mixed cell type originating from fibrous tissue and basal cells of the epidermis.

Fibroblast A cell derived from connective tissue.

Leukaemia A cancer of white blood cells.

Lymphoma A cancer of lymphoid tissue.

Malignant In the context of carcinogenesis, an adjective describing a tumour which grows progressively and invades other tissues. Opposite of benign.

Mesothelioma A tumour of the cells lining the body cavity, surrounding the lungs.

Neoplasm An abnormal new growth, i.e. a cancer.

Oncogenic Tumour-causing.

Sarcoma A cancer developing from fibroblasts.

Telomerase The enzyme that maintains the length of the telomeres at the ends of the chromosomes; it is normally inactive in somatic cells.

Transformation A constellation of phenotypic changes of cells in culture.

Tumour A swelling, due to abnormal growth of tissue, not resulting from inflammation. May be benign or malignant.

---

prove causation is out of the question so other criteria, such as epidemiological data, are also used to make these associations.

What features, if any, do these tumour viruses have in common? Almost all involve the nucleus in their lifecycle and have their genetic material in DNA form at some stage. In fact, representatives of most major families of DNA viruses are associated with cancer, with the exception of the poxviruses. As poxviruses replicate in the cytoplasm, they do not have the same opportunity to affect the function of the nucleus as do the other DNA viruses. It is important also to realize that the genes of a virus can initiate or contribute to tumourigenesis only if the virus infection does not kill the cell. For almost all viruses identified as tumourigenic, a natural mechanism exists whereby cells can potentially survive an infection, giving the chance that they will evolve an altered phenotype subsequently. The herpes- and retroviruses routinely establish latent and non-cytopathic infections respectively in their natural hosts (see Chapters 9 and 17). The papillomaviruses, with their bi-partite lifecycle (see Section 10.3), maintain their genomes in cells which do not support virus production while polyomaviruses can establish persistent productive infections in their natural hosts. Finally, it is important to consider the genetic status of the virus. Within a population of virus particles of any type, there will typically be many defective particles which, through random mutation, lack some functions essential for productive infection. If the genes which are important for transformation or tumour formation remain intact while the ability to replicate productively and kill the cell is lost, then such a particle may be capable of transforming what is normally a permissive cell.

Survival of the infected cell is only part of the story. If viral genes are to contribute actively to an altered cell phenotype which includes uncontrolled cell growth, they must be successfully inherited by all the daughter cells of the originally infected cell. This can be achieved most easily by integration of the viral DNA into the host genome. Of the tumourigenic viruses, only the retroviruses have an integration function as part of their normal lifecycle. Alternatively, the viral DNA may carry sequences in its genome which allow it to be copied during the DNA synthesis phase (the S phase) of the cell cycle and partitioned between daughter cells at subsequent mitosis. Both Epstein-Barr virus (and possibly other oncogenic herpesviruses) and the papillomaviruses can do this, although papillomavirus-induced tumours ultimately contain an integrated portion of the viral DNA. If there is no other mechanism of DNA persistence, the relevant viral genes must become inserted into the host genome by random non-homologous recombination which is a rare event, occurring experimentally in roughly 1 in 10^5 infected non-permissive cells. This is how adenovirus, SV40 or polyomavirus DNA is retained in transformed and tumour cells, and how papillomavirus DNA eventually becomes integrated during tumour development.

Although maintaining key viral genes in the cell to contribute to the altered phenotype on an ongoing basis is the most obvious mechanism by which viruses might cause tumourigenesis, there is an alternative. It is possible to conceive of viruses acting transiently, contributing to tumourigenesis in a 'hit-and-run' mechanism, e.g. by promoting genome instability in the host cell and hence the acquisition of mutations in critical host genes. Once such mutations have arisen, the viral genes would not need to be sustained within the cells all the way to the appearance of disease. This mechanism may be applicable to hepatitis C virus.

For each of the viruses associated with human malignancies, infection does not lead inevitably to cancer. Carcinogenesis is multi-factorial and hence is a rare event even though the associated virus infection may be common. Various factors, such as host genotype, age at infection, diet, environmental carcinogens (other than viruses) and other invading organisms, may all contribute to the process of virus-associated oncogenesis. Thus to call a virus a 'tumour virus' is rather misleading, as the name refers to one relatively infrequent aspect of its lifecycle. A more accurate name would be 'tumour-associated virus'. A tumourigenic interaction with a host offers essentially no advantage to the virus. Usually, it loses the ability to produce progeny and to transmit to other hosts in the course of such an interaction, whilst threatening the life of its existing host.

25.3 Polyomaviruses, papillomaviruses and adenoviruses: the small DNA tumour viruses as experimental models

The polyomaviruses, papillomaviruses and adenoviruses are completely distinct in the molecular details of their gene expression. They also have completely different disease profiles: adenoviruses cause a variety of respiratory, gastrointestinal tract or eye infections; polyomaviruses cause urinary tract infections; and papillomaviruses cause warts on various epithelial surfaces. It is particularly interesting, therefore, that the mechanisms by which these viruses can immortalize and transform appropriate cells in culture should be so similar.

Genetics of viral transformation

SV40, polyomavirus, human adenoviruses and bovine papillomavirus type 1 (BPV1) can transform appropriate non-permissive rodent cell types in culture (human papillomavirus effects on human cells are considered further in Section 25.4). While SV40, polyomavirus and adenovirus DNA become integrated during this process, BPV1 DNA is maintained as an episome. Studies of these transformed cells revealed which were the important viral genes for transformation in each case (Box 25.3); these genes are summarized in Box 25.4 and are the same genes that are required for tumourigenesis. All of these genes are essential for normal virus replication in their natural target cells.

Studies of the adenovirus transforming genes were important in developing the idea that multiple genetic changes are needed to observe transformation or the appearance of a tumour.

Box 25.3

Evidence that defined the transforming gene(s) of BPV1, adeno- and polyomaviruses

- Adenovirus, SV40 and polyomavirus:
  - In cells transformed by these viruses, viral DNA is stably integrated into the host genome, as detected by Southern blotting, without any apparent preference as to the integration site.
  - Typically, only part of the genome is integrated; comparing the retained sequences between multiple cell lines to find the regions in common identified the crucial genome regions for transformation, i.e. the transforming genes.
  - Cloned genomic DNA from these viruses can transform cells; by cloning progressively smaller genome fragments and testing their transforming capacity, the individual transforming genes were identified.
- BPV1:
  - Transformation is achieved with cloned genome in a plasmid, which is maintained episomally.
  - BPV1 transforming genes were identified by testing the transforming capacity of cloned sub-fragments.
  - Once genes essential for BPV1 replication were deleted, transformation required random integration of the transforming genes as for SV40, etc.

Box 25.4

The oncogene products of the small DNA tumour viruses

Virus	Protein(s) implicated in cell transformation
SV40	Early region, large T antigen
polyoma	Early region, large T antigen plus middle T antigen
human papilloma	E6 and E7 proteins
bovine papilloma type 1	E5, E6 and E7 proteins (depending on cell type)
human adenovirus	E1A and E1B proteins

Neither the E1A nor E1B genes alone were sufficient to produce permanently transformed cell lines from baby rat kidney (BRK) cells, but both genes together could efficiently transform these cells (Fig. 25.2). E1A alone produced abortive transformants, cells which appeared transformed but which could not be stably maintained. Thus, at least two separate functions are needed for adenovirus to fully transform BRK cells. Only the fully transformed cells were tumorigenic in congenitally T-cell-immunodeficient mice (known as athymic nude mice). Similarly, polyoma virus large T and middle T genes together are needed for full transformation by this virus.

In contrast, SV40 large T antigen alone is able to fully transform BRK cells. Further experiments showed that it contains three separate transformation functions, one or more of which is needed to achieve cell transformation depending on which cells are used for the experiment. Thus, some T antigen mutants are defective for transformation of one rodent cell line but not another. This result is taken as a reflection of the different level of mutation already present in the genomes of these different cell lines before introducing the T antigen gene. Even the most 'normal' cell line in culture has undergone some genetic changes during its establishment and so the further

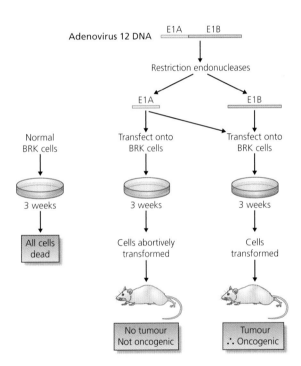

Fig. 25.2 Demonstration of cooperation between gene functions to achieve full oncogenic transformation. While the E1A region alone of adenovirus type 12 is transforming, the cells do not survive indefinitely (abortive transformation). Only cells carrying the adenovirus type 12 E1B gene as well are fully transformed and tumorigenic in syngeneic rats. BRK, baby rat kidney. The E1B gene alone gives no transformation.

changes that are needed before a transformed phenotype is observed will vary.

The role of the immune response in controlling viral oncogenicity

The E1A and E1B genes of a range of different human adenovirus serotypes, e.g. Ad5 and Ad12, can transform rodent cells in culture. However, only a few of these serotypes, such as Ad12 but not Ad5, can cause tumours in normal immunocompetent animals. The difference between these viruses resides in their E1A genes (Box 25.5). In addition to its role in cell immortalization/transformation, Ad12 E1A is able to block T-cell recognition of transformed cells whereas Ad5 E1A is not. It seems that Ad12 oncogenicity resides in the ability of the virus to 'hide' the cells it has transformed from the immune system, a form of specific immunosuppression. However, it is important to remember that this Ad12 E1A function has the primary purpose of optimizing the survival of the virus during natural infections rather than enhancing its tumourigenicity. These

studies of adenovirus oncogenicity demonstrate how important the immune system is in preventing the growth of tumours that would otherwise arise.

The activities of virus-encoded proteins leading to transformation and tumourigenesis

The mechanisms by which SV40 T antigen, adenovirus E1A, E1B proteins, etc. contribute to transformation are now well understood. In each case, the protein achieves its effect(s) through one or more specific interactions with host cell proteins involved in regulating the mammalian cell cycle. Remarkably, these interactions are very similar for each virus, revealing how fundamental the disruption of the cell cycle is to their normal lifecycles. The domain structures of these proteins is summarized in Fig. 25.3. Genetic studies have shown that each of the interactions highlighted is important for transformation by that protein. The following is a simplified account of some of these interactions. For more detail, readers are

---

Box 25.5

Evidence that an adenovirus E1A function determines adenovirus oncogenicity

- BRK cells transformed with Ad12 E1A plus E1B genes generate tumours in syngeneic rats whereas cells transformed with Ad5 genes do not. (Syngeneic means genetically identical to the cells used; such animals are used to avoid rejection of the tumour cells through immunological recognition of foreign rat antigens.)
- When the E1A and E1B genes from Ad5 and Ad12 are mixed and matched, cells transformed with Ad12 E1A plus Ad5 E1B genes are tumourigenic in this assay whereas cells with Ad5 E1A plus Ad12 E1B genes are not.
- Both Ad5- and Ad12-transformed cells express viral antigens but only the Ad5-transformed cells are lysed by activated cytotoxic T lymphocytes (CTL).
- Ad12 E1A functions turn off expression of major histocompatibility complex (MHC) class I antigen, which is essential for antigen presentation to CTL.

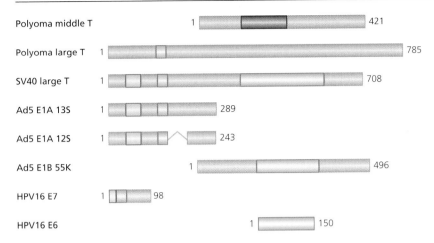

Fig. 25.3 The transforming proteins of SV40, polyoma, papilloma and adenovirus. Horizontal bars represent the proteins indicated (not to scale). The adenovirus E1A 13S and 12S polypeptides are highly related, differing only in the removal (by splicing of the mRNA) of 46 amino acid residues from the shorter protein. For expression and sequence relatedness of the SV40 and polyoma proteins, see Section 10.2. The various shaded blocks indicate functional rather than sequence relatedness, although there is sequence similarity between the indicated Rb-binding sequences. Orange - Rb-binding; turquoise - p53 binding; green - accessory region required to release transcription factors from bound Rb; red - Src protein binding. For definitions of Rb, p53 and Src, see text.

referred to texts dealing with mammalian cell biology.

One of the targets of the viral transforming proteins is the cellular protein called Rb or p105, a tumour suppressor gene product that is a key cell cycle regulator in mammalian cells. The Rb protein prevents unscheduled S-phase (DNA synthesis) in the cell by binding and rendering inactive a transcription factor needed to turn on essential S-phase genes. When a cycle of DNA synthesis and division is correctly signalled, Rb protein is specifically phosphorylated, so releasing its bound transcription factor which allows S-phase to proceed. SV40 and polyoma large T antigens, the adenovirus E1A proteins and papillomavirus E7 protein all inactivate Rb as well. Each one binds to Rb via a protein sequence (Fig. 25.3, orange) that contains a short conserved sequence, Leu-X-Cys-X-Glu. Binding allows a second region in each protein

(Fig. 25.3, green) to release the transcription factor and so enable S-phase progression. Two other relatives of Rb are targeted in the same way. Each of the small DNA tumour viruses needs the cell it infects to enter S-phase to support its own replication, so these viral functions are absolutely crucial to infection in vivo, where newly-infected cells typically will not be dividing rapidly or even at all. However, when allowed to act outside of the normal context of infection, these activities promote transformation because they remove a key negative regulation on continuous cell division. The fact that the Rb gene is a tumour-suppressor means that its inactivation by direct mutation has been observed to lead to tumours, again demonstrating the mechanistic link between transformation and tumourigenesis.

Another mechanism common to the small DNA tumour viruses involves the very important cell protein p53. This protein acts as a sensor of the replicative health of the cell, preventing the survival of damaged cells that might otherwise threaten the health of the organism. Whenever the cell's DNA is damaged, p53 activity is induced and this leads firstly to the cell cycle being halted, to allow DNA repair systems a chance to fix the damage and, if repair fails, then to the induction of apoptosis (Section 13.5). When the restraint on cell proliferation normally imposed by Rb protein is inactivated, p53 activity is induced. Having inactivated Rb, these viruses therefore need a mechanism for blocking the p53 response. SV40 large T antigen (but not polyoma large T), papillomavirus E6 and adenovirus E1B 55K proteins all bind to p53 (Fig. 25.3, turquoise), inactivating and/or targeting it for proteolytic degradation. Also, a second E1B protein (19K) acts independently to block apoptosis. The polyoma middle T antigen interacts with the Src protein (Fig. 25.3, red). The Src protein is encoded by the gene acquired as cDNA by Rous sarcoma virus (see Section 25.7) and its interaction with polyoma middle T antigen leads indirectly to a block on the p53 response. Thus, it appears that all the small DNA viruses target both Rb-mediated and p53-mediated regulation of the cell cycle. The cooperation between adenovirus E1A and E1B genes to achieve full transformation (Fig. 25.2) can now be understood; E1A proteins overcome cell cycle control through effects on Rb, and E1B 55K protein is needed to block the resulting p53 response. p53 is a crucial protection against tumour formation, as is shown by the very high frequency with which naturally-occurring human tumours contain a mutated p53 gene. When these viruses inactivate p53, the infected cell takes an important step towards a transformed or tumourigenic state.

When SV40 transforms growth-arrested rodent cells, the process requires small t as well as large T antigen. This discovery led to the definition of another interaction with host cell proteins which has counterparts in other viruses. SV40 small t antigen, polyoma small and middle T antigens, and adenovirus E4 Orf4 protein all bind to protein phosphatase 2A (PP2A), an enzyme which is important in cell cycle regulation. PP2A is a heterotrimeric protein, with active A and C subunits regulated by a variety of B subunits. These viral proteins act as alternative B subunits, displacing existing B subunits to alter enzyme activity and hence disrupt regulation of cell division.

25.4 Papillomaviruses and human cancer

Human papillomaviruses and carcinogenesis

Certain human papillomaviruses play a causative role in the development of cervical cancer, one of the most common cancers of women worldwide. They are also associated, more rarely, with other cancers of the anogenital tissues in men and women. Vaccines are now in use that it is hoped will protect against the most common of the cancer-causing HPVs and thus also protect people from the tumours that these viruses can cause.

There are around 100 human papillomavirus (HPV) types, as defined by a minimum level of sequence difference between viruses since no system for serotyping exists. These viruses cause warts (which are benign tumours) and can be subdivided into those infecting the skin and those infecting internal epithelial cell layers. Among the latter, the two most prevalent are types 6 and 11, which are associated mainly with genital warts, but also with papillomas of the vocal cords and respiratory tract. The occurrence of genital warts correlates with an early initiation of sexual activity and a high number of sexual partners, reflecting the transmission of HPV by sexual contact. The incidence of cervical cancer also correlates with these indicators and analysis of cervical carcinoma biopsies routinely reveals HPV DNA of one of a small group of types, most commonly HPV16, 18 and 31, that are termed high-risk HPV. These are only minor components of the total load of genital HPV infections but seem selectively to be involved with this serious disease. Vaccines against these high-risk HPVs, comprising virus-like particles formed of the L1 major capsid protein of each virus, are now in routine use in many countries.

Progression from wart to carcinoma involves successive changes in the properties of the HPV-infected cells and is invariably associated with a change from maintenance of the HPV DNA as a free episome to its integration into the host genome and loss of expression of all but the E6 and E7 genes. The selective retention of these genes indicates the importance of E6 and E7 proteins in the development of the tumour. As discussed above (Section 25.3), these two proteins affect key regulators of the cell cycle, but these activities are broadly common to all papillomaviruses. E6 and E7 from high-risk HPVs must have some distinct molecular characteristics that promote progression from a benign wart to an invasive carcinoma. This question is still being studied, aided by the observation that only high-risk E6 plus E7 can transform primary human keratinocytes (the natural target for these viruses), a property that correlates with their tumourigenic potential (Box 25.6).

25.5 Polyomaviruses and human cancer

Merkel cell polyomavirus

Merkel cells are mechanosensory cells in the skin. They are thought to be epithelial in origin but are closely associated with sensory nerve endings. Merkel cell carcinomas occur relatively rarely, usually in skin areas that have high sun exposure and typically in older people. However, there is also a pronounced elevation of risk for this tumour in immunosuppression and this prompted a search in Merkel cell tumours for evidence of an infectious agent. Using techniques which compared the total RNA content of normal and tumour cells, a new virus with all the molecular characteristics of a polyomavirus (see Sections 7.2 and 10.2) was identified in 2008 and named Merkel cell polyomavirus (MCPyV). A similar approach had earlier been used to identify a novel human herpesvirus (see Section 25.6). MCPyV makes

---

Box 25.6

Features unique to E6 and/or E7 from high-risk HPV

- E7 is more effective at releasing transcription factor from Rb than is E7 from low-risk types.
- E6 is more effective at causing degradation of p53.
- E6 binds to a class of host proteins known as PDZ proteins as well as to p53.
- E6 re-activates the expression of the host telomerase gene.
- E6 and E7 together promote mitotic instability (chromosome segregation errors at mitosis).

large T and small t antigens from its early gene, as well as a shorter form of large T. Merkel cell tumours contain integrated MCPyV DNA and retain expression of these early proteins; indeed, if this expression is disrupted experimentally in Merkel tumour cells in culture they undergo apoptosis, which demonstrates the critical role that the MCPyV T antigens play in maintaining tumour cell viability. The detailed mechanisms by which these T antigens cause cell transformation are still being worked out but it appears there are both similarities with and differences from the paradigm established for SV40 (Section 25.3).

SV40 as a possible human tumour virus

SV40 is not naturally an infectious agent of humans. However, in cell culture, human cells are semi-permissive for this virus, indicating a low level of replication. During the 1950s, when the mass vaccination campaign for polio was beginning, the live Sabin vaccine was grown in cultures of African green monkey kidney cells that, it was subsequently realised, contained SV40 as an inapparent infection. Thus, many millions of people were exposed to SV40 during this period when they received the Sabin vaccine. (Current vaccine production methods avoid the risk of such contamination.) At one point, there appeared to be evidence for ongoing transmission of SV40 within the human population but this has now been discounted. There were also studies linking the presence of SV40 DNA to the occurrence of certain human tumours (particularly mesothelioma and certain types of brain tumour). However, these findings are now thought to be due to contamination of samples with SV40 in the laboratory so at this point, SV40 is not believed to be a cause of human cancers.

25.6 Herpesvirus involvement in human cancers

Two human herpesviruses are associated with human cancers. Epstein-Barr virus infection is ubiquitous but the disease outcomes vary greatly. Several types of cancer are apparently caused by EBV in association with different cofactors. The existence and disease-causing potential of the second virus, HHV8 or KSHV, was revealed by the epidemic of immunodeficiency due to HIV-1 infection.

Epstein-Barr virus

EBV infects everyone. Childhood infection is subclinical, but infection of young adults can cause prolonged debilitation (glandular fever or infectious mononucleosis, named after the extensive proliferation of host T lymphocytes in the blood in response to the expanding pool of infected B cells). Whether or not the primary infection is symptomatic, EBV establishes a lifelong latent infection of circulating B cells (see Section 17.4) with reactivations occurring periodically in individual cells. This lifecycle also involves epithelial cell infection, during primary infection and/or reactivations and, in both epithelial and B cell populations EBV infection is associated with specific tumours (Table 25.1). In each of these tumour types, one or more of the EBV gene products that function during latent infection is present (Box 25.7).

The mechanisms by which EBV infection contributes to cancer development are now being unravelled. Natural infection results in a permanent pool of infected B cells, proliferation of which is only held in check through the action of T cells. This explains the incidence of EBV-associated B cell proliferative disease after

Box 25.7

Epstein-Barr virus latency/immortalization functions

EBV encodes up to 9 proteins during latency and in EBV-induced disease states, as indicated [BL – Burkitt's lymphoma; HD – Hodgkin's disease; NPC – nasopharyngeal carcinoma; LPD – lymphoproliferative disease]:

EBNA1: DNA binding protein required for replication and episomal maintenance of EBV DNA in dividing cells. [BL; HD; NPC; LPD]

EBNA2: Transcriptional activator of the viral LMP genes and host genes, which associates with the RBP-Jκ host transcription factor. Variation in EBNA2 sequence defines two types of EBV that may differ in their disease-causing potential. [LPD]

EBNA3A,3B,3C: Modulators of the activity of EBNA2. [LPD]

EBNA-LP: Transcriptional co-activator that cooperates with EBNA2. [LPD]

LMP1: Cell-surface molecule that causes constitutive activation of intracellular signalling to activate NF-κB. [HD; NPC; LPD]

LMP2A,B: Cell-surface molecules that activate Src and Akt signalling. [HD; NPC; LPD]

All latency and EBV disease states also express non-coding EBER RNAs and viral miRNAs.

organ transplantation, since such organ recipients have to receive some immunosuppressive therapy to prevent rejection of their grafts. Inadvertent immunosuppression may also play a part in the development of Burkitt's lymphoma (BL), another B cell tumour, which forms in the lymphoid tissue particularly of the jaw but also other tissues. Endemic BL is found in high incidence in children, particularly boys, aged 6–9 in tropical Africa and New Guinea where the tumours almost universally contain EBV DNA. This geographical distribution is markedly coincident with the distribution of the malaria parasite, *Plasmodium falciparum*, which Dennis Burkitt (who first described the tumour) suggested as a possible disease-co-factor. It has been demonstrated that malaria drastically lowers the control of EBV-infected cells by EBV-specific T lymphocytes, by decreasing the proportion of helper T cells in relation to suppressor T cells. Repeated attacks by the parasite are thought to reduce the immune

response to the level needed to allow a tumour to establish. Exactly how malaria achieves this immunosuppression is not known. As well as endemic BL in these restricted areas of the world, BL also occurs sporadically at low frequency in all populations and here the proportion that is associated with EBV infection is much smaller. This nicely illustrates the complexity of cancer development; although EBV is a significant contributory cause of BL in some populations, the disease can arise without any apparent contribution from the virus.

Besides EBV infection, a feature of Burkitt's lymphoma cells is that they characteristically contain one of three chromosome translocations, which join a specific region of chromosome 8, usually to a site in chromosome 14 or, more rarely, to sites in chromosomes 2 or 22. These chromosomal translocations place a cellular proto-oncogene, *myc* (see Section 25.7) on chromosome 8 under the transcriptional control of the chromosome regions encoding the immunoglobulin heavy and light chains respectively (chromosomes 14, 2, 22). These genes are of course highly active in B-cells, so the effect of the translocations is to deregulate *myc* expression and cause inappropriate production of its protein product, a DNA-binding protein responsible for aspects of transcriptional control in the cell. This then impairs the cell's ability to control its division.

EBV is associated with other tumours too. About 50% of Hodgkin's lymphomas, another B cell tumour, are EBV-positive. There is also a long-established association of the virus with nasopharyngeal carcinoma, an epithelial cell tumour. Again, the disease shows a marked prevalence in a restricted geographic area, in this case South-east Asia. Here, consumption of salted fish from an early age, which is thought to expose the infected cells to various carcinogens, is thought to be a co-factor in disease development. Possession of certain MHC haplotypes also acts as a risk factor. EBV infection is also linked with minor subset of gastric cancers, another epithelial cell tumour.

Human herpesvirus 8

HHV-8 (also known as Kaposi's sarcoma-associated herpesvirus, KSHV) is the most recently recognized human herpesvirus. It was identified through the cloning of sequences found consistently in a tumour, Kaposi's sarcoma (KS), which had become a defining feature of AIDS during the recognition of the HIV epidemic in homosexual men (see Chapter 21). HHV-8 is now thought to be significantly involved in the development of this tumour, which appears only very rarely in the population in the absence of HIV1-induced immunodeficiency. Once the involvement of HHV-8 in KS had been established, it was realized that the virus was also present in the much rarer sporadic cases of KS that occur in HIV-negative people. Interestingly, these cases are geographically concentrated around the

Mediterranean and in East Africa, suggesting other co-factors or genetic factors are present that increase the susceptibility of these populations. Another rare tumour associated with HHV-8 is primary effusion lymphoma, a body cavity tumour of B cell origin. The molecular features of the virus that are responsible for its tumourigenic activity are still being defined. HHV-8 is now known to be present at various levels of prevalence in different populations, but always at a frequency greater than the malignancies believed to be associated with infection. Thus, as with the other examples already described, HHV-8, although suggested to be a cause of certain human cancers, is not sufficient of itself to elicit a tumour.

25.7 Retroviruses as experimental model tumour viruses

The first retroviruses to be discovered were oncogenic and these viruses were much studied subsequently for this reason. It is now clear that few retroviruses are oncogenic in the context of natural infection. However, the study of oncogenic retroviruses led to the discovery of oncogenes in the host genome and these genes are now understood to be of great importance in the development of cancer.

Highly oncogenic retroviruses that also transform cells in culture

Rous sarcoma virus was isolated in 1911 as a filterable agent capable of causing sarcomas at the site of injection in chickens. It was named after its discoverer, Peyton Rous, who later received the Nobel prize. This virus is an example of a subset of the oncogenic retroviruses which transforms immortalized cells in culture and causes tumours in animals with high efficiency and rapid onset. Each of these viruses has, in its genome, sequences that have been acquired from the host and genetic analysis reveals that these are responsible for the cell transformation and tumourigenic phenotypes. The acquired genes represent cDNA copies (reverse-transcribed mRNA) from cellular proto-oncogenes (see Section 25.1); indeed, analysis of such viruses was the way in which the oncogenic potential of these normal components of our genomes was first identified. The study of these oncogenes has been hugely informative as to the mechanisms which control the growth of mammalian cells and how these can go wrong, leading to tumour development.

What causes a normal cell gene to take on the character of an oncogene when it is captured as cDNA by a retrovirus? Deregulated expression from the powerful viral LTR promoter is at least partly responsible for these genes acquiring oncogenic properties (Fig. 25.4a). Another factor is that, when compared to their cellular counterparts, the acquired oncogenes often have mutations which affect protein function; classic examples are the altered *ras* genes of Kirsten and Harvey murine sarcoma viruses. The same *ras* mutations have also been found in the DNA of human bladder carcinomas which have no viral involvement in their development.

With the exception of Rous sarcoma virus, which acquired its oncogene, termed *src*, 3′ to its *env* gene, all highly oncogenic retroviruses have lost essential viral genes (*gag*, *pol* or *env*, see Section 10.8) during the acquisition of their oncogenes. They are therefore defective and can grow only with support from a standard retrovirus, such as the avian (ALV) or murine (MLV) leukemia viruses from which they derived. The oncogenes play no part in the virus life cycle. The detailed mechanism by which these viruses arose is unclear. Models proposed

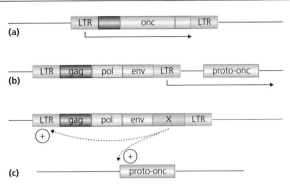

Fig. 25.4 Models of retrovirus oncogenesis involving integration of viral genomes into cellular DNA. (a) Integration of a defective viral genome carrying an oncogene whose transcription is under the control of the strong promoter in the viral LTR. (b) Integration of a virus that lacks an oncogene, so that its right-hand LTR drives expression of a cellular proto-oncogene (proto-onc). (c) Expression of a transcription-enhancing product from viral gene X which affects both viral and cellular transcriptional control, upregulating expression of a proto-oncogene.

involve integration of an intact provirus adjacent to a cellular gene, subsequent production of a hybrid mRNA containing viral and cell sequences and then recombination during reverse transcription to create a provirus with its essential flanking LTRs (see Chapter 9). There is no obvious evolutionary advantage to the virus in acquiring cellular sequences at the expense of its own essential genes, so these events may best be viewed as rare accidents of the retroviral life cycle.

Other oncogenic retroviruses

If immortalized fibroblastic cells in culture are infected with standard retrovirus such as ALV or MLV, very little morphological change occurs. The cells are not transformed nor do they show cytopathology. Despite this, these retroviruses can still produce malignant disease, in this case leukemia, when introduced into a suitable host, but unlike the highly oncogenic retroviruses, the efficiency is low (only a few of the infected animals show disease) and the time taken to observe disease is much greater. Genetic analysis does not reveal any viral oncogene to which the oncogenic properties of these viruses can be allocated. Thus the abilities of the virus to go through a productive cycle and to cause leukemia cannot be separated by mutation.

The clue to understanding how ALV, MLV, etc. cause disease came from an analysis of DNA from the tumours they induced. Although the sites of provirus insertion were not identical, in each case insertion had occurred in the same region of the host genome. Characterization of the surrounding DNA led to the discovery of the *myc* gene, already known from its presence in cDNA form in a highly oncogenic retrovirus. This result led to the concept of retroviral insertional mutagenesis (Fig. 25.4b). By inserting upstream of a proto-oncogene such as the *myc* gene, the right viral LTR, which contains a powerful promoter identical to that in the left LTR (see Chapter 9), drives its high-level expression. This alters the timing and level of expression of the gene, so causing aberrant cell behaviour (activation of *myc* expression is also involved in Burkitt's lymphoma, see Section 25.6). In fact, insertional mutagenesis can work even when the insertion is downstream of the target gene, or in the wrong orientation to drive transcription directly via the LTR promoter. This is because the LTRs also contain enhancer elements that will upregulate expression from a gene's endogenous promoter whenever they are placed in reasonable proximity to it (Fig. 25.5). Why does disease caused by these viruses have a long latency? This is because retroviral integration is a random process. Many rounds of infection will typically be needed to give any probability of a provirus inserting in a region of the genome that can have an oncogenic effect.

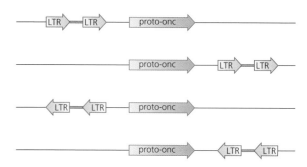

Fig. 25.5 Four variations on the retroviral insertional mutagenesis model shown in Fig. 25.4b. Integrated proviruses are indicated in red with flanking LTRs in green. Transcriptional regulation of a cellular gene will be altered by the provirus in all four scenarios, because the enhancers in each LTR operate in a position and orientation-independent manner to upregulate transcription from the gene's endogenous promoter.

25.8 Retroviruses and naturally-occurring tumours

Mammary tumours in mice

Given the random nature of retroviral integration, natural retroviral infections might be expected to confer a risk of mutational activation of cellular proto-oncogenes, as is seen experimentally with the avian and murine leukemia viruses. One example of this is the disease observed with mouse mammary tumour virus (MMTV), which is endemic in certain mouse strains and is spread from mother to pup in the milk. When mammary tumours are examined for MMTV DNA, the proviral integration sites are in reasonable proximity to one of a small number of genes, known as the *int* genes. As a result, the expression of the *int* gene will come under the influence of the viral LTR (Figs. 25.4b; Fig. 25.5). The MMTV LTR is sensitive to steroid hormone regulation, perhaps explaining why the onset of disease is linked to pregnancy in infected female animals.

The fact that specific targeting of these regions by provirus is seen in tumour cells is evidence that the effect on neighbouring genes is relevant to the disease process.

Adult T cell leukaemia in humans

The first human retrovirus, T cell lymphotropic virus type 1 (HTLV-1), was not recognized until 1980, followed soon after by HIV-1. HTLV-1 is associated with adult T cell leukaemia in a minority of those infected.

Adult T cell leukaemia (ATL) is a tumour of $CD4^+$ T lymphocytes. The geographic distribution of ATL cases across the globe is markedly non-uniform, with hotspots in Japan and the Caribbean Islands. This distribution closely matches that of HTLV-1 infection (seroprevalence), suggesting an involvement of the virus in this disease. Confirming this, ATL cells always contain integrated HTLV-1 DNA. However, other factors are also involved because, as with all the other examples of human viruses associated with cancer, only a minority (about 1%) of those infected show the disease and there is a 20–40-year interval between becoming infected and developing ATL. One of these factors is the nature of the immune response that is mounted to HTLV-1, and in particular to the Tax protein that it encodes, during the initial infection. If immunological tolerance to Tax is established in an individual, due to infection occurring when very young or because their particular MHC antigens cannot adequately present Tax peptides to cytotoxic T cells, then the chance of subsequently developing ATL is increased.

Analysis of the HTLV-1 genome shows that no acquired oncogene is present; all the genes are required for virus replication and none of them is related to a host cell gene. Also, there is no common region of integration of HTLV-1 in

the DNA from different tumours, suggesting that insertional mutagenesis is not involved. A clue to the mechanism is the fact that, despite apparently lacking an oncogene, HTLV-1 can transform T lymphocytes in culture. This activity depends on the viral Tax protein, which is encoded by an additional viral gene to the minimal retroviral *gag*, *pol* and *env* genes. The probable mechanism of Tax action is shown in Fig. 25.4c, and is somewhat similar to that of the adenovirus E1A protein. Tax is a transcriptional activator that interacts with and alters the activity of a number of host cell transcription factors. Together with these factors, Tax acts on the viral LTR and also on host cell genes such as those for interleukin 2 (IL-2) and its receptor; IL-2 is a key growth factor for T cells. Tax also interferes with Rb and p53 functions (see Section 25.3), again promoting loss of control of cell division. These changes alter the cell's growth properties so as to predispose it to accumulate further mutations as it divides, a process which can eventually lead to the cell initiating a tumour. A further HTLV function (HBZ) has also been implicated in HTLV tumourigenesis; HBZ is encoded from the opposite genome strand of the provirus to Tax and the other viral proteins, breaking the paradigm of retroviral gene expression.

25.9 Hepatitis viruses and liver cancer

Hepatitis B virus chronic infection is strongly associated with liver cancer in later life. With 300 million carriers, this makes HBV probably the most potent human carcinogen after cigarette/tobacco smoking. A second hepatitis virus, HCV, is also strongly associated with liver cancer.

Infection by hepatitis B virus (HBV) may either be resolved by the immune response or become chronic (see Section 16.5), the latter being most frequent in people infected in infancy. Those who become chronically infected have a 200-fold increased risk of developing primary (i.e. derived from liver cells rather than metastatic from other locations) hepatocellular carcinoma (HCC) over the rest of the population. In many cases of HBV-positive liver cancer, the tumour cells contain integrated HBV sequences. The integration site is the same in all cells of a tumour, indicating that integration occurred in a single cell which subsequently divided to form the tumour. There is, however, no consistent integration site among different patients. This observation tends to exclude insertional mutagenesis as a mechanism of virus-induced tumourigenesis and instead suggests a direct role of virus function(s) in the development of disease. As the HBV X protein can both transactivate transcription and disrupt p53 function (see Section 25.3), it has been suggested as the likely culprit. However, its expression is usually undetectable in tumour cells and the integrated sequences are usually 'scrambled' by rearrangements preventing most viral gene expression. Also, any proposed mechanism must take account of the long time lag (perhaps 40 years) between primary infection and the appearance of cancer. Therefore, an alternative theory is that the virus exerts an indirect effect through the long-term damage to the liver which it inflicts. In a chronic infection, cells are constantly being lost through immune-mediated destruction of infected cells. The liver responds by regeneration and this persistently increased rate of division among hepatocytes may then predispose the cells to accumulate mutations, ultimately leading by chance to the emergence of a cell with a set of mutations which gives it cancerous properties.

A persuasive piece of evidence for the theory that HBV can cause liver cancer through the indirect effects of chronic liver damage has come from the more recent discovery and characterization of hepatitis C virus (HCV) infection which also predisposes to liver cancer after a long delay. HCV is the only tumour virus listed in Table 25.1 which does not have any DNA involvement in its replication cycle. Because its genome is always in the form of RNA, none of the mechanisms by which genetic material might persist in a dividing population of pre-cancerous cells (see Section 25.2) are available to it. Therefore 'hit-and-run' or indirect mechanisms like that proposed for HBV are the only options. Given the long delay between infection and disease, the latter is the more likely model. This conclusion is also supported by the fact that other, non-virological, agents which cause chronic liver damage, such as alcohol, also give rise to an increased risk of hepatocellular carcinoma. The extent of HCV infection in developed countries is only now becoming apparent and promises to be a major public health problem.

25.10 Prospects for the control of virus-associated cancers

Vaccination against tumour-causing viruses has the potential to reduce substantially the incidence of virus-associated cancers. Vaccines for high-risk human papillomaviruses and for hepatitis B virus already exist and are in widespread use.

The power of vaccines against tumour viruses to prevent the tumours that these viruses cause has been amply demonstrated with Marek's disease virus of chickens, a herpes virus which causes a cancer of cells of the feather follicles, and with successful experimental vaccines against herpesvirus saimiri and EBV. Immunization against tumour viruses is as easy (or as difficult) as immunization against any non-tumour virus. Excellent vaccines for humans against HBV and against high-risk HPVs are already on the market; the HBV vaccine is now used in nationwide vaccination programmes in most countries. In the long term, reducing the numbers of people becoming infected with HBV or HPV should be reflected in declining incidence of HCC and cervical cancer respectively.

Since vaccines rarely succeed in completely blocking a virus infection, only in rendering it subclinical, and virus-initiated tumours may arise from a single infected cell, it can be argued that vaccines that prevent standard disease caused by a virus may be less effective at preventing any long-term oncogenic consequences of infection. However, the probability of an individual infected cell later giving rise to a tumour is very low – the oncogenic potential of the virus comes from the large number of such cells produced during the initial infection, each of which is then at risk of progressing into a tumour. Thus, by reducing the extent of the initial virus infection, a vaccine should substantially reduce the number of infected cells at risk of malignant progression and hence the probability of cancer caused by that virus arising.

To combat other known virus-induced cancers, several new vaccines are needed, for example to EBV and HCV. One problem in applying such vaccines may be that each of these viruses only rarely causes cancer. Even with a disease as emotive as cancer, will there be sufficient interest to persuade people to avail themselves of immunization, or to have their children vaccinated, given that the disease consequences are so uncertain and distant in time, and that any vaccination carries some small risk? Will potential recipients understand that, at best, such vaccines can only protect against virally-induced tumours when many still think of cancer as a single disease? If not, when many

vaccinees still experience cancer, albeit from a non-viral cause, the public may feel that such vaccines have failed and acceptance may decline. Development of cancer vaccines must therefore proceed in tandem with the more traditional anti-cancer approaches involving the elimination of co-factors, education about the risk of co-factors to inform people's lifestyle choices, improved early diagnosis and better treatment. These are as applicable to virus-induced cancers as to other cancers.

Key points

- Normal cells cannot grow and divide indefinitely in culture whereas immortalized, transformed or tumour-derived cells can.
- The underlying basis of the changes in cell behaviour that lead to tumour formation is the acquisition of mutations.
- Some viruses carry genes that have the effect of substituting for one or more of the mutations that would be needed to convert a normal cell into a tumour cell.
- For a virus to contribute to tumourigenesis, it must carry one or more relevant genes that become inherited by daughter cells at division and the cell must not be killed by the infection.
- Key elements of the growth control pathways of cells are targeted by many viruses in order to favour their own growth.
- Diverse viruses are associated with specific human cancers, always in conjunction with one or more cofactors. No virus infection is known that leads inevitably to cancer development in humans.
- Collectively, viruses are causally involved in the development of about 20% of human cancers.
- Vaccines against specific viruses may protect against the development of the cancers with which they are associated.

Questions

- Discuss the mechanisms by which certain viruses are thought to subvert the normal growth controls of cells in culture and to contribute to tumour formation in vivo.
- Write an essay on the significance of virus infections as a cause of human cancer.

Further reading

Butel, J. S. 2000. Viral carcinogenesis: revelation of molecular mechanisms and etiology of human disease. *Carcinogenesis* 21, 405–426.

Dawson, C. W., Port, R. J., Young, L. S. 2012. The role of the EBV-encoded latent membrane proteins LMP1 and LMP2 in the pathogenesis of nasopharyngeal carcinoma (NPC). *Seminars in Cancer Biology* 22, 144–153.

Dilworth, S. M. 2002. Polyoma virus middle T antigen and its role in identifying cancer-related molecules. *Nature Reviews: Cancer* 2, 1–6.

Endter, C., Dobner, T. 2004. Cell transformation by human adenoviruses. *Current Topics in Microbiology and Immunology* 273, 163–214.

Farazi, P. A., DePinho, R. A. 2006. Hepatocellular carcinoma pathogenesis: from genes to environment. *Nature Reviews: Cancer* 6, 674–687.

Zur Hausen, H. 2009. Papillomaviruses in the causation of human cancers – a brief historical account. *Virology* 384, 260–265.

Matsuoka, M., Jeang, K.-T. 2011. Human T-cell leukemia virus type 1 (HTLV-1) and leukemic transformation: viral infectivity, Tax, HBZ and therapy. *Oncogene* 30, 1379–1389.

McLaughlin-Drubin, M. E., Münger, K. 2009. Oncogenic activities of human papillomaviruses. *Virus Research* 143, 195–208.

Mesri, E. A., Cesarman, E., Boshoff, C. 2010. Kaposi's sarcoma and its associated herpesvirus. *Nature Reviews: Cancer* 10, 707–719.

Rawat, S., Clippinger, A. J., Bouchard, M. J. 2012. Modulation of apoptotic signaling by the Hepatitis B virus X protein. *Viruses-Basel* 4, 2945–2972.

Riley, T., Sontag, E., Chen, P., Levine, A. 2008. Transcriptional control of human p53-regulated genes. *Nature Reviews: Molecular Cell Biology* 9, 402–415.

Spurgeon, N., Lambert, P. F. 2013. Merkel cell polyomavirus: a newly discovered human virus with oncogenic potential. *Virology* 435, 118–130.

Young, L. S., Rickinson, A. B. (2004) Epstein-Barr virus: 40 years on. *Nature Reviews: Cancer* 4, 757–767.

Chapter 26

Vaccines and immunotherapy: the prevention of virus diseases

Vaccines prevent infections (prophylaxis). Since prevention is always better than cure, and because really effective antiviral therapy has been difficult to achieve, there is great importance attached to producing safe and effective vaccines against viruses. In recent years, the use of specific antibodies to provide short-term protection has advanced with the production of 'humanized' monoclonal antibodies.

Chapter 26 Outline

26.1 The principles of vaccination
26.2 Whole virus vaccines
26.3 Advantages, disadvantages and difficulties associated with whole virus vaccines
26.4 Subunit vaccines
26.5 Advantages, disadvantages and difficulties associated with subunit vaccines
26.6 Considerations for the generation and use of vaccines
26.7 Adverse reactions and clinical complications with vaccines
26.8 Eradication of virus diseases by vaccination
26.9 Immunotherapy for virus infections
26.10 Adverse reactions and clinical complications with immunotherapy

The devastating effects of infectious disease in people and their animals and plants have been evident since prehistoric times. The death, disease and economic loss caused by virus infections have prompted investigation of ways to protect against them, with the primary desire of preventing infection. Long before the nature of infectious agents was understood it was known that prior infection could, in some circumstances, prevent recurrence of the infection. For example, smallpox virus infection was endemic in some parts of China around 1000 BC and it was observed that survivors of the disease never became infected again. It was also recognized that contact with the pocks on the skin of infected individuals led to infection so the process of *variolation* was established in which people were deliberately exposed by scarification to pock material from an infected person. If the recipient survived the process – and many did not – they were immune to future infection (Box 26.1). This was the first recorded use of a mechanism of vaccination that went on to form the basis for our current approach to prevention of infection which is described in this chapter.

Historically, immunization with vaccines has been far more effective than antiviral chemotherapy, though the application of

Introduction to Modern Virology, Seventh Edition. N. J. Dimmock, A. J. Easton and K. N. Leppard.
© 2016 John Wiley & Sons, Ltd. Published 2016 by John Wiley & Sons, Ltd.

Box 26.1

The eradication of smallpox by vaccination

Smallpox was a scourge dating back millennia in human history. Ancient texts describe the devastating effects of the disfiguring infection which carried a 30% risk of death for those infected. The mummified body of Pharaoh Ramses V, who died in 1196 BC, carries signs of the characteristic pocks on the skin that give the disease its name. The disease remained a major threat with an estimated 300 million fatal cases in the 20th century alone. Smallpox was caused by two closely-related viruses: variola major caused serious illness that could be life threatening, and variola minor caused a milder infection that rarely caused death. Smallpox was spread extremely easily by contact with contaminated clothing or bed linen or direct contact with infected patients.

In China the process of variolation, exposure to infected material by scarification, to protect against smallpox disease was described more than 3000 years ago. This provided protection against subsequent infection to those who survived the process. We now know that this worked because the process established a protective immune response. The practice of variolation spread westwards and was recorded in letters home by Lady Mary Wortley Montague, the wife of the British ambassador to Turkey, who in 1718 had her 5-year-old son treated. Lady Wortley Montague championed this process and introduced it to the British royal family. After tests on condemned prisoners proved successful, their participation leading to their release, King George I had two of his grandchildren inoculated in 1722. Fortunately, both children survived and the technique acquired broad acceptance.

Some time later Edward Jenner, an English country doctor, noted that dairymaids who had contracted the mild disease cowpox never succumbed to smallpox infection. Cowpox infection in humans produces small pocks on the skin which bear some similarity to those of smallpox and this was taken to suggest the two infections were related. In 1796, Jenner made a preparation of material from the skin pocks of the dairymaid Sarah Nelmes and used it to vaccinate 8-year-old James Phipps. Jenner then took the brave step of deliberately exposing the young boy to smallpox – he survived and vaccination was born. The wider introduction of vaccination was hampered by scare stories at the time, with concerns about the introduction of cow material into the body. This led to leading cartoonists of the day representing the process as likely to cause horns and other bovine features to appear on the bodies of the treated people. Despite this, the success of vaccination led to very wide adoption of the process with subsequent reduction in the incidence of disease.

Vaccination against smallpox spread rapidly and in the 1800s many countries required compulsory vaccination for children. At some point during the maintenance of Jenner's original cowpox material across the world, it was supplanted by vaccinia virus. The origin of vaccinia virus is unknown but it shared the ability to protect against variola virus without causing significant disease.

The widespread application of vaccination led to the elimination of smallpox, with the last naturally acquired case in the UK in the 1930s and in the USA in the 1940s. In 1958, Russia proposed that a programme designed to eliminate smallpox through vaccination should be initiated and this was accepted in 1959 by the precursor to the World Health Organisation. The intensity of mass vaccinations was stepped up in 1966. As the number of cases declined, a process of 'ring vaccination' was adopted in which anyone who might have been exposed to a smallpox patient was identified and vaccinated as quickly as possible. This effectively reduced the spread of infection and in 1977 the last-ever recorded natural case of smallpox was identified in Somalia – the patient, Ali Maow Maalin, survived the variola minor infection.

All stocks of variola virus, except one held in Russia and one held in the USA, were destroyed, marking the global elimination of smallpox. Since 1977, the only cases to have arisen came from accidental laboratory exposure.

antiviral drugs is an extremely valuable additional weapon in the treatment of infection when prevention has not worked or is not yet possible (see Chapter 27). It is important to appreciate that not all vaccines work, or work as well as they should, and for some viruses (notably HIV-1) there is still no vaccine available despite years of work. These problems will be discussed below.

There are a number of different approaches available to generate a vaccine, from preparing a single purified antigen to the generation of a virus variant that does not produce disease but can induce an immune response. While the consideration here focuses on viruses, some of the types of vaccines discussed below share features with bacterial vaccines; others are specific for viruses. Alternative approaches, such as toxoid vaccines, are also used for bacteria and will not be considered here.

Vaccination is not achievable in some systems and alternative approaches must therefore be used. For example, since plants do not have an inducible immune system, prevention of plant virus diseases has relied upon other means, such as selective breeding of plants that are genetically resistant to the virus or its vector for transmission, or by control of the vector. Control of animal virus diseases through improvements in nutrition, management of transmission, and education is also a vitally important aspect of prevention, especially in the developing world.

26.1 The principles of vaccination

The aim of vaccination is to generate virus-specific adaptive immunity without the individual having to experience the disease. The adaptive immune response is discussed in Chapter 14. Ideally, this immunity is comprehensive in that it provides as broad a spectrum of protection as possible and is sufficiently long-lasting to provide protection for many years. Following successful vaccination, the memory retained in the primed immune system means that it is able to mount a rapid and strong response to prevent the establishment of the infection when the virus is encountered (Fig. 26.1). This works only if the immune response initiated by vaccination is suitable to inhibit the infection.

Individual versus herd immunity

It might seem that every individual should be immunized to protect against an infection, but 100% immunity is not usually needed to eliminate a virus from the population. What is required is a sufficient percentage of immune individuals to break the chain of transmission of the virus. This is referred to as *herd immunity*, and is the key to successful implementation of a vaccine programme. A natural situation that

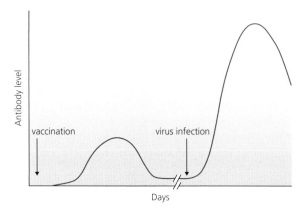

Fig. 26.1   Antibody levels following vaccination and subsequence infection. Following vaccination there is a rise in specific antibodies directed against the antigen. The memory of this is retained. There is a delay of 7 days or more following vaccination before a rise in antibody level which peaks and then gradually declines to a basal level. Following infection of a previously vaccinated person the specific antibody levels against the same antigen rise very rapidly and achieve very much higher levels.

illustrates this point is the incidence of measles virus in island populations. Natural measles infection leaves people immune to reinfection, and a population of at least 500,000 is needed to generate a sufficient supply of susceptible children to prevent the virus from dying out (see Section 4.5). When fewer than the minimum required number of susceptible individuals are present the virus cannot survive in the population. The exact percentage of immune individuals required for effective herd immunity can be calculated and is different for each infectious agent; knowledge of this important parameter can used to establish targets and to inform implementation of immunization programmes.

   The principle of herd immunity is key to a successful vaccination programme but it also represents a major hurdle for those directing the programmes. As a vaccine programme begins to succeed, the level of the disease threat in the population reduces. This also reduces the public perception of the seriousness of the disease and may lead to a reduction in uptake of the vaccine. If uptake drops below a critical level, herd immunity is insufficient to limit transmission effectively and the disease incidence will rise again. In addition, as discussed below, with almost every vaccine approach there is a risk of adverse reactions occurring in a few vaccinees. At the beginning of a vaccine programme the seriousness of the disease caused by the virus ensures that low levels of adverse reactions do not lead to significant public concern. However, as disease incidence drops the level of adverse reactions becomes an important factor in public acceptance of the vaccine. Consequently, as the perception of the disease as a threat diminishes, so the perception of the vaccine as a hazard can often increase, sometimes without justification. This can lead to lower uptake of the vaccine, a rise in the proportion of susceptible individuals and a consequent rise in disease incidence. Such a situation arose in the UK following the publication of a paper in 1998 suggesting that the measles vaccine was linked to autism in some children. Despite the discrediting of the science in the paper, the public perception of a problem led to a dramatic reduction in vaccine acceptance in some areas of the UK which led in turn to a dramatic rise in measles incidence in those areas some years later.

26.2 Whole virus vaccines

Whole virus vaccines can take two forms: infectious (live) virus or inactivated, non-infectious (killed) virus. Live virus vaccines must:

- cause less disease than the natural infection
- have no significant clinical side effects

- stimulate effective and long-lasting immunity in the individual and/or stimulate sufficient herd immunity to deprive virus of the susceptible hosts needed for its survival
- be genetically stable

Killed virus vaccines must:

- retain their full immunogenicity

The first vaccines to be developed comprised either infectious ('live') or non-infectious ('killed') virus particles and these remain the mainstay for prevention of both human and animal virus infections (Table 26.1). Both types of whole virus vaccines involve the growth of infectious virus under carefully controlled conditions.

Infectious (live) virus vaccines

A vaccine should produce both effective and long-lasting immunity. Different viruses are susceptible to different parts of the immune system and it is therefore necessary for a vaccine to stimulate immunity of the correct type, in the correct location, and of sufficient magnitude for it to be effective. This is most readily achieved with a live virus vaccine as its interaction with the immune system will most closely mimic that of the virulent virus. A key requirement for a live vaccine is that the virus must cause a reduced amount of disease. Ideally, the virus would cause no disease at all but this is very hard to achieve. The first recorded application of this principle was by Edward Jenner who used cowpox virus to vaccinate against smallpox. The principle here

Table 26.1
Examples of human and veterinary whole virus vaccines.

Infecting virus	Family	Live or killed vaccine	Disease (if named differently from the virus) and other comments
Human viruses			
Hepatitis A	Picornavirus	Killed	Hepatitis
Influenza A and B	Orthomyxovirus	Live	
Measles	Paramyxovirus	Live	
Mumps	Paramyxovirus	Live	
Polio	Picornavirus	Killed, live	Poliomyelitis
Rabies	Rhabdovirus	Killed	
Rubella	Togavirus	Live	German measles
Variola	Poxvirus	Live vaccinia virus	Smallpox
Yellow fever	Flavivirus	Live	
Varicella-zoster	Herpesvirus	Live	Chicken pox
Veterinary viruses			
Canine distemper	Paramyxovirus	Live	
Rinderpest	Paramyxovirus	Live	
Equine influenza	Orthomyxovirus	Killed	
Foot-and-mouth	Picornavirus	Killed, live	Affects cattle and pigs
Newcastle disease	Paramyxovirus	Live	Fowl pest
Parvo	Parvovirus	Killed, live	Causes death in young dogs and cats
Pseudorabies	Herpesvirus	Killed, live	Aujeszky's disease of pigs

was that cowpox, a natural strain of an animal virus, caused only a mild disease in humans but was antigenically related to variola virus that caused life-threatening smallpox. The antibodies raised against cowpox were able to neutralize the infectivity of variola virus, thus protecting people from disease. Vaccinia virus, a related virus of unknown origin, was subsequently used for the same purpose and proved to be sufficiently successful that it led to the eradication of smallpox virus from the world (Box 26.1). The use of vaccinia virus in this way gave rise to the term 'vaccine', which is now employed when any immunogen is used against an infectious disease.

However, the use of closely-related viruses is possible in very few cases and in most situations the only alternative for the development of a live virus vaccine is to use a variant that causes little or no disease. If a natural variant is not available it is necessary to generate one. The process of producing a virus which causes a reduced amount of disease for use as a live vaccine is called *attenuation*. The disease-causing virus is referred to as *virulent* and the attenuated strain as *avirulent*. It is important to appreciate that these are only relative terms – there will always be a risk, however small, that a live vaccine may cause a very mild disease. The important factor is that the risk of disease that may rarely be seen after vaccination must be considerably less than the risk of disease during infection with the target virus.

Avirulent vaccines have been obtained by selecting for naturally-occurring avirulent variants during cycles of growth under laboratory conditions. Attenuation is achieved empirically, and experience has shown that it can be helped by replication in cells unrelated to those of the normal host, by growth at a sub-physiological temperature, or by recombination with an avirulent laboratory strain. The yellow fever virus 17D vaccine is an example of a highly successful attenuated live vaccine that was produced by an empirical approach. Its

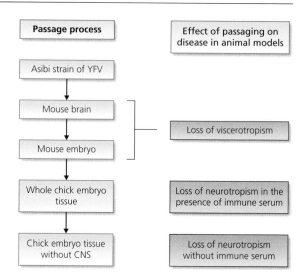

Fig. 26.2 Passage history of the yellow fever virus 17D vaccine. After passage in mouse brain and mouse embryo tissues the virus lost its ability to readily spread to other tissues of the body (loss of viscerotropism). Subsequent growth in chick embryo tissue significantly reduced the efficiency of growth in the CNS (neurotropism). Successful vaccination initially required the addition of immune serum but further growth in chick embryo tissue from which neural tissue had been removed eliminated the requirement for the immune serum.

isolation illustrates how long and complicated this procedure can be (Fig. 26.2). The process of attenuation of the Asibi strain of yellow fever virus was begun in the 1920s and the candidate vaccine was finally generated more than seven years later. An indication of the effort required is that the final stage of attenuation alone required a minimum of 227 successive growth cycles (termed passage) in chick embryo tissue without the brain and spinal column, to generate material that could be tested in humans. This material formed the basis for the vaccine that is still used today in one of the most successful and safe vaccines available. The success achieved with the yellow fever virus vaccine led the way in the production of attenuated virus vaccines.

The ability to produce genetically-modified viruses has opened the possibility of using this technology to generate attenuated viruses. This is not trivial as the mechanisms of attenuation are not well understood. However, many studies on several different viruses have tackled this problem and two approaches have so far emerged. The first is to generate attenuated viruses by generating specific mutations in the virus genome. Whilst highly attractive, this remains problematic due to the complexity of viral pathogenicity which is often due to the effect of several genes acting in concert in a complex way. A more simple, and to date more successful, approach is to produce hybrid viruses in which the genetic 'backbone' is derived from a virus for which an attenuated vaccine is already available and to introduce into this backbone a gene encoding a major antigen from the virus of choice. A successful example of this is recombinant vaccinia virus expressing the major glycoprotein of rabies virus being used as a rabies vaccine. When animals eat food baits 'seeded' with the recombinant vaccinia virus, they become infected and generate antibodies that protect them from later rabies virus infection.

Non-infectious inactivated ('killed') virus vaccines

For some viruses, no attenuated mutant is available and there is no alternative live virus that shares antigenic characteristics with the target virus. In these circumstances, a strategy that has been used successfully is to treat the target virus in a way that leads to inactivation of the virus infectivity, generating what is commonly referred to as a killed virus vaccine. The most common treatments used are to heat the virus or to treat it with chemicals such formaldehyde or β-propiolactone to destroy its infectivity. The use of inactivated virus vaccines is attractive because, despite being derived from

a virulent virus, killed vaccines should cause no disease at all. However, for this to be true it is essential to kill every infectious particle present. The process of killing the virus must be established from first principles for each virus to ensure that all infectivity is removed. Whilst simple in principle, this requirement represents a major issue for killed vaccines (see below).

An example of the implementation of a highly successful killed virus vaccine was seen with the use of the inactivated polio virus vaccine developed by Jonas Salk and introduced in the USA in the 1950s. The effect of the vaccination programme was an almost immediate and dramatic reduction in the incidence of poliomyelitis (Fig. 26.3). A similar approach has subsequently been taken with many other viruses (Table 26.1).

26.3 Advantages, disadvantages and difficulties associated with whole virus vaccines

A major advantage of using a live virus vaccine is that the immune response that is stimulated is likely to closely mimic the response generated by the natural infection. This means that the live vaccine is likely to stimulate all of the aspects of immunity which are most likely to provide protection from subsequent infection such as systemic and mucosal B cell and T cell responses. This immunity is frequently found to be long-lasting. A further advantage is that, as the vaccine virus is able replicate, only a very small dose is required and this usually need be delivered only once, making it more acceptable to patients and easier to distribute to isolated communities where repeated access to healthcare professionals may be difficult. Associated with this is that the vaccine may not require an injection and can be delivered by a

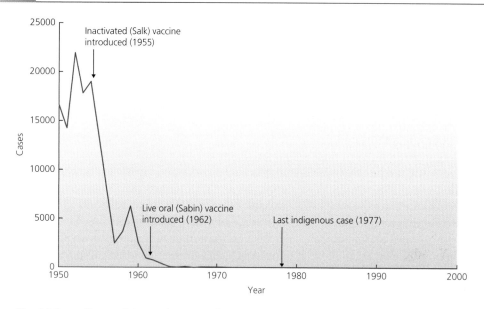

Fig. 26.3 Poliomyelitis incidence in the USA. The times of introduction of the inactivated and live attenuated poliovirus vaccines are indicated. The time of the last indigenous case of poliomyelitis in the USA is also indicated.

more 'natural' route such as inhalation for an influenza virus vaccine or by oral ingestion for poliovirus. Finally, they are typically cheaper per dose than a killed viral vaccine though this is offset by the greater costs of maintaining the correct conditions during transportation and storage. These factors make live virus vaccines extremely attractive. However, they do have disadvantages.

A constant concern with the use of attenuated live virus vaccines is the risk that the virus may revert to virulence when it multiplies in the immunized individual. This risk is specific for each vaccine strain, and is related to the number and nature of the mutations present that give its attenuated phenotype; even the excellent Sabin live poliovirus vaccine is not perfect in this regard (Box 26.2). As discussed below, even attenuated viruses may still cause some mild disease and this can be considered unacceptable by the general population. Whilst it is usually the case that an attenuated virus is likely to generate good levels of protective

immunity, there is a possibility that the attenuation process may alter the immunogenicity such that immune response is not entirely complete, leaving a small window of opportunity for the virulent virus to establish an infection. This must be rigorously tested before the vaccine can be used but is difficult to establish until large-scale testing has been performed. Coupled with this is the risk of interaction of the live vaccine with an ongoing infection in the recipient. This concern has two aspects. It is important that vaccines which are formed by attenuating virulent strains of viruses cannot be restored to infectivity by genetic interactions with viruses occurring naturally in the recipient, or with each other if the vaccine contains more than one type of virus. The presence of ongoing infections in the recipient is also of concern as it poses a potential risk of interference with the natural immune response which might compromise the protection that would otherwise be generated; for this reason, patients presenting with an ongoing infection

Box 26.2

Evidence for the basis of poliovirus vaccine attenuation and its reversion to virulence

There are three poliovirus serotypes and immunity to each is required for protection against poliomyelitis. This requires that the vaccine is trivalent. The avirulent type 3 component of poliovirus live oral vaccine is known to spontaneously revert to virulence, with the result that there is a very, very low incidence of vaccine-associated paralytic poliomyelitis (approximately one case per 10^6 doses of vaccine for the first dose and diminishing to zero by the third dose). Viruses isolated from such patients are neurovirulent when inoculated into monkeys in the test normally applied to ensure the safety of new batches of vaccine.

Long after the vaccine had been made, sequencing techniques were developed, and comparison of the wild-type 'Leon' strain of type 3 poliovirus with the vaccine strain that was derived by Albert Sabin found that only 10 of the 7431 nts that comprise the poliovirus genome had mutated during the attenuation process. There were two mutations in the 5′ non-coding region, one in the 3′ non-coding region, three leading to amino acid changes and four silent mutations in the coding regions (Fig. 26.4a). Comparison of the vaccine strain with the virulent revertants found seven nucleotide changes associated with reversion, one in each of the terminal non-coding regions and five in the coding regions. Of the latter, one was silent (at position 6034) and four resulted in amino acid changes in structural proteins (Fig. 26.4b). Analysis of genomes of viruses that have reverted to neurovirulence with the attenuated genome showed that they had one nucleotide change in common, at residue 472 in the 5′ non-coding region (changing a cytosine to uridine and back to cytosine). The $472U \rightarrow C$ was the only change that occurred in all viruses responsible for cases of vaccine-associated illness, demonstrating its role in attenuation.

The observation that a nucleotide change in a region of the virus genome that does not code for a protein associated with serious disease was initially very confusing. Subsequent molecular modelling has shown that the $472U \rightarrow C$ mutation greatly alters the secondary structure of the viral RNA, but exactly how this affects virulence is not fully understood. Viruses with 472U have reduced replication and translation and decreased interaction of the RNA with a cellular RNA-binding protein in neuronal cells, but not in cells of non-neuronal origin. Secondly, the mutation at nucleotide 472 increases but does not fully restore neurovirulence to equivalence with that of the original Leon strain. Thus, one or more of the other residues may contribute to virulence.

are usually not given live vaccines. A further important concern is the risk of contaminating microorganisms being present in live virus vaccine preparations; it is not possible to treat the vaccines in ways that will kill all microorganisms as that would also kill the virus, rendering the vaccine ineffective. A final consideration is that of maintaining the viability of the virus. Viruses are very fragile and long-term storage usually requires ultra-low temperatures such as -70 °C and lower. Viability can be maintained at more accessible temperatures such as 4 °C but even this can be difficult to maintain. Maintenance of a 'cold chain' to ensure delivery of effective live vaccines in remote areas of the world is a major

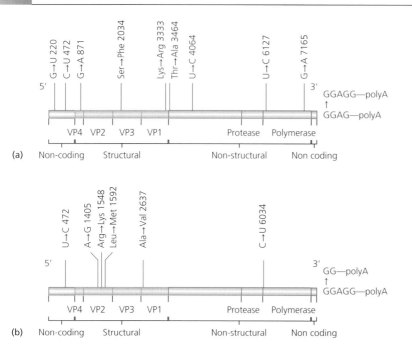

Fig. 26.4 Mutations linked with attenuation in the Sabin live oral poliovirus vaccine. (a) Sequence changes when the original type 3 poliovirus wild-type neurovirulent Leon strain was attenuated to form the Sabin vaccine, and (b) additional sequence changes when the vaccine strain reverted to the neurovirulent 119 isolate. (Reproduced from *Annual Review of Microbiology* 41, 153–180 (1987) with permission.)

challenge that can limit the accessibility to life saving treatments.

A significant advantage of killed vaccine preparations is that any other unknown, contaminating viruses or bacteria will probably also be killed in the same process. Also, because inactivated vaccines are not able to replicate, it is possible to deliver multiple inactivated vaccines simultaneously without concerns of genetic interaction leading to disease, expanding the utility of the approach. A practical and commercial advantage of inactivated vaccines is that they do not pose the same problems of stability seen with live vaccines. Inactivated vaccines are generally stable for long periods and they do not require elaborate storage conditions; in many cases, the material is prepared in a freeze-dried form that can be transported at normal temperatures.

However, a major practical consideration with inactivated vaccines is that the process of inactivation for each virus is unique and must be determined empirically. To be effective, a killed vaccine must not be infectious, but must be sufficiently immunogenic to stimulate protective immunity. For this reason, inactivating agents should target the viral nucleic acid and not just attack virion proteins, since nucleic acid alone can be infectious and could be released by the action of the cell on virion coat proteins. Formaldehyde and β-propiolactone are two inactivating agents which have been used. The former reacts with the amino groups of nucleotides and cross-links proteins through ε-amino groups of lysine residues. β-propiolactone inactivates viruses by alkylation of nucleic acids and proteins. To prepare a killed vaccine, every one of the

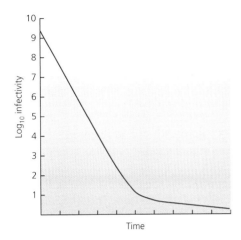

Fig. 26.5 Kinetics of inactivation of a virus to prepare a killed vaccine. In the early stages of treatment there is a dramatic loss of infectivity but as the treatment proceeds the rate of inactivation declines with a residual level of infectivity that is difficult to eradicate.

possibly 1,000,000,000 (10^9) infectious particles in an immunizing dose of virus must be rendered non-infectious. The process of inactivation usually follows a pattern shown in Fig. 26.5. There is a rapid loss of infectivity at early times after treatment has begun but the rate of inactivation slows with a small residue of viable virus that is frequently very difficult to eliminate. This means that the whole virus preparation has to be kept in contact with the inactivating agent for much longer than is predicted by the initial rate of inactivation. In turn, this brings a greater risk that immunogenicity will be altered or destroyed by the treatment. The 'resistant' fraction results from inefficient inactivation of particles trapped in aggregates or clumps within the virus preparation being treated. This poses a very severe concern and requires rigorous testing. Failure to detect remaining infectivity can have devastating consequences (Box 26.3).

The inactivation of viruses must be carefully controlled and excessive inactivation may

---

Box 26.3

The consequences of failure to eliminate virus infectivity in a killed vaccine: the Cutter incident with the Salk inactivated poliovirus vaccine

The inactivated poliovirus vaccine generated by Jonas Salk was prepared by chemical inactivation of virulent poliovirus. The importance of total inactivation of the infectious virus in inactivated vaccines was illustrated by a catastrophic consequence of a failure in testing systems early in its production. One of the companies commissioned to generate the vaccine when it was introduced in 1955 was the Cutter Laboratories. Following initial reports of poliomyelitis associated with the vaccine almost immediately after its release, the company recalled its product. Unfortunately, by that time 120,000 of the suspect vaccine doses had been used to immunize young children. It was subsequently shown that, despite company checks indicating that no infectious virus was present, the vaccine preparation contained a low level of virulent virus. Of the 120,000 children given the vaccine, 40,000 developed disease. Fortunately, in most cases this did not lead to damage of the central nervous system. However, 56 children developed classical paralytic poliomyelitis, and five died. Furthermore, the virus was able to spread and this led to an epidemic of poliovirus in the families and friends of the infected children, with a further 113 cases of paralytic poliomyelitis and a further five deaths.

generate problems such as alteration of the immunogenic characteristics of the virus that compromises the efficacy of the vaccine. Under these circumstances, immunity acquired from administration of a killed vaccine may even potentiate the disease that is experienced when the wild virus is contracted. An extreme example of this was seen in a clinical trial of a potential vaccine against respiratory syncytial virus (a paramyxovirus that causes a lower respiratory tract infection in very young children). The problem was due to the inappropriate stimulation of subsets of CD4$^+$ T helper cells, which are classified on a functional basis into T helper type 1 (T$_H$1) cells and T$_H$2 cells (see Section 14.2). Any immunogen stimulates a balance of these two cell types. T$_H$1 cells support cell-mediated immunity (including CD8$^+$ cytotoxic T cells and IgG2a) that has evolved to deal with intracellular infections), and T$_H$2cells drive inflammatory responses. It is thought that the batch of killed vaccine used in the trial was altered by the inactivation process in such a way that it produced a response biased to T$_H$2 cells, rather than the T$_H$1 cell response seen in natural infection. When the immunized children contracted a natural respiratory syncytial virus infection, only the T$_H$2 arm had immunological memory, and this led to an enhanced inflammatory response that enhanced the pathology caused by the virus. Unfortunately, this was so severe that some of the children died.

A practical disadvantage of killed viruses is that they do not multiply, so an immunizing dose has to contain far more virus than a dose of live vaccine, and repeated doses may be required to induce adequate levels of immunity. This increases both the cost and the amount of non-viral material injected; the latter raises concerns over hypersensitivity reactions when these substances are experienced again. Additionally, achieving the necessary multiple treatments may be hard in areas of the world where access to medical professionals is

difficult. Inactivated vaccines must be delivered by injection, which is not attractive to recipients, and the site of injection is important and may play a role in the efficiency of the subsequent immunity that is induced. Linked with this, another problem is that killed vaccines do not reach and stimulate mucosal immunity in the intestinal and respiratory tracts where virus normally gains entry to the body, and may thus not stimulate the full range of immunity needed for the greatest levels of protection. However, providing that the killed vaccine is a potent immunogen, the high levels of serum IgG which result from vaccination can often serve in place of local immunity, presumably because there is a sufficient concentration of IgG to diffuse to the extremities.

26.4 Subunit vaccines

Subunit vaccines can take several forms. They may consist of:

- one or a subset of the virus proteins
- non-infectious virus-like particles containing a subset of the virus structural proteins.

A subunit vaccine, as the name suggests, consists of a component of the virus which is used to stimulate an antibody response. The subunit may be one or more proteins that are known to contain epitopes which stimulate an immune response. These usually include a protein(s) that forms part of the virus particle but may also include virus proteins which are normally found only within infected cells. Alternatively, such a vaccine may consist of a peptide or collection of peptides representing some of the epitopes contained within virus proteins. These components must be highly

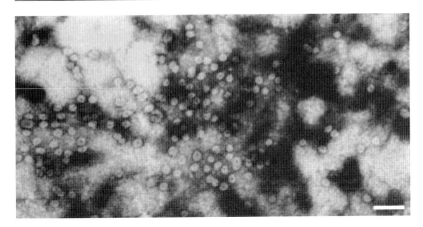

Fig. 26.6   Virus-like particles (VLP) consisting of hepatitis B virus surface antigen. The virus protein was expressed in yeast cells in the absence of any other virus material. These VLP are used as a vaccine. The bar indicates 100 nm. (Image from Howard et al., 1986, *Journal of Virological Methods* 14, 25–35. Reproduced with permission.)

purified to avoid risks of contaminating material causing adverse responses. This approach also requires a great deal of knowledge about the nature of the virus antigens and epitopes to ensure that an appropriate protective immune response is elicited by immunization.

As described in Section 12.1, very few viruses are able to spontaneously assemble into infectious particles. However, it has been found for some viruses that incomplete particles, lacking any nucleic acid, can be generated spontaneously if one or more structural proteins are expressed in large quantities in eukaryotic cells. These particles will, of course, contain only a subset of the virus proteins normally found in a particle and they are not infectious; they are known as virus-like particles (VLP). A highly successful example of this was seen with the development of a VLP vaccine against hepatitis B virus (Fig. 26.6). It was observed that incomplete virus-like particles were present in the bloodstream of patients infected with hepatitis B virus. These particles consisted of the hepatitis B virus surface antigen (HBsAg; originally called the Australia antigen) formed into small, highly immunogenic particles. The gene for the HBsAg

was cloned and when expressed in yeast cells this generated VLPs that could be purified and used as a vaccine. A similar approach has been taken more recently with a VLP vaccine for human papilloma virus in which the virus L1 protein, which forms the major component of the virus capsid, is expressed in cells.

26.5 Advantages, disadvantages and difficulties associated with subunit vaccines

Improvements in the techniques of protein production using genetically-engineered bacteria, yeast and animal cells have made the preparation of subunit vaccines much simpler and cheaper. A major advantage of subunit vaccines is that they utilize highly purified material, significantly reducing the risk of contamination with infectious agents or other materials that may lead to adverse effects. Because the material is inert they often do not need the cold-chain for delivery that is required for live vaccines, and this makes supply considerably more simple and convenient, especially in isolated parts of the world that may lack a high technology infrastructure. The production of the subunit components can be carefully controlled and is usually very cost-effective.

The primary difficulty with subunit vaccines is that they generate an adaptive immune response which is restricted to B cell derived antibody production with little or no T cell involvement. This means that the immune response when the vaccinee meets the real virus is limited only to antibody production; this potentially limits the scope of protection to strains that have the same sequences in the proteins which form the vaccine. If a strain is encountered that differs in sequence, there may be limited or no protection. This problem arises with the influenza A virus subunit vaccines which must be reformulated every year to induce immunity against the new strain associated with each annual epidemic (see Chapter 20). Two other potential problems with this approach in terms of ease of administration and acceptance by patients are, as with killed whole viral vaccines, that the subunit vaccines must be administered by injection, the site of injection may be important for the quality of the subsequent response, and multiple doses (and therefore repeat injections) may be required.

26.6 Considerations for the generation and use of vaccines

Several factors must be considered when undertaking a vaccination programme. In vaccinated individuals the successful establishment of protective immunity is referred to as the vaccine 'take' and the rate of take differs with different vaccines. Many elements play a role in determining the optimal approach such as the dose, the use of potentiaters of the immune response, the age of the vaccinee and whether multiple vaccines can be used simultaneously. For example, vaccination with a low dose of the live attenuated oral monovalent G1 human rotavirus vaccine shows seroconversion (production of neutralizing antibodies) in 81.5% of vaccinees while a high dose generates seroconversion in 88% of recipients. Though relatively small, this difference can determine whether a vaccination campaign is successful or not. In most cases, a combination of several factors determines the success of a programme.

Age of vaccinees

All vaccines are tested exhaustively for safety and efficacy before they are adopted for general use. However, the immune responses and susceptibility to some infections of infants and the elderly differ significantly from the norm, and these groups require separate testing. The susceptibility of infants to infection and disease is different from that of mature adults in that they are less susceptible to, e.g. poliovirus, cytomegalovirus and Epstein-Barr virus, but more susceptible to, e.g. rotavirus and human respiratory syncytial virus, so live vaccines must be introduced with caution. In addition, the immune system of infants is not completely mature until 2 years of age, with different parts developing at different rates. As a result, they may not be able to generate the response needed to resist a particular infection. At the other end of the spectrum, the immune response gradually declines with age, so the elderly may not respond as vigorously to the same vaccine dose as younger adults. Thus, a more aggressive vaccination regimen may be required to achieve the same level of immunity. In addition, the elderly are more susceptible to certain infections (e.g. rotavirus, human respiratory syncytial virus, influenza A virus), which again has implications for the safety of live vaccines. For these reasons, new live attenuated vaccines may be licensed only for certain age groups.

Multivalent vaccines

The immune system can respond to more than one antigen at a time. This allows the use of multiple viruses or antigens simultaneously in a vaccine 'cocktail'. The use of such a multivalent vaccine with killed viruses or subunits has the advantage of reducing the total number of injections necessary to provide immunity to a number of pathogens, which is usually welcomed by the recipients. This approach generates good immunity as seen with a UK vaccine that comprises five elements which protect against diphtheria, tetanus, whooping cough, *Haemophilus influenzae* B and poliovirus. Multicomponent vaccines containing live viruses are more problematic as there may be mutual interference in multiplication, possibly as a result of the induction of interferon (see Section 13.1). However, the live poliovirus vaccine contains the three serotypes, and a triple live vaccine of measles, mumps and rubella viruses (MMR) is in routine use.

Post-exposure use of vaccines

Once signs and symptoms of an infection are apparent, it is usually too late to immunize. This is because the development of an adaptive immune response takes several days to reach a significant level, and in most cases the virus will already have established an infection which must take its natural course by that time. However, a notable exception is seen with rabies virus vaccine. The approach taken here is that the vaccine is delivered as close as possible to the time of infection. The reason that this vaccine works after exposure when others do not is that, following the initial infection, rabies virus travels along peripheral nerves to its target cells in the spinal cord and brain in a process

that can take several weeks. The disease does not start until the virus reaches the central nervous system by which time the vaccine has induced the necessary immune responses to provide protection. Other viruses begin the infection and disease process much more rapidly than rabies and the immune system cannot catch up.

Use of adjuvants

A vaccine should produce both effective and long-lasting immunity. However, not all antigens stimulate a sufficiently robust immune response that provides the necessary level of protection from infection. Immunogenicity can be enhanced by mixing virus with an *adjuvant*, which stimulates the immune system in ways that are poorly understood. The antigen binds to the adjuvant or, with oil-based adjuvants, is emulsified. Emulsions act as a depot that releases the immunogen over a prolonged period, which can result in a greater immune response. However, many adjuvants cause local irritation and currently only alum (aluminium phosphate and aluminium hydroxide) and MF59 (a squalene oil in water emulsion) are licensed for use in humans. Adjuvants cannot be used with live virus vaccines and are restricted to inactivated or subunit vaccines.

26.7 Adverse reactions and clinical complications with vaccines

Vaccines have been enormously successful in reducing the burden of infectious disease. However, their use is always associated with a small risk of adverse reactions in some of the vaccinated individuals. These adverse reactions can range from mild to extremely severe and potentially life-threatening, though the most severe reactions are very rare. Whilst the impact of these adverse reactions on the affected individuals is of great concern, they have to be balanced by consideration of the risk to the population from the infectious disease being targeted. For example, the risk of poliomyelitis following infection is approximately 1 in 100 infections. This contrasts with a risk of acquiring poliomyelitis due to reversion of the live attenuated vaccine virus of between 1 in 1,000,000 to 1 in 25,000,000 doses of vaccine.

One of the best-studied vaccines in terms of risk of adverse reactions is vaccinia virus, as this was used in a worldwide immunization campaign that ultimately led to the elimination of smallpox. Vaccinia virus is considered to be one of the most successful vaccines to have been used but even this has some side effects. In a study of 14 million people who received the vaccine, mortality was estimated at one death per million resulting from complications following primary vaccination and one death per four million following subsequent revaccination. This compares with a mortality rate of 30% for those infected with smallpox virus. As with all vaccines, there was a range of adverse reactions with vaccinia virus and these are shown in Box 26.4.

Several other factors play a role in the appearance of adverse reactions to vaccines. A major consideration is the presence of other materials in the vaccine preparation arising

---

Box 26.4

Adverse reactions seen in a study of 14 million vaccinia virus recipients

Four primary adverse reactions were seen.

1. *Eczema vaccinatum* was seen in people with a history of eczema. Infants were more severely affected with potentially large areas of affected skin.
   (74 cases, no deaths; 60 additional cases of eczema vaccinatum occurred in the contacts of vaccinated persons, with 1 death)

2. *Progressive vaccinia* was seen only in immunodeficient people. In the most severe cases progressive disease led to death, approximately 2–5 months after vaccination.
   (11 cases, with 4 deaths)

3. *Generalized vaccinia* was seen in otherwise healthy individuals. The first appearance was a rash 6–9 days after vaccination.
   (143 cases, no deaths)

4. *Postvaccinial encephalitis* was the most severe consequence. In infants less than 2 years of age this was associated with convulsions. Recovery was often incomplete, leaving cerebral impairment and paralysis. In children greater than 2 years, the symptoms had a rapid onset, with fever, vomiting, headache, and malaise, followed by symptoms of neurological dysfunction such as loss of consciousness, amnesia, confusion, restlessness, convulsions and coma.
   (16 persons, with 4 deaths)

Data from the World Health Organisation

from the manufacturing process. For example, the yellow fever 17D vaccine is considered to be extremely safe; however, as the live virus is prepared by growth in embryonated chickens' eggs it cannot be given to anyone with an allergy to egg products. Vaccines also frequently contain additional materials such as adjuvants that are present to enhance the immunogenicity of the vaccine (see Section 26.6), to improve the stability, or to provide some level of protection from contamination. A list of some of the possible constituents is shown in Table 26.2. The presence of these materials

Table 26.2

Materials found in certain vaccine preparations. Only a limited number of these materials would be found in any single vaccine preparation.

Material	Function
Aluminium gels or salts of aluminium	adjuvant approved for use in human vaccines
Water:mineral oil emulsion	to enhance antigen delivery
Microspheres	polymers of lactic or glycolic acids which encapsulate the antigen to enhance delivery
Cholesterol and phospholipids	to enhance delivery
Liposomes	to enhance delivery
Antibiotics	to protect against bacterial contamination
Egg protein	a by-product of the production process
Formaldehyde	used to inactivate some viruses
Monosodium glutamate	stabilizer to protect against changes in heat, light, acidity or humidity
2-phenoxy-ethanol	stabilizer to protect against changes in heat, light, acidity or humidity
Thimerosal (also known as Thiomersal)	preservative to prevent contamination and growth of potentially harmful bacteria

must always be considered before administration of a vaccine in case of any allergic or other reaction they may generate in a small number of recipients. Some of the constituents, such as thimerosal, have generated considerable controversy amongst some concerned groups.

26.8 Eradication of virus diseases by vaccination

For the individual, the aim of a vaccine is to provide adequate protection from infection. However, the application of vaccines can, in principle, be used to eradicate the virus in question from the world, so providing escape from infection for the entire population. To date, only two viruses have been declared eradicated from the world by vaccination: smallpox and the cattle virus rinderpest. The reason there have been so few successful eradications is that this is only achievable if a virus fulfils a number of specific criteria (Table 26.3).

It can be seen that most of the factors are determined by characteristics of the virus but the nature of the immune response established by the vaccine is also important. An important feature is whether the vaccine is *sterilizing* or *non-sterilizing*. These terms refer to whether or not the vaccine induces an immune response that completely prevents replication of the virus in the vaccinee. Thus a sterilizing vaccine ensures that the virus cannot replicate in the vaccinated individual, while a non-sterilizing vaccine generates an immune response that suppresses virus replication sufficiently to prevent disease but a virus infection still takes place and virus may still be shed and be able to be transmitted to a susceptible contact. An example of non-sterilizing vaccines is seen with most avian influenza virus vaccines which prevent disease but permit the virus to grow at a

Table 26.3
Criteria for the world-wide eradication of human viruses.

Criterion	Variola virus	Measles virus	Poliovirus types 1–3	Comment
There must be overt disease in every instance so that infection can be recognized	+	+	−	99% of poliovirus infections are subclinical
The causal virus must not persist in the body after the initial infection	+	+	+	People whose B cell system is deficient can have persistent poliovirus infections and excrete virus for years
There must be no animal reservoir from which reinfection of humans can occur	+	+	+	Great apes are susceptible to poliovirus
Viral antigens must not change	+	+	+	
Vaccination must provide effective and long-lasting immunity	+	+	+	

low level. Clearly, non-sterilizing vaccines cannot lead to the eradication of a virus – indeed, some argue that they ensure the survival of the virus by effectively creating a reservoir of infection that cannot be seen, allowing it to persist in the population. The use of the term 'sterilizing' in the context of vaccines has generated some unfortunate misunderstandings in some communities where the term has been misinterpreted to indicate an effect on recipients rather than the virus. This has compromised some vaccine programmes in certain areas of the world.

The planned eradication of poliovirus

In 1988, the WHO launched the Global Polio Eradication initiative with the plan to use the available vaccines to eliminate polio from the planet by 2000. However, based on the criteria in Table 26.3, poliovirus falls down on one of the key criteria for eradication as 99% of infections are subclinical. In addition, although there is no natural animal reservoir, old-world primates are susceptible to infection and chimpanzees in the wild have been infected. This would suggest that eradication of poliovirus is unlikely to be successful. This goal appears even more unlikely when it is appreciated that the live polio vaccine efficiently reverts to neurovirulence which means that the use of the vaccine has the unwanted consequence of constantly introducing infectious virus capable of causing disease back into the environment (Box 26.2). However, eradication is a major target to which huge resources have been devoted. The progress in elimination of poliovirus has been dramatic with great success seen very rapidly (Fig. 26.7). The early successes have been further extended by focussed vaccination campaigns and, other

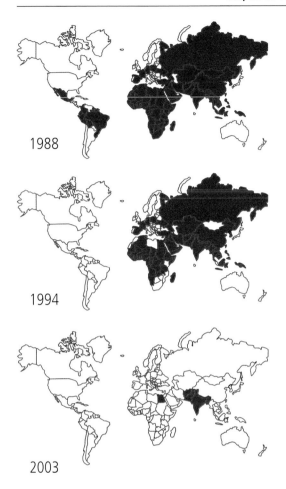

1988

1994

2003

Fig. 26.7 Progress of the WHO eradication programme for poliovirus. The incidence of poliovirus infection across the world at the time of beginning the poliovirus eradication programme (1988) and in two intervening years (1994 and 2003) are highlighted. This demonstrates the early successes in reducing disease incidence. In recent years attention has focussed on the few countries where poliovirus outbreaks occur. Egypt and Niger were declared polio non-endemic in 2006 and India was declared polio-free in 2014. The virus is now restricted to areas of conflict or low socioeconomic status of rural communities.

than in parts of the world where armed conflict prevents the vaccination programme from being implemented, the initiative has been successful.

The keys to the success of the poliovirus eradication scheme to date, and what allows us to be optimistic that total eradication from the world is possible, are the infrastructure that underpins the programme and the availability of both a live and a killed vaccine. Laboratories have been set up in all countries, and there are events such as national vaccination days in which the entire population of a country is encouraged to take up the offer of vaccination – a massive undertaking. This is linked with intensive vaccination whenever there is an outbreak. The aim is to establish immunity in as large a number of the population as possible over a short period. The herd immunity that is established will break the chain of transmission of the virus. With no susceptible hosts the virus cannot survive. When the number of cases is reduced sufficiently the original killed Salk vaccine can be used to ring vaccinate potential contacts of infected individuals. This removes the possibility of a vaccine-associated infection and poses no risk of introduction of live infectious virus into the environment. The campaign has been very successful.

There are still other problems to overcome. Analysis of sewerage (where all polioviruses end up) has shown that neurovirulent poliovirus has been replicating and circulating in some countries where there has been a highly vaccinated community for years. While this virus is clearly derived from the live vaccine, it has recombined with other enteroviruses to make a brand new chimeric virus. This raises concerns that a virus like this may be able to fill the niche left by poliovirus.

The final steps of poliovirus eradication will include the destruction of all other sources of the virus from which reintroduction might occur. These include frozen stocks of poliovirus held in many laboratories where poliovirus was (and still is) studied for research purposes, and frozen faecal samples in hospital pathology laboratory freezers all over the world that may contain poliovirus.

26.9 Immunotherapy for virus infections

For the many virus infections for which there are no vaccines the options are limited. One possibility is to introduce antibodies directly into the body, either into the bloodstream or intramuscularly from where they enter the circulatory system. This is a process referred to as immunotherapy or passive immunization. This approach was more widely used in the past and is less common now, though the use of monoclonal antibodies has revived the practice. It is important to appreciate that this is not equivalent to vaccination as no immune response is stimulated in the recipient and the immunity lasts only as long as the antibody survives in the body. Typically, the half-life of an antibody delivered in this way is 30 days though this can be prolonged in some circumstances. A significant advantage of using immunotherapy is that the recipient is protected from infection immediately following the treatment rather than having to wait for the immune system to respond to an immunogen.

Whilst in theory the antibody used in such treatments can come from any animal source, the risk of an adverse reaction makes it more desirable to use human antibodies. The material may consist of a 'pool' of immunoglobulins taken from a number of humans who are immune to the particular virus, combined together. Alternatively, and increasingly commonly, the antibodies are monoclonal antibody from human hybridomas, cloned human antibody, or mouse immunoglobulin that has been humanized by replacement of all but its paratope with human IgG sequences (Box 26.5). The latter types are likely to give fewer problems with adverse reactions. Immunotherapy with humanized mouse monoclonal antibodies (MAbs) or human MAbs is not intended to replace the stimulation of

---

Box 26.5

Humanising mouse monoclonal antibodies (MAbs)

Despite their huge impact on virology, the use of MAbs in immunotherapy has been frustrated by technical limitations. MAbs are made by immunizing an animal, extracting primed B cells from the spleen and then immortalizing the B cell by fusion with a myeloma (B cancer) cell. While the system works for mice, it is not applicable to humans as no human myeloma cell has proved effective in this process, though there has been some success by immortalizing human B cells with Epstein-Barr virus. The problem of treating people with a mouse MAb is that their immune system very rapidly reacts to the mouse antigens present in the foreign protein molecule. At best, this destroys the antibody and its beneficial effects; at worst, it initiates a potentially harmful, or even lethal, anaphylactic response. One solution is to humanize a mouse MAb by using recombinant DNA technology to replace the mouse sequences with sequences from a human antibody, except short sequences responsible for binding to antigen (the paratope) that are composed of complementarity-determining regions (CDRs) of the hypervariable parts of the variable (V) regions of the light (L) and heavy (H) chains. By minimizing the amount of foreign (mouse) protein, the problem of rejection is overcome.

One of the first humanized MAbs, palivizumab, is licensed for use against respiratory syncytial virus.

immunity with vaccines but, instead provides immunity in circumstances where no other treatment is possible or effective. It is used routinely to protect infants who have underlying conditions that render them seriously at risk from infection, or otherwise healthy infants who have contracted a life-threatening infection. Immunotherapy can also be considered for the congenitally immunodeficient, or those whose immune systems are compromised by accident or deliberately in the course of transplant surgery or cancer chemotherapy.

Efforts continue to improve the antibodies available for immunotherapy. Most of the current effort consists of identifying mouse monoclonal antibodies with more desirable characteristics such as high avidity to the target antigen. These can then be humanized for clinical use. Alternative strategies are utilizing recombinant DNA technology to clone human antibody genes for production of human Ig in cell culture systems and then to subject these to mutagenesis to develop increased avidity of binding to antigen. It is also possible to manufacture *microantibodies* which are peptides with a sequence derived from just one of the six 8–23 amino acid residue CDRs (complementarity-determining regions) of an antibody that form the epitope/antigen-binding site. The small size of microantibodies reduces the problems of accessibility to target tissues seen with the large antibodies from which they are derived (see below).

26.10 Adverse reactions and clinical complications with immunotherapy

The primary concern with immunotherapy is the risk of generating an immune response directed against the therapeutic antibody. This is a particular concern when the antibody has been derived from an animal. The antibody preparations are usually purified immunoglobulin so the risk of contamination with other material is reduced, but not completely eliminated. However, the immunoglobulin is detected by the immune system of the recipient as 'foreign' and this can lead to stimulation of an immune response. At the least, this means that subsequent treatments with the same preparation or with a different immunoglobulin from the same animal species will be eliminated rapidly from the circulation before they can exert a protective effect. With each successive treatment with the same foreign immunoglobulin the risk of an anaphylactic response is increased – in the most extreme circumstances this can be life-threatening. The risk of an immune reaction to the treatment is reduced if human immunoglobulin, either single or pooled from several people, is used and the risk is further reduced when humanized monoclonal antibodies are used.

Key points

- The purpose of a vaccine is to stimulate immunity without having to experience the infection or the disease.
- Live vaccines are made by attenuating virulent, wild-type virus; they cause an infection in the vaccinee that is inapparent or less severe than the wild-type virus infection.
- Killed or non-infectious vaccines are made by inactivating virulent virus.
- Subunit vaccines can be prepared using individual components or groups of components of the virus particle.
- Like all medical procedures, vaccination carries a risk, but this is far less than the risk of contracting the wild-type virus infection.
- Vaccination offers the hope of eradicating some human virus diseases, but so far only smallpox has succumbed.

- Immunotherapy offers the possibility of providing immediate, but short-term protection by injection of antibodies directly into the body.

Questions

- Compare and contrast the properties of live attenuated and killed whole viral vaccines, commenting on the advantages and limitations of each in controlling human viral infections.
- What factors need to be considered when deciding policy on vaccination against viral infections, and attempts to achieve global eradication of named viral infections?
- Compare the advantages and disadvantages of subunit versus whole virus vaccines.

Further reading

Barrett, A. D. T., Higgs, S. 2007. Yellow fever: a disease that has yet to be conquered. *Annual Review of Entomology* 52, 209–229.

Both, L., Banyard, A. C., van Dolleweerd, C., Wright, E., Ma, J. K., Fooks, A. R. 2013. Monoclonal antibodies for prophylactic and therapeutic use against viral infections. *Vaccine* 31, 1553–1559.

Coffman, R. L., Sher, A., Seder, R. A. 2010. Vaccine adjuvants: putting innate immunity to work. *Immunity* 33, 492–503.

Graham, B. S. 2013. Advances in antiviral vaccine development. *Immunology Reviews* 255, 230–242.

Henderson, D. A., Klepac, P. 2013. Lessons from the eradication of smallpox: an interview with D. A. Henderson. *Philosophical Transactions of the Royal Society of London B. Biological Sciences* 368, 20130113.

Jin, H., Chen, Z. 2014. Production of live attenuated influenza vaccines against seasonal and potential pandemic influenza viruses. *Current Opinions in Virology* 6, 34–39.

Lauring, A. S., Jones, J. O., Andino, R. 2010. Rationalizing the development of live attenuated virus vaccines. *Nature Biotechnology* 28, 573–579.

Stiehm, E. R. 2013. Adverse effects of human immunoglobulin therapy. *Transfusion Medicine Reviews* 27, 171–178.

Verardi, P. H., Titong, A., agen, C. J. 2012. A vaccinia virus renaissance: new vaccine and immunotherapeutic uses after smallpox eradication. *Human Vaccines and Immunotherapeutics* 8, 961–970.

Zhao, Q., Li, S., Yu, H., Xia, N. and Modis, Y. 2013. Virus-like particle-based human vaccines: quality assessment based on structural and functional properties. *Trends in Biotechnology* 31, 654–663.

The World Health Organisation web pages on immunization, vaccines and biologicals: http://www.who.int/immunization/diseases/en/

Chapter 27
Antiviral therapy

The introduction of antibiotics in the 1940s to control bacterial infections prompted the hope that similarly broad-spectrum treatments for virus infections (antivirals) might also be found. However, even today with nearly 50 antiviral drugs approved for human use, this hope has not been realized. Most antivirals are specific for just one or a few viruses. Half of the available drugs are specifically for HIV-1.

Antiviral therapy is the use of drugs to intervene at some point following infection to alter the course of infection and so improve the outcome for a patient. This is in contrast to the use of vaccination, which is almost always deployed prophylactically, to prevent infection in healthy people. The use of antivirals is also almost entirely restricted to treatment of human infections. Whilst the same principles could be applied to treatment of infection in animals, either veterinary or agricultural, the costs and practicalities are usually prohibitive; treatment of plant infections in the same way would be confounded both by these considerations and by issues of access to the infected tissues due to the different architecture of the organism. This chapter considers the development of antivirals for use in treating significant infections in humans.

27.1 Scope and limitations of antiviral therapy

Given the obvious clinical impact of virus infections, it is not surprising that considerable efforts have been devoted to the development of effective therapies. This effort stretches back over 50 years. However, the potential of antiviral therapy has always been limited by practical considerations of timing. The aim of specific drug therapy is, just as in antibiotic therapy for bacterial infections, to slow or stop the growth of the pathogen so that the host immune system can gain the upper hand and eradicate the infection. Logically, this therapy can only be started once a patient has recognized that they are ill and a clinician has diagnosed their problem. However, for many if not all acute virus infections, the onset of overt signs and symptoms occurs only after the peak of virus replication has passed, reflecting the fact that

Introduction to Modern Virology, Seventh Edition. N. J. Dimmock, A. J. Easton and K. N. Leppard.
© 2016 John Wiley & Sons, Ltd. Published 2016 by John Wiley & Sons, Ltd.

these signs of infection are actually due to the action of various parts of the host immune system. So the time window within which even a highly effective antiviral can be deployed against an acute infection and achieve clinical benefit is narrow. It is for this reason that the majority of successful antiviral drugs that have been developed are used against long-lasting persistent, chronic or latent infections.

The development of antiviral drugs has also been constrained by the recognition that viruses do not possess the same broad-spectrum targets for therapy as are found in bacteria. Antibiotics can be deployed with success without having a precise understanding of which organism is causing the infection because these drugs target essential biochemical processes common to many different species of bacteria. These drugs are also generally safe to use because the targeted processes are either absent from the host or differ greatly from those that operate in the host, so that the drug has no direct effects on essential host functions. In contrast, viruses replicate with the use of many essential host processes. For example, protein synthesis is the target for many successful antibacterial drugs, but with viruses this process is always provided by the host as also is energy metabolism. Some viruses also use host systems for mRNA synthesis and/or genome replication, though many provide their own enzymes for one or both of these processes. Finally, viruses lack a cell wall, synthesis of which is again a target for many antibacterial therapies. Thus, rather than targeting generic processes common to large numbers of viruses, antivirals tend to be highly specific. They are developed to target a process that is essential to the replication of a particular virus and will typically be active only against that virus or others closely related to it.

27.2 Antiviral therapy for herpesvirus infections

Herpesviruses establish lifelong persistent or latent infection, with periodic reactivations that can be very debilitating. Reactivation is also a serious problem in patients undergoing immunosuppressive therapy for organ transplantation. Drugs such as aciclovir that target herpesvirus DNA synthesis are very effective in blocking the clinical manifestations of virus reactivation.

The field of antiviral drug research was born out of the search for anticancer drugs, where interest was directed towards compounds that inhibited DNA synthesis. Attention focused on nucleoside analogues, modified forms of the deoxyribose-base structures that are phosphorylated and used as substrates for DNA synthesis throughout all life. These analogues inhibit DNA synthesis either by binding in the enzyme active site and being incapable of reaction, thus blocking the enzyme, or by reacting with the growing DNA chain to produce a product that is not able then to accept any further nucleotide additions, thus blocking further synthesis. Researchers seeking treatments for the symptoms associated with reactivation from latency of various herpesviruses (see Chapter 17) began to test the effect of such compounds as inhibitors of herpesvirus replication. Even though these viruses use their own DNA polymerases, compounds such as idoxuridine (IDU) that were initially intended to be active against cellular DNA synthesis still displayed activity against the virus in cell culture. However, attempts to move such drugs into the clinic highlighted the significant issues of toxicity and side effects which remain key problems in antiviral drug development today. In this case, because DNA synthesis is an essential process in all of our tissues, IDU also affected cell DNA synthesis and, hence, caused damage as well as benefits. The therapeutic window for IDU (the range of drug concentrations that is high enough to give therapeutic benefit without unacceptable side effects) was too narrow for it to be used systemically; it is, however, still licensed for use in an ointment to treat herpesvirus infection of the eye.

IDU is a nucleoside analogue, comprising a modified uracil base linked to an unmodified deoxyribose sugar ring. Another nucleoside analogue that emerged as a potential therapy from early attempts to inhibit herpes simplex virus (HSV) DNA synthesis was adenosine arabinoside, in which the modification from a standard nucleoside lies in the sugar ring. Although well-tolerated in transfer to the clinic, the drug was ineffective because of the speed with which it was removed from the system by a deamination reaction catalyzed by a host enzyme; it was consequently impossible to maintain an effective concentration of the drug with any reasonable dose regime. However, while searching for a drug that would block this deamination and so make adenosine arabinoside more effective, the first real breakthrough in antiviral research was made. A compound, aciclovir (or acyclovir), was discovered in these screens that turned out to have intrinsic and highly potent activity directly against replication of the virus, again by inhibiting DNA synthesis (Fig. 27.1). Aciclovir works as a chain-terminator, becoming incorporated into viral DNA and then lacking the essential 3′ OH group for further chain extension to occur. A key feature of aciclovir is its high therapeutic index, which arises from it requiring activation by an enzyme that is encoded by HSV and is therefore present only in infected cells. Thus, uninfected cells are relatively untouched by the drug and the potentially toxic side effects of inhibiting host DNA synthesis are largely eliminated. These

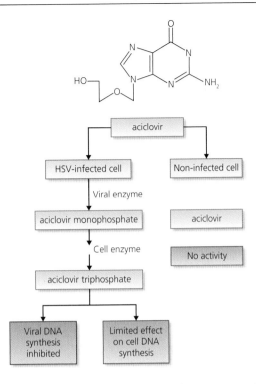

Fig. 27.1 Aciclovir is active against recurrent HSV infection because it inhibits viral DNA synthesis. It has an excellent therapeutic index for two reasons. Firstly, the drug requires activation by phosphorylation and only the viral thymidine kinase enzyme can do this: no cell enzymes will act on it. Secondly, the active form of the drug, aciclovir triphosphate, is bound and used preferentially by the viral DNA polymerase in comparison with cell DNA polymerases.

features make aciclovir very safe to use and it has gone on to be marketed worldwide in a number of formulations. In the UK, it is the only antiviral drug considered safe enough to be sold without prescription from a doctor as a treatment for cold sores caused by reactivation of HSV.

Two other compounds, ganciclovir and penciclovir, have also been developed as anti-herpesvirus drugs (Fig. 27.2). Although they have a common mechanism of action with

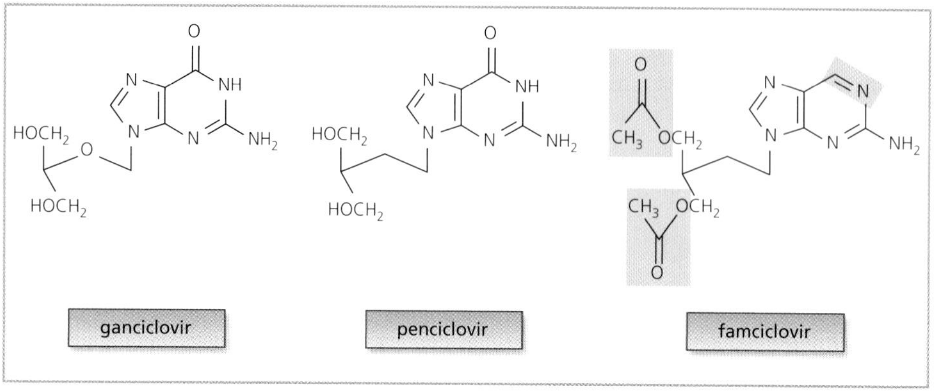

Fig. 27.2 Like aciclovir, ganciclovir and penciclovir are guanosine analogues that are active against various herpesviruses once activated to a triphosphate form. Famciclovir is a pro-drug derivative of penciclovir that is converted readily to penciclovir when ingested; the pro-drug features of the molecule are highlighted.

aciclovir, they differ in their potency against the various herpesviruses because of differences in the way these drugs are activated in the body. Ganciclovir is used against human cytomegalovirus infections in immunosuppressed patients. Finally, prodrug forms of these drugs have been developed. Prodrugs are processed into the standard form of the drug during absorption from the gut; they are developed to give better bioavailability than the original active compound so that oral administration can give effective therapeutic drug levels. Examples here are valaciclovir and famciclovir (Fig. 27.2).

27.3 Antiviral therapy for influenza virus infections

As an acute and usually self-limiting infection, the opportunity for treating influenza virus infections with specific antiviral drugs is limited. The key value of drugs to treat these virus infections lies in their potential to limit the effects of severely pathogenic pandemic strains of influenza A virus that may arise in the future.

The first specific antiviral compounds active against influenza viruses were amantadine and rimantadine, molecules containing the unusual bridged ring structure of adamantane. They were first licensed in the 1960s for the prevention of infection by the circulating strains of Hong Kong flu (influenza A virus). These molecules target the M2 protein of the virus, which is present in the viral envelope as a tetramer where it functions as an ion channel. After virus attachment and endocytosis, this channel must allow protons into the viral core for fusion and release of the genome segments into the cytoplasm to occur. The amantadine ring structure binds in the lumen of the M2 channel (Fig. 27.3), effectively blocking the flow of protons through it and so preventing these key steps in the infectious cycle from taking place. Unfortunately, influenza A viruses can mutate very readily, and the selective pressure of use of these drugs has meant that, essentially, all circulating strains of human influenza A virus are resistant to them. Thus the use of amantadine as a front-line treatment for these viruses is no longer recommended.

More recently, two drugs that target the action of influenza virus neuraminidase (NA) have been approved widely for use; these are

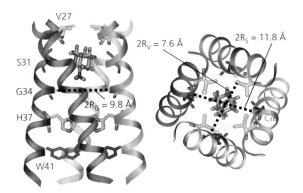

Fig. 27.3 Amantadine is a small, cross-bridged cyclic hydrocarbon molecule with an attached amino group. It binds in the lumen of the influenza A virus M2 ion channel, which is a tetrameric protein, blocking the flow of protons through the channel and hence preventing the initiating events in infection. (Reprinted by permission from Macmillan Publishers Ltd: Cady, S.D. et al. *Nature* 463, 689–692, copyright 2010.)

zanamivir and oseltamivir, which are active against both influenza A and B viruses. Two further drugs in this class have recently received approval in some countries. All these drugs were produced by rational design, using the known structure of the enzyme active site to model potential inhibitors that were then synthesized and tested. The neuraminidase is one of two proteins displayed on the surface of an influenza virus particle (Chapter 20). Its key role in the infectious cycle is thought to be in aiding virus release from an infected cell by breaking down non-productive re-attachment to the cell during exit. During infection, viral replication in the respiratory tract is thought to peak 1–3 days after the initial recognition of symptoms, creating a short time window in which a drug that inhibits replication might have an effect if promptly administered. Data in support of the licensing of zanamivir and oseltamivir suggest that their use shortens the symptomatic period by about one day, but it is unclear how effective they are at reducing

severity of disease in those who are worst affected.

It is known that resistance can arise to neuraminidase inhibitors through as little as one amino acid change in the NA protein. However, each drug has a different structure and so mutations generally confer resistance to just one drug rather than the whole class. Prior to the 2009 H1N1 pandemic (Chapter 20), the circulating H1N1 virus developed resistance to oseltamivir to the point where almost all isolates tested were no longer inhibited by the drug, but this strain has been replaced by the 2009 version and, so far, the 2009 H1N1 has remained sensitive, though sporadic isolates found in many parts of the world have been shown to be resistant. The world continues to be fearful of a future influenza pandemic involving a highly pathogenic strain, perhaps arising out of transmission and adaptation of an avian influenza A virus to a human host. One of the counter-measures that many countries have put in place is to stockpile neuraminidase inhibitors to deploy in the face of such an event.

27.4 Antiviral therapy for HIV infections

HIV-1 infection progresses to AIDS unless treated. Specific antiviral compounds are now available that target several different aspects of the virus replication cycle. By combining three drugs with different targets into a therapeutic regime, the disease process can be held in check and emergence of resistant viral mutants greatly reduced.

Nucleoside reverse transcriptase inhibitors (NRTI)

When the HIV-1 pandemic was first recognized in the 1980s (see Chapter 21), there was an immediate focus on developing antiviral therapies that might be effective against this virus. Because HIV-1 is a retrovirus, establishing infection in a cell requires the synthesis of a DNA copy of its genetic material by the viral enzyme reverse transcriptase (RT; see Chapter 9). The spread of infection within a host should thus be blocked by DNA synthesis inhibitors. Given the experience in developing these both as anticancer drugs and as antivirals targeting herpesviruses, there was rapid progress. The first drug to be approved for clinical use was azidothymidine (AZT), marketed as Zidovudine, which was launched in 1987. Initial experience of the use of AZT was very promising. Virus loads in the blood of patients were reduced and there was a marked restoration in CD4+ T cell counts, both indicators of a very favourable response. However, this promise turned to disappointment when, within a few months of starting this therapy, both these indicators reversed back to their pretreatment levels and the therapeutic benefit was lost. The reason for this turned out to be the evolution of viral drug escape mutants in the patients. An example of such an effect is shown in Fig. 27.4, which also illustrates the improvement in therapy when using drugs in combination. The emergence of drug-resistant pathogens was not a new phenomenon, but the pace of evolution of resistance in HIV-1 was remarkable. Even today, with the use of triple drug therapy, the emergence of resistant HIV-1 strains is a continuing clinical problem.

AZT is a nucleoside analogue: it targets the active site of reverse transcriptase and is incorporated into DNA as a chain-terminator (Fig. 27.5), similar to the action of aciclovir in herpesvirus therapy. However, there is no specific activation of AZT by a viral enzyme, meaning that AZT is more toxic to healthy cells than aciclovir. In fact, the favourable therapeutic index of AZT comes from the fact that HIV RT enzyme has a rather higher affinity

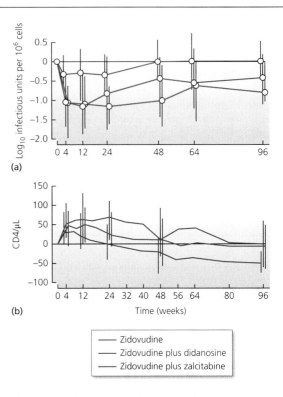

(a)

(b)

— Zidovudine
— Zidovudine plus didanosine
— Zidovudine plus zalcitabine

Fig. 27.4 The course of HIV-1 infection over 96 weeks, measuring change in virus load (a) and change in CD4+ cell counts (b), following commencement of AZT (zidovudine) monotherapy or either of two regimes of dual therapy including AZT and didanosine or zalcitabine in previously untreated patients. Patients were enrolled in the trial with CD4+ cell counts of between 50 and 350 per μl. (Reprinted from *The Lancet*, vol. 350, Brun-Fezenet F. et al., HIV-1 viral load, phenotype, and resistance in a subset of drug-naive participants from the Delta trial, 983–990, Copyright (1996), with permission from Elsevier.)

for AZT triphosphate than host DNA polymerases, meaning that at appropriate doses, inhibition of the viral enzyme can be achieved whilst having acceptably low levels of effect on host tissues. Even so, the side effects made the required dose regime difficult for some patients to tolerate. Subsequently, a large

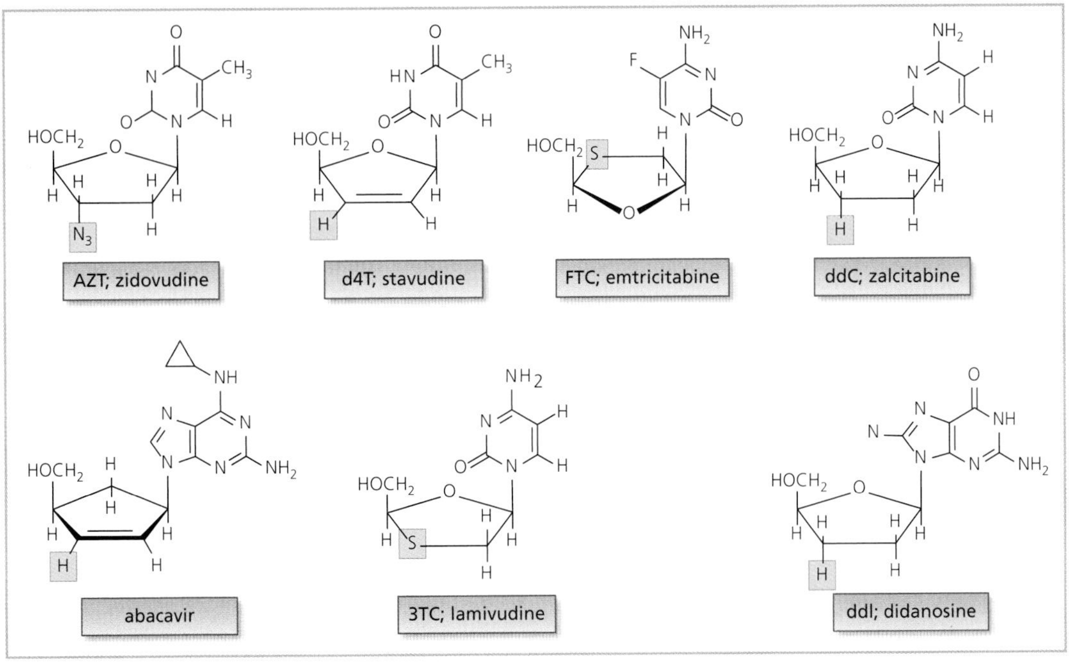

Fig. 27.5 Comparison of the formulae of nucleoside analogue inhibitors of HIV-1 reverse transcriptase approved for clinical use. All these molecules require phosphorylation by cellular enzymes before being incorporated into DNA. Note that all analogues shown share a loss of the 3′ hydroxyl group on the sugar rink (highlighted). Thus they cannot form a bond with the next nucleotide, and hence act as chain terminators when incorporated into a growing DNA strand.

number of other nucleoside analogues have been developed for use in HIV therapy. These drugs, known collectively as the nucleoside RT inhibitors (NRTI) are shown in Fig. 27.5. Each has its own particular characteristics and, importantly, the mutations in RT that give resistance to one of these drugs do not necessarily impart resistance to others. Thus patients can be moved from one drug to another if resistance emerges.

Other HIV drug targets

The RT active site is not the only HIV molecular target to have been exploited successfully as a drug target. A second class of RT inhibitors was found that binds outside the active site to inhibit its catalytic activity, an example being nevirapine. This class of drugs is only active against HIV-1 because the RT of HIV-2 lacks the drug binding site. Since the site is not crucial to the activity of RT, resistance to these non-nucleoside RT inhibitors (NNRTI) emerges very readily.

Another important target is the HIV protease (PR). This enzyme is required to process the viral polyproteins during the maturation of virus particles to gain infectivity (see Section 12.8). The first drug to target PR was saquinavir. It is a non-cleavable isostere of the polypeptide substrate of the enzyme, meaning that it is a structural analogue of the substrate that fits into the active site (Fig. 27.6). Since it cannot be cleaved to products it remains tightly bound, so preventing the further processing of

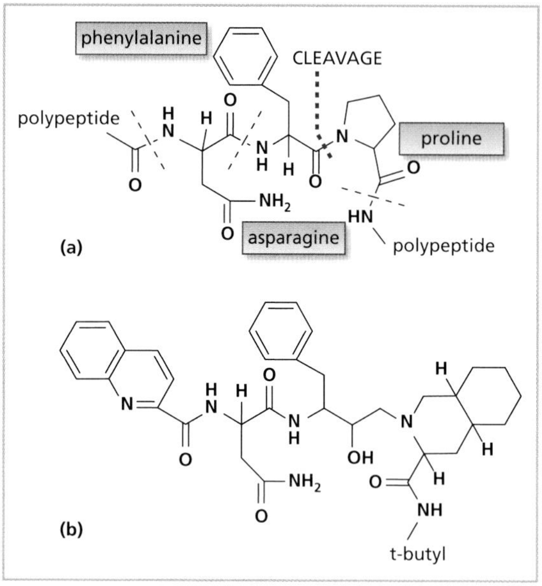

Fig. 27.6 Comparison between the structures of a natural peptide substrate for the HIV-1 protease (a) and the specific HIV-1 protease inhibitor, saquinavir (b). Saquinavir is a non-cleavable analogue of the transition state through which the natural substrate passes when being cleaved by the enzyme.

substrate by the enzyme. Saquinavir was developed by successive systematic structural modification and evaluation from an initial lead compound, which is the classical approach to drug development. Later, drugs such as nelfinavir and amprenavir were produced by rational drug design using the 3D structure of PR to model inhibitors that would bind the active site. Subsequently, other new drugs have targeted either the integrase enzyme (raltegravir) or the fusion process that occurs during delivery of the virus particle into the cell (enfuvertide). Of all the drugs discussed in this chapter, enfuvertide is the most chemically complex, being a 36 residue peptide derived from the HIV-1 gp41 transmembrane protein.

Combinations of three drugs drawn from the classes of protease inhibitors, NRTIs and NNRTIs formed the basis of what has become a successful therapy for HIV, known as highly active anti-retroviral therapy (HAART; Box 27.1). Drugs against other newer targets have since been incorporated into this regime as required. The principle is that, by facing the virus simultaneously with drugs against different

---

Box 27.1

Key points about HAART for HIV infection

- Uses three drugs simultaneously; these bind different parts of the viral reverse transcriptase or protease molecules or sometimes now other HIV molecular targets
- In use since 1994
- Reduces virus load in plasma
- Restores lost immune functions
- Halts progression to AIDS
- Does not clear latently infected CD4+ memory cells
- Lapses in therapy result in virus rebound to normal levels
- Is not tolerated by all people
- Has to be taken indefinitely
- Not known if therapy can be tolerated for a lifetime
- Uncertain if resistant virus will eventually break through

targets, the probability of a variant virus emerging in the population that has the sequence changes necessary to evade all the different drugs in the cocktail is dramatically reduced. Patients on HAART are monitored regularly for their blood virus load and, if any signs of resistant variants are seen, then the drug cocktail is varied to counteract this. With HAART properly applied, the life expectancy of HIV-positive individuals is substantially increased towards what is normal for the population.

27.5 Antiviral therapy for hepatitis virus infections

The long-term liver damage associated with chronic infections by hepatitis B and C viruses can be limited or avoided by extended periods of combination therapy with a stabilized form of type 1 interferon and an appropriate specific antiviral compound. For both viruses, the probability of successful therapy is related to virus genotype.

As discussed in Chapter 22, viral hepatitis is caused by a series of unrelated pathogens. Antiviral therapy is only relevant in the context of those of these viruses that establish persistent infections, hepatitis B and C viruses. An effective vaccine exists for HBV, which therefore forms the main focus of efforts to limit the future health impact of this virus (see Section 26.4). However, there are many people who are already chronically infected by this virus and in some cases long-term treatment is needed to prevent or minimize liver damage. Treatment is particularly indicated if, after the acute phase of the infection is over, a patient still has serum markers of ongoing liver damage and there is detectable virus in the blood. Available treatments do not generally lead to the complete loss of virus from the system (a

cure) but they can have a substantial beneficial effect on the long-term clinical outcome.

First line therapy is usually a form of type 1 interferon (IFN), PEG-IFNα2a, given by weekly injection for a year. IFN is a natural antiviral, produced by the body as part of the innate immune response (Chapter 13); conjugating IFN with PEG (polyethylene glycol) increases its half-life in the body, so reducing the frequency of dosing required for effective therapy. Some genotypes of the virus respond better than others to treatment, as determined by reduction in virus load and the production of antibodies to the HBV e antigen (Section 22.4). Other treatment options include a selection of nucleoside reverse-transcriptase inhibitors (NRTI). Initially lamivudine, a drug already mentioned as a treatment for HIV-1 (Fig. 27.5), was used but this has now been supplanted for HBV treatment by other drugs such as tenofovir and entecavir (Fig. 27.7). Despite the close structural similarity between the various NRTIs, they can target HBV and HIV-1 very differently: for example, entecavir is not active against HIV-1 whilst having good activity against HBV.

Hepatitis C virus has been the focus of intense research since the full scale of the infection in the global population has become apparent. Since there remains no vaccine

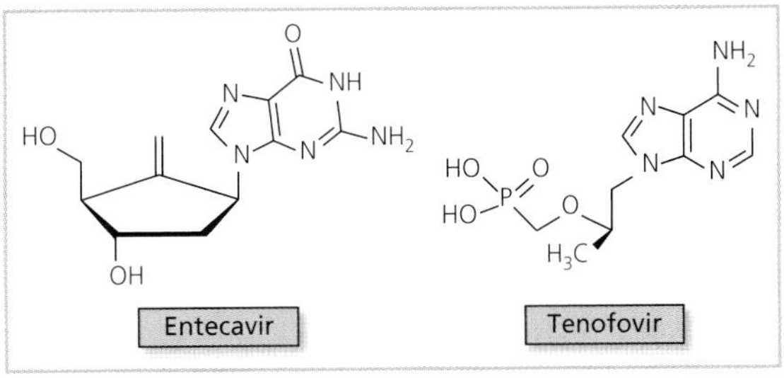

Fig. 27.7 Reverse transcriptase inhibitors used to treat chronic HBV infections. Tenofovir is now used exclusively in a prodrug form where the hydroxyl groups of the phosphonic acid element are derivatized to neutralize their charge and improve oral bioavailability.

against HCV, the development of effective antiviral compounds is particularly important. For a number of years, the standard therapy was a lengthy course of treatment with PEG-IFN plus an antiviral compound, ribavirin, a drug that is active against a variety of RNA viruses (see below). This combination proved curative in about half of patients in Western countries, i.e. it eradicated the virus from the body. This relatively low success rate is now realized to be linked to genetic variation, both in the virus and the human population (Section 22.6). The established practice, given the side-effects of interferon therapy, was to monitor patients over the first few weeks of treatment and, if virus loads reduced, to continue with the full course whilst ceasing therapy if no effect was seen.

More recently, a series of drugs that target specific processes in the HCV replication cycle have either been licensed or are approaching the market. Telaprevir and Boceprevir are the first of a growing series of drugs that act on the viral NS3/4A protease. This enzyme is encoded within the viral polyprotein that is the primary product of viral protein synthesis, and is necessary for the cleavage of the polyprotein into its active components. The drugs are similar in concept to the successful inhibitors of HIV-1 protease: they are structural analogues of the section of polypeptide that naturally binds in the enzyme's active site. They have been shown to substantially improve the cure rate in HCV infections, particularly in those for whom PEG-IFN plus ribavirin is less likely to be effective or has failed. However, given the high mutation rate of HCV – another virus that establishes a quasispecies of variant genome sequences in a host (see Section 4.2), it is unsurprising that use of these drugs alone leads rapidly to the emergence of resistance. Therefore, they are being used in combination with PEG-IFN and ribavirin. Other drugs coming to market target either NS5A or the viral RNA polymerase. It is hoped that these

new drugs, used in combination, will transform the treatment of those infected with HCV, similar to the approach used to give successful maintenance therapy against HIV-1 disease.

27.6 Therapy for other virus infections

Among the various virus infections of humans, those that can cause the most severe acute disease are typically RNA viruses – i.e. viruses that replicate using an RNA-dependent RNA polymerase. Perhaps because they often have a high mutation rate, such viruses form the majority of zoonotic infections and these often cause severe, even life-threatening, pathogenesis. In such circumstances, any antiviral drug that can have even a marginal effect on the replication of the virus might have a significant clinical impact, albeit only in the rare instances of such infections. Ribavirin has been used to treat several exotic infections, including Lassa fever (an arenavirus) and various bunyavirus infections though the effect on outcome is not clear. It has also been used widely in aerosol form to treat infants hospitalized with respiratory syncytial virus infection, a paramyxovirus (Fig. 27.8). Ribavirin is another nucleoside analogue, in this case with the structure of the purine base disrupted, and would be expected to interfere with RNA synthesis. Its precise mechanism of action is unclear however, particularly in light of evidence that it is also active against some DNA viruses. In fact, it may function through multiple pathways, including a disruptive effect on purine nucleotide metabolism that would affect all nucleic acid synthesis.

One final area where antiviral treatment is proving increasingly important is in the area of transplantation medicine. After receiving a tissue or organ transplant, patients must be immunosuppressed to prevent immunological

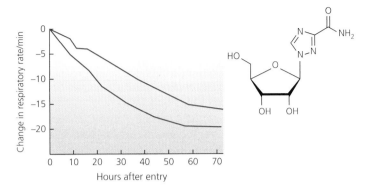

Fig. 27.8 Ribavirin is active against a number of different viruses, including respiratory syncytial virus. In infants hospitalized with severe RSV infection, the use of aerosolized ribavirin (red line) significantly shortened the time taken to reduce the excessively rapid respiration rate at admission (compared with placebo, blue line) that characterizes such infants. (Graph reproduced from *Archives of Disease in Childhood*, W. Barry, F. Cockburn, R. Cornwall, J.F. Price, G. Sutherland and A. Vardag, vol 61, 593–597 ©1986, with permission from BMJ Publishing Group Ltd.)

rejection of the graft. This leaves them vulnerable to many infections, particularly reactivation of viruses being harboured latently in either the patient or the transplanted tissue. Cytomegalovirus (CMV; a herpesvirus) and various adenoviruses are a particular problem. For CMV, drugs such as ganciclovir are useful whilst for adenovirus, cidofovir has shown good effects. Cidofovir is yet another nucleoside analogue and inhibits a number of different viral DNA polymerases with some selectivity over the DNA polymerases of the host. Unlike most of the other drugs considered in this Chapter, it requires delivery by intravenous injection.

Key Points

- Antiviral drugs have been slow to arrive, but nearly 50 have been approved for human use.
- Most antiviral drugs are specific for one or a few similar viruses; only ribavirin shows properties suggestive of broader spectrum activity.

- The most numerous class of antiviral drugs are nucleoside analogues that inhibit genome replication of particular viruses.
- Other molecular targets within virus lifecycles that have been exploited to develop useful antiviral drugs include site-specific viral proteases and aspects of virus entry into and exit from the cell.
- New antiviral drugs are now being made by rational design through understanding of the molecular details of viral proteins and replication processes.
- Given their natural antiviral activity, interferons are surprisingly ineffective at treating infections in general, but they have specific uses.
- The emergence of drug-resistant viral strains is a problem for many different viruses and can be limited by using combinations of two or more drugs together.

Questions

- Discuss the mode of action of antiviral therapeutics, using selected examples, and explain why broad-spectrum antiviral therapeutics are so difficult to achieve.

- Explain how nucleoside analogues can inhibit virus replication, using specific examples.

Further Reading

Broder, S. 2010. The development of antiretroviral therapy and its impact on the HIV-1/AIDS pandemic. *Antiviral Research* 85, 1–18.

De Clercq, E. 2004. Antivirals and antiviral strategies. *Nature Reviews: Microbiology* 2, 704–720.

De Clercq, E. 2011. A 40-year journey in search of selective antiviral chemotherapy. *Annual Review of Pharmacology and Toxicology* 51, 1–24.

Martinez-Picado, J., Martinez, M. 2008. HIV-1 reverse transcriptase inhibitor resistance mutations and fitness: a view from the clinic and ex vivo. *Virus Research* 134, 104–123.

Pawlotsky, J.-M. 2014. New hepatitis C therapies: the toolbox, strategies, and challenges. *Gastroenterology* 146, 1176–1192.

Samson, M., Pizzorno, A., Abed, Y., Boivin, G. 2013. Influenza virus resistance to neuraminidase inhibitors. *Antiviral Research* 98, 174–185.

Chapter 28

Prion diseases

Prion diseases are the subject of widespread interest and concern as threats to public health following an epidemic of bovine spongiform encephalopathy (BSE) in the United Kingdom in the 1980s and 1990s that spread to other countries, and the associated emergence of a new human disease, variant Creutzfeld-Jakob disease (variant CJD).

Chapter 28 Outline

This chapter deals with an intriguing group of diseases that affects a variety of animals, including man. These diseases are known collectively as *spongiform encephalopathies* because of the characteristic pathology which they display in regions of the brain. Historically, these were known as slow virus diseases following the demonstration of an infectious basis for the sheep and goat disease, *scrapie*, the first identified disease of this type, and in recognition of the fact that the disease course was prolonged. However, extensive attempts to characterize a classical viral agent associated with these diseases have failed, and instead there is a considerable body of evidence which ascribes infectivity in these diseases to a protein.

This notion, the *prion hypothesis*, which includes the concept of a replicating protein, has now gained widespread, although not universal, acceptance. Its principal champion, Stanley Prusiner, received the 1997 Nobel Prize for his work on this subject.

28.1 The spectrum of prion diseases

Several diseases of animals and humans are now believed to be caused by prions. These are summarized in Table 28.1. All show a characteristic pathology in the central nervous system, with parts of the brain becoming vacuolated or spongy, in many cases also with extensive deposits of extracellular protein fibrils or plaques (Fig. 28.1), but with no sign of inflammation, i.e. no invasion of immune cells. Different brain regions are affected by the various spongiform encephalopathies (Fig. 28.4 shows an example). For most of the diseases listed, the presence of infectivity has been shown by transmission of disease to an experimental animal (typically a mouse or hamster) by intracranial injection of a homogenate of diseased tissue and subsequent

Introduction to Modern Virology, Seventh Edition. N. J. Dimmock, A. J. Easton and K. N. Leppard.
© 2016 John Wiley & Sons, Ltd. Published 2016 by John Wiley & Sons, Ltd.

Table 28.1
Prion diseases in the date order that transmissibility was demonstrated.

Disease and occurrence	Host species	Date
Scrapie Common in several countries throughout the world	Sheep, goats	1936
Transmissible mink encephalopathy (TME) Very rare, but adult mortality nearly 100% in some outbreaks	Mink	1965
Kuru Once common among the Fore-speaking people of Papua New Guinea, now rare	Humans	1966
Creutzfeld–Jakob disease (CJD) Occurs in iatrogenic*, familial and sporadic forms. The latter has uniform worldwide incidence of 1 per million per annum	Humans	1968
Gerstmann–Sträussler–Scheinker (GSS) syndrome An inherited disease; less than 0.1 per million per annum	Humans	1981
Chronic wasting disease (CWD) Colorado and Wyoming, USA	Mule-deer, elk	1983
Bovine spongiform encephalopathy (BSE)	Cattle	1988
Feline spongiform encephalopathy (FSE)	Domestic cat	1991
Fatal familial insomnia (FFI)	Humans	1995
Variant Creutzfeld-Jakob disease (variant CJD)	Humans	1997

*Relating to medical intervention

serial passage to further mice/hamsters. Hence, these diseases have been termed *transmissible spongiform encephalopathies* (TSE).

The symptoms of TSEs vary in detail but generally share the features one would predict for a progressive loss of function in the brain. Early symptoms include personality changes, depression, memory problems and difficulty in coordinating movements. Later, the deficits become profound, with patients losing the ability to move or to speak. Characteristically, all the TSEs have long incubation periods (e.g. 1 year for a mouse; 4–6 years for a cow). For the human TSEs, the age at which signs and symptoms first appear depends on the disease type. Sporadic classical CJD symptoms start at a mean age of 60 years while symptoms of inherited forms emerge somewhat earlier (45 years). One of the defining features of variant CJD (see Section 28.6) is its early age at onset, below 30 years. To date, all TSEs have been invariably fatal; no effective treatment has been established.

28.2 The prion hypothesis

The infectivity that causes prion diseases is thought to reside in a protein that adopts a stable abnormal conformation rather than in a conventional, nucleic acid-containing, entity.

The essence of the prion hypothesis is that the infectious agent which causes TSEs is a specific structural conformation of a standard cell protein, PrPc (the cellular form of PrP, encoded by the *prnp* gene), which is a glycoprotein found on the surface of cells in

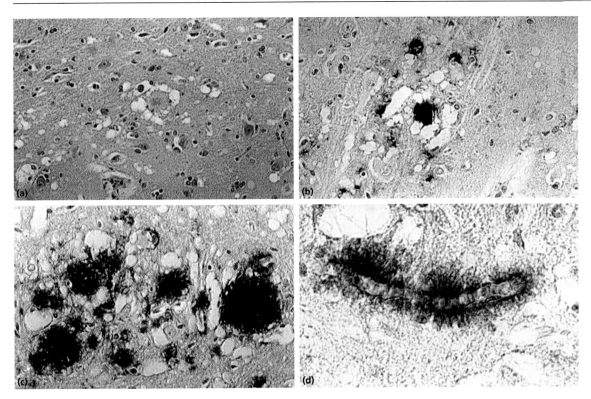

Fig. 28.1   Spongiform degeneration and protein (PrPsc) plaque deposition in CWD-infected elk brain. The images show sections from brain cortex either haematoxylin and eosin stained (A) or with detection of PrPsc deposits by immunohistochemistry (brown stain, B–D). Magnifications: A - x180; B - x280; D - x720. (Reproduced with permission from Liberski et al., (2001) *Acta Neuropathologica* 102, 496–500. © Springer-Verlag.)

brain and other tissues. In its unusual *prion* conformation, termed PrPsc (meaning the scrapie form of PrP), the protein has pathogenic properties. For an agent to be infectious, it must be able to replicate itself. An incoming prion is thought to achieve this by driving the conversion of host molecules that are in the standard structural conformation (i.e. PrPc) into the altered, prion conformation. In short, the infectious aberrant form is thought to act as a template on which existing normal protein molecules are converted into new altered structures which precisely resemble the original infectious agent and are themselves infectious (Fig. 28.2).

Some of the evidence in favour of the prion hypothesis is summarized in Box 28.1. Although now widely accepted, one problem with the hypothesis is the difficulty in accounting for scrapie strain variation (Box 28.2). In essence, when different isolates of sheep scrapie are grown in genetically identical laboratory animals, they show distinct properties that are stable on serial passage. For the prion hypothesis to accommodate these data, PrPc of a specific and defined amino acid sequence must be able to adopt not just one aberrant conformation but several, in each case resembling exactly the conformation of the PrPsc in the infecting scrapie strain, so that its unique pathogenic properties are faithfully

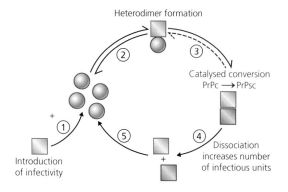

Heterodimer formation

Catalysed conversion
PrPc ⟶ PrPsc

Introduction
of infectivity

Dissociation
increases number
of infectious units

Fig. 28.2 A model for the propagation of TSE infectivity according to the prion hypothesis. PrPc (circles) and PrPsc (squares) represent the normal and an abnormal conformation respectively of the *prnp* gene product. Host PrP in the normal and abnormal conformation are coloured in red. The initiating infectivity (blue) is proposed to have the same structural features and properties as the progeny PrPsc molecules (although see discussion of the species barrier, Section 28.4). Steps 2 and 3 in the propagation cycle may be reversible. Alternative models differ principally in suggesting that polymerisation of the altered structural form is important to its potential for catalyzing further structural conversions.

Box 28.1

Evidence supporting the prion hypothesis for TSE diseases

- A series of observations argues that the TSE agents cannot be conventional viruses or viroids. Although the scrapie agent passes through filters which admit only viruses and can be titrated in mice (often reaching concentrations of over 10^7 infectious units/ml), it is much more resistant than typical viruses to inactivation by heat, radiation, and chemicals such as formaldehyde. The rate at which infectivity is destroyed by radiation indicates that any nucleic acid present would be tiny, at most 250 nucleotides of single-stranded nucleic acid. Moreover, infectivity is not susceptible to nucleases and no unique nucleic acid molecule, even of this minimal size, has been found to co-purify with infectivity.

- Several observations directly support the prion hypothesis. Infectivity co-purifies with PrPsc, which forms the plaques seen in some TSE-affected brain tissues. The sequence of this protein is identical to the host protein PrPc but its tertiary structure is very different, which renders the protein core insoluble and resistant to protease digestion. PrPsc can catalyze the conversion of PrPc molecules into the PrPsc structure in the test tube. Also, transgenic mice that lack the *prnp* gene are completely resistant to experimental transmission of TSEs, as you would expect if this gene provides the substrate for prion replication, while mutations in this gene in humans are associated with inherited forms of TSE (Section 28.3). Finally, efficient in vitro conversion of PrPc to PrPsc has been achieved in which new infectivity is generated over many cycles of in vitro replication.

- A prion-type replication mechanism has precedents in yeast. The yeast *PSI*+ and *URE3* phenotypes result from the presence of aberrant conformations of two normal proteins, Sup35p and Ure2 respectively. Interestingly, a molecular chaperone (a protein which catalyzes changes in the conformation of other proteins) is involved in generating the *PSI*+ trait in a previously negative (*psi*−) cell population, and in subsequent elimination of the trait from that population. Mutant forms of the normal proteins have also been shown to increase the rate of conversion to give the abnormal phenotypes.

Box 28.2

Evidence for scrapie strain variation

- Different scrapie isolates are characterized by their very precise, but different, incubation periods in laboratory animals of a given type or inbred strain. A single scrapie strain will produce disease onset at times that vary by only two or three days among a group of animals whereas the mean incubation periods for different scrapie strains range from less than 100 days to more than 300 days. Scrapie strains also differ reproducibly in details of their pathogenesis.
- Scrapie strains adapted to one rodent species can be introduced into another species where, after crossing the species barrier when the incubation period is extended and more variable (see Section 28.4), they again show reproducible and discrete incubation periods (Fig. 28.3).
- When scrapie strains, serially passaged in mice and then in hamsters, are passaged back to the original mouse strain they may either display their original properties (incubation time, pathogenic details) or they may 'mutate' to a new set of properties that are again stable on further passage.

---

reproduced. This idea is difficult to reconcile with our understanding of how stable protein folding is achieved whereas strain variation would be easy to understand if the infectivity had a nucleic acid component.

28.3 The aetiology of prion diseases

The first TSE to be identified was scrapie, a natural infection of sheep, which takes its name from the tendency of diseased animals to scrape themselves against fence posts, presumably to relieve itching of the skin. The disease is usually spread maternally but can also be spread horizontally between unrelated animals and its infectious basis was further confirmed by transmission into experimental animals (see Section 28.2). There is also clear evidence for an infectious basis to other prion diseases of animals, such as BSE (see Section 28.5).

Among the human prion diseases listed in Table 28.1, Kuru clearly has an infectious basis.

It was spread by ritual cannibalism and, since its identification, its incidence has declined along with this practice. *Iatrogenic* (meaning related to medicine) CJD also shows infectious aetiology. It has arisen through various medical treatments that involve transfer of tissues/tissue extracts between individuals. For example, the treatment of dwarfism, a pituitary growth hormone deficiency, with intramuscular injections of this hormone led to a number of cases of CJD. Disease occurred because the growth hormone was extracted from pituitary glands obtained from cadavers, some of which were harbouring inapparent prion disease at the time of death. The risk of transmission was heightened by the need to pool material from a large number of individuals to obtain sufficient hormone for use.

On the other hand, GSS, FFI and familial CJD are autosomal dominant inherited genetic disorders while sporadic CJD, as the name suggests, occurs at a random low frequency with no known cause. How can these be accounted for within the prion hypothesis? For the inherited diseases, the explanation lies in the discovery that each is linked with one or more specific mutations in the *prnp* gene (Box 28.3). Studies of these mutations show that subtle differences in the amino acid sequence of PrP can affect the time course and nature of disease that is observed following conversion to a PrPsc form, as the prion hypothesis predicts. These inherited disease-associated PrPc sequence differences are proposed to destabilize the protein so that spontaneous conversion into an altered, PrPsc, form is of such high probability that it inevitably occurs within a normal lifetime. After this, replication of the aberrant structure can take place in the same way as following infection with exogenous prions. Again, the precise nature (shape and structure) of this altered form must determine the disease phenotype. Supporting this idea, mice made transgenic for the human GSS mutant *prnp* gene spontaneously develop disease. Following the same reasoning, sporadic CJD is thought to occur when a molecule of the normal PrPc undergoes this same spontaneous conversion to initiate replication of the relevant PrPsc form. In the absence of a destabilizing mutation, this spontaneous conversion would be a rare event, accounting for the low frequency of sporadic CJD over a normal human lifespan and for the typical late onset of this disease.

28.4 Prion disease pathogenesis

Following infection, PrPc expression is required in the central nervous system for brain degeneration to occur. If infection occurs peripherally (e.g. in the gut), cells of the immune system that express PrPc are also necessary for neuroinvasion.

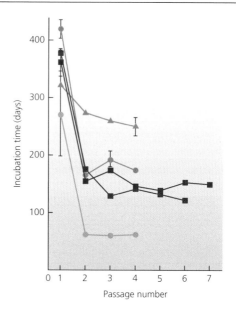

Fig. 28.3 The incubation time to disease of five distinct, mouse-adapted, scrapie strains upon serial passage in hamsters. Each symbol type represents data from a different scrapie strain. Passage 1 shows a longer and more variable incubation time for each strain because of the species barrier to transmission (see Section 28.4). Subsequent passages show remarkably constant, but strain-specific, incubation times. Data are taken from Kimberlin et al. (1989). J. Gen. Virol. 70, 2017–2025. Standard errors, except where shown, were insignificant.

If formation of altered conformations of PrP underlies prion diseases of all aetiologies, how does infectivity reach the brain in each case? Sporadic and inherited TSEs might arise directly from pathogenic conversion of PrP being initiated within the central nervous system (CNS). Some cases of iatrogenic CJD, which have arisen through direct introduction of infectivity into the CNS (comprising the brain and spinal chord), for example by grafting of contaminated donor dura mater (the covering of the brain) during brain surgery or transplant of contaminated donor corneas, could similarly arise through propagation of infectivity from the

Box 28.3

Human *prnp* mutations associated with inherited disease

- Specific mutations in human PrP are associated with either CJD, GSS or FFI. For example, replacement of glutamic acid by lysine at position 200 results in CJD while replacing proline with leucine at position 102 results in GSS.
- Polymorphisms at other positions in human PrP (not themselves pathogenic) modify the disease pattern associated with pathogenic mutations. A key polymorphism is at position 129, coding either for valine or methionine.
- Almost all disease due to specific CJD and GSS mutations occurs in individuals with methionine at position 129 on both the normal and mutated *prnp* alleles (methionine 129 homozygotes) even though they represent only 40% of the population.
- Replacing aspartic acid at position 178 with asparagine causes FFI when residue 129 in that allele is a methionine but CJD when residue 129 is a valine (Fig. 28.4). Note that these diseases target different regions of the brain. In each case, disease onset and progression are exacerbated when the amino acid at position 129 in the normal allele is the same as that at position 129 in the mutated allele, i.e. the person is homozygous at the polymorphic position in *prnp*.

site of exposure through the CNS. Experiments with PrPc-null transgenic mice (i.e. they make no endogenous PrP) carrying a graft of PrP$^+$ brain tissue show that expression of PrP in the recipient CNS is essential if pathogenesis is to be seen, leading to the idea that, within the CNS, infectivity is propagated through the tissue from the site of infection by the progressive conversion of normal PrP to a pathogenic form.

Other cases of iatrogenic CJD have come from peripheral exposure to infectivity, for example via injection of contaminated pituitary growth hormone (see Section 28.3), while Kuru, BSE and variant CJD are, with varying degrees of certainty, known to have been transmitted through ingestion of contaminated food (see Sections 28.5, 28.6). How, in all of these various circumstances, does infectivity reach the CNS? Information on this subject has come mostly from studies of infection in various transgenic mouse lines. These show that elements of the immune system are crucially important to amplify the infectivity and

transmit it on to the CNS. Indeed, the spleen has long been known as a site of replication of these agents and infectivity accumulates there earlier than it does in the brain following peripheral infection (Fig. 28.5). Both B cells and follicular dendritic cells (FDC) have been implicated in TSE agent peripheral replication and, possibly, different types of TSE require different immune cell types for the agent to penetrate the CNS. Amplification of infectivity in FDC seems to be crucial if disease is to be manifested in the brain and the FDC must, as predicted from the prion hypothesis, be expressing PrPc for them to fulfil this role.

A second key aspect of pathogenesis concerns the species barrier to infection. Experimentally, this is observed as a very much greater difficulty (lower frequency of transmission and longer duration to overt disease) in transmitting infection between species than within a species (Fig. 28.3). This is thought to reflect the need for structural similarity (ideally sequence identity) between the infecting prion and the

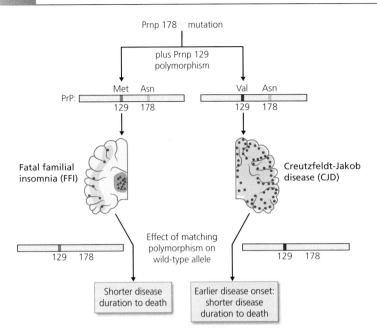

Fig. 28.4 The interaction between pathogenic mutations at position 178 of human PrP and the amino acid present at the polymorphic position 129 on either the mutant or normal allele. Both the nature of the disease pathology, its time of onset and rate of progression are determined by the position 129 polymorphism. Light green - regions of spongiform degeneration; dark green - regions of neuronal loss and replacement with astrocytes and glial cells. See Box 28.3 for further details. (Redrawn, with permission, from Gambetti (1996) *Curr. Topics Microbiol. Immunol.* 207, 19–25.)

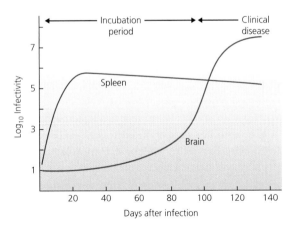

Fig. 28.5 Multiplication of the scrapie agent in the spleen and brain of mice and the time course of disease.

endogenous PrPc for efficient structural conversion of the latter protein to a prion. Thus human prions transmit disease to standard laboratory mice very poorly, more easily to mice carrying transgenic copies of the human *prnp* gene in addition to the equivalent mouse genes, and most easily to mice carrying the human gene but not the mouse gene. The disease-enhancing effects of homozygosity at position 129 (see Box 28.3) may reflect a similar effect; the presence of two sequence variants in a heterozygote would be expected to inhibit infectious propagation. However, animals which have apparently not been infected in cross-species transfers, in that they remain asymptomatic, can nonetheless harbour considerable infectivity. This calls into question the simple molecular view of the species barrier to disease transmission as a reduced probability of PrP structural conversion.

28.5 Bovine spongiform encephalopathy (BSE)

BSE (popularly known as mad cow disease) appeared suddenly in dairy cattle in the UK in the 1980s, was formally recognized in 1986 and developed into a large-scale epidemic involving nearly 185,000 animals to 2005, by which time the epidemic was essentially over (Fig. 28.6); this had huge economic consequences. Much smaller numbers of cases occurred in other European countries, about 5200 over the same period, with a very few in other parts of the world such as in North America. Two possible sources for BSE emergence are considered possible. Either the agent resulted from the scrapie agent of sheep adapting to cattle or else a rare case of sporadic disease in a cow, equivalent to sporadic CJD in humans, was transmitted to other cattle. In both cases, transmission is thought to have been made possible by the practice of giving cattle artificial food concentrates which contained protein rendered from carcasses of animals, including sheep and cattle. It is likely that a single event originated the epidemic because all BSE isolates appear to represent a single prion strain (Box 28.4). This was then propagated by successive cycles of rendering of cattle residues from the abattoir into cattle food concentrates, which gave optimum conditions for a cattle-adapted scrapie variant to emerge and/or for propagation of a pre-existing cattle prion. Once this disease etiology was recognised, a ban was imposed on feeding of ruminant animal protein to ruminants in the UK from 1988, soon followed elsewhere, which brought the epidemic under control.

The low number of BSE cases detected in North America is surprising given (a) the importation of potentially-infected animals into North American herds that took place before the nature of the problem was appreciated; (b) the history of risk-prone cattle feeding practices in the USA and Canada similar to those in Europe; and (c) the evidence of feed-related transmission in the few cases that have been seen. To some extent, the low case count may be a reflection of the relatively low intensity of testing of cattle being applied. It may also be the

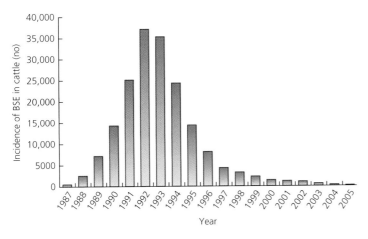

Fig. 28.6 Annual incidence of BSE in the U.K. cattle herd, 1987–2005. Graph drawn from data taken from the U.K. Department for Environment, Food and Rural Affairs (DEFRA) website, July 2005.

Box 28.4

Evidence that BSE isolates are a single strain

- When BSE is transmitted experimentally into mice, different isolates appear to be remarkably homogeneous with regard to incubation period; each of four mouse strains tested shows a characteristic BSE incubation time (Fig. 28.7a). This is a classical definition of strain identity in studies of scrapie.
- BSE isolates share a PrPsc molecular 'signature' on Western blots. These signatures reflect the number and location of glycosylation sites and the size of the protease-resistant core of the protein (Fig. 28.7b). In contrast, when scrapie strains are analyzed they differ widely in all these properties.
- Different BSE isolates show similar distribution of brain pathology upon experimental transmission to mice, again in contrast to different strains of Scrapie.

---

case that the amounts of infectivity in the cattle food chain in these countries did not reach the levels needed to produce an epidemic before measures were put in place to block this route of transmission.

Scrapie had been endemic in sheep in the UK for nearly 300 years, and the practice of feeding cattle with animal protein was in place well before the 1980s, so why did transmission to cattle only occur then? It is likely that this

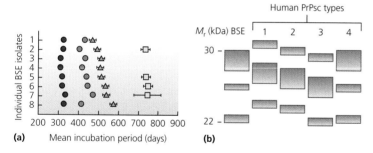

(a) Mean incubation period (days)

(b)

Fig. 28.7 The biological and molecular properties of different BSE isolates transmitted to mice are very similar to each other, and to those of vCJD isolates. (a) The incubation times of BSE isolates from cattle upon transmission to mice. Each line on the figure represents a different BSE isolate and the four types of symbol show the mean incubation time for that isolate in four different strains of mouse. (Adapted, with permission, from Bruce *et al.* (1997) *Nature* 389, 498–501).
(b) Molecular signatures of TSE isolates. A schematic representation of a Western blot analysis of polypeptides from the brains of mice infected with different TSE isolates that have been separated by SDS-PAGE. The bands represent protease-resistant core fragments of PrPsc, differing in length and glycosylation pattern, that have been detected with antibody to PrPsc. The thickness of each band represents its intensity in the original analysis. Human PrP types 1, 2 and 3 are seen in various sporadic and iatrogenic CJD isolates; the type 4 pattern is seen in variant CJD isolates and is very similar to the pattern consistently generated by BSE isolates in this type of analysis. (Reprinted from *Current Opinion in Genetics and Development*, vol. 9, JDF Wadsworth, GS Jackson, AF Hill & J Collinge, Molecular *Biology of Prion Propagation*, 338–345 (1999) with permission from Elsevier Science, using original data from Collinge et al. (1996) *Nature* 383, 685–690.)

resulted from a change that took place between 1980 and 1983 in the way meat-and-bone-meal cattle food was prepared from animal carcasses. During this period, the use of organic solvent extraction in the process decreased by nearly 50%. This extraction had involved heating for 8 hours at 70 °C and then the application of superheated steam to remove traces of solvent, which would have inactivated scrapie infectivity. Such extreme measures are needed as TSE infectivity is very stable, far more so than typical micro-organisms, partly because of its unique composition among infectious agents and also because the PrPsc tertiary structure is exceptionally stable among proteins.

28.6 BSE and the emergence of variant CJD

Following the BSE epidemic in the UK, an outbreak of a new human TSE, variant CJD, occurred. This represents human infection by the BSE agent, transmitted through the food chain. By early 2014, around 220 variant CJD cases had occurred globally, 177 of them in the UK.

Once the likely cause of the BSE epidemic had been established, a ban was imposed on using meat and bone meal in cattle feed to prevent further spread in cattle. As a precaution, a ban was also imposed in the UK in 1989 on the human consumption of specified bovine offals (brain, spinal cord and lymphoid tissues) that were considered potentially infectious (see Section 28.4). However, it was recognized that people had been eating potentially contaminated cattle meat products for a number of years prior to this point. There were also other risks to human health to be considered; for example, calf serum is widely used in the pharmaceutical industry for the growth of cultured cells to make virus vaccines

and other cattle by-products are used in cosmetics.

At the time, the risk of transmission of BSE through the food chain was considered to be low (but not zero). It was reasoned that scrapie had been endemic in the UK sheep population for nearly three centuries and there was no evidence of transmission to humans, despite the large amount of sheep meat consumed. There was also evidence of substantial species barriers to transmission of scrapie infectivity (see Section 28.4). Nonetheless, the disease TME (Table 28.1) had previously emerged in captive mink through feeding them scrapie-contaminated sheep which showed such cross-species transfer through feed could occur. More worrying was the subsequent emergence of FSE in domestic cats in the early 1990s, presumably as a result of consuming contaminated food, which suggested that this species barrier could also be breached by the BSE or scrapie agent; later, FSE was firmly linked to BSE transmission.

This view of the limited risk to human health from BSE was changed dramatically in 1996 by the recognition of a novel human disease in the UK, now known as variant CJD (vCJD). Its clinical characteristics differed somewhat from those of previously described CJD (however acquired), but what was particularly striking was the difference between the ages of the victims. vCJD patients were typically much younger (less than 30 vs. an average 60 years) than were sporadic CJD patients. Experiments quickly showed that vCJD represented human infection by the BSE agent (Box 28.5).

By early 2014, 177 people had died of definite or probable vCJD in the UK and about 40 more had succumbed in other European countries. Isolated cases in other parts of the world have been linked with probable exposure to infection in Western Europe. All the cases so far have been in people homozygous for methionine at the polymorphic position 129 in their *prnp* genes (129MM). This polymorphism

---

Box 28.5

Evidence that vCJD represents BSE infection in humans

- Infection of mice with material from either BSE or vCJD samples occurs with very similar incubation times and with similar characteristic pathogenic signatures in the brain. Both these features are distinct from those occurring when classical CJD strains infect mice.
- The molecular signatures of the BSE and vCJD agents are similar (Fig. 28.7b).
- The pathogenic lesion profile, incubation time and molecular signature of vCJD in mice (all isolates are identical) differ from those of all known scrapie strains.

---

is already known to influence human disease in the context of pathogenic *prnp* mutations (see Section 28.3). The annual UK death rate from vCJD peaked in 2000 with 28 deaths and then declined; 2012 was the first year since vCJD emerged when there were no UK deaths from the disease though in 2013 there was a single case. The number of deaths from sporadic CJD also increased over the period of the vCJD outbreak in the UK, from a base of around 30 per year in the 1980s to a peak of 77 in 2003. This increase is thought to be due simply to an increased number of referrals for diagnosis of TSE following the heightened awareness of these diseases that the BSE/vCJD epidemics generated, rather than any real increase in incidence of sporadic CJD linked to BSE.

28.7 Concerns about variant CJD in the future

The eventual size of the vCJD outbreak is hard to estimate and depends greatly on whether those so far affected are typical of the rest of the population in their disease susceptibility or incubation period. Individuals who have subclinical vCJD infections may be a source of further iatrogenic infections, whether or not they eventually display disease symptoms.

For obvious reasons, there is intense interest in predicting the eventual size of the vCJD problem. All the time the annual UK vCJD death rate was holding steady, uncertainty about the incubation time from infection to disease made such prediction very difficult. Was this a plateau, or just the front end of a very gradually increasing curve? However, the firm downturn in the death rate to very low levels now seen has led to estimates that the final number of people affected in the UK will be only a few hundred. However, the 129MM homozygotes so far found to be susceptible to vCJD are only about 40% of the population. A major uncertainty was therefore whether other genotypes were completely resistant to infection by BSE or simply had a longer incubation time to disease. A partial answer to this question came with the detection of a clinically inapparent vCJD infection in a 129MV heterozygote who died of unrelated causes, suggesting that, as in genetically engineered mice, infectivity can propagate in such people. If they can also progress to disease within a natural lifespan, the eventual size of the outbreak will be significantly greater.

Of course, finding one infected person says nothing about the frequency with which 129MV people became infected by ingesting BSE-infected meat, especially since this single individual was actually infected iatrogenically. Indeed, it was also not known whether those few 129MM people who succumbed to vCJD were part of a small subset of this 40% of the population who became infected or whether infection was widespread but pathogenesis is a rare consequence. Resolving these issues required unbiased surveys of the prevalence of infection in the general population. For several years, the problem was getting an assay that was sufficiently sensitive for this purpose. However, a 2013 survey of more than 30,000 UK appendix samples stored over several decades suggested about 1 in 2000 of the population was infected. It is important to note, though, that neuroinvasion from such gut-associated tissue is a further hurdle for prion infectivity to overcome before disease could even potentially occur, so this is certainly not an estimate of the likely future scale of vCJD in the UK.

Another uncertainty in predicting the future trend of vCJD cases is whether or not there will be significant human-to-human transmission. Although the outbreak originated from consuming BSE-infected meat, iatrogenic transmission from people infected by this route could amplify the number of cases considerably. This risk exists because people harbour significant prion infectivity, particularly in lymphoid and neural tissue, long before they show disease symptoms. Furthermore, as discussed above, the pool of those harbouring infectivity probably includes 129MV heterozygotes who have been exposed to BSE in addition to presently asymptomatic 129MM homozygotes, and these two genotypes comprise a large majority of the population.

Infection may be passed on from people harbouring prion infectivity if they donate blood or organs, or through instruments contaminated during routine surgical or dental procedures carried out on them. This is more than a theoretical risk as three symptomatic secondary cases of vCJD have been seen in blood transfusion recipients; the asymptomatic 129MV carrier already mentioned was also infected by this route. To minimize this risk, blood donations in the UK are not accepted from people who have themselves received a donation in the past. Depletion of white cells from blood before transfusion also limits transmission risk substantially since it is these cells that potentially harbour infectivity. In the USA, blood donations are not accepted from people who have lived for extended periods in the UK. In hospitals, the re-use of certain surgical instruments is a concern, since standard cleaning and sterilizing procedures are not sufficient to eliminate infectivity from them.

Finally, it is worth noting that infected cattle are not the only source of TSE infectivity in human food. As already noted, scrapie-infected sheep meat has been consumed for years in the UK and elsewhere, apparently without human disease consequences as the frequency of sporadic CJD is similar in countries with and without endemic sheep scrapie. However, since it has been shown that the BSE prion is experimentally transmissible to sheep, there is concern that, during the silent phase of the BSE epidemic before the ruminant protein feed ban was imposed, this agent could have infected sheep and that this might then be infectious for humans consuming sheep meat. There is also the condition CWD that affects elk and mule deer in an area of the western USA (Table 28.1). The possibility has been considered that hunters and others who eat meat from these animals might also be at risk of TSE infection. However, in a number of cases of human TSE-like disease investigated for a possible connection to CWD, no convincing evidence of a link has so far been obtained.

28.8 Unresolved issues

Despite the explosion of research interest in the TSEs, there remain several significant unanswered questions about them. Firstly, it is unclear why the accumulation of a protein in an aberrant conformation should cause the mass death of neurones and spongiform degeneration of the brain, although it is notable that other neurodegenerative conditions, such as Alzheimer's disease, are also characterized by insoluble protein deposition in the brain. Secondly, it is unclear how the different pathologies associated with each human prion disease, which include different effects on specific brain areas in each case, can be generated through the structural conversion of the same normal protein. In other words, why does conversion of PrP to an abnormal form damage area A of the brain and spare area B, while conversion to another abnormal form damages area B but not A when the substrate for conversion, normal PrPc, is present in both areas? Finally, what is the normal function of PrP, aberrations in which seem to have such disastrous consequences? The fact that *prnp-*null mice are viable and healthy tells us that the function is not essential, at least under the conditions of laboratory animal maintenance. Prevention and cure for TSEs will surely depend on a much fuller understanding of these diseases.

Key points

- The prion diseases are a collection of delayed onset neurodegenerative diseases that are mostly transmissible – the transmissible spongiform encephalopathies.
- The infectious entity in these diseases is a stable aberrant conformer of a cell protein called PrP.
- According to the prion hypothesis, protein molecules in the aberrant form catalyze the conversion of further molecules into that form, so replicating the infectious agent.

- The principal human TSE is Creutzfeld-Jakob disease. CJD can arise spontaneously, through inheritance of a mutation in PrP, through infected tissue transplant/injection or through eating infected material.
- A major epidemic of bovine spongiform encephalopathy occurred in cattle in the United Kingdom through the feeding of material from recycled animal carcasses.
- BSE infectivity in meat entering the human food chain has transmitted into humans, causing variant CJD in around 220 people by 2014.

Questions

- Discuss the evidence for and against the hypothesis that all spongiform encephalopathy diseases are caused by the propagation of an altered conformation of the prion protein.
- Discuss the mechanisms by which transmissible spongiform encephalopathy infectivity is believed to be spread within an individual, between individuals and between species.
- Describe the prion hypothesis and discuss how it can account for the existence of human diseases with both an infectious and an inherited basis.

Further reading

Aguzzi, A., Nuvolone, M., Zhu, C. 2013. The immunobiology of prion diseases. *Nature Reviews: Immunology* 13, 888–902.

Bruce, M. E. 2003. TSE strain variation. *British Medical Bulletin* 66, 99–108.

Collinge, J., Clarke, A. R. 2011. A general model of prion strains and their pathogenicity. *Science* 318, 930–936.

Grassmann, A., Wolf, H., Hofmann, J., Graham, J. & Vorberg, I. 2013. Cellular aspects of prion replication in vitro. *Viruses-Basel* 5, 374–405.

Head, M. W. 2013. Human prion diseases: molecular, cellular and population biology. *Neuropathology* 33, 221–236.

Head, M. W. & Ironside, J. W. 2012. Review: Creutzfeldt-Jakob disease: prion protein type, disease phenotype and agent strain. *Neuropathology and Applied Neurobiology* 38, 296–310.

Hill, A. F., Collinge, J. 2003. Subclinical prion infection. *Trends in Microbiology* 11, 578–584.

Kraus, A., Groveman, B. R., Caughey, B. 2013. Prions and the potential transmissibility of protein misfolding diseases. *Annual Review of Microbiology* 67, 543–564.

Lloyd, S. E., Mead, S., Collinge, J. 2013. Genetics of prion diseases. *Current Opinion in Genetics & Development* 23, 345–351.

Pattison, J. 1998. The emergence of bovine spongiform encephalopathy. *Emerging Infectious Diseases* 4, 390–394.

Prusiner, S. B. 2013. Biology and genetics of prions causing neurodegeneration. *Annual Review of Genetics* 47, 601–623.

Ricketts, M. N. 2004. Public health and the BSE epidemic. *Current Topics in Microbiology & Immunology* 284, 99–119.

The UK Creutzfeldt-Jakob Disease Surveillance Unit web site. http://www.cjd.ed.ac.uk/

The World Health Organisation TSE web pages. http://www.who.int/topics/spongiform_encephalopathies_transmissible/en/

Part V

Virology – the wider context

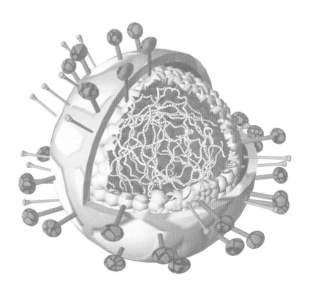

Chapter 29

The economic impact of viruses

The impact of viruses can be assessed in many ways. The economic effects of virus disease play an important role in the medical, animal and plant spheres. Consideration of the economic impact can often inform decisions about investment in exploring treatments for virus infections.

Chapter 29 Outline

29.1 The economics of virus infections of humans

29.2 The economics of virus infections of animals

29.3 The economics of virus infections of plants

29.4 The Netherlands tulip market crash

It is easy, and understandable, to consider viruses only in terms of the severity of disease they cause – mild through to fatal. Such considerations are important but only indicate part of the real impact of viruses and can lead to potentially problematic conclusions. For example does a virus that kills most of the people it infects but which infects relatively few people, such as Ebola virus, have more impact than a virus that causes serious and possibly life-threatening illness in only a very small proportion of those infected but which infects most people in the world, such as the currently dominant H1N1 strain of influenza virus? For those seriously affected by a virus infection the difference is hard to appreciate and there are no easy answers to such questions but it indicates the difficulty in trying to quantify the impact of a disease. This approach is also very anthropocentric in that it is focussed primarily on disease in humans. It is important to appreciate that this is a very narrow view and that viruses affect us in many different ways. Some of these are explored in Chapter 31.

Inevitably, a common method of assessment of the impact of an infection is to try to determine the economic consequence of the disease. However, while this might seem to be straightforward there are problems when trying to identify what should be counted as an economic consequence of a disease. For example, self-diagnosis of mild respiratory disease leads to loss of productivity if workers are unable to attend work through illness but because such events are not officially documented, the economic consequences are extremely difficult to assess. As a result all economic impact determinations are, of necessity, estimates with some degree of inaccuracy and decisions to include or exclude certain parameters can have significant effects on the final figures that are produced. For many, the thought of assigning an economic cost to a disease is distasteful and it can be legitimately argued that it is irrelevant in many circumstances. However, an estimate of the economic impact of infections can be informative in some situations. In particular,

Introduction to Modern Virology, Seventh Edition. N. J. Dimmock, A. J. Easton and K. N. Leppard.
© 2016 John Wiley & Sons, Ltd. Published 2016 by John Wiley & Sons, Ltd.

it can be used to formulate an estimate of a cost-benefit analysis when deciding on the distribution of limited resources, though it is always important to question if an economic impact has captured the real importance of a disease, as can be seen particularly with some plant virus infections discussed below.

29.1 The economics of virus infections of humans

Estimating the accurate economic impact of human infections is complicated by a number of factors. The first is that it may be difficult to obtain what appears to be the most basic piece of information: how many people have been infected? This is particularly true in developing countries or where there is a limited healthcare infrastructure from which the data might be obtained. However, it is often difficult even in the most developed countries if the infection causes a range of symptoms that generate illness across the spectrum from mild to severe. The next problem is how to decide what constitutes economic cost. The simplest approach is to determine the cost of healthcare associated with the infection, but again this may not apply equally to all parts of the world; the healthcare spend in some countries is very limited compared to others and this will affect the estimated cost. Also, it is clear that treatment costs alone do not reflect the full economic cost as infected people may not be able to work effectively, or possibly at all, during the infection, so affecting national productivity. Of necessity, any figure relating to lost productivity must be an estimate and the accuracy will depend on whether the goods being produced are tangible (commercial goods such as cars, textiles, etc.) or intangible (such as office based activities), together with the part of the world being considered, as lost value of equivalent goods and services will vary greatly

between countries. These considerations mean that when comparing the economic costs of different infections it is important to be sure that the assessments are truly comparable and that they have used similar approaches to generate the final figures. Without this, comparisons may be very misleading. Inevitably, most economic assessments refer to developed parts of the world where the infrastructure and information to allow such assessment is more readily available; this generates some bias in the data and risks underestimating the accurate costs.

The economic impact of Dengue virus infection

As described in Section 23.2, Dengue virus infections have been increasing over many years with current estimates of up to 400 million infections and hospitalization of possibly 500,000 people, predominantly children, every year. The two areas of the world most affected by Dengue virus infection for which economic data are available are the Americas in the western hemisphere of the globe and Southeast Asia in the eastern hemisphere (Fig. 29.1).

Estimates of the economic impact of dengue virus infection in the Western hemisphere are complicated because of the significant variety of economic infrastructures throughout the Americas. While large areas have good healthcare and reporting systems, many others have variable or limited systems and this will inevitably lead to underreporting of disease which in turn will affect the estimates of economic impact. The most recent estimates suggest that in the Americas the cost of dengue virus infections in 2010 was in the region of $2.1 billion per year. The level of uncertainty about various factors suggested that the true figure lay between $1 billion and $4 billion per year. During this period there were an estimated 2–3 million Dengue virus cases per

Fig. 29.1 Map showing the countries in the Americas and Southeast Asia included in assessment of the economic cost of Dengue virus infections.

year in this area. While 40% of these costs were estimated from direct medical and non-medical costs the remaining 60% was estimated as indirect costs, predominantly lost productivity in the countries affected.

Estimates of the cost of Dengue virus infection in Southeast Asia are complicated by the definition of what constitutes the area. In an analysis published in 2013, a total of 12 countries were studied from which data could be obtained. These countries are indicated in Fig. 29.1. In this area, there were approximately 2.9 million Dengue virus cases per year over the study period. The considerations of economic impact suggested that Dengue virus infections cost the combined economies of these countries approximately $950 million per year (with a range of $600 to $1.4 billion). This figure included hospital and outpatient medical care which represented approximately 48% of the total costs, with the remainder being an estimate of the indirect costs of the infections. The economic burdens of the disease were not evenly distributed and some countries, such as Indonesia, bore a higher cost than others.

While these costs carry some caveats as discussed, even the lowest estimates indicate

that the economic burden associated with the disease is significant. These estimates, in association with the direct human impact of Dengue disease, have served to emphasize the importance in tackling this growing disease threat.

The economic impact of West Nile virus infection in the USA

The consideration of the impact of Dengue virus infection is for an endemic virus infection in the affected areas and gives an indication of the 'steady-state' that can occur. The appearance of a new infection in a defined economic area allows a more accurate estimate of the economic cost. As described in Section 23.4, West Nile virus was introduced into the USA in 1999. Since then, the virus has spread throughout the country, bringing with it the costs associated with treatment, community-level prevention and lost economic production. While the costs of healthcare have been determined with some degree of accuracy to date, the estimates of indirect costs remain elusive and can only be extrapolated from limited studies in specific states. However, the figures illustrate the potential scale of the financial effects of such new infections.

From its first appearance to the end of 2013 it is estimated that West Nile virus has infected over 3 million people in the USA. Fortunately, almost 80% of infections are asymptomatic or generate only relatively mild symptoms that do not require treatment. However, a total of 39,557 people were hospitalized or required some level of treatment and 1667 people died from infection over the time-frame studied. Current estimates indicate that the total cost to date has been approximately $780 million (with a range of $670 million to $1.1 billion), giving costs of $56 million per year (a range of $48 million to $79 million per year). This estimate is primarily based on treatment costs but includes an assessment of the indirect costs to the economy. These costs were not evenly distributed throughout each year since the appearance of the virus but have fluctuated to reflect the changing incidence. The estimates indicate the potentially significant impact of the introduction of a new virus to a country.

The economic impact of SARS

The examples of Dengue virus and West Nile virus above provide an indication of the potential economic costs of either an endemic or a newly introduced virus to specific regions of the world. With the outbreak of SARS coronavirus infection in 2003, it was possible for the first time to determine the financial cost of the emergence of a new virus that infected almost every continent, albeit with relatively few cases (8098 people infected of whom 774 died, see Section 24.3).

The extremely dramatic appearance of SARS and its rapid international spread posed several problems for many countries across the world. The human impact of the outbreak was clear but the economic effects of the outbreak and its aftermath were significant. These included the costs associated with treatment and preventative measure introduced in most countries but there were also effects on travel, tourism and international trade. Although the SARS outbreak lasted only a few months and affected fewer than 10,000 people, the overall assessment of the global economic impact of SARS is $50 billion to $60 billion. At first sight, it seems difficult to accept such a vast cost for such a limited event, but it becomes more understandable in the light of some further detail. Prior to the outbreak, the Chinese economy showed a growth of 9.9% in the first financial quarter of 2003 but this dropped to only 2% in the second quarter following the outbreak. At the same time, the predicted growth rate of the Hong Kong economy dropped from 3.5% to 0.5%. Together these effects amount to very large sums of lost productivity. An indication of the impact on global travel can be seen by the numbers of tourists visiting Indonesia which dropped from 5 million prior to the outbreak to 2 million following the announcement of only two suspected cases in that country and no associated deaths. These effects were also seen in many other parts of the world and cumulatively explain the estimated costs.

The example of SARS indicates the potentially devastating effects of emerging viruses on the world's economy. As the world continues to become ever more interconnected, the potential costs of such virus emergences will increase.

29.2 The economics of virus infections of animals

We rely on domesticated livestock as part of the human food supply and the production of farmed animals plays a major role in the economies of many countries. All animals are subject to infection by viruses, and the diseases that result from these infections can have devastating effects for the animals. With animal diseases it is very common to consider the impact in purely economic terms, either directly in financial terms or by assessing loss of yield. However, this can be very misleading in certain circumstances. In some parts of the world, one or a few animal species may represent a significant component of subsistence living for a population and any large loss in productivity can have a huge impact on that population whilst presenting a relatively small financial cost. The considerations in determining how to deal with animal diseases are also quite different to those of humans, and in extreme cases 'at risk' animals may be slaughtered as a mechanism to prevent further spread, leading to another aspect of impact.

The economic impact of porcine reproductive and respiratory syndrome virus in the USA

Porcine reproductive and respiratory syndrome (PRRS) virus was first reported in 1987 in the USA and Europe. PRRS virus affects pigs and causes serious respiratory disease in young pigs and reproductive failure in older sows. In the early years of the infection, it was referred to as 'blue ear' disease, which reflected one of the frequent side effects of the respiratory symptoms –cyanosis that appeared as discolouration in the ears of light-coloured animals. The virus spreads very rapidly in pig herds and has been responsible for significant losses in the USA pig industry. More recently, a highly-pathogenic strain of the virus appeared in China and this has been responsible for significant disease in pig herds in Asia. There is no treatment for PRRS virus and slaughter is frequently the only option adopted to eliminate the virus from an infected area.

Estimates of the cost of PRRS virus infection in the USA in 2011 suggest that the direct financial costs due to lost productivity amounted to approximately $660 million per year. This was made up of $300 million due to losses in breeding herds and $360 million due to losses in herds raised for food production. While this figure is extremely high and represents a significant burden for the industry, it is recognized as being incomplete as it does not include indirect costs incurred by the industry. These additional costs include veterinary bills and the provision of enhanced biosecurity to reduce the future risk of introduction and subsequent spread of the virus. These additional costs are estimated to amount to approximately $480 million annually, bringing the cost of the infection to over $1 billion per year.

The economic impact of foot and mouth disease virus in the UK 2001 outbreak

In February 2001, the UK government reported an outbreak of foot and mouth disease virus in pigs identified at an abattoir. In the following months, the infection spread across a large area of the country, directly affecting in excess of 2000 farms and indirectly affecting more than 10,000 other animal facilities until the outbreak was declared over in September 2001 (Fig. 29.2). The infection affects cattle, sheep and pigs and was the worst outbreak so far recorded in the UK. The 7-month outbreak has been estimated to have cost the UK economy a total of £8 billion. The costs were comprised of £2.5 billion paid by the Government in compensation for slaughtered animals, coupled with payment for the costs associated with disposal of slaughtered animals and subsequent cleaning of premises, and an estimated £3.1 billion in losses to agriculture and food supply chains. The remainder of the costs was due to a

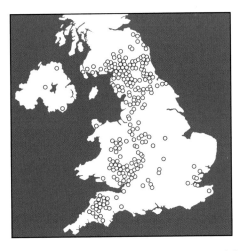

Fig. 29.2 Map showing the locations of the premises infected with the 2001 UK foot and mouth disease outbreak. (Copyright DEFRA, reproduced with permission.)

wide range of effects, such as effects on rural tourist and other industries from large-scale access restrictions in the countryside and indirect, subsequent effects of international trading restrictions and significant reductions in tourism throughout other parts of the country.

The effect of the outbreak was devastating in financial terms but also from the perspective of animal losses as the UK government regulations required slaughter of all cattle, sheep and pigs within 10 km of any confirmed site of infection. The consequence of this policy was that, within the 12,000 directly or indirectly affected farms, a total of almost 5 million sheep, 700,000 cattle and 400,000 pigs were killed. Following the outbreak the farming community had to restock the animal populations and comply with new regulations covering the transport of animals across the country, all of which also came with additional costs.

The UK foot and mouth outbreak indicates very clearly the potential impact, both financial and in terms of animal welfare, that can arise from a single event leading to the introduction of a highly infectious virus into a valuable livestock population.

The economic impact of Nipah virus in Malaysia

The appearance of Nipah virus in Malaysia in 1998 is described in Section 24.2 where the effect on humans is considered. The virus was shown to have originated in fruit bats and spread to pigs. All of the human infections were in people who had been in close contact with infected pigs. While there are no clear data for the economic impact of Nipah virus infections in humans it has been possible to determine the economic impact of the infection on the pig industry in Malaysia.

A very detailed assessment of the economic losses in Malaysia has been presented that puts the total cost at US$582 million for the short-

lived outbreak. The sum is made up of $97 million in compensation for the 1.1 million pigs slaughtered as a direct result of infection but primarily as a precaution against spread, and $120 million in lost pork production. The indirect costs of $229 million included lost tax revenue to the Government and losses in international trade. The remaining $136 million was due to the implementation of a control programme, providing biosecurity and slaughter facilities across the affected areas. As with the UK foot and mouth disease outbreak, this episode indicates the very significant economic impact of even a small outbreak of a new virus.

29.3 The economics of virus infections of plants

It is ironic that, although virology as a science began with an investigation of a plant virus (tobacco mosaic virus) by Dmitri Iwanowsky, the impact of virus infections in plants is now frequently ignored, probably because such infections do not induce welfare concerns in the same way as human or animal infections. However, plant infections can have considerable impact on our wellbeing by reducing the yield of important crops. This is particularly true in areas of the world that rely on a limited number of staple crops for food. In many of these areas, farming may be carried out at a subsistence level rather than for large-scale commercial production and in these circumstances the impact of plant disease may not be great in financial terms but can be catastrophic for those affected by the loss. Many crops are grown throughout the world, and of these the main food crops are rice, wheat, maize, cassava and sweet potatoes. Any infections of these crops will have a very serious consequence. A consideration of the main crops will show that these frequently form a major part of the diet of the poorest parts of the world where it is difficult to make an assessment of economic impact.

Virus infections in plants are transmitted by insects (see Section 18.2) and the prevalence of the insect vector, climatic and other environmental conditions all play a role in determining the outcome of the infection. The viruses may also exist within a reservoir plant species which is not used for food but which provides a continuous source of infection. The outcome of infection with many plant viruses is also affected by the stage of growth of the plant when the virus is introduced. A further complicating factor when considering the impact of a specific virus on a crop is that diseases may be the result of infection by a virus complex – a combination of several different viruses rather than a single agent – and the make-up of the complex, with the presence of each member determined by multiple factors, determines the outcome. These factors interact with each other in a complex way that makes it difficult to predict the appearance of an outbreak in a crop at any specific time, which in turn makes it difficult to implement preventative measures in many cases.

For the purposes of this discussion we will consider three examples of the effect of virus infection on crops: two of the primary food crops, maize and sweet potatoes, and a different type of crop, grapes for wine production. These examples illustrate different elements and processes of analysis that form part of the economic consideration of the impact of plant diseases.

The economic impact of virus infections of maize

Maize streak virus (MSV) is one of the most destructive virus diseases of cultivated maize and is considered to be one of the major threats to food supply in Africa. The virus is transmitted by leafhoppers and is limited to Africa and neighbouring large islands. MSV infection can completely destroy an entire crop (Fig. 29.3). Maize was introduced as a crop plant into Africa and provided a new host for MSV which also infects a number of grass species in the continent. The grasses act as a reservoir for the virus from where it can be transmitted to maize. A critical factor in maize streak disease is that the greatest impact occurs in plants infected as seedlings before the production of the second leaf. In these circumstances, the plant will completely fail to produce the seed that is the food product. Infection at later times after this developmental stage produces progressively less effect on seed (food crop) yield as the plant grows.

Despite its clear importance, no accurate figures on the economic impact of MSV on crop yields are available, and the assessments have therefore taken the approach of estimating the potential improvement in yield in the theoretical absence of the virus. In other words, if the virus was eradicated what would the value of increased productivity be? On this basis, current estimates place the value of the loss of maize yield due to MSV to be between $120 million and $480 million per year. While it is not possible to eliminate MSV, an alternative approach has been the development of disease-resistant maize strains. The use of these strains reduces the burden of disease and it is estimated that their large-scale use would potentially reduce losses due to MSV infection by up to 50%. Control of the leafhopper vector may also help reduce infection rates, particularly at key stages in the growth of the plant.

The economic impact of virus infections of sweet potatoes

Sweet potatoes are grown in all tropical and sub-tropical parts of the world. They are a major

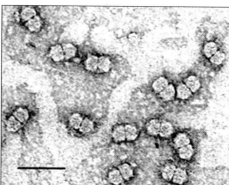

Fig. 29.3   (Left) A maize plant infected with maize streak virus showing the characteristic streaking phenomenon in front of uninfected healthy plants (copyright D. Shepherd, reproduced with permission). (Right) An electron micrograph showing the classical 'geminate' particle structure of maize streak virus. (Copyright E. Rybicki, reproduced with permission.)

food crop in many parts of the world and losses through infection represent a significant problem. Almost 90% of sweet potato production occurs in China where a range of different viruses can cause disease. The viruses are transmitted either by aphids (sweet potato feathery mottle virus (SPFMV), sweet potato latent virus (SwPLV) and sweet potato vein virus (SPVMS)) or by whiteflies (sweet potato mild mottle virus (SPMMV) and sweet potato yellow dwarf virus (SPYDV)). The most important of these is SPFMV which particularly affects sub-Saharan African sweet potato production. Whilst SPFMV infection alone causes some degree of damage to infected plants, it is when SPFMV infection is accompanied by sweet potato chlorotic stunt virus (SPCSV) in a complex that there are serious symptoms of disease; estimates suggest

that the combination virus infection can lead to reduction in yields of up to 50%.

The importance of sweet potato to subsistence farming in poor parts of the world makes it difficult to determine the economic impact of infection. Whilst it is possible to estimate reduction in yields by extrapolation from laboratory and small-scale field experiments, these estimates are necessarily inaccurate. An alternative approach is to assess the cost of implementation of virus-prevention measures. For sweet potatoes, this is achieved by vegetative propagation of virus-free vines and tubers and using these as the source of the annual planting root stock for farming. This approach has been successfully implemented in China and has led to significant reductions in losses due to infection. The generation of the root stock requires a robust infrastructure to support the farming community in providing the virus-free material. This approach led to an increase in yield of approximately 40% in early trials conducted in Shandong province in China in the late 1980s. The success of the trials led to more extensive use of virus-free planting with an overall increase in sweet potato yield in excess of 25%, representing an increase of global yield by 2.6% with a commercial value of almost $340 million in 2014 prices.

The economic impact of virus infections of grapevines

In the examples discussed above, the crop plants are annual (maize) or perennial but treated as annual (sweet potato) crops. Thus, they are crops that are replanted each year which makes it relatively easy to replace infected plants on a yearly basis. This is not possible with plants that are cultivated as perennials, such as for various fruit crops. The investment in these plants requires many years of cultivation for efficient and cost-effective cropping and replacement can represent a significant economic burden. These considerations play a role in determining the economic impact of virus infection in such plants that is absent for annual crops.

Grapevines are grown in many parts of the world. The grapes are a high-value product that are not primarily used as a food crop but rather in the production of wine where the quality of the product is a major part of the value. Grapevine leafroll disease is found in almost every part of the world and is caused by a complex of several different insect-transmitted viruses of which grapevine leafroll-3 virus is a particular concern. Grapevine leafroll disease can cause yield losses of 30–70% but it also has additional effects, including delay in fruit ripening, which affects harvest time and increases the acidity of the extracted grape juice, both of which compromise the quality of

the product. Such is the ubiquity of the infection that it is difficult to provide an assessment of the economic impact by comparison with an uninfected situation. However, just as for the sweet potato it is possible to determine the likely benefit obtained from using virus-free stock for replanting following removal of diseased plants. Such an assessment has been performed using the grape-growing area of Northern California, where, even for such a relatively small but high-production area, the use of this approach would bring benefits valued at between $53.5 million per year, potentially rising to $330 million. The figure of $53.5 million represents approximately 6.5% of the value of the average annual grape juice yield in Northern California.

29.4 The Netherlands tulip market crash

The impact of virus infections of plants has not always come from negative effects of the consequent plant disease. In the late 17th century, the Netherlands was experiencing a huge economic boom due to international trade through their colonies. This prompted increased speculation in a variety of luxury commodities, one of which was tulips. The Netherlands has been a provider of cut flowers for centuries and tulips have been a particularly important component of the market. In the 17th century, tulips were found in which the normally solid coloured petals instead exhibited a highly-attractive swirled pattern with regions of light or white colour (Fig. 29.4). This is referred to as a broken pattern. The tulips appeared with this elegant pattern in an unpredictable way and for no obvious reason. The element of apparent chance in the generation of these tulips led to an intense market in their trade. This was exhibited by bidding to purchase bulbs derived from plants with the broken appearance in the

Fig. 29.4 Tulip exhibiting a broken petal colouring consistent with that seen with tulip breaking virus infection. (Copyright B. Hoover, reproduced with permission.)

hope of obtaining a particularly attractive specimen which could then command high prices. The costs paid in this form of gambling rapidly became immense; the largest sum paid for a single *Semper augustus* bulb was equivalent to the price at the time of three houses with gardens. A sense of the intensity of the bidding for these bulbs can be seen with the final winning bid for one bulb in 1636, which was: 4 tons of wheat, 8 tons of rye, 2 tons of butter, 1000 pounds of cheese, 4 tons of beer, 2 barrels of wine, 4 oxen, 8 swine, 12 sheep, a suit of clothes, a bed, and a silver drinking cup. Fortunes could be made or lost depending on the quality of the resulting flower. Sadly, but perhaps inevitably, the tulip market suffered a catastrophic crash in 1637 during which many lost their entire fortunes and the Netherland economy was seriously damaged – one of the earliest examples of a market crash through inappropriate speculation.

We now know that the entire tulip craze was established due to the effect on the plant of tulip breaking virus. This virus infection causes alteration in the distribution of anthracyanin which is responsible for pigmentation. The random distribution is not predictable but can produce the swirling patterns particularly sought after by the early speculators. Ultimately, the market was likely to fail not least because, although the plant appears healthy initially following infection, the virus causes damage that accumulates and eventually the plant becomes seriously weakened and cannot survive. This is a very rare example where an infection was positively sought for financial gain. Tulips with broken petal colouration are now readily obtained but these have been generated by selective breeding and plants infected with tulip breaking virus are not traded.

Key points

- Determination of the economic costs of virus infections include a large variety of factors.
- Virus infections can be assessed in terms of their economic impact. For infections in humans these assessments may be used to decide on the assignment of limited health care resources.
- The economic cost of human infections includes both direct, medical, and indirect, social and employment, costs to the national economies of the world.
- Virus infections of animals have both welfare and economic implications. The economic costs are generally calculated in terms of lost productivity but can also include other direct and indirect costs.
- Virus infections of plants have most impact when they involve food crops. However, the economic costs of these infections can be matched or exceeded by the costs of infection of non-food crops.

Further reading

Clark, C. A., Davis, J. A., Abad, J. A., Cuellar, W. J., Fuentes, S., Kreuze, J. F., Mukasa, S. B., Tugume, A. K. Tairo, F. D., Valkonen, J. P. T. 2012. Sweetpotato viruses: 15 years of progress on understanding and managing complex diseases. *Plant Disease* 96, 168–185.

Martin, D. P., Shepherd, D. N. 2009. The epidemiology, economic impact and control of maize streak virus. *Food Security* 1, 305–315.

Rybicki, E. P., Pietersen, G. 1999. Plant virus disease problems in the developing world. *Advances in Virus Research* 53, 127–175.

Shepard, D. S., Undurraga, E. A., Halasa, Y. A. 2013. Economic and disease burden of dengue in Southeast Asia. *PLoS Neglected Tropical Diseases* 7, e2055.

Shepard, D. S., Coudeville, L., Halasa, Y. A., Zambrano, B., Dayan, G. H. 2011. Economic impact of dengue illness in the Americas. *American Journal of Tropical Medicine and Hygiene* 84, 200–207.

Staples, J. E., Shankar, M. B., Sejvar, J. J., Meltzer, M. I., Fischer, M. 2014. Initial and long-term costs of patients hospitalized with West Nile virus disease. *American Journal of Tropical Medicine and Hygiene* 90, 402–409.

Chapter 30

Recombinant viruses: making viruses work for us

As well as providing potent and constantly-evolving threats to human health, and to animals and plants, viruses also offer the possibility of being used for human benefit through the construction and appropriate use of recombinant viruses. These are viruses where the genome has been altered in a planned way by experimental manipulation.

Chapter 30 Outline

30.1 Recombinant viruses as vaccines
30.2 Recombinant viruses for gene therapy
30.3 Retroviral vectors for gene therapy
30.4 Adenovirus vectors for gene therapy
30.5 Parvovirus vectors for gene therapy
30.6 Oncolytic viruses for cancer therapy
30.7 Recombinant viruses in the laboratory

Recombinant viruses have been invaluable in the study of virus replication cycles, allowing us to study the effect on function of precisely-targeted changes in the viral genome. They are also powerful tools in the study of fundamental cell processes in the laboratory, where they can be used to carry cDNA for specific host genes into cells in culture and hence to cause expression of the protein products. Another use of recombinant viruses is as vaccines. Finally, recombinant viruses have been of great interest as potential gene therapy agents. This last topic has been the focus of intense research interest over the past 20 years.

The starting point for any programme to generate a recombinant virus is to clone its genome. After genetic material has been isolated from virus particles, it can be manipulated in exactly the same way as any other RNA or DNA molecule. Thus, DNA virus genomes may be cloned directly while RNA virus genomes may be cloned as cDNA. These cloned molecules can then be modified by site-specific alteration or, more drastically, portions may be removed and replaced with foreign DNA sequences. In this way, it is quite straightforward to generate modified forms of either an intact viral genome or, if the genome is large, a portion of it, cloned in a bacterial plasmid. What is far more difficult is to complete the process by using these cloned sequences to recreate infectious virus particles, a process often termed *virus rescue*. This requires techniques specific to each type of virus and is not yet possible for all virus types. At its simplest, all that is needed is to transfect the purified recombinant genome (DNA or RNA transcribed from cloned DNA as appropriate) into susceptible cells. However, many viruses also need some of their proteins to be supplied

Introduction to Modern Virology, Seventh Edition. N. J. Dimmock, A. J. Easton and K. N. Leppard.
© 2016 John Wiley & Sons, Ltd. Published 2016 by John Wiley & Sons, Ltd.

with the genome in order for any recombinant virus to be recovered. Some of the double-stranded RNA viruses remain refractory to all attempts at generating recombinant virus from cloned cDNA.

30.1 Recombinant viruses as vaccines

Harmless or attenuated viral strains can be used as carriers for gene sequences derived from pathogenic viruses. When a host is infected with such a recombinant virus, it produces a protective immune response against the pathogenic virus.

As discussed in Chapter 26, there are two broad classes of viral vaccine: live agents that have been attenuated to eliminate their pathogenicity and killed preparations of either whole virus or purified components. Work over many years has led to the isolation of a number of genetically stable, safe live viral vaccines that are routinely deployed to protect against, e.g. measles and yellow fever. Vaccinia virus and its modern more attenuated derivatives such as *modified vaccinia Ankara* (MVA) also belong in this vaccine class, although vaccinia is no longer in widespread use because its target, smallpox virus (variola), has been eradicated (Box 26.1). Live vaccines have a number of practical and scientific advantages but when attempting to expand the range of vaccines to include new targets it is often not possible to achieve a suitable live vaccine derivative of the pathogenic virus. Recombinant viral vaccines offer a path to achieving the benefits of live vaccines for such difficult cases. The principle is to embed one or more genes from the target pathogen into the genome of an established attenuated viral vaccine strain or other non-pathogenic viral genome so that, when the

recombinant vaccine is administered, its genes encode proteins from the target pathogen and hence elicit protective immune responses against it.

The only live virus recombinant vaccine to have been widely used to date is a vaccinia-rabies recombinant that has been deployed to protect wildlife against rabies virus infection (Section 26.2) and hence, indirectly, to reduce human exposure to rabies virus. Vaccinia virus (a poxvirus) has a large double-stranded DNA genome approaching 200 kbp in length. The technology to manipulate its genome is explained in Box 30.1. What results is a recombinant virus where all the genes encoding structural proteins of the vaccinia virion remain. Thus the recombinant virions retain their normal protein composition but the added genes from the target, in this case the surface glycoprotein (G) gene of rabies virus, are expressed to produce protein within infected cells in the vaccine recipient. These are then presented as antigens to elicit a protective immune response against the target.

A rather different strategy underlies the development of candidate vaccines to protect against dengue fever (Section 23.2). In this case, the successful yellow fever virus (YFV) vaccine strain (17D) has been used as the basis for recombinant viruses in which the genes for the YFV surface-displayed proteins (M and E) are replaced with the exactly homologous genes from dengue fever virus (DEN; Fig. 30.1). In this way, the recombinant vaccine particles encode and display DEN antigens instead of YFV antigens. Because there are four serotypes of DEN in circulation, four equivalent recombinants have been made. Whilst the viability of such recombinant viruses should never be assumed, in this case they can be grown successfully, reflecting the fact that DEN and YFV are related members of the Flavivirus family with equivalent genome organizations and modes of replication. Following earlier phase II trials, results of a large phase III trial of

---

Box 30.1

Isolation of recombinant vaccinia virus

The genome of vaccinia virus (VV) is too large to be conveniently cloned in its entirety and several viral proteins that enter a cell within the particle are necessary for productive infection. Hence, reconstituting recombinant VV (rVV) simply by transfecting cloned manipulated genome DNA is not feasible. Instead, a recombination–selection strategy is employed that results in rVV carrying an insertion within a target non-essential gene, e.g. thymidine kinase (*tk*).

The processes that are carried out are:

i. Clone the region of the VV genome containing the *tk* gene and flanking DNA.
ii. Manipulate the cloned plasmid DNA to insert the gene of interest, as cDNA, under control of a VV promoter, in place of *tk* sequence using standard DNA handling techniques.
iii. Transfect cells with the manipulated plasmid DNA clone, superinfect with VV and apply selection for the absence of *tk* function by addition of, e.g. bromodeoxyuridine (BrDU) to the medium. Cells infected with *tk*⁺ VV will incorporate the mutagen BrDU into virus which will lose viability whilst rVV lacking functional *tk* will survive. Isolate rVV plaques and screen for correct genotype.

An alternative development of this approach is to incorporate a positive selection marker such as the guanine phosphoribosyltransferase (*gpt*) gene into the recombinant gene cassette. By adding mycophenolic acid to the medium, plus the purines hypoxanthine and xanthine, only rVV containing this gene can survive.

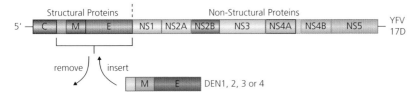

Fig. 30.1 The genome organization of yellow fever virus (YFV, a flavivirus) showing the gene replacement strategy for converting the YFV vaccine strain 17D into a recombinant vaccine for dengue (DEN) viruses.

the quadrivalent recombinant vaccine were published in July 2014, showing that a course of three vaccinations gave good efficacy in protecting against dengue virus disease, especially against the more severe dengue haemorrhagic fever. This approach will hopefully form the basis of a useable vaccine in due course.

30.2 Recombinant viruses for gene therapy

The ability of viruses to carry gene sequences into target cells can be used to deliver functional gene sequences into host cells for purposes other than production of immune responses. If immune responses can be avoided, such cells can survive and display altered properties that correct or ameliorate a disease process.

Gene therapy is simply the introduction of DNA sequences into the cells of a patient with the aim of achieving a clinical benefit. It was originally conceived as a way of restoring normal function in patients with specific inherited single gene defects, such as cystic fibrosis. In such conditions, absence of a normal copy of the gene means that a specific protein function is missing, with severe physiological consequences. By putting back a 'good' copy of the gene (normally as cDNA) into the patient's cells, all these consequences should, in theory, be corrected. From this beginning, the gene therapy concept has now grown to encompass a variety of possible applications, with the greatest number of clinical trials being in the field of cancer therapy. When it comes to treating cancer by gene therapy, the goal is totally different from gene therapy for inherited conditions. Rather than trying to restore normal function to the tumour cells, the objective is to kill them, or to cause them to undergo apoptosis – but to do so specifically, so that normal cells are not damaged.

All gene therapy experimentation is carefully regulated. In concept, gene therapy could be applied to either the germ cells or the somatic cells of an individual, but, for ethical reasons, only the latter is allowed. Any attempt to modify the DNA that is passed on to subsequent generations is currently illegal. Thus, gene therapy is restricted in scope to the treatment of the health problems of the individual. It cannot be used to attempt to eliminate 'bad' genes from the human gene pool.

For all the exciting possibilities of gene therapy to become reality, DNA has to be carried across the membranes of cells and ultimately to reach the cell nucleus. This is something that naked DNA is very poorly equipped to achieve. The cell membrane is a huge barrier to such a long, highly-charged molecule and, if it is endocytosed into a cell, it is then vulnerable to degradation in lysosomes. By contrast, nucleic acid that is inside an infectious virus particle avoids these problems. Firstly, viruses have evolved specific interactions with cell surface molecules that lead to their efficient entry into the cell and, secondly, if that entry involves arrival in the cytoplasm within an endocytic vesicle, then viruses have mechanisms to allow efficient escape into the cytoplasm (see Chapter 6). These properties make viruses important tools for gene therapy. A recombinant virus (the vector) carrying a foreign gene (often referred to as a transgene) is rather like a Trojan horse, being able to penetrate the defences of the cell before unloading its cargo. What that cargo does to the cell subsequently is totally dependent on the gene(s) chosen; having carried the gene into the cell, the role of the virus is over. This process of virus-mediated delivery of a gene into a cell is known as transduction.

To be potentially useful as a gene therapy vector, a virus should have a number of features (Box 30.2). Bearing these factors in mind, the viruses that have seen extensive development as vectors so far include mouse retroviruses, human adenovirus and the parvovirus, adeno-associated virus (AAV). However, no virus meets all of the criteria for an ideal gene therapy vector and, as considered below, there are some significant drawbacks to the vectors that have received extensive attention so far. Thus, there is no such thing as the perfect viral gene therapy vector and each gene therapy application is likely to need its own vector, chosen and then tailored to meet the precise circumstances.

Box 30.2

Necessary or desirable features in a virus to be used as a gene therapy vector

- The virus must be able to enter the desired human cell (i.e. receptors for the virus must be present on the cell surface, see Chapter 6).
- It must be possible to generate recombinant forms of the virus that are deleted for one or more essential functions and so do not replicate in the patient. Equally, it must be possible to grow these replication-defective recombinants in cell cultures in the laboratory.
- In its deleted, replication-defective form, the virus must be safe to use in people. In particular, it should not be able to recover replication-competence by recombining with other viruses that might be present in vivo.
- The virus should deliver its genetic material to the nucleus in the form of DNA.
- It should be possible to grow recombinant forms of the virus to high concentrations.
- The replication-defective form of the virus should be able to accommodate a sufficient length of foreign DNA to be useful (most viruses have quite stringent upper limits on the amount of nucleic acid that they can include in a particle).
- If therapy is to be 'permanent', the transgene needs to persist in the cells indefinitely and to be inherited by daughter cells if the cell divides. To achieve this, a viral vector DNA needs to achieve either integration into a chromosome of the cell or else be capable of autonomous replication of its genome and partition during cell division.

30.3 Retroviral vectors for gene therapy

Very promising results have been obtained using retroviral vectors in clinical trials of gene therapy, but serious safety issues emerged during follow-up of patients treated for one particular condition. Now there are much improved retroviral vectors which seek to address these safety issues.

Retrovirus vectors have all their normal coding sequences removed and replaced with a cDNA of interest. The recombinant virus is then grown in a special cell line (packaging cells) which provides the viral proteins that are needed to form particles containing the recombinant genome (Fig. 30.2). When used therapeutically, the virus particles reverse-transcribe their genome and integrate the DNA copy, including the transgene, randomly into the host cell genome using enzymes included in the particle (see Chapter 9), but the absence of all viral genes from the particle means that no further events of the infectious cycle can occur. Hence, the transgene that the vector has carried into the cell is present for the life of the cell and will be passed on if the cell divides.

Retroviral vectors grow only to relatively low concentrations, which means that they have typically been used only to transduce cells taken from a patient; these must then be reimplanted later (ex vivo gene therapy). These vectors are thus well suited to treating cells of the lymphoid system because, firstly, the cells can be easily explanted and re-implanted and, secondly, transgene integration is required for persisting therapy in these cells as lymphoid cells are being constantly replaced from progenitor cells.

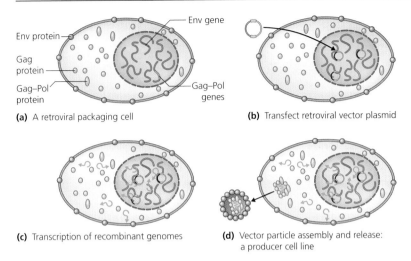

Env protein

Gag protein

Gag–Pol protein

Env gene

Gag–Pol genes

(a) A retroviral packaging cell

(b) Transfect retroviral vector plasmid

(c) Transcription of recombinant genomes

(d) Vector particle assembly and release: a producer cell line

Fig. 30.2 Production of recombinant retroviruses by transfection of cloned recombinant genome plasmid into a packaging cell line. Cells may be cultured on as a producer cell line, with recombinant virus being harvested from the growth medium. The essential retroviral protein coding genes, gag, pol, and env (see Section 10.8) are integrated in the packaging cell line in two distinct segments, to minimize the risk that recombination between these sequences and the vector could produce a viable retrovirus again.

Several diseases have been the target of clinical trials of retrovirus-mediated gene therapy, in particular two inherited immunodeficiencies: adenosine deaminase (ADA) deficiency and X-linked severe combined immunodeficiency disease (X-SCID). Both conditions are characterized by a profound lack of T lymphocytes and hence a failure of both humoral and cell-mediated arms of the adaptive immune response (see Chapter 14). In both cases, following re-implantation of bone marrow-derived stem cells treated ex vivo there was clear evidence of clinical improvement. In the case of ADA deficiency, the therapy has proved safe and is considered an accepted treatment for patients where other options are not available. However, in the case of X-SCID, although there was early optimism that the treatment had cured the disease, serious adverse effects quickly emerged in the form of acute leukemia in around 25% of treated children. The ability of retroviruses to affect the expression of host growth control genes when

they integrate their DNA into the genome and so cause a tumour (see Section 25.7), was always recognized as a risk with retroviral vectors but thought to be of very low probability. However, this unfortunately became reality in several children suffering from X-SCID who received retroviral gene therapy.

The results of the X-SCID gene therapy trials showed that safety of retrovirus vectors needed to be improved: two important developments have been made. Firstly, an understanding of the precise mechanism of reverse transcription (Chapter 9) has been used to create *self-inactivating* (SIN) retroviral vectors (Fig. 30.3). These have improved safety because the LTRs are no longer available to cause insertional activation of neighbouring genes. Secondly, alternative retroviruses are being used as the basis for the design of the vectors. The initial wave of retroviral vector research was done using murine leukemia viruses (MLV). When these, or vectors derived from them, infect cells,

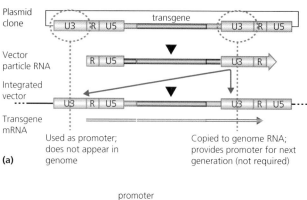

Used as promoter;   Copied to genome RNA;
does not appear in   provides promoter for next
genome   generation (not required)

(a)

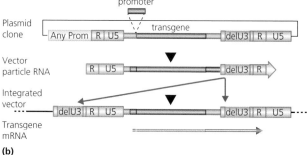

(b)

Fig. 30.3 Comparison of standard retroviral vectors (a) and self-inactivating (SIN) retroviral vectors (b). Retroviral LTRs are formed during reverse transcription at either end of the proviral DNA that is then inserted into the host genome. The U3 element of the LTRs has enhancer-promoter activity. By eliminating U3 from the downstream LTR in cloned vector plasmid DNA, the resulting vector particles contain RNA that can be packaged like retroviral RNA but which lacks a functional U3 that would be necessary for the creation of active LTRs during insertion in the next infected host.

the integration process – although nominally random with respect to target sequence in the host cell genome – actually has a propensity to target the promoters of active genes. In this way, such vectors have an elevated risk of affecting the expression of genes that are important for the normal biology of the cells they infect. The alternative, which is being used increasingly for vector research, is rather surprising at first glance: it is none other than HIV-1, the cause of the AIDS pandemic (Chapter 21). Although a dangerous pathogen, the basis of HIV-1 pathogenicity lies ultimately in its genes. Stripped of these, it is capable of providing a harmless gene carrier that has two advantages over vectors based on MLV: firstly,

the integration process does not target promoter regions which gives a safety advantage, especially when combined with SIN technology, and secondly, infection by an HIV-based vector particle leads to the establishment of an integrated DNA copy in non-dividing cells as well as dividing cells, whereas MLV vectors are restricted to the latter.

30.4 Adenovirus vectors for gene therapy

Adenovirus vectors have been widely developed for gene therapy applications. However, immune recognition of cells that have taken up the vector remains a problem. Their most promising application so far is in cancer gene therapy where such responses are an advantage.

There are many different human adenoviruses. The one that has been most widely studied as a gene therapy vector is adenovirus type 5, a mild respiratory pathogen. As adenoviruses encode a much larger number of proteins than retroviruses (see Section 10.4), it has not been possible yet to make packaging cells that will allow the growth of vectors where all the viral genes have been deleted (gutless vectors, Fig. 30.4b). Such vectors can be grown using an intact virus (helper virus) to provide the necessary viral proteins, but most applications to date have used vectors where most of the viral genes are retained so the vectors can be grown in cells that complement their selective

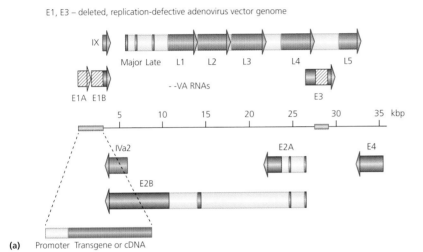

(a)

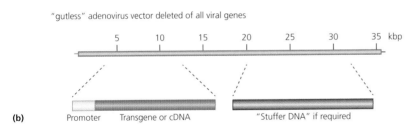

(b)

Fig. 30.4 The genomes of adenovirus gene therapy vectors, either (a) deleted for the essential E1A and E1B genes, plus non-essential E3 genes (to make more space for the transgene), or (b) deleted for all adenovirus sequences except the termini, which provide origins of replication and the signal for packaging the genome into particles. In (a), the maximum capacity for the transgene is around 7 kbp while in (b), the maximum is around 35 kbp and, for short transgene sequences, additional 'stuffer DNA' is needed to achieve the minimum genome length necessary for stable particle formation. The viral genes are shown in simplified outline in (a), with early genes coloured purple and late genes coloured brown (see Section 10.4). Viral sequences deleted during vector construction are shown as green boxes in the genome and the affected genes are shown striped in (a).

gene deficiencies. A widely-tested example is a vector lacking the E1A and E1B genes, which encode proteins required for the expression of the other viral genes, and the E3 gene, which encodes several proteins that are not needed for growth in cell culture but which modulate the host response to infection in vivo (Fig. 30.4a); these are known as *first generation Ad vectors* and can accommodate about 7 kbp of foreign DNA. Very high concentrations of vector particles can be obtained, suitable for direct administration to patients for in vivo therapy.

The major problem which has been encountered with adenovirus vectors deleted for E1 and E3 genes has been the adaptive immune response directed against vector proteins that develops when they are used. This effectively eliminates cells that have acquired the transgene over the course of just a few weeks and ensures that any subsequent administration of the same vector is dealt with even more rapidly. The immune response arises in part from some residual expression of the viral genes that remain on the vector and so it can be reduced by using gutless vectors. However, such immune responses may be an advantage in the context of cancer gene therapy if the virus can be targeted specifically to tumour cells. A further difficulty is that receptors for adenovirus 5 are not as widely distributed as first thought. Initially, cystic fibrosis was considered to be a promising target for therapy with adenovirus 5 vectors since this gene defect affects respiratory epithelial cells which are the natural target of the virus. However, the virus receptors are on the basolateral membranes of these cells, not the apical surface, so infection is not efficient. Work is continuing to modify adenovirus 5 receptor specificity and to use other adenoviruses as vectors. However, using adenovirus vectors for cancer therapy, where the aim of transgene delivery is to cause the death of the cell, remains a very promising approach. Here, transgenes are typically immunological modulators designed to increase immune recognition of tumour cells and the vectors are modified to target tumour cells specifically by altering their receptor-binding activity. Finally, it should be noted that, as with retroviral vectors, safety issues have emerged during clinical trials using adenovirus vectors. One patient died in 1999 following administration of a very high dose of vector particles, apparently as a result of an unusual response to the virus in that individual. Such variation at the level of the individual patient is a major problem in the development and safety testing of novel therapies.

30.5 Parvovirus vectors for gene therapy

The parvovirus AAV (see Sections 7.6 and 10.7) has only a small genome, so even by removing all its genes its capacity for foreign DNA is limited to about 4 kbp. However, a key advantage of AAV over other vectors is that cells transduced by it seem to elicit minimal destructive immune responses. It was also thought initially that the ability of AAV to integrate at a specific chromosomal location (Section 15.3) might give it an advantage over the random integration of retroviral vectors, but it now appears that this property of AAV depends upon expression of the viral Rep proteins (Fig. 10.8), and these genes have to be deleted when converting it to a vector in order that there should be sufficient space for a transgene.

30.6 Oncolytic viruses for cancer therapy

Oncolytic viruses offer a particular avenue for cancer therapy that is distinct from gene therapy, though the two approaches increasingly come together. The aim of oncolytic therapy is to harness the intrinsic ability of virus infection to kill infected cells and to adapt and target the virus so that killing ability is enhanced and focused onto tumour cells whilst sparing normal cells in the patient. Thus a key distinction from gene therapy is that the ability of the virus to replicate is not disrupted before giving it to the patient. However, it is crucial that this replication ability should be made selective for tumour cells. Ways of achieving this are summarized in Box 30.3. Often, such adaptations are specific to particular tumour types so that an oncolytic virus would be an appropriate therapy only for one or a few tumour types.

There has been considerable interest in making oncolytic adenoviruses tumour-selective based upon the function of wild-type virus in inhibiting p53 activity (Section 25.3).

Since many human tumours are p53-deficient it was argued that a mutant adenovirus that was unable to inactivate p53 should replicate selectively in p53-deficient tumours relative to the surrounding normal tissue. Subsequent analysis showed that the replication ability of such a mutant virus in fact did not correlate with p53 status in a variety of normal and tumour cell lines. However, these oncolytic adenoviruses, which lack E1B 55 K protein function, have shown some benefit for head and neck cancers when injected directly into the tumour in conjunction with *cis*-platin and 5-fluorouracil chemotherapy and a version of this virus has been licensed for use in China. Other viruses on which experiments have been conducted to develop candidate oncolytics

---

Box 30.3

Targeting oncolytic virus activity to tumour cells

The ability of a virus to kill the cell it infects, either directly or by activating the cell's own apoptosis pathways (Section 13.5), can be focused on tumour cells specifically to create a targeted therapeutic when given to patients. This can be achieved by:

- Modifying the viral surface protein responsible for attachment and entry into a target cell so that it recognizes a cell surface molecule that is found only on tumour cells. The resulting virus should only be able to enter and infect those tumour cells.
- Identifying a gene within the virus genome whose products are essential for virus replication and cell death, and place it under the control of a tumour-specific promoter, i.e. a promoter from a cell gene that is found to be strongly activated in a particular tumour type and not in normal cells. The resulting recombinant virus should only grow in and kill tumour cells.
- Identifying a gene within the virus genome whose products are essential for virus replication in normal cells but which are dispensable for growth in tumour cells. A recombinant virus carrying loss-of-function mutations in that gene should replicate in and kill tumour cells specifically.
- Isolating virus strains that are attenuated for spread in normal tissue because they lack the ability to counteract innate immunity. These responses are typically defective in tumour cells, meaning that such a virus may replicate selectively in tumour tissue.

Once made tumour-specific, oncolytic viruses may then be armed with one or more transgenes designed to actively kill tumour cells, linking together oncolytic and gene therapy.

include reovirus, Newcastle disease virus and measles virus. Importantly, all of these are RNA viruses, demonstrating that the field of recombinant viruses is not limited to those that have DNA genomes or a DNA provirus.

30.7 Recombinant viruses in the laboratory

All the viral vector systems discussed in the earlier sections of this chapter first proved themselves in a laboratory research setting. Modern cell biology research has at its heart the testing of how function is altered when particular proteins are either added to the contents of cells (transient or stable over-expression) or else removed from the cell (gene knock-out, or transient or stable RNA knock-down). For stable over-expression, a favoured approach is to clone the relevant promoter-cDNA into a retroviral vector, transduce the selected cell line with the vector and then to select for cells that have incorporated the retrovirus using a dominant selectable marker (such as puromycin resistance) which is also included in the vector. In the same way, stable knock-down at the RNA level can be achieved by transducing cells with a retroviral vector that encodes an appropriate shRNA construct to down-regulate the target gene via the miRNA pathway (see Section 13.7). When only transient gene delivery is desired, adenovirus vectors have also been widely used.

Viral vectors have also been crucial in fundamental studies of the immune response. Immune cells harvested from an experimental animal following infection with a virus or other form of immune challenge can be stimulated in vitro to determine which particular cell types are responsive and how numerous they are. To do this requires particular antigens from the virus concerned to be expressed by antigen-presenting cells in vitro and both adenovirus

and vaccinia virus vectors have proved very useful for this.

Finally, recombinant viruses generally, i.e. any virus in which its natural genome has been manipulated experimentally, have been absolutely essential to the study of fundamental virology. By making targeted changes to the genome of a virus and then comparing its biologic properties with those of the parental wild-type virus, it is possible to draw precise conclusions about the function of the element that has been altered. This is the technique of *reverse genetics* and it has underpinned a great many of the advances that have been detailed elsewhere in this book.

Key points

- Many virus genomes can be manipulated in the laboratory to produce recombinant viruses.
- Recombinant viruses are powerful tools for the study of virus biology.
- Recombinant viruses can efficiently carry foreign genes into cells, either in the laboratory or in vivo.
- Foreign gene sequences within recombinant viruses may serve to produce antigens that elicit immune responses (recombinant vaccines) or may express useful functions that alter the host biology for therapeutic purposes (gene therapy).
- A recombinant vaccinia virus is used as a rabies vaccine for wild animal populations.
- Retrovirus, adenovirus and parvovirus vectors have been widely developed and tested for gene therapy.

Questions

- Consider the ways in which viruses can be used as tools for the benefit of humankind.
- Explain how a retrovirus must be modified to convert it to a useful vector and how such vectors work in practice.

Further reading

Brunetti-Pierri, N., Ng, P. 2008. Progress and prospects: gene therapy for genetic diseases with helper-dependent adenovirus vectors. *Gene Therapy* 15, 553–560.

Capasso, C., Garofalo, M., Hirvinen, M., Cerullo, V. 2014. The evolution of adenoviral vectors through genetic and chemical surface modifications. *Viruses-Basel* 6, 832–855.

Cavazza, A., Moiani, A., Mavilio, F. 2013. Mechanisms of retroviral integration and mutagenesis. *Human Gene Therapy* 24, 119–131.

Lentz, T. B., Gray, S. J., Samulski, R. J. 2012. Viral vectors for gene delivery to the central nervous system. *Neurobiology of Disease* 48, 179–188.

Mercedes Segura, M., Mangion, M., Gaillet, B. et al. 2013. New developments in lentiviral vector design, production and purification. *Expert Opinion on Biological Therapy* 13, 987–1011.

Rollier, C. S., Reyes-Sandoval, A., Cottingham, M. G. et al. 2011. Viral vectors as vaccine platforms: deployment in sight. *Current Opinion in Immunology* 23, 377–382.

Walsh, S. R., and Dolin, R. 2011. Vaccinia viruses: vaccines against smallpox and vectors against infectious diseases and tumors. *Expert Review of Vaccines* 10, 1221–1240.

Chapter 31

Viruses: shaping the planet

Viruses can kill large numbers of hosts, so exerting an extraordinary evolutionary pressure for change. They also affect all forms of cellular life. In organisms alive today, there is evolutionary evidence for a crucial role of viruses in their biology. All the evidence indicates that in major ways viruses have shaped, and continue to shape, the planet that we share with them.

Chapter 31 Outline

Through the preceding chapters we have discussed viruses as discrete entities, considering the principles by which they can 'live' with their hosts in order to reproduce and maximize their evolutionary success. In so doing, we have shown how viruses can be agents of harm to their hosts, causing disease and even jeopardizing their individual survival. In this final chapter, we consider some of the less obvious ways in which viruses influence the nature of other organisms.

31.1 Virus infections can give a host an evolutionary advantage

The interaction of virus and host can become symbiotic to the point where the existence of the host is absolutely dependent on the virus.

Viruses of certain parasitic wasps are telling us something that some virologists have suspected for many years – that one reason for viruses continuing to exist is that they can provide their host with an evolutionary edge. For the past 74 million years, ichneumonid and braconid wasps have been carrying integrated copies of the genome of a bracovirus in their DNA. There is virus replication and production of free virus particles, but only in special cells of the ovary. The larvae of these wasps are parasitic on caterpillars, such as that of the tobacco

Introduction to Modern Virology, Seventh Edition. N. J. Dimmock, A. J. Easton and K. N. Leppard.
© 2016 John Wiley & Sons, Ltd. Published 2016 by John Wiley & Sons, Ltd.

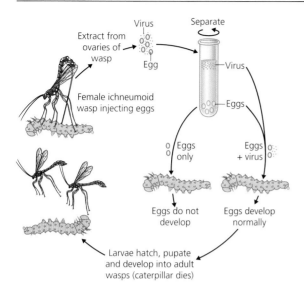

Fig. 31.1 Eggs of the ichneumonid wasp only develop in their host caterpillar when injected together with a polydnavirus.

hornworm moth; an infected female wasp deposits eggs into the caterpillar by injection, along with bracovirus particles. The eggs hatch out and the wasp larvae feed on the living caterpillar until the caterpillar dies, whereupon the wasp larvae pupate and emerge in due course as new adult wasps. However, if the wasp eggs are separated from the bracovirus by centrifugation and then injected alone into caterpillars, they fail to develop. Adding back purified virus allows normal wasp development to take place (Fig. 31.1). On examination, it was found that the virus selectively suppresses the caterpillar's immune system in favour of the wasp larva, and alters the developmental regulation of the caterpillar so that it cannot pupate and so remains a caterpillar – and hence a suitable host for the parasite. One of the virus proteins responsible is an analogue of cellular cystatins, which are inhibitors of cysteine proteases. Thus, virus is needed to suppress (by some unknown mechanism) the immune responses of the caterpillar, producing the

situation where a virus has become essential to the successful lifecycle of its host.

31.2 Endogenous retroviruses and host biology

A significant amount of the mammalian genome is comprised of endogenous retrovirus sequences. These are the residue of ancestral retroviral infections that have invaded the germ line. Whilst most of these residual sequences are inert, some may have deleterious effects on human health while others are beneficial or even essential to our wellbeing.

When retroviruses infect a cell, a DNA copy of their genome is inserted at random into the genome of the host, where it remains for the life of the cell (see Chapter 9). Whilst the vast majority of infections affect somatic cells, occasionally a retrovirus infects a germ cell. If gametes formed from that cell subsequently give rise to offspring, then every cell in the body of the new individual will contain a copy of the inserted retroviral sequences. These are known as endogenous retroviruses, as distinct from exogenous viruses that infect an individual from its environment.

Examination of the human genome reveals large numbers of endogenous retrovirus (HERV) sequences that comprise perhaps as much as 8% of the total DNA. The vast majority of these sequences are the result of ancient insertion events that predate human speciation, but some are more recent. Examination of the sequences of these retroviral inserts reveals that almost all are defective – i.e. unable to produce virus particles. Many have suffered gross deletion of the genome to leave just a single inserted long terminal repeat (LTR; see

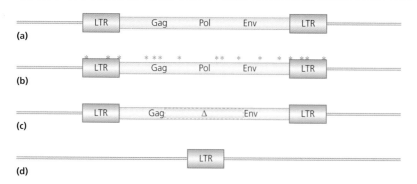

Fig. 31.2 Endogenous retroviral genomes decay by mutation. (a) The initial product of integration following horizontal infection of a germ cell; (b–d) possible organizations of the endogenous virus sequence after many generations within the host germline. (b) Random point mutations can degrade the protein-coding capacity of the genome and/or alter the function of the two LTRs. (c) Large random deletions within the coding region can remove some or all of the genes. (d) Homologous recombination between the two LTRs of the original integrated copy can precisely excise the entire coding region to leave a single LTR embedded in the genome.

Section 9.4), while most of the others have deletions and/or frame-shift mutations affecting some or all of the *gag*, *pol* and *env* genes which encode the internal structural proteins, viral particle enzymes and envelope glycoproteins respectively (Fig. 31.2). This is presumably the result of strong negative selection pressure on intact functional endogenous genomes; the high rate of insertional mutagenesis that would result from virus production and re-infection would likely be extremely harmful. However, retained expression of Env proteins may have the selection benefit of providing protection against superinfection by exogenous retroviruses that have the same receptor specificity.

The most recently acquired human endogenous retroviruses belong to a sequence subgroup known as HERV-K. Many of these are found only in chimpanzees and humans, reflecting the recent emergence of this HERV lineage, while around 70 are human-specific and so are very recently acquired. Even for these viruses, most of their genes are inactivated by deletion or mutation, but one or two of the sequences are only minimally

altered, suggesting the possibility of functional expression. In addition, there is the possibility that the complete set of proteins needed for particle formation might be provided by different defective endogenous viruses (i.e. complementation). Indeed, HERV-K gene activity is detected in testis tissue and particles have been seen in teratocarcinoma cells (cells from germ cell tumours). Some of the human-specific HERV-K loci are so recently inserted that they are still polymorphic – i.e. some people possess the insertion while others do not; these loci are a potential source of individual biological variability in our population.

There has been a lot of interest in the possibility that HERVs might not be biologically inert. Either the production of virus, or of mRNA transcripts that are then reverse transcribed and inserted elsewhere on the genome, or the effects of the integrated sequences or their protein products on the host cell, might be expected to have adverse effects on the biology of the host. Many attempts have been made to link HERV activity with complex conditions such as rheumatoid arthritis, multiple sclerosis, and schizophrenia. However, it is very difficult to prove a causal connection because any activity detected is coming from multiple HERVs and it might be that only one specific HERV is relevant to a disease; these diseases are themselves heterogeneous and so probably have multiple causes, further confusing the picture. One disease connection with fairly strong evidence is the activity of a subgroup of HERV-K in men with germ cell tumours. Such patients have a high frequency of antibodies to HERV-K proteins compared to those without the disease and, as already noted, cell lines from such tumours can form HERV-K particles. However, this might be a consequence of disease rather than a causal relationship. Studies of these potential disease connections of HERVs continue.

As well as being considered as agents of disease, HERVs also have positive roles in mammalian biology. A protein known as syncytin is produced from a gene that has evolved from the *env* gene of a HERV-W integration in the human genome. Syncytin is involved in the formation of syncytiotrophoblasts, which are multi-nucleated cells in the placenta that form the boundary between maternal and foetal tissue. These giant cells are created by fusion of trophoblasts to one another, and the role of syncytin in this process is shown by its ability to fuse trophoblast cells in culture. The generation of syncytiotrophoblasts is critical in the formation of a functional placenta. Mice contain two syncytin genes, both derived from retroviruses up to 25 million years ago. Deletion of the mouse syncytin A gene leads to failure to form a functional placenta with early death of the embryo. Deletion of the mouse syncytin B gene also leads to placental abnormality. Phylogenetic analysis of the syncytin genes from a wide range of mammalian species has shown that syncytin-like genes have been acquired independently at least three times during the evolution of different branches of

the mammalian tree, within which they probably play similar roles to those described in humans and mice. This may be viewed as an adaptation of the ancestral viral function of this protein in mediating fusion between the viral and target cell membranes. The expression of retroviral Env-derived proteins in the placenta has also been suggested to be immunosuppressive and, hence, an important means of achieving tolerance of the antigenically foreign foetus by the maternal immune system. Clearly, the involvement of HERVs in something as fundamental to mammalian biology as reproduction illustrates very well how HERVs can be acquired by the host in a functional as well as a physical sense, and so become critical genes within the host genome.

Whether or not they retain any functional genes, endogenous retrovirus insertion loci have one or two LTRs that originally contained active promoter/enhancer sequences. Although the functions of these sequences may have been degraded or modified by mutation in the time since acquisition, they clearly have the potential to alter the expression of nearby genes. Whilst this may be harmful, as in the tumourigenic activation of oncogenes (Chapter 25), it can also be beneficial. One good example of this is the production of salivary amylase. Amylase, an enzyme required to break down starch, is produced in humans both by the parotid gland for secretion into the mouth and by the pancreas for secretion into the gut. There is a cluster of five closely-related amylase genes on human chromosome 1, two of which are transcribed only in the pancreas and the other three only in the parotid gland. The basis for this difference has been shown to be the presence of an inserted endogenous retrovirus immediately 5′ to the parotid-specific genes, that creates a novel specific promoter active in this tissue; these three genes arose by gene duplication after an ancestral retroviral insertion. Acquisition of the ability to begin breaking down starch in the mouth may have been selected for because of benefits in the selection of nutritious food sources by taste or the more rapid absorption of nutrients through predigestion before reaching the gut.

In summary, endogenous retroviruses are not simply the historic byproducts of germline retroviral infection events, they are an integral part of our genomes and our biology. This reality blurs the distinction between virus and host: for retroviruses, at least, they are both separate from and part of their hosts.

31.3 Bacteriophage can be pathogenicity determinants for their hosts

Bacteria are susceptible to their own set of viruses, known as bacteriophage (phage). The effects that phage have on their hosts range from benign co-existence as a latent infection (lysogeny) to acute cell death. Lysogeny, exemplified by the behaviour of the *E. coli* phage λ (Chapter 17), involves the integration of a phage genome into that of the host. Once such a lysogen has been established, when a cell divides both daughters will acquire a copy of the phage genome and so will also be infected.

Perhaps the best documented example of a lysogenic phage conferring pathogenicity on its host is in *Corynebacterium diphtheria*, the cause of diphtheria. The primary determinant of pathogenicity resides in the ability of the bacterium to produce a toxin and the gene for this is carried on a lysogenic phage. When the phage is not carried, the bacterium is normally a harmless commensal but, once the phage is acquired, it becomes a dangerous pathogen. This interaction can be viewed as the phage altering the biology of its host so that it becomes a more ruthless exploiter of the resources of its own host so as to favour phage replication. The

exotoxin secreted by *Vibrio cholerae* is also encoded on a phage. For several pathogenic bacterial species, comparative genome analysis of multiple isolates has shown that their disease-causing capability resides in so-called pathogenicity islands (PI) that are mobile between strains. In *Staphylococcus*, these PIs mobilize by hijacking the DNA excision and packaging systems of helper phages so that defective phage particles form that carry the PI instead of the phage genome. These then move the PI DNA to new host cells where it can integrate again. Phages are therefore integral to the disease-causing capacity of many bacterial species.

31.4 Cyanophage impacts on carbon fixation and oceanic ecosystems

The world's oceans are host to an immense range and quantity of flora and fauna. Amongst these are viruses which represent the most abundant form of organism found in the sea. These viruses can only exist by infecting suitable hosts. The cellular and multicellular organisms within the oceans that serve as hosts for marine viruses, particularly the cyanobacteria, themselves play a vital role in many biogeochemical cycles on which life on the planet depends; as a consequence, virus infection of these organisms has a major impact on the global ecosystem. Key processes affected by the action of phage include the carbon cycle, in which organic carbon can be retained deep in the oceans in the form of particulate organic matter, particularly that derived from dead organisms. When marine viruses kill their hosts, the living organic matter is converted into inanimate particulate and dissolved organic matter, both of which contribute differently to the carbon cycle. Thus, the level of cell death due to virus infection has a direct effect on the amount of carbon available as a nutrient for other organisms. The precise scale of the effect of virus infection on the carbon and other biogeochemical cycles is not known, but it has been estimated that virus infection leads to the death of between 20% and 40% of marine prokaryotes in the surface layers of the world's oceans each day: this is a cycling of organic carbon on a truly vast scale.

The effect of virus infections in the ocean is perhaps most visibly observable with phytoplankton (algal) blooms. These arise periodically as a population in a defined region and flourish due to an abundance of nutrients and suitable environmental conditions. The appearance of an algal bloom can also be seen as the consequence of a relative failure of phage infection to control the local algal population. Such blooms can grow to immense sizes covering hundreds of square miles, sufficiently large to be seen from orbiting satellites (Fig. 31.3). Algal blooms represent a hazard in many ways, but particularly as many of the species that form them can produce harmful toxins which pose a threat to marine animals, particularly whales and dolphins. Blooms usually grow over a period of time and then rapidly disappear. In some cases, the disappearance is due to the spread of a lytic virus throughout the phytoplankton population. One factor involved in the frequency of algal blooms may be diminished phage virulence due to increased mutagenesis of phage genomes by ultraviolet light reaching the surface water layers because of depletion of the ozone layer in the upper atmosphere over the past few decades.

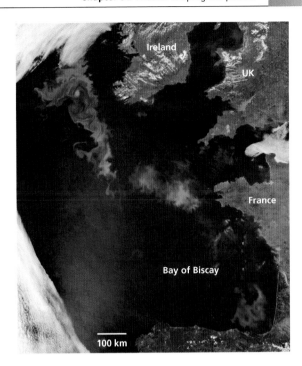

Fig. 31.3 Algal bloom seen from space. The bloom stretches from Ireland into the Bay of Biscay. (Image from NASA.)

31.5 Virology and society: for good or ill

We began this book with a comment about the extraordinary number of virus particles on our planet. The cellular life we see on Earth today is both dependent on and threatened by these myriad viruses that parasitize it. Undoubtedly, viruses are an evolutionarily ancient form of life, examples of which have been evolving with their hosts and moving between hosts through much, if not all, of evolutionary history. It may even be that the viruses we see today represent the descendants of a non-cellular lineage which developed in parallel with cellular life out of the era of developing organic chemistry that preceded life on Earth: two different forms of life, each dependent on the other. Rather than being something we should always strive to overcome, viruses are in reality a necessary part of the ecosystem which we would do well to understand and to come to terms with, even as we try to mitigate their less desirable effects on us.

Key points

- Virus infections can be beneficial to a host as well as harmful.
- Endogenous retroviruses, the residue of past integrations into the germline, are a major part of mammalian genomes.
- Endogenous retrovirus sequences carry out essential functions for the host.
- The pathogenicity of well-known bacterial pathogens is often determined by viruses that infect the bacteria.
- The health of the global ecosystem depends on the balance between cyanobacteria and phage in the world's oceans.

Questions

- Discuss the significance of viruses in the biology of cellular life.

Further reading

Boyd, E. F. 2012. Bacteriophage-encoded bacterial virulence factors and phage-pathogenicity island interactions. *Advances in Virus Research* 82, 91–118.

Gundersen-Rindal, D., Dupuy, C., Huguet, E., Drezen, J.-M. 2013. Parasitoid polydnaviruses: evolution, pathology and applications. *Biocontrol Science and Technology* 23, 1–61.

Stoye, J. P. 2012. Studies of endogenous retroviruses reveal a continuing evolutionary saga. *Nature Reviews: Microbiology* 10, 395–406.

Suttle C. A. 2007. Marine viruses – major players in the global ecosystem. *Nature Reviews: Microbiology* 5, 801–812.

Index

Page numbers in italics refer to figures; those in bold refer to tables or boxes. The index is organized in letter-by-letter order, i.e. spaces and hyphens are ignored in the alphabetical sequence. Since the major subject of this book is viruses, few index entries are listed under headings beginning with 'virus' or 'viral'. Readers are advised to seek more specific references.

Introduction to Modern Virology, Seventh Edition. N. J. Dimmock, A. J. Easton and K. N. Leppard.
© 2016 John Wiley & Sons, Ltd. Published 2016 by John Wiley & Sons, Ltd.